Foreland Basins

Foreland Basins

EDITED BY P. A. ALLEN
AND P. HOMEWOOD

SPECIAL PUBLICATION NUMBER 8 OF THE
INTERNATIONAL ASSOCIATION OF SEDIMENTOLOGISTS
PUBLISHED BY BLACKWELL SCIENTIFIC PUBLICATIONS
OXFORD LONDON EDINBURGH
BOSTON PALO ALTO MELBOURNE

© 1986 The International Association of Sedimentologists

Published by
Blackwell Scientific Publications
Editorial offices:
Osney Mead, Oxford OX2 0EL
8 John Street, London WC1N 2ES
23 Ainslie Place, Edinburgh EH3 6AJ
52 Beacon Street, Boston
  Massachusetts 02108, USA
667 Lytton Avenue, Palo Alto
  California 94301, USA
107 Barry Street, Carlton
  Victoria 3053, Australia

First published 1986

Typeset by CCC, printed and bound in Great Britain by
William Clowes Limited, Beccles and London

DISTRIBUTORS

USA and Canada
  Blackwell Scientific Publications Inc
  PO Box 50009, Palo Alto
  California 94303

Australia
  Blackwell Scientific Publications
    (Australia) Pty Ltd
  107 Barry Street,
  Carlton, Victoria 3053

British Library Cataloguing in Publication Data

Foreland basins.—(Special publication/
  International Association of Sedimentologists
  ISSN 0141-3600; 8)
  1. Sedimentation and deposition
  2. Sedimentary structures
  I. Allen, P. A.   II. Homewood, P.   III. Series
  551.3'04   QE571

  ISBN 0-632-01732-5

Library of Congress Cataloguing-in-Publication Data

Foreland basins.

  (Special publication no. 8 of the International
Association of Sedimentologists)
  Includes bibliographies and index.
  1. Orogeny.   2. Geology, Stratigraphic.
3. Sedimentation and deposition.   I. Allen, P. A.
II. Homewood, Peter.   III. Series: Special Publication
of the International Association of Sedimentologists; no. 8.
QE621.F57   1986   552.8   86-23265
ISBN 0-632-01732-5

Contents

Preface

The papers assembled in this volume on Foreland Basins are the outcome of a symposium sponsored by I.A.S. and S.E.P.M. The meeting was held at Fribourg, Switzerland, in 1986 and was accompanied by four field trips investigating various aspects of Alpine and Apenninic tectonics and sedimentation in south-central Europe.

The conference was conceived initially because we felt that the geology carried out in the Molasse basin over a number of years (in particular the sedimentology) warranted a public airing, and that the subject of Foreland Basins in general was sufficiently topical to justify an international meeting. Although we were optimistic, with enthusiastic encouragement from both Harold Reading (I.A.S.) and Dick Moiola (S.E.P.M.), we hardly imagined the overwhelming response from all sides. Perhaps we underestimated the tourist attraction of Fribourg, a very pretty medieval town located in the heart of the Swiss Molasse.

Our approach to the topic is founded on the idea that the main advances in the understanding of sedimentary basins will come from the integration of Earth Science sub-disciplines rather than from more pronounced specialization. With this in mind, we tried to attract geophysicists, structural geologists, petrographers, stratigraphers and economic geologists, as well as sedimentologists, to create the widest possible range of discussion. We were understandably satisfied with the outcome. However, the papers presented here are mostly from the sedimentological fraternity, many of whom it must be said have dabbled quite seriously in regional and local tectonics. This is entirely appropriate for an I.A.S. special publication.

The papers themselves fulfil two aims. Firstly they represent a collection of case-studies embracing a wide range of basin types and tectonic and stratigraphic settings; from Andean ocean–continent systems to Alpine continent–continent systems, from Palaeozoic to present-day and from small Mediterranean compartments like the Taranto Gulf to colossal zones of subsidence like the Rocky Mountain basin. Secondly, they highlight a number of specific themes such as the petrographic signature of foreland basin deposits, their subsidence history and its relation to orogenesis and the stratigraphic architecture of the basin fill.

The authors were very cooperative and speedy in submitting both their manuscripts and their revisions as required. Our reviewers were most obliging and thorough. Thanks to all of them, the editorial task was considerably lightened. The secretarial staff at the Universities of Oxford and Fribourg are also gratefully acknowledged.

P.A.A. & P.H.

Spec. Publs int. Ass. Sediment. (1986) **8**, 3–12

Foreland basins: an introduction

PHILIP A. ALLEN*, PETER HOMEWOOD† *and* GRAHAM D. WILLIAMS‡

* *Department of Earth Sciences, University of Oxford, Parks Road, Oxford OX1 3PR, U.K.*
† *Institut de Géologiè, Université de Fribourg, Pérolles, CH-1700 Fribourg, Switzerland*
‡ *Department of Geology, University College, PO Box 78, Cardiff CF1 1XL, U.K.*

Foreland basins are fairly easily defined as the sedimentary basins lying between the front of a mountain chain and the adjacent craton. Bound up with foreland basins is the concept of the geosyncline and the idea of 'polarity' of developing mountain belts (Aubouin, 1965), whereby orogenesis advances from the deeper outer 'eugeosyncline' towards the shallower inner 'miogeosyncline' bordering the foreland. Aubouin's type-geosynclines were those of the European Alpine system. Dickinson (1974) formally introduced the term 'foreland basin' and proposed two genetic classes: (1) peripheral foreland basins such as the Indo–Gangetic basin and the north Alpine Molasse basin, situated against the outer arc of the orogen during continent–continent collision (A-type subduction of Bally & Snelson, 1980), and (2) retro-arc foreland basins such as the Late Mesozoic–Cenozoic Rocky Mountain basins situated behind a magmatic arc and linked with subduction of oceanic lithosphere (B-type subduction of Bally & Snelson, 1980). Foreland basins in broad terms correspond to the class of perisutural basins on continental lithosphere associated with major compressional zones of deformation or megasutures. Foreland basins, however, vary greatly, particularly in the state of the lithosphere which supports them, and one should certainly be aware of potential classification and terminological problems where the foreland basin evolves from, for example, a previous trench-accretionary complex or, more commonly, from a highly attenuated passive continental margin. The Alpine foreland basins of Europe are illustrated in Fig. 1.

For many geologists the classic example of a foreland basin is the Molasse basin, skirting the Swiss Alps, and its lateral continuation into Bavaria and Austria in the east and into French Savoy in the southwest (Homewood, Allen & Williams; Pfiffner, this volume). The terrigenous sediments of the Molasse basin, and the coarser detrital facies in particular, have led to the christening of coarse, predominantly continental sediments as 'Molasse' facies. The term

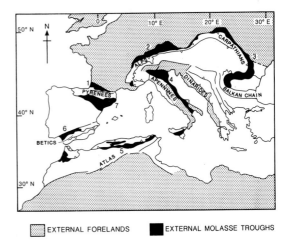

Fig. 1. Distribution of foreland basins and major tectonic elements of Alpine Europe and adjacent regions. Modified from Dewey, Pitman *et al.* (1973). 1, Aquitaine Basin; 2, north Alpine (Molasse) Basin; 3, Carpathian foreland basin; 4, Apenninic basins and Po Basin; 5, Atlas foreland basin; 6, north Betic foreland basin; 7, South Pyrenean and Ebro Basin.

itself has been traced further back to the fourteenth century, referring to building stone in the Rhône valley near St Maurice, Switzerland (Rütsch, 1972).

Since the last century (Bertrand, 1897) the notion of 'Molasse' has been related to the later stages of orogeny, and has been incorporated within the successive paradigms of geosynclines and plate tectonics. Much confusion has arisen over the years concerning the term 'Molasse' and its companion 'Flysch' (Hsü, 1970), largely because they have been used both for lithofacies and tectofacies. It is not our intention to debate the use of the term 'Molasse' here, for it will become apparent that within this volume alone it is used in a number of different senses (cf. Homewood, Allen & Williams; Ricci Lucchi, this volume). A fuller discussion can be found in van Houten (1974) and Mitchell & Reading (1986).

LITHOSPHERIC BEHAVIOUR

Geophysical investigations of the oceanic lithosphere at seamounts, mid-ocean ridges and bordering trenches gave an appreciation of the way in which emplaced loads are supported by the lithosphere (Parsons & Sclater, 1977; Watts, 1978; Cochran, 1979). In the ocean the lithosphere appears to behave elastically and its deflection under a superimposed load is dependent only on its thermal age at the time of loading. An elastic model was subsequently applied (Watts, Karner & Steckler, 1982), to the continental lithosphere. However, the purely elastic model has a number of drawbacks:

(1) it necessitates extremely high elastic bending stresses (Turcotte & Schubert, 1982);
(2) it does not conform to the laws of deformation of rock-forming minerals (Goetze, 1978);
(3) it fails to account for the observed rapid relaxation of the lithosphere to an asymptotic elastic thickness following a loading event (Bodine, Steckler & Watts, 1981).

In contrast, a lithosphere which behaves visco-elastically (Maxwell rheology) would respond very differently under an emplaced load (Walcott, 1970; Beaumont, 1981). The main feature of the visco-elastic model is that following loading the lithosphere should progressively soften, so that its deflection is dependent principally on the time since loading. However, the visco-elastic model made the unacceptable prediction that flexural stresses should eventually vanish to zero over geological time (Karner, Steckler & Thorne, 1983).

From a sedimentologist's or stratigrapher's point of view, there has been a largely futile but nevertheless understandable search for basin geometries that support an elastic lithosphere at one extreme or a visco-elastic lithosphere at the other. Whilst there may have been some real hope of testing the geophysical models in sedimentary basins resulting largely from flexure due to a thermal contraction load (and this is indeed where most effort was expended), the geological complexities of foreland basins made them very poor candidates for this sort of approach. Nevertheless, Jordan (1981) found that subsidence in the foreland basin of the Rockies in western U.S.A. could be adequately explained by a purely elastic model, while Beaumont (1978) preferred a visco-elastic behaviour to account for the stratigraphy and subsidence of the Alberta basin, Canada.

The importance of lithospheric composition and temperature structure in determining its response to loading has been emphasized by Kusznir & Karner (1985). The temperature and stress dependency of lithospheric material has been used to construct a thermo-rheological model which satisfactorily accounts for the long-term finite rigidity of the lithosphere as well as the short-term rapid relaxation. Such a model, involving an initial softening but followed by a period of near-constant or asymptotic rigidity, has yet to be tested against well documented case-studies.

Since the effective elastic thickness and therefore flexural rigidity of a lithospheric plate is a function of its thermal state (Courtney & Beaumont, 1983), the geological events which take place before collision and flexure are of considerable interest. Foreland basins are commonly superimposed on an already thinned continental margin. Stockmal, Beaumont & Boutilier (1986) provide an analysis of the implications of overthrusting a previously rifted passive margin. Of particular importance is the fact that as overthrusting continues, the orogenic wedge loads progressively more rigid lithosphere. All of these factors, and others, contribute to the preserved stratigraphy of a foreland basin. Clear documentations of an early passive margin history are provided in this volume by Houseknecht in the Carboniferous Arkoma basin, south-central U.S.A., Wuellner, Lehonten & James from the Permo–Carboniferous Val Verde and Marathon basins, west Texas, U.S.A. and Hiscott, Pickering & Beeden from the Palaeozoic of eastern Canada. In Europe, the Mesozoic to Palaeogene Helvetic passive margin of the Alpine orogen is noted by Homewood, Allen & Williams and also by Pfiffner. Finally, the deposits of the western Taiwan foreland basin overlie the mainland China passive margin, as described by Covey. Cross describes the subsidence and deformation history of the Sevier fold-thrust belt which was imposed upon a lithosphere which had previously been heated and stretched in the arc and back-arc region. A similar evolution of back-arc to foreland basin is envisaged by Biddle, Uliana *et al.* for the Palaeozoic Magallanes basin in southern South America and by Audley-Charles for the Banda orogen in SE Asia. Clearly the tectonic 'grain', thickness and the thermal structure of the lithosphere prior to loading can be highly variable.

The Bouguer gravity anomaly over orogenic belts typically shows a major gravity 'low', generally displaced towards the foreland from the greatest topographic relief and, in some cases, an associated gravity 'high' displaced towards the overriding plate

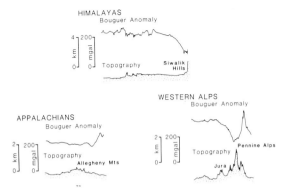

Fig. 2. Gravity profiles across (A) Alps, (B) Appalachians and (C) Himalayas, after Karner & Watts (1983). The Alps and Appalachians both show a large asymmetrical negative Bouguer anomaly coupled with a symmetrical positive Bouguer anomaly. The Himalayas lack the gravity 'high'.

(Fig. 2). The first detailed interpretations viewed the negative anomalies as simply due to an overthickened crust with a mass deficiency and the positive anomalies to the presence of high density rocks at depth. Later interpretations, with the benefit of advances in the field of plate tectonics, placed the gravity anomalies in the context of plate convergence, collision and overthrusting (Brooks, 1970; Daignières *et al.*, 1982; Fountain & Salisbury, 1981). Karner & Watts (1983) specifically interpreted the gravity field in terms of lithospheric flexure. Foreland basins, the sediment mass of which contributes to the negative Bouguer anomaly, can be viewed simplistically as the flexed depression in front of a mountain belt due to the lateral transfer of mass consequent upon plate collision. Because, in the case of the Alps and Appalachians, there was not a good match between the dimensions of the foreland basin and the observed topographic loads, Karner & Watts (1983) postulated the existence of an additional 'hidden' or subsurface load as an extra driving force for foreland flexure and subsidence. Such a hidden load could take the form of density variations within the lithosphere or could be due to transmitted compressional horizontal stresses caused by plate collision. A similar situation exists in the Apennines and Carpathians where the topographic load of the mountain belt is dramatically insufficient to create the observed foreland basin (Royden & Karner, 1984). If the foreland plate has previously been extended and a deep oceanic bathymetry exists prior to overthrusting, very thick overthrust wedges can develop with only modest topographic expressions. Stockmal *et al.* (1986) regard this as sufficient to make the existence of subsurface

hidden loads unecessary. In the Himalayas the topography (or surface load) more closely matches the observed Bouguer anomaly over the foreland basin (Wang, Shi & Zhou, 1982). However, the topographic load of the Himalayas is too great to be supported purely by the elastic stresses of the overridden Indian plate, suggesting that an external force system is present (Lyon-Caen & Molnar, 1983). The external force has been interpreted as gravity acting on the cold mantle component of the broken and weakened Indian lithosphere.

Whereas the main Bouguer anomaly pattern is often a correlatable feature along the strike of mountain belts, significant lateral variations in the flexural rigidity of the Indian plate have been postulated (Lyon-Caen & Molnar, 1985). Flexural rigidities in two profiles just 200 km apart were different by a factor of up to 10. Such large variations cannot be explained by an elastic lithosphere whose rigidity depends only on its thermal age.

SUBSIDENCE HISTORY AND MOBILITY OF DEPOCENTRES

The primary cause of the formation of sedimentary basins by widespread subsidence lies within the lithosphere. A number of mechanisms exist, including continental extension related broadly to upwelling of mantle in the form of hot rising jets or sheets (and subsequent cooling), flexure caused by supra- or subcrustal loading, and downdrag by cold underplated lithosphere.

The subsidence history of rift-sag and continental margin basins has been extensively documented as comprising an initial rapid fault-controlled subsidence followed by an exponentially decreasing subsidence caused by thermal contraction and sediment loading. Although there may be debate as to the precise nature of the initial thermal, basin-forming event, there seems little doubt that the exponential segment of the subsidence curves is a response to lithospheric cooling. Much less is known of the typical subsidence histories of foreland basins. Kominz & Bond (1982) analysed the foreland basins of western North America which resulted from thrusting in the Cordilleran orogenic belt, Beaumont (1981) modelled the Mesozoic to Palaeogene Alberta Basin, Canada and Jordan (1981) studied the Idaho–Wyoming thrust belt and adjacent foreland basin. Kominz & Bond (1986) provided an analysis of the Denver, Green River and Alberta basins from the viewpoint of thermal modelling.

The conclusion of Kominz & Bond (1982, 1986) that the magnitude of subsidence decreases with distance from the orogenic belt is hardly unexpected. It is also documented in this volume by Homewood, Allen & Williams using decompacted geohistory plots. A second feature is the common *relatively* gradual onset of foreland basin subsidence giving a convex-up (accelerating) subsidence curve, although the rate is considerably greater than the previous exponential segment. This is a feature which characterizes the Denver Basin, Green River Basin and Alberta Basin (all summarized in Kominz & Bond, 1986) as well as the Molasse Basin of western Switzerland and the Hoback and south-west Montana zones of the Sevier fold-thrust belt illustrated by Cross. It is certainly not a universal feature, since Cross's geohistory plot for Utah, for example, contains a sharp rather than gradual acceleration in subsidence rate at the initiation of Sevier thrust loading, giving an angular rather than convex-up initial subsidence curve. Sediment accumulation rates are also given for the Oligocene to recent Venetian and Northern Apennine basins in Italy by Massari, Grandesso *et al.* and Ricci Lucchi respectively. Burbank, Raynolds & Johnson provide sediment accumulation rates from the north-western Himalayan foreland basin based essentially on a palaeomagnetic time-scale covering the last 5 Myr. Strong temporal and spatial changes in subsidence rates appear to be correlated with the detailed hinterland tectonics in the case of the north-western Himalaya, and similar features can be discerned from the Cenozoic eastern Andean foreland basin of Argentina. Here, at about 9 Ma, the sediment accumulation rate increased by a factor of 5 in response to uplift of the central Precordillera (Johnson, Jordan *et al.*). An entirely different method, a stratigraphic approach, of estimating differential subsidence using punctuated aggradation cycles is suggested by Anderson, Goodwin & Goodmann.

The migration of depocentres and of feather-edge pinch-outs is a feature common in foreland basins. Ricci Lucchi describes a stop-start style of migration in the northern Apennines. The Apenninic foreland basin climbed the Adriatic margin at a speed of 5–10 mm yr^{-1} during the Oligo–Miocene 'flysch' stage but migration was erratic and slower during the following Plio–Pleistocene 'molasse' stage. Homewood, Allen & Williams, using the palinspastically restored positions of pinch-outs on the European foreland in western Switzerland, speculate on a gradual slowing down in the rate of migration of the depositional realm from 9 mm yr^{-1} in the Oligocene

to just 2 mm yr^{-1} in the Miocene. A comparable approach in the Himalayan foreland basin yielded rates of 10–15 mm yr^{-1} averaged over the last 15–20 Myr (Lyon-Caen & Molnar, 1985). A similar rate for depocentre migration is evident from the progressive southward progradation of depositional sequences in the Catalonian foreland basin of the eastern Pyrenees described by Puigdefabregas, Muñoz & Marzo. Burbank, Raynolds & Johnson evaluate the rate of south-westerly progradation of coarse conglomeratic facies from the Himalayan ranges into the foreland basin as 30 mm yr^{-1}, but this represents an almost instantaneous figure that would be dangerous to extrapolate too far back in time. Audley-Charles reports a migration of the axis of the Timor trough towards the Australian craton at a spectacular rate of 75 mm yr^{-1} from mid- to late Pliocene. All of the rates reflect a mobile hinterland so characteristic of the foreland basin type.

Can foreland basin subsidence be confidently discriminated from styles of subsidence due to other mechanisms? The fault-controlled subsidence caused by lithospheric extension is given by McKenzie (1978) and Sclater & Christie (1980) and graphically by Dewey (1982). Its magnitude is controlled by a number of lithospheric thickness, density and temperature terms but also by the stretching factor β. The principal factor for a constant β is the ratio of crustal (seismic) to lithospheric (thermal) thickness, which will vary greatly between, for example, island arcs and old cratons. In North Sea studies the fault-controlled initial subsidence was 38% of the total subsidence, the remainder being attributable to thermal contraction.

Decompacted subsidence curves for the north Alpine Molasse basin are superimposed on the thermal subsidence curves for North Sea Wells (Sclater & Christie, 1980) in Fig. 3. Evidently there is an order of magnitude difference in subsidence rate between the two basin types. In terms of fault-controlled initial subsidence, hypothetical stretching factors in the range 2 to 3 would be required to produce the total subsidence observed in the Molasse basin, assuming the depression to be continuously filled with sediment to the brim and the crustal and lithospheric thicknesses to be 31·2 and 125 km respectively. However, in order to form a 10 km deep sediment-filled basin, stretching factors in excess of those allowable for a lithosphere with no flexural strength (unacceptable) and perfect Airy compensation ($\beta = \infty$) are predicted. The deepest sediment-filled basin permitted by uniform stretching at $\beta = \infty$ is 7·63 km (McKenzie, 1978). In other words, some foreland basins may be confused in terms of

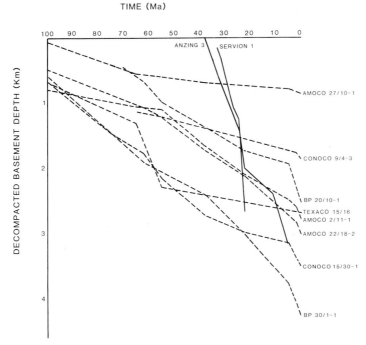

TIME (Ma)

Fig. 3. Decompacted subsidence curves for the north Alpine foreland basin and selected wells from the central North Sea. Stratigraphic columns decompacted using technique outlined by Sclater & Christie (1980). Post-mid-Cretaceous gradient of North Sea wells is determined by amount of extension according to Sclater & Christie. In Molasse basin, Servion 1 is in western Switzerland at front of Subalpine zone, Anzing 3 is close to Münich, southern Germany.

their subsidence histories with the initial extensional subsidence associated with highly stretched continental lithosphere, but not with the longer period of slower thermal subsidence. The close correspondence of Molasse basin decompacted subsidence with that of the southern Rhine graben and Valence basin (Fig. 4) emphasizes this fact.

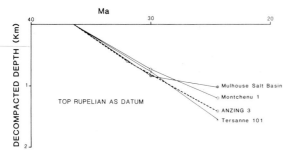

Fig. 4. Decompacted subsidence curves for the Molasse basin (Anzing 3), southern Rhine graben (Mulhouse salt basin) and Valence basin (Tersanne 1 and Montchenu 1) (Debrand-Passard, 1984), showing the similarity of extensional fault-controlled subsidence and foreland basin subsidence.

STRUCTURAL EVOLUTION

In the simplest terms foreland basins develop at the front of active thrust belts where the bulk transport direction is towards the evolving basin. Because the thrust load is inherently mobile the foreland basin itself becomes involved in deformation (Fig. 5). To what extent the basin becomes dissected or becomes completely detached depends on a number of variables including thrust tip propagation rate, availability of subsurface easy-slip horizons underlying the basin and the angle of convergence. Where a basin is accumulating sediment ahead of an active thrust system it can be termed a toe-trough or foredeep *sensu stricto*. Where deformation has progressed under the basin so that it rests on moving thrust sheets it is termed a thrust-sheet-top (Ori & Friend, 1984) or piggy-back basin. These different tectonic settings for sedimentation are conspicuous in the Alpine chains of Europe.

In the Molasse basin of Switzerland sedimentation was generally taking place ahead of the thrust front

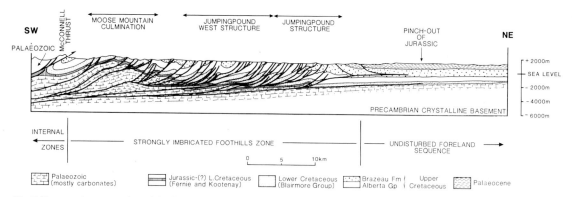

Fig. 5. Structural cross-section of the frontal zone of the Canadian Rocky Mountains, Alberta, showing the intense imbrication of foreland basin deposits in the foothills region and the undisturbed sequence in the NE (after Ollerenshaw, 1975).

according to Homewood, Allen & Williams. Although it is believed that the advancing Alpine thrust sheets were partly buried by erosional detritus, depocentres appear to have always lain close to but ahead of thrust tip lines. Post-depositional tectonics subsequently detached the entire basin in western and central Switzerland as deformation propagated into the Jura province, whilst in the eastern part of the country and further east into Bavaria and Austria there is no evidence of detachment. The décollement of the basin in the west was made possible by the presence of thick Triassic salt in the subsurface (Rigassi, 1977) (Fig. 6). The Apenninic chain presents a very different picture. Here, major basins ahead of the thrust front were coupled with important piggy-back basins located on top of Apenninic thrust sheets (Ricci Lucchi). As the entire system migrated towards the craton, depositional events in both piggy-back and toe-trough basins were synchronized. Ricci Lucchi calls this the 'flysch-stage'. They become decoupled at the end of the Miocene when the thrust belt was uplifted and cannibalized to provide detritus for the molasse of the Po basin. Ori, Roveri & Vannoni provide the Plio–Pleistocene continuation of this sequence of events by seismic stratigraphic and structural studies of the central Adriatic Sea. The Apenninic foreland basin in the Adriatic Sea was strongly affected by contemporaneous Plio–Pleistocene thrust deformations, giving structural culminations subsequently denuded by slope failure and submarine erosion and, in external areas, subtle fold drapes over blind thrusts. In the southern Apennines the same linkage of Miocene piggy-back basins and toe-troughs exists in the Irpinian basin described by Pescatore & Senatore. Eastward and southward migration of deformation in the Neogene

and Quaternary has caused sedimentation to shift to the Taranto Gulf. The Venetian basin in northern Italy is a complex poly-history basin (Massari, Grandesso et al.). In the Oligo–Miocene it was a toe-trough to the Dinaric range in the south-east, but became incorporated in the south Alpine kinematic system in the late Miocene. Subsidence and thrusting then migrated from west to east along the strike of the chain in response to oblique convergence.

The southern Pyrenees, Spain are another example of linked piggy-back and toe-trough basins. Those of the central sector of the southern Pyrenees have previously been documented (Rosell & Puigdefabregas, 1975; Mila & Rosell, 1985 for sedimentological and stratigraphic accounts, Williams, 1985 for a recent structural summary). Puigdefabregas, Muñoz & Marzo provide an integrated structural, stratigraphic and sedimentological study of the eastern sector in this volume.

Syntectonic unconformities in basin-margin sequences demonstrate contemporaneous tectonic activity and sedimentation. Anadon, Cabrera et al. show that syntectonic intraformational unconformities in Palaeogene sediments of the eastern Ebro Basin, Spain develop in a number of tectonic settings: (1) at thrust fronts, (2) along uplifting structures related to basement-involved strike-slip faulting and (3) linked to syntectonic folding of Mesozoic basement. Tankard's syntectonic unconformities in the Carboniferous of Kentucky, U.S.A. are interpreted as due to uplift in a flexural forebulge region on the cratonward or feather-edge margin of the Appalachian foreland basin. This forebulge, represented by the Cincinnati–Waverly Arch complex, migrated eastward (basinward) during periods of tectonic quiescence, which

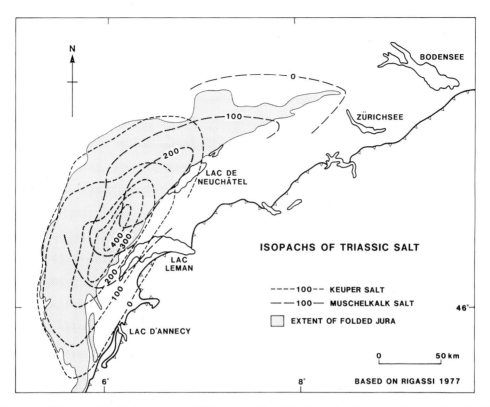

Fig. 6. Isopachs of Triassic salt in the Swiss foreland. The extent of the folded Jura closely matches the subsurface distribution of salt acting as an easy slip horizon (after Rigassi, 1977).

Tankard believes is due to relaxation (softening) of a viscoelastic lithosphere.

Tectonics have a primary influence on sediment dispersal patterns. Uplifting thrust fronts may not act as major sediment suppliers but may instead form barriers to basinward sediment transport. This effect is clearly demonstrated by Hirst and Nichols in the southern Pyrenees. The apices of major fluvial distributary systems are located at structural lows in the thrust front; these systems derived sediments from catchments reaching far into the Pyrenean orogen. By contrast, very small but exquisitely exposed alluvial fans, such as at Aguero (Nichols, 1986), with limited drainage basins abut the thrust front of the External Sierras, the frontal thrust system of the southern Pyrenees. Frontal thrust systems perform the same blocking action on sediment dispersal in the marine sequences of the Apennines (Ricci Lucchi).

The enhanced submarine slopes caused by shortening in the orogenic wedge may, however, cause the shedding of slides, slumps and other gravity flows. The wildflysch and schistes à bloc of the Swiss and French Alps are examples and the denudation complexes of the central Adriatic Sea are another.

Transverse faults, whether they be lateral and oblique ramps, strike-slip faults unrelated to the thrust system or extensional structures, serve as conduits or feeders for the removal of erosional detritus from the orogenic belt to the foreland basin, and if they traverse the basin itself are responsible for large thickness variations in the sedimentary fill. Burbank, Raynolds & Johnson, Ricci Lucchi, Massari, Grandesso *et al.*, Homewood, Allen & Williams and Hirst & Nichols all comment on the influence of such lineaments. Transverse faults developed very early in the Apenninic orogeny. The Sillaro Line, which splits the Marnoso arenacea into two parts, played a predominantly vertical role, the Oligo–Miocene step faulting being inverted in the late Miocene. The Forli Line on the other hand is a neotectonic lineament with active seismicity, yet it was active in the Messinian when it separated different evaporitic basins and at earlier times when it favoured the localization of submarine palaeochannels. The Forli Line may be entirely unrelated to Apenninic thrusting and more akin to the important transversal elements segmenting the southern Alps described by Massari, Grandesso *et al.*

To date, the inversion of old extensional faults either related to continental margin rifting or to other mechanisms such as extension induced by flexure, has not been widely documented. Jackson (1980) has, for example, shown that old listric normal faults are presently being rejuvenated by thrusting in the Zagros Mountains, and Stoneley (1982) describes an example in the Wessex Basin, Britain. Certainly, extensional tectonics are not uncommon in the early stages of foreland basin development. Houseknecht shows in his seismic interpretations across the Arkoma basin normal faults which are truncated by low angle thrusts. In the Alps of eastern Switzerland the earliest deposits (late Eocene to early Oligocene) of the foreland basin fill extensional basins defined by basement faults. Later compression has been concentrated at shallower crustal levels according to the detailed restorations of Pfiffner. His interpretation would apparently preclude the possibility of substantial inversions utilizing the same fault planes in this example.

PETROGRAPHY OF FORELAND BASIN DEPOSITS

Molasse deposits have long been recognized as important in the analysis of late orogenic histories. The occurrence of different clast types or diagnostic heavy mineral assemblages in the sedimentary succession allow the evaluation of unroofing history as deeper structural levels are brought to the surface (e.g. Füchtbauer, 1967). Can light fraction petrography differentiate by itself, foreland basins from other basin types? Schwab tackles this problem by sampling ancient assemblages from the Ouachita, Appalachian, Cordilleran and Alpine belts. They are compared with one another and with modern deep sea and continental margin deposits. Schwab's review suggests that the early fill of foreland basins is quartz-rich and felspar-poor, originating essentially from cratonic sources, whilst later deposits are richer in rock fragments derived from orogenic sources. Only small amounts are derived from uplifted subduction complexes or from magmatic arcs. Lawton and Graham *et al.* give detailed case-studies, concentrating on clast-types. The compositional trend in the Indianola Group of the Cretaceous foreland basin in Utah (Lawton) is of quartz clast enrichment upwards—a reversal of the general trend suggested by Schwab. This reflects the

weathering and erosion of first a Mid- to Late Palaeozoic carbonate sequence in the Sevier fold thrust belt, and then at deeper stratigraphic levels, a Precambrian to Cambrian quartzite and argillite unit. These distinct influxes are linked to major ramp uplift along different thrust faults. As deformation continued, the Indianola Group was folded and subsidence terminated. Graham *et al.*'s study of the Maastrichtian Sphinx Conglomerate in south-west Montana highlights the importance of changing provenance through time. Cambrian through Cretaceous clasts are found in an inverted stratigraphy, and the bedding trends (coarseness and thickness) closely correlate with the availability of resistant clasts in the Laramide foreland thrust belt. Graham *et al.*'s main thesis is therefore that evolution of source terrains plays a fundamental and sometimes dominant role in determining the style of foreland basin sedimentation in addition to the effects of tectonics and climate.

FORELAND BASIN EVOLUTION

The oldest deposits of foreland basins are commonly predominantly fine-grained, often turbiditic sediments which accumulated in sub-shelf water depths. The Murrees of the Pakistan sub-Himalaya, the Taveyannaz and Val d'Illiez sandstones of the north Alpine foreland basin, the Marnoso-arenacea of the northern Apennines and the Hecho Group of the southern Pyrenees (Labaume, Séguret & Seyve, 1985) are all well known examples. Further, but less well known examples documented in this volume are the early deposits of the Pliocene–Pleistocene Taiwan foreland basin (Covey), while the sedimentary fills of the Palaeozoic (Taconic) foreland basin of Quebec (Hiscott, Pickering & Beeden) and the Cretaceous–Tertiary Magallanes Basin, southern South America (Biddle *et al.*) remained essentially deep water throughout. The later deposits of foreland basins are, in contrast, predominantly shallow-water or continental and typify the term 'Molasse'. The Siwaliks of the sub-Himalayas (Graham, Dickinson & Ingersoll, 1975; Parkash, Sharma & Roy, 1980) and the freshwater Molasse of the Alpine perimeter in Europe are excellent examples. The significance of this evolutionary pattern, already recognized by Miall (1978), is discussed at some length by Covey. The early deep water stage of the western Taiwan foreland basin was accompanied by growth of the Taiwan orogen but topography was relatively subdued and sediment delivery rates were relatively low. The later shallow water stage occurred when the mountain belt had grown to a 'steady-state' size and rapid erosion was counterbalanced by uplift. During this phase the basin filled with detritus and thereafter any excess sediment was removed from the foreland basin by fluvial and shallow marine processes. A constant basin geometry was therefore established. Some of the following 24 papers in this volume may provide a substantial data base for those who wish to test this hypothesis.

On a general note, the early underfilled stage of foreland basins may be a natural consequence of loading an initially stretched lithosphere. For a normal unstretched crust, emergence and the shedding of clastic wedges accompanies the onset of shortening, whilst for progressively thinner crusts the onset of rapid clastic sedimentation is delayed more and more in the orogenic cycle, awaiting the emergence of the orogenic belt above sea-level (Dewey, 1982, p. 400). This heightens the need to investigate a prolonged period of lithospheric history before the response to orogenesis by foreland basin development can be fully appreciated.

CONCLUDING STATEMENT

Foreland basins now represent a well-established model with respect to sedimentation and tectonics. The case-histories presented at the Fribourg conference in September 1985 and those presented in this volume suggest a common theme, whether they be recent to present-day, Tertiary, Mesozoic or Palaeozoic examples. This apparent uniformity may be partly due to uncritical obedience to the ruling dogma and we should beware of this possibility.

We like to think that this volume comes at an appropriate time for an overview of the state-of-the-art of research concerning foreland basins. The subject is emphatically not closed and continuing studies and future research will certainly provide different concepts and interpretations. We hope that this volume will play its part in stimulating that progress.

ACKNOWLEDGMENTS

We gratefully acknowledge colleagues John Platt and Harold Reading for reviewing the manuscript, Mary Marsland for typing and Claire Pope for draughting.

REFERENCES

ÀUBOUIN, J. (1965) Geosynclines. In: *Developments in Geotectonics*. Elsevier, Amsterdam, 335 pp.
BALLY, A.W. & SNELSON, S. (1980) Realms of subsidence. In: *Facts and Principles of World Petroleum Occurrence* (Ed. by A. D. Miall). *Mem. Can. Soc. Petrol. Geol.* **6**, 9–75.
BEAUMONT, C. (1978) The evolution of sedimentary basins on a viscoelastic lithosphere: theory and examples. *Geophys. J. R. astr. Soc.* **55**, 471–497.
BEAUMONT, C. (1981) Foreland basins. *Geophys. J. R. astr. Soc.* **65**, 291–329.
BERTRAND, M. (1897) Structure des Alpes françaises et récurrence de certains facies sédimentaires. *Vle. Int. geol. Congr.* (Zurich), pp. 161–177.
BODINE, J.H., STECKLER, M.S. & WATTS, A.B. (1981) Observations of flexure and the rheology of the oceanic lithosphere. *J. geophys. Res.* **86**, 3695–3707.
BROOKS, M. (1970) Positive Bouguer anomalies in some orogenic belts. *Geol. Mag.* **111**, 399–400.
COCHRAN, J.R. (1979) An analysis of isostasy in the world's oceans 2. Mid-ocean ridge crests. *J. geophys. Res.* **84**, 4713–4729.
COURTNEY, R.C. & BEAUMONT, C. (1983) Thermally activated creep and flexure of the oceanic lithosphere. *Nature*, **305**, 201–204.
DAIGNIÈRES, M., GALLART, J., BANDA, E. & HIRN, A. (1982) Implications of the seismic structure for the orogenic evolution of the Pyrenean range. *Earth planet. Sci. Lett.* **57**, 88–100.
DEBRAND-PASSARD, S. (1984) Synthèse géologique du sud-est de la France: Stratigraphie et Paléogeographie. *Mém. Bur. Réch. geol. Min.* **125**, 615 pp.
DEWEY, J.F. (1982) Plate tectonics and the evolution of the British Isles. *J. geol. Soc. London*, **139**, 371–412.
DEWEY, J.F., PITMAN, W.C. III, RYAN, W.B.F. & BONNIN, J. (1973) Plate tectonics and evolution of the Alpine system. *Bull. geol. Soc. Am.* **84**, 3137–3180.
DICKINSON, W.R. (1974) Plate tectonics and sedimentation. In: *Tectonics and Sedimentation* (Ed. by W. R. Dickinson), pp. 1–27, *Spec. Publ. Soc. econ. Paleont. Miner.*, Tulsa, **22**.
FOUNTAIN, D.M. & SALISBURY, M.H. (1981) Exposed cross-sections through the continental crust: implications for crustal structure, petrology and evolution. *Earth planet. Sci. Lett.* **56**, 263–277.
FÜCHTBAUER, H. (1967) Die Sandsteine in der Molasse nordlich der Alpen. *Geol. Rdsch.* **56**, 226–300.
GOETZE, C. (1978) The mechanisms of creep in olivine. *Phil. Trans. R. Soc. A*, **288**, 99–119.
GRAHAM, S.A., DICKINSON, W.R. & INGERSOLL, R.V. (1975) Himalayan–Bengal model for flysch dispersal in the Appalachian–Ouachita system. *Bull. geol. Soc. Am.* **86**, 273–286.
HOUTEN, VAN F.B. (1974) Northern Alpine molasse and similar Cenozoic sequences in southern Europe. In: *Modern and Ancient Geosynclinal Sedimentation* (Ed. by R. H. Dott Jr & R. H. Shaver), pp. 260–273, *Spec. Publ. Soc. econ. Paleont. Mineral.*, Tulsa, **19**.
HSÜ, K.J. (1970) The meaning of the word flysch—a short historical search. In: *Flysch Sedimentology in North America* (Ed. by J. Lajoie), pp. 1–11. *Spec. Publ. geol. Soc. Can.* **7**.
ILLIES, J.H. (1975) Recent and palaeo-intraplate tectonics in

stable Europe and the Rhinegraben rift system. *Tectono-phys.* **29**, 251–264.

JACKSON, J.A. (1980) Reactivation of basement faults and crustal shortening in orogenic belts. *Nature*, **283**, 343–346.

JORDAN, T.E. (1981) Thrust loads and foreland basin evolution, Cretaceous, western United States. *Bull. Am. Ass. Petrol. Geol.* **65**, 291–329.

KARNER, G.D., STECKLER, M.S. & THORNE, J.A. (1983) Long term thermo-mechanical properties of the continental lithosphere. *Nature*, **304**, 250–253.

KARNER, G.D. & WATTS, A.B. (1983) Gravity anomalies and flexure of the lithosphere at mountain ranges. *J. geophys. Res.* **88**, 10,449–10,477.

KOMINZ, M.A. & BOND, G.C. (1982) Tectonic subsidence calculated from lithified basin strata. *Abstr. Progr. geol. Soc. Am.* **14**, 534.

KOMINZ, M.A. & BOND, G.C. (1986) Geophysical modelling of the thermal history of foreland basins. *Nature*, **320**, 252–256.

KUSZNIR, N. & KARNER, G.D. (1985) Dependence of the flexural rigidity of the continental lithosphere on rheology and temperature. *Nature*, **316**, 138–142.

LABAUME, P., SÉGURET, M. & SEYVE, C. (1985) Evolution of a turbiditic foreland basin and analogy with an accretionary prism: example of the Eocene South Pyrenean basin. *Tectonics*, **4**, 661–686.

LYON-CAEN, H. & MOLNAR, P. (1983) Constraints on the structure of the Himalaya from an analysis of gravity anomalies and a flexural model of the lithosphere. *J. geophys. Res.* **88**, 8171–8191.

LYON-CAEN, H. & MOLNAR, P. (1985) Gravity anomalies, flexure of the Indian plate, and the structure, support and evolution of the Himalaya and the Ganga Basin. *Tectonics*, **4**, 513–538.

MCKENZIE, D. (1978) Some remarks on the development of sedimentary basins. *Earth planet. Sci. Lett.* **40**, 25–32.

MIALL, A.D. (1978) Tectonic setting and syndepositional deformation of molasse and other nonmarine-paralic sedimentary basins. *Can. J. Earth Sci.* **15**, 1613–1632.

MILA, M.D. & ROSELL, J. (1985) *6th European Regional Meeting, int. Ass. Sediment., Exc. Guidebook*, Lleida, Spain.

MITCHELL, A.H.G. & READING, H.G. (1986) Tectonics and sedimentation. In: *Sedimentary Environments and Facies* (Ed. by H. G. Reading), 2nd edn, pp. 471–519, Blackwell Scientific Publications, Oxford.

NICHOLS, G.J. (1986) Syntectonic alluvial fan sedimentation, southern Pyrenees. *Geol. Mag.* (in press).

OLLERENSHAW, N.C. (1975) *Map, Geol. Surv. Can. 1457A*, Calgary.

ORI, G.G. & FRIEND, P.F. (1984) Sedimentary basins formed and carried piggyback on active thrust sheets. *Geology*, **12**, 475–478.

PARKASH, B., SHARMA, R.P. & ROY, A.K. (1980) The Siwalik Group (molasse)—sediments shed by collision of continental plates. *Sedim. Geol.* **25**, 127–159.

PARSONS, B. & SCLATER, J.G. (1977) An analysis of the variation of ocean floor bathymetry and heat flow with age. *J. geophys. Res.* **82**, 803–827.

RIGASSI, D. (1977) Genèse tectonique du Jura: une nouvelle hypothèse. *Paleolab. News* **2**, Terreaux du Temple, Geneva.

ROSELL, J. & PUIGDEFABREGAS, C. (1975) The sedimentary evolution of the Paleogene South Pyrenean basin. *9th Int. Sed. Congr.*, Nice, France, *Guide Book Exc. 19.*

ROYDEN, L. & KARNER, G.D. (1984) Flexure of the lithosphere beneath the Apennine and Carpathian foredeep basins. *Nature*, **309**, 142–144.

RÜTSCH, R.F. (1971) Région-type et faciès de la Molasse. *Arch. Sci. Genève*, **24**, 11–15.

SCLATER, J.G. & CHRISTIE, P.A.F. (1980) Continental stretching: an explanation of the post-Mid Cretaceous subsidence of the central North Sea Basin. *J. geophys. Res.* **85**, 3711–3739.

STOCKMAL, G.S., BEAUMONT, C. & BOUTILIER, R. (1986) Geodynamic models of convergent margin tectonics: transition from rifted margin to overthrust belt and consequences for foreland basin development. *Bull. Am. Ass. Petrol. Geol.* **70**, 181–190.

STONELEY, R. (1982) The structural development of the Wessex Basin. *J. geol. Soc. London*, **139**, 543–554.

TURCOTTE, D.L. & SCHUBERT, G. (1982) *Geodynamics: Applications of Continuum Physics to Geological Problems.* Wiley, New York, 450 pp.

WALCOTT, R.I. (1970) Flexural rigidity, thickness and viscosity of the lithosphere. *J. geophys. Res.* **75**, 3941–3954.

WANG, C.Y., SHI, Y. & ZHOU, W.H. (1982) On the tectonics of the Himalaya and the Tibet plateau. *J. geophys. Res.* **87**, 2949–2957.

WATTS, A.B. (1978) An analysis of isostasy in the world's oceans, 1, Hawaiian–Emperor seamount chain. *J. geophys. Res.* **83**, 5989–6004.

WATTS, A.B., KARNER, G.D. & STECKLER, M.S. (1982) Lithospheric flexure and the evolution of sedimentary basins. *Phil. Trans. R. Soc. A*, **305**, 249–281.

WILLIAMS, G.D. (1985) Thrust tectonics in the south-central Pyrenees. *J. struct. Geol.* **7**, 11–17.

Circum-Pacific and Western Americas

Spec. Publs int. Ass. Sediment. (1986) **8**, 15–39

Tectonic controls of foreland basin subsidence and Laramide style deformation, western United States

TIMOTHY A. CROSS

Department of Geology, Colorado School of Mines, Golden, CO 80401, U.S.A.

ABSTRACT

A variety of Late Mesozoic to Early Cenozoic tectonic events in the Rocky Mountains region are temporally and spatially coincident with inferred variations in kinematics of plate interactions and subduction geometries. This coincidence suggests that these disparate events are the products of a single causal mechanism. The following tectonic features are regarded as genetic expressions of variations in subduction modes and geometries: (1) the history of igneous activity in the western United States; (2) the contrasting styles and loci of deformation along the foreland fold and thrust belt (Sevier style) and the basement-cored uplifts (Laramide style) bordering the northern and eastern margins of the Colorado Plateau; (3) the development and maintenance of the Colorado Plateau as a relatively rigid tectonic block; (4) the timing and geometry of subsidence in the foreland basin; and (5) the disjunct history of subsidence and subsequent uplift of the Colorado–Wyoming region beyond the foreland basin.

During a period of normal subduction (from before 92 to about 80 Ma), thin-skinned décollement style deformation occurred along the Sevier fold and thrust belt opposite the convergent margin. Coeval subsidence of the foreland basin was confined to a relatively narrow zone to the east of the Sevier belt. This subsidence is attributed to lithospheric flexure induced by supracrustal loading of thrust plates and sediment. Geohistory analyses of strata along the axis of the foreland basin indicate that foreland basin subsidence began as early as 115 Ma, with a major episode of rapid subsidence initiated by about 90 Ma.

During an ensuing period of low-angle subduction, the Colorado Plateau was underpinned by subducted lithosphere and behaved as a mechanically rigid block, a consequence of the doubled lithospheric thickness. From about 80 to 67 Ma, rapid subsidence occurred over an anomalously broad region centred about Colorado and Wyoming. This episode of subsidence is attributed to sublithospheric loading and cooling induced by the shallowly subducted oceanic plate. To the north and south of the Colorado–Wyoming locus, foreland basin subsidence continued without interruption coincident with continued foreland folding and thrusting. Another effect of low-angle subduction was the transfer of deformation from the Sevier belt (termination about 75 Ma) to the eastern and northern margins of the Colorado Plateau, coincident with the position of greatest contrast in mechanical properties of the lithosphere. This initiated Laramide style basement-cored uplifts at about 69 Ma. Decoupling of subducted lithosphere from overlying lithosphere at about 50 Ma caused regional uplift and erosional stripping of the Colorado–Wyoming region, lithospheric flexure to the east, and sediment accumulation on the High Plains following a long period of non-deposition.

INTRODUCTION

The Late Mesozoic foreland basin of the western United States displays an unusual discordance in timing, areal limits and geometry of subsidence. In the first phase of its development, the foreland basin existed as a north–south trending, relatively narrow, asymmetric structural trough adjacent and parallel to an eastward advancing foreland fold and thrust belt.

This simple pattern of subsidence adjacent to a linear foreland fold and thrust belt was modified during the Late Cretaceous with the development of a second mode of subsidence in Colorado and Wyoming. In that area subsidence occurred in a broad, sub-circular region centred well to the east of the fold and thrust belt. This second mode of subsidence was coeval with continued development of the foreland fold and thrust belt/foreland basin pair to the north and south of the Colorado–Wyoming region.

In the quest for an explanation of these two spatially disjunct modes of foreland basin subsidence, a coincidence was recognized in the temporal and spatial occurrences of several other seemingly unrelated tectonic events and elements. This coincidence invites an explanation that unites these disparate elements and events by a single causal mechanism. In this study, a large number and variety of geological data were assembled, evaluated and integrated. These constituted the basis for an analysis of the spatial and temporal occurrences and the causes of Late Mesozoic through Early Cenozoic deformation in the western United States. This report presents a synopsis of these data, reviews their significance, and proposes a unified explanation for the genesis of several major tectonic features and events in the Rocky Mountains region.

During the Late Mesozoic and Early Cenozoic, oceanic plates of the Pacific Ocean were converging with and subducting beneath the western margin of North America. Calculated and inferred reconstructions of plate motions indicate that kinematics of subduction and geometries of subducted oceanic plates varied in time and space. Several published models relate absolute plate motions, kinematics of plate interactions, and varying modes of subduction to states of stress, styles and positions of deformation, and emplacement of magmas within the overriding continental plate (e.g. Coney, 1972; Sykes, 1978; Dewey, 1980; Zoback & Zoback, 1980; Cross & Pilger, 1982; Engebretson, Cox & Thompson, 1984, among others). With a knowledge of plate interactions, such models offer predictions about the nature and occurrence of deformation and magmatism which may be tested against geological observations. Alternatively, in the absence of reliable plate reconstructions, they provide a basis for interpreting the causes of observed deformational and magmatic events.

Objectives and outline

The primary intention of this report is to demonstrate temporal and spatial coincidence between the history of plate interactions and the occurrence of major deformational and magmatic events in the western United States. The second intention is to propose a unified explanation for these events which relates them genetically to the kinematics and dynamics of plate interactions and the mechanical properties of the lithosphere within which they occurred. The

occurrence and genesis of the following tectonic elements and events are specifically addressed:

(1) the history of igneous activity in the western United States;
(2) the contrasting styles and loci of deformation along the foreland fold and thrust belt (Sevier style) and the basement-cored uplifts (Laramide style) bordering the northern and eastern margins of the Colorado Plateau;
(3) the development and maintenance of the Colorado Plateau as a relatively rigid tectonic block;
(4) the timing and geometry of subsidence in the foreland basin; and,
(5) the disjunct history of subsidence and subsequent uplift of the Colorado–Wyoming region beyond the foreland basin.

Some of the data and interpretations presented in this report have been published previously. Nonetheless, they are summarized briefly in order to provide an integrated and coherent assessment of the spatial and temporal association among some of the tectonic events and elements listed above. Following the summaries of previous work, new information and compilations of data are presented. These augment the range of tectonic elements and events in the western United States that are temporally and spatially associated. This documentation of coincidence in timing and spatial occurrence among these events provides the basis for suggesting a genetic relationship and for proposing a mechanism that accounts for the observed associations.

Published plate reconstructions provide the temporal, kinematic and dynamic frameworks with which deformational and magmatic events in the western United States may be compared and related. These reconstructions and their geological implications are discussed first. Next, the history of Late Mesozoic to Early Cenozoic magmatism in the western United States is reviewed. Because this history has been thoroughly documented in a number of previous reports, it is summarized cursorily in order to demonstrate consistency and probable genetic relations between the nature of plate interactions and resultant igneous activity.

The history of deformation in the Rocky Mountains region is considered next. This discussion focuses on the disjunct modes of foreland basin subsidence, on the development and maintenance of the Colorado Plateau as a rigid tectonic block, and on the contrasting styles and loci of deformation along the fold and thrust belt (Sevier style) and the basement-cored uplifts

(Laramide style) bordering the northern and eastern margins of the Colorado Plateau. In contrast with the excellent summaries of magmatic history, there does not exist a comparable synthesis of deformation in the Rocky Mountains region for the same time period. Documentation of the varying styles and histories of deformation are contained either in studies of specific, local structural elements or in reviews more limited in regional and temporal scope than is attempted in this report. From a critical review of pertinent literature, a set of geological data that constrains the temporal and spatial limits of deformation in the Rocky Mountains region was assembled. Only those tectonic elements for which geological evidence provides an accurate age of initiation, duration, and/or cessation of deformation are included in this synopsis. These data demonstrate spatial and temporal coincidence between specific modes and geometries of subduction and the major tectonic events and elements listed above.

Finally, a genetic relationship is proposed among the several, seemingly independent, tectonic events and elements and the history of plate interactions along the western United States. From theoretical considerations and empirical observations of contemporary subduction systems, the geological expressions of particular modes of subduction have been recognized (as summarized by Cross & Pilger, 1982, for example). These provide a rationale for explaining the observed histories of deformation and magmatism in the western United States by a particular evolution of plate interactions. Moreover, the changes in geometry of subducted lithosphere associated with this evolution provide a mechanism that accounts for the temporal and spatial occurrence and the nature of deformation in the Rocky Mountains region. It appears that several major, seemingly unrelated, tectonic events and elements in the western United States are the united consequences of a particular evolution of plate interactions.

HISTORY OF PLATE INTERACTIONS FROM PLATE RECONSTRUCTIONS

The history of interaction between North America and the oceanic Kula, Farallon and Pacific plates originally was described by Atwater (1970) and Atwater & Molnar (1973). This history, as determined by rotations of magnetic signatures of oceanic crust, is reasonably well constrained for the past 80 Myr. Refinements and temporal extensions of their plate reconstructions have been accomplished by inclusion of the hot spot, or absolute motion, reference frame in reports by Engebretson, *et al.* (1984), Engebretson, Cox & Gordon (1984), Henderson, Gordon & Engebretson (1984) and Jurdy (1984), among others. Absolute motion models extend the reconstructions to 150 Ma. Although these are non-unique solutions, their validity may be assessed by comparison with the temporal and spatial occurrence of geological events in the western United States that reflect plate interactions. Plate reconstructions for times earlier than 150 Ma rely on palaeomagnetic data from continents and uniformitarian arguments relating varying modes of subduction to observed temporal and spatial patterns of magmatism and deformation in the western United States.

The absolute motion of North America and the relative convergence between North America and the Farallon plate, as derived by Engebretson and co-workers, are shown schematically in Figs 1 and 2. More graphic, but inferential, representations of the history of plate interactions and variations in subduction geometries are presented in Figs 8 and 9. From about 135 to 127 Ma, the Farallon plate was converging with North America at a rate of about 70 km Ma^{-1}, and from 127 to 100 Ma the convergence velocity decreased slightly to about 55 km Ma^{-1}. From 100 to 75 Ma, convergence was oblique to the subduction zone, but at an increased velocity of about 100 km Ma^{-1}. During the entire 145–85 Ma period, North America was moving to the NW at velocities of about 30 km Ma^{-1} in the hot spot reference frame, or oblique to the inferred N30°W trend of the subduction zone separating the North American and Farallon plates. These rates and orientations of relative convergence combined with the oblique absolute motion of North America toward the trench should result in normal (moderate- to steep-angle) subduction and development of a volcanoplutonic arc along the western margin of North America. Moderate-angle subduction also should be espressed as Cordilleran style crustal shortening and foreland fold and thrust deformation as the hot, ductile, isostatically uplifted back-arc region is compressed against the colder, more rigid craton (Armstrong & Dick, 1974; Cross & Pilger, 1982).

At about 75 Ma, the North American plate changed direction in the absolute motion frame and moved south-westward, toward and normal to the trench. Between about 65 and 47 Ma, the absolute motion of North America attained a maximum rate of nearly 50 km Ma^{-1}. From 75 to 44 Ma the relative conver-

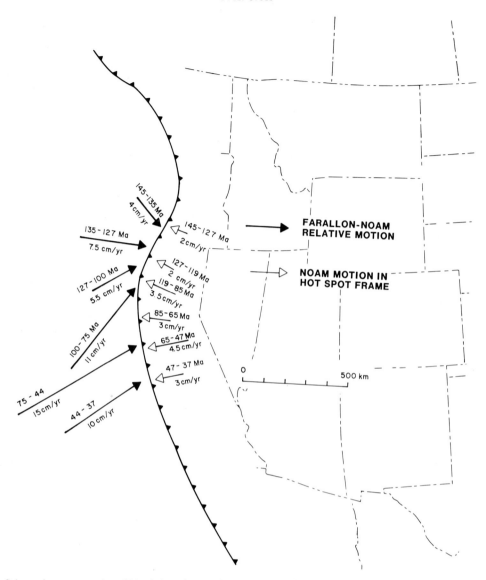

Fig. 1. Schematic representation of North American and Farallon plate motions from 145 to 37 Ma. The inferred position of the subduction zone along the western United States is indicated by the thrust symbol. Open arrows show orientation and velocity of North America in the absolute motion or hot spot reference frame. Closed arrows show relative convergence velocities and orientations of the North American and Farallon plates. Arrows are scaled to the velocities of absolute and relative motions, respectively. Plot derived from plate reconstruction parameters of Engebretson (1972).

gence velocity between the Farallon plate and western United States reached a maximum of 150 km Ma^{-1} and the orientation of convergence was perpendicular to the subduction zone. This combination of high convergence rate and rapid overriding of the trench by North America in the absolute motion frame should result in a shallower angle of subduction. With

shallow subduction there should be a concomitant shift of the volcanoplutonic arc away from the trench or, alternatively, cessation of magmatism (Cross & Pilger, 1982).

Cross & Pilger (1978b) and Livaccari, Burke & Sengor (1981) speculated that, during approximately the same time period, an aseismic ridge on the

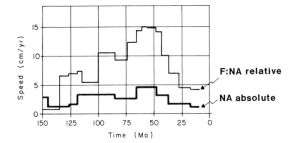

Fig. 2. Velocities of North American and Farallon plate motions from 150 to 10 Ma. Heavy line represents the velocity of North America in the hot spot reference frame. Light line represents the relative convergence velocity between the Farallon and North American plates in a direction perpendicular to an assumed plate boundary trending N40°W. Figure redrawn from Engebretson *et al.* (1984).

Farallon plate was subducted beneath the central part of western United States. They argued that the increased crustal thickness beneath the aseismic ridge and consequent decreased average lithospheric density made that segment of the Farallon plate more buoyant and either contributed to or caused the development of an anomalously low-angle subduction segment. Subsequent refinements in techniques and approaches of plate reconstructions have provided the means of testing these speculations. Henderson *et al.* (1984) reconstructed the positions of aseismic ridges on the Farallon plate by assuming a mirror imagery of aseismic ridge formation on the Pacific (observable) and Farallon (subducted and not observable) plates. Their reconstructions show that, if it existed, an aseismic ridge intersected the trench near Cape Mendocino, California by 65 Ma. During the following 15 Myr the intersection of the postulated aseismic ridge and the North American plate margin moved rapidly to the south. As a consequence of the relative motions between North America and the Farallon plate, the subducted ridge was beneath portions of the central part of western United States from at least 65–55 Ma. Owing to uncertainties in the plate reconstructions, the location of the thermal anomaly that created the ridge, and the maximum age of the ridge, a 10 Myr error in timing of intersection is permitted. From their discussion of these uncertainties, it is evident that any errors are most likely to shift the time of initial intersection to an earlier date. Subduction of an aseismic ridge should decrease the angle of subduction, extinguish the arc above the subducted ridge, and increase the effectiveness of coupling and stress

transmission between the subducting and overriding plates. (Shallow-angle subduction from any other cause would produce the same effects.)

At about 45 Ma, the velocity of the North American plate decreased to about 30 km Ma^{-1} and, concurrently, the relative convergence velocity between North America and the Farallon plate dropped to about 100 km Ma^{-1}. Relative convergence velocity continued to decrease rapidly to a minimum of about 50 km Ma^{-1} by 25 Ma. The anticipated result of these plate interactions is renewed normal subduction and gradual resumption of a coastal volcanoplutonic arc. Slow absolute motion of the North American plate toward the trench combined with slow convergence should create an extensional stress regime in the lithosphere of the upper plate. This regime would be expressed by regional extension in the weak, ductile portions of the crust previously heated by widespread magmatism.

PLATE INTERACTIONS INFERRED FROM HISTORY OF MAGMATISM

The Late Mesozoic and Cenozoic history of igneous activity in the western United States has been described and summarized by Christiansen & Lipman (1972), Lipman, Prostka & Christiansen (1972), Armstrong & Suppe (1973), Snyder, Dickinson & Silberman (1976), Cross & Pilger (1978a) and Lipman (1980). Subsequently published data confirm the space–time patterns of magmatism described in those reports.

These and other studies, based primarily on compilations of isotopic age determinations and compositions of igneous rocks, invoke certain assumptions about genetic relationships between the nature of plate interactions and the generation and composition of magma. From these assumptions and the observed space–time pattern of magmatism, there has emerged a coherent picture of the evolution of plate boundaries and the history and nature of plate interactions along the western United States. The assumptions and the inferred histories of plate interactions have been corroborated in at least two important ways. First, the observed distribution of Neogene, subduction-related igneous rocks is coincident with that predicted from independently derived plate reconstructions (e.g. Snyder *et al.*, 1976; Cross & Pilger 1978a). Second, the inferred plate motion history accounts for a variety of tectonic features and events not directly related to

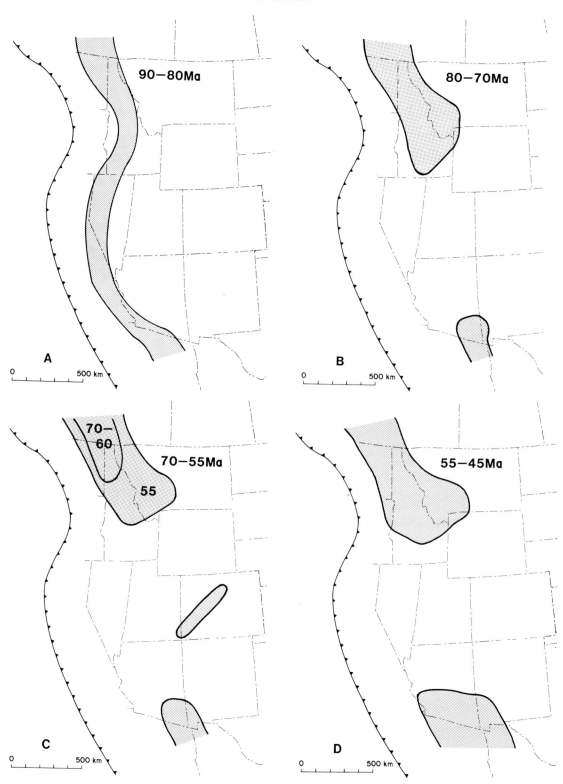

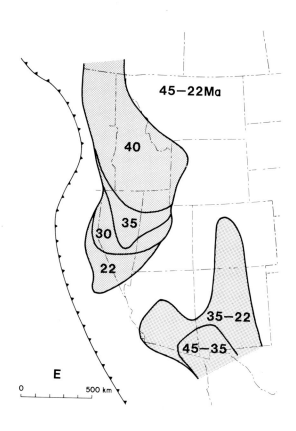

45−22Ma

40

35

30

22

35−22

45−35

E

0   500 km

Fig. 3. Summary of history of magmatism in the western United States. Stipple patterns show positions of volcanoplutonic arc as inferred from isotopic ages of calc-alkaline igneous rocks of silicic and intermediate composition. Ages in million years (Ma) indicated on each figure show geographic limits of the inferred arc for the respective times. Palinspastic map of western United States with Late Cenozoic extension and strike-slip faulting removed is from Hamilton (1978). The inferred position of the subduction zone is indicated as in Fig. 1. (A) From Late Jurassic to about 80 Ma, the Sierra Nevadan volcanoplutonic arc was active along the western margin of the North American plate, indicating a period of normal subduction. (B) During the 80–70 Ma period, a broad swath across the central portion of western United States was devoid of magmatism. Magmatism within the Andean-type volcanoplutonic arc continued to the north and south of this magmatic gap. The magmatic gap is inferred to represent a period of shallow subduction, with normal subduction continuing to the north and south of the gap. (C) The magmatic gap in the central part of western United States continued from 70 to 55 Ma, with the exception of minor volcanism along the Colorado Mineral Belt. A genetic association of the Colorado magmatic episode with subduction is not established. (D) The magmatic gap continued in essentially the same position until about 45 Ma. (E) Beginning at 45 Ma, magmatic centres migrated from the north and south into the region previously occupied by the magmatic gap. Thus, the magmatic gap was progressively diminished in size and by 22 Ma magmatism was renewed along essentially the entire margin of western United States. This is inferred to reflect renewed normal subduction. Magmatism in Colorado and New Mexico from 35 to 22 Ma may reflect either regional extension or partial melting of the former shallowly subducted plate after it detached from the upper plate and sank under gravitational force.

magmatism (e.g. Armstrong, 1974; Cross & Pilger, 1978a).

Because the history of igneous activity and its implications for the history and nature of plate interactions are treated in detail in other reports, only a synopsis is given here. Throughout the Jurassic and most of the Cretaceous, igneous activity was confined to a narrow belt along the western margin of the North American plate. Here was developed an Andean-type volcanoplutonic arc above a steeply dipping, eastward verging subduction zone (Fig. 3A; Hamilton, 1969).

At 80 Ma, magmatism waned dramatically in the Sierra Nevadan province of California, Nevada and southern Oregon, and ceased by 70 Ma (Evernden & Kistler, 1970; Chen & Moore, 1982). With the extinction of the Sierra Nevada arc in this region, a broad swath devoid of magmatism occupied the central part of western United States from about 80 to 72 Ma (Fig. 3B). To the north and south of this magmatic gap, coastal volcanoplutonic arcs extended

from Canada into northern Washington and Idaho, and from Mexico into southern Arizona and New Mexico. Between 80 and 70 Ma, the volcanoplutonic arc in the Pacific Northwest expanded and migrated south-eastward into south-western Montana. Active magmatism in southern Idaho and Montana ceased at 70 Ma. A similar, but less pronounced, expansion and eastward shift occurred along the southern volcanoplutonic arc in Mexico and southern Arizona. Coney (1972) recognized that these major shifts in the loci of igneous activity were coeval with equally fundamental changes in tectonism and attributed these changes to reoriented and accelerated motion of the North American plate.

Minor plutonism and associated volcanism occurred along the Colorado mineral belt from 72 to 55 Ma (Fig. 3C). However, a genetic relationship between these small magmatic centres and subduction is not established. With the exception of this minor episode of igneous activity in Colorado, the magmatic

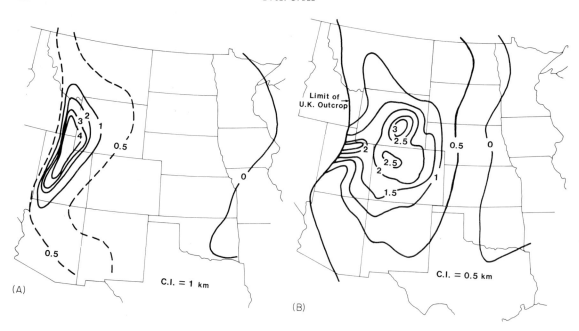

Fig. 4. Two spatially disjunct modes of Cretaceous subsidence are recognized in the foreland of the western United States. (A) One mode was confined to a north–south trending, relatively narrow, asymmetric structural trough. This was colinear with and immediately adjacent to the Sevier fold and thrust belt. Subsidence of the foreland basin (strict sense) is attributed to loading by thrust sheets and derived sediments as discussed in the text. Isopachs show restored thickness of Upper Albian through Santonian strata. (B) The other mode of subsidence occurred during the Campanian and Maastrichtian (84–66 Ma). It was confined to a broad region centred about Wyoming and Colorado. Supracrustal loading by thrust sheets and sediment is insufficient to cause the observed subsidence. Instead, subsidence is attributed to subcrustal loading and cooling by a shallowly subducted oceanic plate. The geographic position and timing of this mode of subsidence in the foreland is coincident with the location and timing of shallow subduction inferred from the history of igneous activity and the development of the magmatic gap. Isopachs show restored thickness of Campanian and Maastrichtian strata. Figure modified from Cross & Pilger (1978b).

gap in the central part of western United States continued to about 45 Ma (Fig. 3D). During the 60–45 Ma period, the northern limit of the magmatic gap is clearly defined by centres of volcanism and plutonism in southern Idaho and south-western Montana. The southern limit of the magmatic gap was essentially stationary from 80 to 45 Ma as defined by continuous igneous activity in southern Arizona and south-western New Mexico. Cessation of magmatism in the Sierra Nevadan arc and development of the magmatic gap has been interpreted in many reports as a consequence of low-angle subduction (e.g. Hyndman, 1972; Lipman *et al.*, 1972; Coney & Reynolds, 1977; Snyder *et al.*, 1976; Cross & Pilger, 1978a; Keith, 1978).

After 45 Ma, magmatism migrated from the north and south to fill the previously amagmatic area (Fig. 3E). Southward expansion of the arc from southern Idaho was rapid, such that by 35 Ma a volcanoplutonic

arc extended continuously from the Pacific Northwest to southern Nevada. Northward expansion of the arc from southern Arizona was slower and less pronounced. By 25 Ma, a volcanoplutonic arc again occupied the western margin of the United States, except for a narrow swath in southernmost Nevada and adjacent regions to the east and west. During the same period (45–25 Ma), magmatism was renewed in the southern Rocky Mountains in Colorado and New Mexico. This episode of igneous activity was geographically and volumetrically restricted from 45 to 35 Ma, but expanded rapidly between 35 and 25 Ma. Expansion of magmatism into the central part of western United States has been interpreted in numerous reports (for example, references cited in the preceding paragraph) as a consequence of renewed normal subduction, or flexure and steepening of the previously shallowly subducted lithosphere. Cross & Pilger (1978a) attributed this episode of igneous activity in the southern Rocky Mountains to decoupling of the shallowly subducted Farallon plate segment from the overlying North American plate, and consequent gravitational sinking, heating and magma generation.

PLATE INTERACTIONS INFERRED FROM SUBSIDENCE IN THE FORELAND

Cross & Pilger (1978b) recognized two spatially disjunct modes of subsidence in the foreland basin of the western United States. One mode, considered the foreland basin in the strict sense, was confined to a north–south trending, relatively narrow, asymmetric structural trough colinear with and immediately adjacent to the Sevier fold and thrust belt (Fig. 4A). Cretaceous to Early Cenozoic subsidence of the Wyoming and Utah sector of the foreland basin was described by Jordan (1981). During evolution of the basin, the depositional and structural axis migrated eastwards in response to the eastward advance of thrust sheets and sediment loads. Through numerical modelling, Jordan demonstrated that flexure of lithosphere in response to loading by thrust sheets and derived sediments is sufficient to explain the observed subsidence history and the asymmetry of this sector of the foreland basin.

The other mode of subsidence occurred during the Campanian and Maastrichtian and was confined to a broad region, well to the east of the foreland basin proper, in southern Wyoming and western Colorado (Fig. 4B). Isopachs of Campanian and Maastrichtian strata are sub-circular in map pattern and depict centres of maximum thicknesses (3 km) displaced approximately 300 km to the east of the foreland basin structural axis. Subsidence in the Colorado–Wyoming locus was coeval with subsidence of the foreland basin (strict sense) to the north and south, and Jordan's (1981) study showed that supracrustal loading by thrust sheets and sediment is insufficient to cause the subsidence. (The Campanian and Maastrichtian stages span approximately 24 Myr, a time span greater than is desirable for directly comparing the history of igneous activity, particularly the development of the magmatic gap, with the history of subsidence in the foreland. However, complexity in stratigraphic nomenclature and correlation, and insufficient numbers of isotopic age dates of stratigraphic units over the region have precluded making isopach maps of stratigraphic units representing shorter time intervals. In general, lower Campanian units are thinner than upper Campanian units. Therefore, it is likely that this second mode of subsidence in the foreland developed late in the Campanian and continued through the Maastrichtian.)

Cross & Pilger (1978b) discussed the temporal and spatial correspondence between the Colorado–Wyoming locus of subsidence and the postulated occurrence of a shallowly subducted Farallon plate segment. The inferred geometry of this episode of shallow subduction placed cold oceanic lithosphere in close contact with overlying continental lithosphere, thus displacing hot, less dense asthenosphere. As a consequence of the changed geometry and vertical density structure, Cross & Pilger proposed that two mechanisms, sublithospheric loading and sublithospheric cooling, were sufficient to cause the observed subsidence. Their calculations, assuming perfectly elastic Airy compensation, indicated isostatic (rapid) subsidence of 3 km due to sublithospheric loading. Subsidence due to cooling of the base of the continental lithosphere (slow) was an additional 2 km.

Bird (1984) conducted a rigorous, quantitative test of these proposed mechanisms by Fourier-transform solution of the plate bending equation and by finite-difference thermal modeling. His calculations confirmed that the additional weight of the subducted plate would have depressed the region by at least the observed amount of subsidence. The magnitude of subsidence calculated from his model is as much as 1 km greater than that recorded by stratal thicknesses, and the area of maximum subsidence predicted by the model is broader than that observed. (Note, however, that the stratal thicknesses reported by Cross & Pilger (1978b) were uncorrected for compaction. Conservatively assuming an average loss of 40% porosity during burial of the predominantly shale and mudstone sedimentary section, the cumulative sedimentary thickness of a 3 km column of strata would have been 4.2 km.) Nonetheless, a remarkable correspondence exists between the modelled and the observed subsidence, given that the region of greatest modelled subsidence occurs west of the Sevier fold and thrust belt, an area of initially high elevation that lacks a sedimentary record. In particular, Bird (1984) noted that the 0 and 1 km subsidence contours of his model are essentially coincident with the observed 0 and 1 km isopach contours, and the patterns of the modelled and observed contours are essentially conformable. Bird also noted that subcrustal cooling would augment the subsidence due to subcrustal loading, although its effect would have been temporary and the additional amount of subsidence was not reported.

SEVIER AND LARAMIDE OROGENIES

Background

Late Mesozoic and Early Cenozoic deformation in the Rocky Mountains region traditionally has been categorized as two temporally distinct compressional events termed the Sevier and Laramide orogenies. The former traditionally is regarded as a dominantly intra-Cretaceous event characterized by thin-skinned thrusts and folds of décollement style, which usually, but not exclusively, did not involve basement. Deformation along the Sevier foreland fold and thrust belt was confined to strata of the westward thickening Palaeozoic and Mesozoic sedimentary prism. The most eastern deformational limits of the Sevier fold and thrust belt correspond to the approximate positions of the Palaeozoic craton-margin hingelines. By contrast, the Laramide orogeny traditionally is regarded as a latest Cretaceous to Early Cenozoic event that affected the craton. Laramide structures are characterized by basement-cored uplifts and asymmetric anticlines, typically bounded by high-angle reverse and thrust faults.

Armstrong (1974) and Burchfiel & Davis (1975), among others, drew attention to the lack of a clear temporal distinction between the two events, if they are regarded from the perspective of differing kinematics and styles of strain. They noted that basement-involved deformation of Laramide age was confined to the southern and central Rocky Mountains of New Mexico, Colorado, Wyoming and southern Montana. In this sector of the Rocky Mountains, deformation along the Sevier fold and thrust belt ceased shortly before initiation of Laramide basement-involved deformation. To the north (in Montana, Idaho and Canada) and to the south (in west Texas, New Mexico, Arizona and Mexico), thin-skinned deformation of the fold and thrust belt was continuous from latest Jurassic(?) to Eocene. In summary, basement-involved deformation was latitudinally restricted to the central and southern Rocky Mountains and was contemporaneous with continued thin-skinned deformation to the north and south. Although the terms Laramide and Sevier are traditionally used as temporal distinctions of deformational events throughout the Rocky Mountains (and, the terms often are used as temporal designations for events in other parts of the world as well), it is clear that such usage should be abandoned. Instead, the terms should be restricted to spatially and temporally restricted orogenic events characterized by differing styles of deformation in the Rocky Mountains.

Armstrong (1974) noted the coincidence among three major events: the cessation of Sevier style deformation in Utah and Wyoming; the initiation of Laramide style deformation in the central and southern Rocky Mountains; and, the history of igneous activity in the western United States. He attributed the abrupt change from thin-skinned to basement-involved deformation and the attendant eastward displacement of deformation to a ductility contrast within the continental lithosphere. Sevier style deformation was coeval and latitudinally conterminous with Andean type volcanism and plutonism. The responses to compressive stresses were predominantly ductile shortening within the crust heated by magmatism, and detachment and internal imbrication of the overlying, westward thickening sedimentary prism. Cessation of igneous activity in the Sierra Nevadan arc and formation of the magmatic gap at 80 Ma caused the previously high heat flow of the arc and back-arc region to dissipate. Consequently, the lithosphere in the vicinity of the magmatic gap cooled, thickened and became resistant to deformation as a consequence of its increasing strength. Thereafter, compressive stresses were relieved by buckling and shear of the more brittle upper crust. By contrast, the lithosphere to the north and south of the magmatic gap, opposite the volcanoplutonic arcs, remained hot and more ductile. Armstrong's explanation adequately accounts for the differences in styles of deformation and their regional distributions. However, it neither accounts for the abrupt eastward displacement of deformation, nor for the curvature and specific loci of Laramide structures.

Laramide structures of the southern Rocky Mountains are oriented approximately north–south, coincident with the eastern margin of the Colorado Plateau (Figs 5 and 9D, E). These occupy a band extending about 200 km east from and parallel to the physiographic limit of the Plateau. As Laramide structures are traced into the central Rocky Mountains, structural trends rotate anticlockwise first to NW–SE and then to east–west. Orientations of these Laramide structures conform closely to the northern and northeastern margins of the Plateau. With the exception of the Uinta Mountains, Laramide structures of the central Rocky Mountains also form a band about 200 km in breadth, but which is displaced about

300 km north and NE from the physiographic margin of the Plateau.

Hamilton (1978, 1981) revived earlier speculations about a possible genetic relationship between Laramide structures and their spatial association with the margins of the Colorado Plateau. He explicitly noted the parallelism between orientations of the margins of the Colorado Plateau and orientations of Laramide structures. He reasoned that the relatively undeformed Plateau behaved as a mechanically rigid body, and that Laramide deformation was an expression of compression between a Plateau microplate and the larger plate of the continental interior. He described kinematically the relative motion between the two plates as a rotation 2°–4° about an Euler pole located to the SE in the Texas Panhandle. Finite rotation of two plates about a geometric pole only may describe the sum of their relative motions over a discrete time interval. However, as recognized by Hamilton, the geometric pole need not have remained fixed and the actual relative plate motions through time (as potentially described by instantaneous rotations about moving poles) may have differed in important ways from the sum of the finite rotation.

Gries (1983) criticized the kinematic description of Hamilton on the basis that it failed to account for perceived differences in the amount and timing of crustal shortening along east–west trending (greater and later, respectively) and north–south trending (lesser and earlier, respectively) Laramide structures. Instead, she proposed an alternative kinematic description of relative motion between the Colorado Plateau microplate and the continental interior. Gries suggested that relative motion during Late Cretaceous through early Palaeocene was east–west, whereas, during the late Palaeocene through late Eocene, it was north–south.

Regardless of the eventual resolution of finite or instantaneous rotations describing the motions of the Plateau relative to the continental interior, the more basic questions of causation and mechanism remain unaddressed. If, as seems probable, Laramide structures are geometrically and kinematically related to the Colorado Plateau, what was responsible for the origin and maintenance of the Plateau as a relatively rigid tectonic block? What caused the abrupt eastward displacement and change in style of deformation from the Sevier to the Laramide belts in the central portion of western United States? What controlled the development of Laramide structures along a 200 km broad zone parallel with the northern and eastern margins of the Plateau? What other Late Mesozoic and Early Cenozoic tectonic events and elements may be related genetically to the change from thin-skinned to basement-involved deformation?

Duration and geographical limits of Sevier deformation

Resolution of these questions ultimately requires specific knowledge of the nature, position and timing of deformation. Consequently, a set of geological data that places temporal and spatial limits on deformation in the Rocky Mountains was compiled from pertinent literature. From among the scores of major structural elements that have been described in a very large literature, only a few dozen were selected. Only those structural elements for which the initiation, duration, and/or cessation of deformation are well defined within narrow time limits by cross-cutting relations and palaeontologic or isotopic age determinations are included in the compilation. All stratigraphic, biostratigraphic, and isotopic ages were converted to the 1983 Decade of North American Geology geological time-scale (Palmer, 1983). In all instances, primary literature was examined. Results of this compilation are presented in the map of Fig. 5. Comparison of this map with other reviews of more generalized or more limited temporal and geographic scope will reveal occasional discrepancies in both the structural elements selected as defining times of deformation and the ages of deformation assigned to specific structures. These discrepancies are attributed to repeated usage of derivative, rather than primary, literature and, in a few instances, to less than critical analysis of the data from which interpretations were derived.

The limits of initiation and cessation of Sevier deformation in portions of the Rocky Mountains remain elusive, particularly in the critical region adjacent to the belt of Laramide structures. Ages of thrusting are best documented along the Overthrust Belt of Wyoming, south-eastern Idaho and northern Utah (see review and analysis of Wiltschko & Dorr, 1983). There is good evidence for thrusting of the Crawford and Meade allochthons at 88 Ma. Earlier movement along the Paris and Willard thrusts, beginning at 144 Ma and continuing episodically until 90 Ma, is indicated by conglomeratic strata interpreted as synorogenic clastic wedges. Stratigraphic relations in central Utah (Fouch *et al.*, 1983; Lawton & Mayer, 1982) provide substantial evidence for initiation of thrusting as early as Cenomanian (97·5 Ma). Armstrong (1968) reviewed more specula-

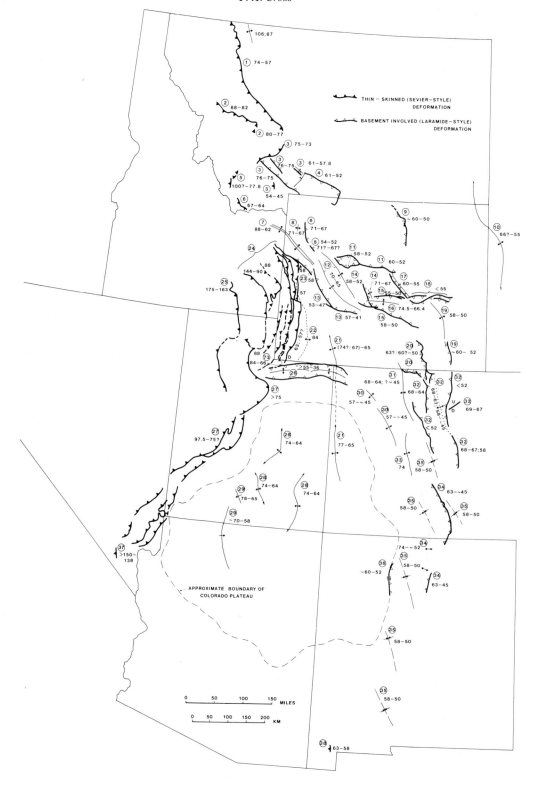

tive evidence that suggested that deformation along the central and southern Utah portion of the Sevier belt began in the Early Cretaceous. Even earlier Middle Jurassic (175–163 Ma) dislocation along the Manning Canyon Shale detachment in northern Utah and southern Idaho was inferred by Allmendinger & Jordan (1981). The earliest siliciclastic sediments definitely derived from the west occur within the Stump Formation (Oxfordian; 156–163 Ma) in western Wyoming, SE Idaho and northern Utah (Brenner 1983). The earliest clastic wedge in the same area occurs within the Morrison Formation (Ephriam Conglomerate) of Oxfordian to Kimmeridgian age (Brenner, 1983). Although these strata were derived from the west, their association with deformation of the Sevier fold and thrust belt has not been established.

In Montana, the earliest well documented occurrence of thin-skinned deformation is thrusting of the Sapphire plate between 88 and 82 Ma (Ruppel, 1963;

Ruppel *et al.*, 1981). Earlier deformation beginning by 100 Ma in SW Montana is suggested by ages of the Beaverhead Formation interpreted as a synorogenic deposit (Ryder & Scholten, 1973). More recent palynologic studies of the Beaverhead Formation by Nichols, Perry & Haley (1985) suggest that deformation did not begin until early Campanian (84 Ma). Lorenz (1982) inferred from an episode of rapid subsidence along the Sweetgrass Arch that initiation of deformation in NW Montana began about 106 Ma.

Along the southernmost extent of the Sevier belt in Nevada, ages of thrusts (Davis, 1973) and of inferred syntectonic conglomerates (Carr, 1980) indicate that deformation began in latest Jurassic (≥ 150 Ma) and continued episodically to 138 Ma.

In summary, structural and stratigraphic relations document that thin-skinned deformation was occurring throughout the Sevier fold and thrust belt by about 90 Ma. Although evidence for earlier initiation

---

Fig. 5. Palaeotectonic map showing positions and age limits of deformational events that are well constrained by available evidence. Sources of information for localities indicated on the map are as follows. **Locality 1** (Montana Disturbed Belt; Glacier National park; NW Montana): Hoffman, Hower & Aronson (1976); Lorenz (1982); Mudge (1972a,b, 1980, 1982); Robinson, Klepper & Obradovich (1968); Schmidt (1978). **Locality 2** (Saphire plate): Ruppel (1963); Ruppel *et al.* (1981). **Locality 3** (SW Montana): Schmidt & Garihan (1983). **Locality 4** (Beartooth Mountains): Foose, Wise & Garabarini (1962). **Locality 5** (Grasshopper and Medicine Lodge plates): Hammons (1981); Perry, Ryder & Maughan (1981); Ruppel *et al.* (1981); Thomas (1981). **Locality 6** (Tendoy thrust): Hammons (1981); Perry & Sando (1982); Perry *et al.* (1981); Ryder & Scholten (1973). **Locality 7** (Targhee Uplift; Ancestral Teton-Gros Ventre Highland): Dorr, Spearing & Steidtmann (1977); Love (1973); Love, Leopold & Love (1978); Wiltschko & Dorr (1983). **Locality 8** (SW flank, Washakie Range): Keefer (1965a, b); Love (1973); Winterfeld & Conard (1983). **Locality 9** (east flank, Bighorn Mountains): Curray (1971). **Locality 10** (Black Hills): Lisenbee (1978); Love (1960). **Locality 11** (southern Bighorn Mountains; southern Owl Creek Mountains; northern Wind River Basin; Casper Arch): Gries (1983); Keefer (1965a, b); Keefer & Love (1963); Love (1978). **Locality 12** (Wind River Range; Wind River Basin): Berg (1963); Gries (1983); Keefer (1965a,b, 1970); Keefer & Love (1963); Love (1960, 1970). **Locality 13** (south flank, Wind River Range): Gries (1983); Love (1970); Steidtmann, McGee & Middleton (1983). **Locality 14** (west flank, Granite Mountains; Emigrant Trail thrust): Keefer (1965a, b); Love (1970, 1971); Reynolds (1978). **Locality 15** (north and south flanks, Granite Mountains; north flank, Great Divide Basin): Gries (1983); Love (1970, 1971); Reynolds (1976, 1978). **Locality 16** (Sweetwater Arch): Love (1960, 1970); Reynolds (1971). **Locality 17** (Casper Arch): Keefer (1965a, b); Keefer & Love (1963); Love (1978). **Locality 18** (north flank, Laramie Range; Casper Mountain): Keefer (1965a, b, 1970). **Locality 19** (NW and east flanks, Laramie Range): Blackstone (1975); Tweto (1975). **Locality 20** (Medicine Bow Mountains): Blackstone (1975); Knight (1953); Tweto (1975). **Locality 21** (Ancestral Rock Springs Uplift; Douglas Creek Arch): Ritzma (1955); Roehler (1961). **Locality 22** (Moxa Arch): Thomaidis (1973); Wach (1977); Wiltschko and Dorr (1983). **Locality 23** (Cache thrust): Kopania (1983); Wiltschko & Dorr (1983). **Locality 24** (northern Sevier fold and thrust belt): Wiltschko & Dorr (1983, and references therein). **Locality 25** (Manning Canyon detachment): Allmendinger & Jordan (1981). **Locality 26** (Uinta Mountains): Hansen (1984); Ritzma (1955, 1971); Standlee (1982). **Locality 27** (central Sevier fold and thrust belt): Armstrong (1968); Standlee (1982); Lawton (1983); Lawton & Mayer (1982). **Locality 28** (San Rafael Swell; Circle Creek Uplift; Monument Upwarp): Lawton (1983). **Locality 29** (East Kaibab and Dutton Mountain monoclines): Bowers (1972); Kelley (1955). **Locality 30** (Axial Arch; White River Plateau): Tweto (1975). **Locality 31** (Park Range): Chapin & Cather (1981); Tweto (1975). **Locality 32** (Front Range): Blackstone (1975); Chapin & Cather (1981); Tweto (1975); Weimer & Davis (1977). **Locality 33** (Sawatch Uplift): Tweto (1975). **Locality 34** (Sangre de Cristo and San Luis Highlands): Chapin & Cather (1981); Dickinson, Leopold & Marvin (1968); Tweto (1975). **Locality 35** (eastern margin of Colorado Plateau): Chapin & Cather (1981). **Locality 36** (Nacimiento Uplift; east flank, San Juan Basin): Baltz (1967); Kelley (1955); Woodward, Kaufman & Anderson (1972). **Locality 37** (southern Sevier fold and thrust belt): Carr (1980); Davis (1973). **Locality 38** (Little Hatchet Mountains): Loring & Loring (1980).

of Sevier deformation is either geographically restricted or more speculative, available data strongly suggest that deformation was widespread by 100 Ma.

An alternative method of defining initiation and duration of deformation is that of geohistory analysis (Van Hinte, 1978). Lithologies, thicknesses and ages of strata at nine locations along the northern and central portions of the Sevier belt were compiled for geohistory analyses. Results of four of these analyses, representative of the set, are presented in Fig. 6. Interpretation of geohistory curves is ambiguous, because the curves only describe the history of subsidence, not the mechanism which induced the subsidence. The shape of the subsidence history curves is typical for foreland basin subsidence (e.g. Jordan, 1981). Because of this observation and because no evidence exists for an alternative subsidence mechanism, it is assumed that a major increase in subsidence rate (the point of inflection on the subsidence history curve) reflects the initial and subsequent flexure of the foreland basin in front of thrust loads. A limitation of these analyses is that the earliest period(s) of thrusting may not be recorded if the thrust load was more than 200 km west of the selected sites.

A convenient starting point for each curve is Early Jurassic, because Late Triassic and Early Jurassic strata in the Rocky Mountains dominantly comprise aeolianite sediments which are assumed to represent deposition at or near sea-level. At one locality in NW Montana (Fig.6A), the geohistory curve shows one inflection point at 108 Ma and a second inflection point occurs at 89 Ma. The subsidence history curve at one location in SW Montana (Fig. 6B) shows identical positions of inflection points. In the Hoback basin of north-western Wyoming (Fig. 6C), the age of initial loading inferred from the geohistory analysis is

114–107 Ma. The most southerly location along the Sevier belt possessing a stratigraphic record sufficiently complete for geohistory analysis is the Gunnison Plateau region of central Utah (Fig. 6D). The inflection point of this curve is at 97 Ma, although substantial subsidence and sediment accumulation (approximately 5 km) occurred between Early Jurassic and 97 Ma. These analyses support observations from field relationships that deformation throughout the northern and central portion of the Sevier belt was initiated in the late Early Cretaceous.

As previously discussed, cessation of deformation is diachronous along the Sevier belt. In Montana and along the Overthrust Belt of Wyoming, Idaho and northern Utah, thrusting was episodic through the early Eocene. Ages of deformation obtained from most well documented structures indicate that thrusting ceased by 57 Ma, although one thrust in Montana (locality 3, Fig. 5; Schmidt & Garihan, 1983) is bracketed as post-54 and pre-45 Ma. By contrast, structural and stratigraphic relations in the central Sevier belt in Utah indicate that the youngest deformation is late Campanian ($\geq$ 75 Ma). This is supported by the geohistory analysis of strata from central Utah (Fig. 6D). An inflection point at 74 Ma marks a major decrease in subsidence rate from 134 m Ma^{-1} during the 92–74 Ma interval to 25 m Ma^{-1} during the 73–23·7 Ma interval.

Duration and geographical limits of Laramide deformation

The age of cessation of Sevier deformation in central Utah marks the initiation of Laramide deformation (compare Figs 7A and 7B). The earliest Laramide

---

Fig. 6. Geohistory analyses at four locations along the Sevier fold and thrust belt. Location of each plot is shown on the small inset map. The origin of each curve in the Early Jurassic is assumed to be at sea-level, because Late Triassic and Early Jurassic strata were deposited as widespread, aeolianite sediments. Solid curves represent total subsidence. Dashed curves represent subsidence of the Early Jurassic surface corrected for the incremental load induced by the weight of sediment through time and, thus, the subsidence due to tectonic loading by thrust sheets. Dots on curves mark time/thickness positions at which calculations were made.

Isostatic corrections were based upon Airy's model of isostasy which assumes perfectly elastic response to loading and ignores the flexural strength of the lithosphere (see, e.g. Watts & Steckler, 1979). Corrections for compaction were based upon lithology and follow the exponential porosity functions presented by Sclater & Christie (1980) with the following modifications: grain-supported limestones were treated as sandstones and mud-supported limestones were treated as shaley sand. No corrections for eustatic variation or palaeobathymetry were made.

Sources of data for each curve are: (A) NW Montana (Mudge, 1982, 1972a); (B) SW Montana (Richards, 1957); (C) Hoback Basin, NW Wyoming (Dorr *et al.* 1977; Wanless, Belknap & Foster, 1955); and (D) Gunnison Plateau–Cedar Hills region, central Utah (Jefferson, 1982; Lawton, 1983; Standlee, 1982; Stanley & Collinson, 1979).

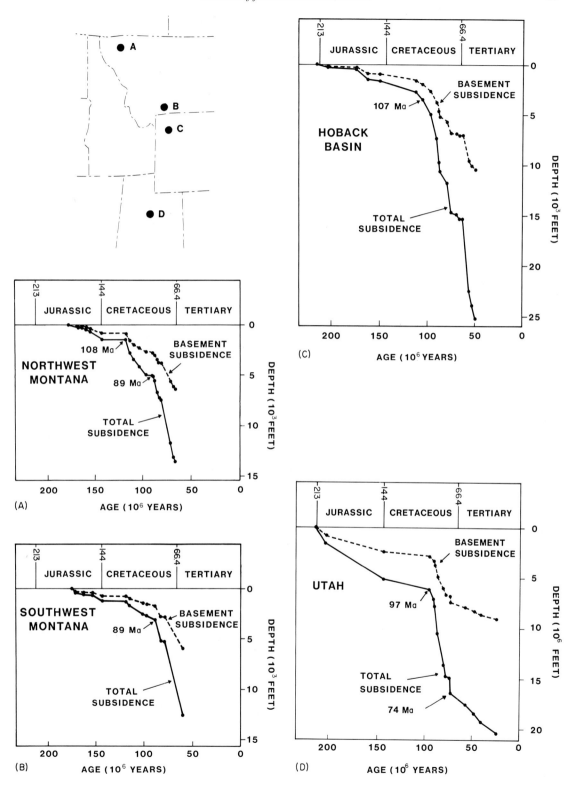

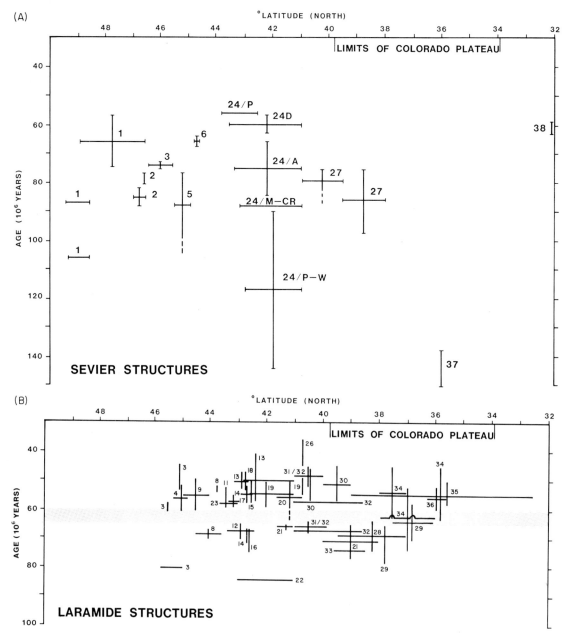

Fig. 7. The times of initiation and cessation of deformation of structures shown on the palaeotectonic map of Fig. 3 are plotted with respect to contemporary latitude. Latitudinal limits of structures are drawn perpendicular to a line of cross-section parallel to and bisecting the trend of the Rocky Mountains (approximately N10°W). The latitudinal limits of the physiographic margins of the Colorado Plateau are identified on each cross-section for geographic reference. (A) Structures associated with Sevier deformation characterized by thin-skinned thrusts and folds of décollement style which typically did not involve basement. (B) Structures associated with Laramide deformation characterized by basement-cored uplifts and asymmetric anticlines, typically bounded by high-angle reverse and thrust faults. The apparent time of hiatus in Laramide deformation is highlighted by the pattern (65–60 Ma).

Laramide structures occupy a space/time frame that differs from that of Sevier structures. Sevier deformation begins earlier than Laramide deformation and initially is continuous along the length of the Rocky Mountains. The earliest Laramide structures developed at or near the beginning of the Maastrichtian (74.5 Ma) at the same time that Sevier deformation in Utah, Wyoming and Idaho ceased. To the north and south of the belt of Laramide style deformation, Sevier style deformation was continuous and coeval with Laramide deformation.

structures developed at or near the beginning of the Maastrichtian (74.5 Ma). Most consist of gentle folds of the Colorado Plateau, the ancestral Rock Springs uplift and Douglas Creek arch, the San Luis Highlands, the Sawatch uplift, and the Sweetwater arch (Fig. 5). An exception to this style of deformation is exhibited by two basement-involved thrusts of earliest Maastrichtian age in SW Montana (Schmidt & Garihan, 1983). Generally, these earliest Laramide structures have a north–south orientation, but some trend east–west and others NW–SE. Several more structures developed later in the Maastrichtian beginning about 69 Ma. These occur along the northern and eastern margins of the Colorado Plateau and again exhibit the same three structural orientations, with the north–south trend being dominant. Two-thirds of these structures are gentle folds and the remainder are basement-involved reverse faults.

If the structures selected in this compilation are representative of the style and timing of deformation of other Laramide structures, the following observations may characterize the early phase of Laramide deformation. Initial deformation was widespread and essentially synchronous throughout the Colorado Plateau and along its margins. Early Laramide deformation is typified by gentle folds and flexures, with minor fault-bounded basement-cored uplifts. By the close of the Cretaceous, the general geographic limits of the Laramide belt were outlined and three structural orientations were defined, the most prominent of which was north–south. Data compiled in this report do not support the speculation of Dickinson & Snyder (1978) that there was a southward migration in the initiation of Laramide deformation. The widespread distribution of early Laramide structures suggests an equally broad distribution and transmission of stress. As discussed subsequently, this widespread distribution stands in marked contrast to the concentration of strain within a relatively narrow curvilinear belt that typifies later Laramide deformation.

There is an apparent hiatus in Laramide deformation during the early Palaeocene. Given the restrictions of the compilation and of the field relations from which temporal limits of deformation were determined, scrutiny of the palaeotectonic map (Fig. 5) reveals that the early phase of Laramide deformation was over by the end of the Cretaceous and that deformation was renewed at 60 Ma. This pause between tectonic episodes is more clearly displayed in the compressed cross-sectional plot of time versus geographic position of deformation (Fig. 7). Because

the selection criteria eliminated many structures generally regarded as 'Laramide' in age, this apparent hiatus of about 5 Myr duration may be an artefact of a small sample size. Nonetheless, two factors argue for its reality. First, the few structures which extend into the 65–60 Ma period (right side of Fig. 7B) are those with a relatively long range between their established limits of initiation and cessation of deformation; the actual period of deformation may have been shorter. Second, as discussed below, the apparent early Palaeocene hiatus separates structures of different styles and locations of deformation.

Whether or not the apparent tectonic pause is a reality, a second phase of Laramide tectonism began at 60 Ma and continued through the early Eocene. Two phases of Laramide tectonism are distinguished by differing styles and locations of deformation. After the early Palaeocene, Laramide structures are dominated by basement-cored, fault-bounded uplifts. This style of deformation contrasts with the broad, ductile folds characteristic of the early Laramide tectonic episode. Structures formed during the late Laramide phase of deformation are confined to a 200 km-wide belt parallel to the margins of the Colorado Plateau. This also contrasts with the more widespread distribution of early Laramide structures. As noted previously, in the southern Rocky Mountains this belt of late Laramide structures is coincident with the eastern margin of the Colorado Plateau. In the central Rocky Mountains the 200 km-wide belt is displaced about 300 km north and NE from the physiographic margin (but not necessarily the palaeotectonic margin) of the Plateau. In addition, orientations of individual late Laramide structures closely parallel the orientation of adjacent Plateau margins. The parallelism between the physiographic margins of the Colorado Plateau and orientations of structures suggests that the latter were formed by compression at the juncture of a relatively rigid Plateau tectonic block and the continental interior, as suggested most recently by Hamilton (1978, 1981).

Inspection of the palaeotectonic map (Fig. 5) and the compressed cross-section (Fig. 7) reveals that Laramide deformation was renewed at 60 Ma along the eastern and northern margins of the Colorado Plateau. By contrast, the end of Laramide deformation is less well established because few structures possess cross-cutting relations, such as strata deposited upon thrust contacts, that narrowly constrain the cessation of deformation. Many structures have limits of cessation that group around 52–50 Ma; others have limits that group around 45 Ma. Otherwise, deformation apparently was continuous along the curvilinear Laramide belt from 60 Ma to middle Eocene.

Gries (1983) postulated a late Palaeocene lull in Laramide deformation which she regarded as reflecting a change from east–west compression during the Late Cretaceous to early Palaeocene to north–south compression during the Eocene. She proposed that north–south trending structures were formed during the early phase of deformation and east–west trending structures were formed during the late phase of deformation. In this view, the Eocene phase of deformation resulted in greater strain than the earlier phase, and produced significant overlap of basement along east–west oriented structures bounded by reverse faults. The data compiled for this report do not show the late Palaeocene lull in deformation postulated by Gries. Rather, deformation appears continuous from late Palaeocene to the early Eocene. Further, the selected data do not support Gries' conclusion that north–south trending structures were formed first and that east–west trending structures were formed last; structures of both orientations are coeval. However, owing to the nature of this compilation, it is not possible to address whether strain was greater along east–west oriented structures than along north–south oriented structures.

Summary

In summary, the Sevier and Laramide orogenies reflect differing kinematic and dynamic responses to compressive stresses that existed within western North America during the Late Mesozoic and Early Cenozoic. Deformation associated with these events was diachronous along the Rocky Mountains region. The Sevier foreland fold and thrust belt (and its equivalents in Canada and Mexico) occurred opposite an Andean-type volcanoplutonic arc in lithosphere rendered ductile by high heat flow in the arc and back-arc region. Laramide deformation occurred within a region devoid of magmatism and above an inferred shallowly-inclined, subducted oceanic plate. Initiation of Laramide deformation was essentially coeval with cessation of Sevier deformation within the same latitudinal belt; thin-skinned deformation of the fold and thrust belt continued to the north and south and was contemporaneous with Laramide deformation.

Available data on the timing of formation of Laramide structures indicate two phases of deformation, the early phase (about 74–65 Ma) characterized by more ductile behaviour of the crust and the late phase (about 60–45 Ma) characterized by more brittle

behaviour. Structures of the early phase are widely distributed throughout the Colorado Plateau as well as along its margins, suggesting that stresses were widely distributed and evenly transmitted. Structures of the late phase, however, are restricted to a 200 km-wide belt parallel with the margins of the Plateau. During both phases of deformation, structures developed along the northern and eastern peripheries of the Plateau have orientations parallel to the Plateau margins. This suggests that the Plateau behaved as a rigid block and that strain occurred between it and the continental interior.

DISCUSSION

The coincidence in timing and spatial occurrence of several disparate tectonic events acts as a strong enticement to provide an explanation which links them by a common genesis. The explanation proposed here is an extension of Armstrong's (1974) insight and suggestion that a ductility contrast within the North American lithosphere was responsible for the localization and differing styles of deformation characteristic of the Sevier and Laramide orogenies. Armstrong attributed the ductility contrast to the precursor history of magmatism; crust in the area of high heat flow behind the active volcanoplutonic arc (Sevier belt) was ductile, whereas crust in the area formerly occupied by the magmatic gap (Laramide belt) was more brittle. The principal deficiency of Armstrong's explanation is that it fails to account for the rapid eastward displacement of deformation (from Sevier to Laramide) and for the localization of Laramide structures within a narrow belt adjacent to the margins of the Colorado Plateau. Also, it does not address nor account for the episode of anomalous subsidence in the foreland, an observation which post-dated Armstrong's paper.

The timing and geographic occurrence of the major events previously discussed are shown in relation to the inferred history of plate interactions as a series of cartoons in Figs 8 and 9. These serve as an illustrative guide to the following proposal that genetically links these events by a single causal mechanism.

During the period of normal subduction, prior to about 80 Ma, décollement style deformation occurred along the Sevier fold and thrust belt in a region of high heat flow behind the Sierra Nevadan arc. Coeval subsidence of the foreland basin was confined to a relatively narrow belt adjacent to the foreland fold and thrust belt. Supracrustal loading by thrust sheets

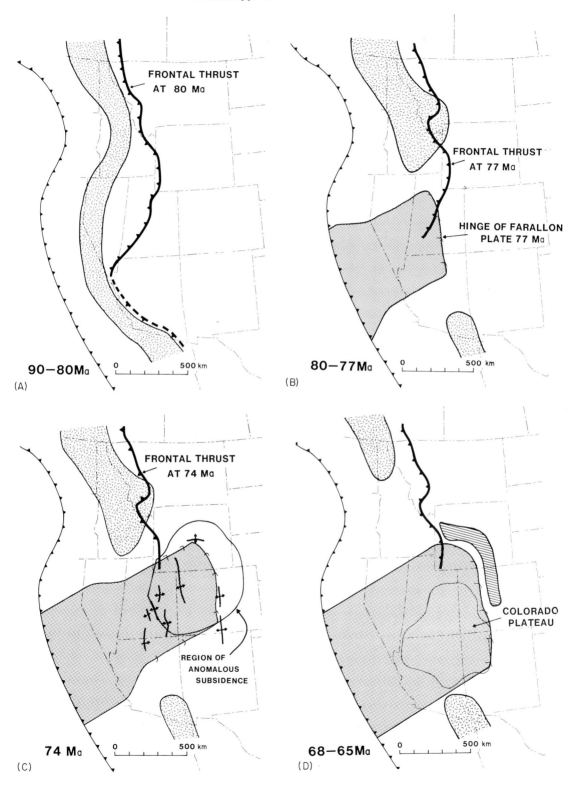

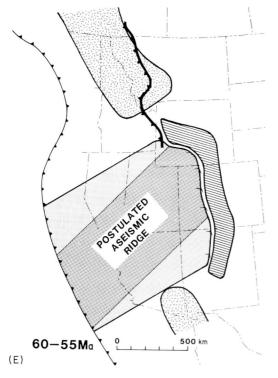

60—55Ma

(E)

Fig. 8. Cartoons showing the inferred history of plate interactions, the evolution of subduction geometries, and the occurrence of the major tectonic and magmatic features discussed in the text. The inferred position of the subduction zone is indicated as in Fig. 1. The eastern deformational limit of the Sevier foreland fold and thrust belt is indicated by a thrust symbol with barbs on the west. The area occupied by Laramide deformation is indicated by the hachured fields and by the anticline symbols. Inferred position of subducted Farallon plate indicated by stipple pattern.

and derived sediment satisfactorily accounts for the location and amount of subsidence in the foreland basin.

Plate reconstructions indicate that North America changed direction and moved, in the hot spot reference frame, perpendicularly towards the trench by about 75 Ma. This change in the absolute motion of North America occurred during a period of rapid Farallon–North America relative convergence. An anticipated consequence of this change in absolute motion is that North America would override the trench and induce low-angle subduction. Inferences from geological data indicate that this episode of low-angle subduction was initiated before 75 Ma. Magmatism in the Sierra Nevadan arc ceased by 80 Ma, suggesting the termination of normal subduction and the initiation of low-angle subduction by then. In addition, if the genetic

correlation of anomalous subsidence in the Colorado–Wyoming region with low-angle subduction is valid, the subducted Farallon plate must have reached Colorado by the mid-Campanian.

During the episode of low-angle subduction, western United States was directly underlain by the Farallon plate and a broad amagmatic area replaced the coastal volcanoplutonic arc. Heat flow in the region of the magmatic gap gradually dissipated, and the lithosphere cooled and became less ductile. Cessation of deformation along the Sevier foreland fold and thrust belt occured at about 75 Ma. At the same time, deformation within the same latitudinal zone was transferred to the broad region of the Colorado Plateau. The first phase of Laramide deformation may be attributed partially to a ductility contrast within the lithosphere of the North American plate, as suggested by Armstrong (1974). However, it is more likely that this effect was augmented by widespread transmission of stress into the overriding plate by the shallowly subducted Farallon plate. Broadly distributed open folds developed within the Colorado Plateau beginning at 74 Ma. The relatively ductile and widespread deformation during this period is inferred as the response to shear(?) coupling between the overriding and subducted plates and the concomitant transmission of stress over a broad area.

Toward the end of the first phase of Laramide deformation, from about 69 to 65 Ma, basement-cored, fault-bounded uplifts developed within a 200 km-wide belt along the north-eastern margin of the Colorado Plateau. Previously, this area had been amagmatic and, presumably, the crust was relatively cool and brittle. Localization of deformation within this narrow belt may be explained by substantial areal differences in the strength of continental lithosphere that were induced by the subducted plate. To the west of the early Laramide belt, under the Colorado Plateau region, North America was underpinned by the Farallon plate. To the east of this belt, the North American plate was underlain by asthenosphere. It is likely that the belt of early Laramide deformation was localized along the juncture between a double and a single thickness of lithosphere, that is, above the hingeline of the Farallon plate where it detached from North America and descended into the asthenosphere. This juncture would represent the position of greatest contrast in mechanical properties of the continental lithosphere and, correspondingly, the position of least lithospheric strength. Stress transmitted into the overriding plate by the Farallon plate would be relieved along the zone of least strength, coincident

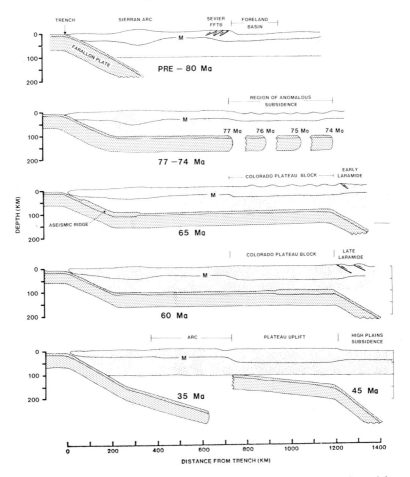

Fig. 9. Cross-sectional representation of Fig. 8 showing the inferred evolution of subduction geometries and the major tectonic and magmatic events considered to be genetic consequences of that evolution.

with the juncture between a single and a double thickness of lithosphere.

This suggestion also explains a long-standing enigma: 'Why did the Colorado Plateau behave as a tectonically rigid block, while regions to the north and south were intensely deformed?' The answer may be simply that the Plateau was underpinned and supported by the shallowly subducted Farallon plate. Because of the doubled lithospheric thickness, the Plateau behaved in a mechanically rigid fashion and intense deformation was restricted to that portion of the crust just beyond its margins.

As previously discussed, plate reconstructions suggest that an aseismic ridge had entered the subduction zone by, or more likely before, 65 Ma. The late

Laramide episode of deformation may correspond to the time when the leading edge of the aseismic ridge reached the hingeline of the Farallon plate and began its descent into the asthenosphere. Presumably, lithosphere containing an aseismic ridge is more buoyant and, during low-angle subduction, would transmit more stress into the overlying lithosphere. In effect, the aseismic ridge would have 'bumped' the overlying lithosphere more intensely than before, but the strain still would have been concentrated along the zone of least strength. Increased transmission of stress by coupling between an aseismic ridge and the overriding plate also may account for the possible greater intensity of late Laramide deformation and its occurrence over a longer curvilinear belt. Although intellectually unsatisfying, no explanation for the pause between early and late Laramide deformation, if it existed, is given at this time.

The end of Laramide deformation at about 50 Ma corresponds to the time that the Farallon plate decoupled from North America. Because the Farallon plate no longer was in contact with overlying continental lithosphere, it no longer could generate and transmit stress into the North American plate. This decoupling event is inferred from the history of igneous activity in which the previously amagmatic area was filled gradually from the north and south by, respectively, southward and northward migration of magmatism. The source of these magmas is not established. It is possible that they were derived by partial melting of the detached Farallon plate as it descended into the asthenosphere. Alternatively, the Farallon plate may have broken, forming an eastern and a western segment, and renewed magmatism was a product of normal subduction of the western segment.

The preceding analysis and the proposed genetic relationship among several tectonic and magmatic events in the western United States is based upon two forms of argument. The first is their coincidental occurrence in time and space. The second involves a logic that links their development to the kinematics of plate interactions and the consequent evolution of subduction geometries. Although the expanations that have been advanced seem compelling at this time, they may be further strengthened through prediction of other events and subsequent verification of such predictions. To this end, two predictions are offered; verification has not been attempted and is beyond the scope of this report.

In analogy with the argument that low-angle subduction causes regional subsidence, decoupling should cause regional uplift of the area previously underpinned by the subducted plate. A potentially fruitful investigation would be to examine the evidence that, after 50 Ma, the Colorado Plateau and the area of Laramide deformation were sites of regional uplift. Some evidence that this occurred is summarized in reports by Mackin (1947), Epis et al. (1980) and Trimble (1980). A second prediction is that the lithosphere to the east of the regional uplift should have flexed downward in response to the supracrustal load induced by the uplift, and the High Plains to the east of the Laramide front should have become a site of deposition during the late Eocene and Oligocene. Again, some support for this prediction is found in summaries of Cenozoic events in the High Plains. A

profound regional hiatus in the High Plains was developed during the Laramide orogeny. During this period, the High Plains were tectonically static and represented a surface of sediment transport by fluvial systems and widespread intense soil development. Thirty million years elapsed between deposition of the last sediments (late Maastrichtian) of the Cretaceous Interior Seaway and the next depositional event beginning in the late Eocene or early Oligocene. Renewed deposition during the middle Cenozoic may represent this predicted episode of lithospheric flexure, but a thorough analysis is warranted.

ACKNOWLEDGMENTS

Over the years that the ideas presented here have been gestating, I have benefited from many stimulating discussions pertaining to the tectonic evolution of the western United States. But, I would like to single out those with Rex Pilger (Louisiana State University) and Mark Baker (University of Texas, El Paso) which have proved particularly challenging and fruitful. The manuscript was reviewed by Philip Allen (University of Oxford) and Bob Dott, Jr (University of Wisconsin, Madison), and I thank them for their many comments that improved the clarity and technical aspects of this report.

REFERENCES

ALLMENDINGER, R.W. & JORDAN, T.E. (1981) Mesozoic evolution, hinterland of the Sevier orogenic belt. Geology, 9, 301–313.
ARMSTRONG, R.L. (1968) Sevier orogenic belt in Nevada and Utah. Bull. geol. Soc. Am., 79, 429–458.
ARMSTRONG, R.L. (1974) Magmatism, orogenic timing, and orogenic diachronism in the Cordillera from Mexico to Canada. Nature, 247, 348–351.
ARMSTRONG, R.L. & DICK, H.J.B. (1974) A model for the development of thin overthrust sheets of crystalline rock. Geology, 2, 35–40.
ARMSTRONG, R.L. & SUPPE, J. (1973) Potassium-argon geochronometry of Mesozoic igneous rocks in Nevada, Utah, and southern California. Bull. geol. Soc. Am. 84, 1375–1392.
ATWATER, T. (1970) Implications of plate tectonics for the Cenozoic tectonic evolution of western North America. Bull. geol. Soc. Am. 81, 3513–3535.
ATWATER, T. & MOLNAR, P. (1973) Relative motion of the Pacific and North American plates deduced from sea-floor spreading in the Atlantic, Indian, and South Pacific Oceans. In: Proc. Con. Tectonic Problems of the San Andreas Fault System. (Ed. by R. L. Kovach & A. Nur). Stanford Univ. Publ. geol. Sci. 13, 136–148.

BALTZ, E.H. (1967) Stratigraphy and regional tectonic implications of part of Upper Cretaceous and Tertiary rocks east-central San Juan Basin, New Mexico. *Prof. Pap. U.S. geol. Surv.* **552**, 101 pp.

BERG, R.R. (1963) Laramide sediments along the Wind River Thrust, Wyoming. In: *Backbone of the Americas. Mem. Am. Ass. Petrol. Geol.* **2**, 220–230.

BIRD, P. (1984) Laramide crustal thickening event in the Rocky Mountain foreland and Great Plains. *Tectonics*, **3**, 741–758.

BLACKSTONE, D.L., Jr (1975) Late Cretaceous and Cenozoic history of Laramie Basin region, southeast Wyoming. In: *Cenozoic History of the Southern Rocky Mountains. Mem. geol. Soc. Am.* **144**, 249–279.

BOWERS, W.E. (1972) The Canaan Peak, Pine Hollow, and Wasatch Formations in the Table Cliff region, Garfield County, Utah. *Bull. U.S. geol. Surv.* **1331-B**, 39 pp.

BRENNER, R.L. (1983) Late Jurassic tectonic setting and paleogeography of western interior, North America. In: *Symposium 2—Mesozoic Paleogeography of West-Central United States*, pp. 119–132. Society of Economic Paleontologists and Mineralogists, Rocky Mountain Section.

BURCHFIEL, B.C. & DAVIS, G.A. (1975) Nature and controls of Cordilleran orogenesis, western United States: extensions of an earlier synthesis. *Am. J. Sci.* **275-A**, 363–396.

CARR, M.D. (1980) Upper Jurassic to Lower Cretaceous (?) synorogenic sedimentary rocks in the southern Spring Mountains, Nevada. *Geology*, **8**, 385–389.

CHAPIN, C.E. & CATHER, S.M. (1981) Eocene tectonics and sedimentation in the Colorado Plateau–Rocky Mountain area. In: *Relations of Tectonics to Ore Deposits in the Southern Cordillera. Ariz. geol. Soc. Digest*, **14**, 173–198.

CHEN, J.H. & MOORE, J.G. (1982) Uranium-lead isotopic ages from the Sierra Nevada batholith, California. *J. geophy. Res.* **87**, 4761–4784.

CHRISTIANSEN, R.L. & LIPMAN, P.W. (1972) Cenozoic volcanism and plate-teconic evolution of the western United States. 2. Late Cenozoic. *Phil. Trans. R. Soc.* **271**, 249–284.

CONEY, P.J. (1972) Cordilleran tectonics and North American plate motion. *Am. J. Sci.* **272**, 603–628.

CONEY, P.J. & REYNOLDS, S.J. (1977) Cordilleran Benioff zones. *Nature*, **270**, 403–406.

CROSS, T.A. & PILGER, R.H., Jr (1978a) Constraints on absolute motion and plate interaction inferred from Cenozoic igneous activity in the western United States. *Am. J. Sci.* **278**, 865–902.

CROSS, T.A. & PILGER, R.H., Jr (1978b) Tectonic controls of Late Cretaceous sedimentation, western interior, USA. *Nature*, **274**, 653–657.

CROSS, T.A. & PILGER, R.H., Jr (1982) Controls of subduction geometry, location of magmatic arcs, and tectonics of arc and back-arc regions. *Bull. geol. Soc. Am.* **83**, 545–562.

CURRAY, W.H., III (1971) Laramide structural history of the Powder River Basin, Wyoming. In: *Symposium on Wyoming tectonics and their structural significance*, pp. 49–60. *Wy. geol. Ass. 23rd Field Conf. Guidebook.*

DAVIS, G.A. (1973) Relations between the Keystone and Red Springs thrust faults, eastern Spring Mountains, Nevada. *Bull. geol. Soc. Am.* **84**, 3709–3716.

DEWEY, J.F. (1980) Episodicity, sequence and style at convergent plate boundaries. In: *The Continental Crust and its Mineral Deposits. Spec. Pap. geol. Ass. Can*, **20**, 553–573.

DICKINSON, R.G., LEOPOLD, E.B. & MARVIN, R.F. (1968) Late Cretaceous uplift and volcanism on the north flank of the San Juan Mountains, Colorado. In: *Cenozoic Volcanism in the Southern Rocky Mountains. Col. School Mines Q*, **63**, 125–132.

DICKINSON, W.R. & SNYDER, W.S. (1978) Plate tectonics of the Laramide orogeny. In: *Laramide Folding Associated with Basement Block Faulting in the Western United States. Mem. geol. Soc. Am*, **151**, 355–366.

DORR, J.A. Jr, SPEARING, D.R. & STEIDTMANN, J.R. (1977) Deformation and deposition between a foreland uplift and an impinging thrust belt: Hoback Basin, Wyoming. *Spec. Pap. geol. Soc. Am.* **177**, 82 pp.

ENGEBRETSON, D.C. (1972) *Relative motions between oceanic and continental plates of the Pacific Basin*. Ph.D. Dissertation, Stanford University. 211 pp.

ENGEBRETSON, D.C., COX, A. & GORDON, R.G. (1984) Relative motions between oceanic plates of the Pacific basin. *J. geophys. Res.* **89**, 10 291–10 310.

ENGEBRETSON, D.C., COX, A. & Thompson, G.A. (1984) Correlation of plate motions with continental tectonics: Laramide to Basin-Range. *Tectonics*, **3**, 115–119.

EPIS, R.C., SCOTT, G.R., TAYLOR, R.B. & CHAPIN, C.E. (1980) Summary of Cenozoic geomorphic, volcanic and tectonic features of central Colorado and adjoining areas. In: *Colorado Geology*, pp. 135–156. Rocky Mountain Association of Geologists.

EVERNDEN, J.F. & KISTLER, R.W. (1970) Chronology of emplacement of Mesozoic batholithic complexes in California and western Nevada. *Prof. Pap. U.S. geol. Surv.* **623**, 42 pp.

FOOSE, R.M., WISE, D.U. & GARABARINI, G.S. (1962) Structural geology of the Beartooth Mountains, Montana and Wyoming. *Bull. geol. Soc. Am.* **72**, 1143–1172.

FOUCH, T.D., LAWTON, T.F., NICHOLS, D.J., CASHION, W.B. & COBBAN, W.A. (1983) Patterns and timing of synorogenic sedimentation in Upper Cretaceous rocks of central and northeastern Utah. In: *Symposium 2—Mesozoic Paleogeography of West-Central United States*, pp. 305–336. Society of Economic Paleontologists and Mineralogists, Rocky Mountain Section.

GRIES, R. (1983) North-south compression of Rocky Mountain foreland structures. In: *Rocky Mountain Foreland Basins and Uplifts*, pp. 169–179. Rocky Mountain Association of Geologists.

HAMILTON, W. (1969) The volcanic central Andes—a modern model for the Cretaceous batholiths and tectonics of western North America. *Bull. Oregon Dept. Geol. Mineral Ind.* **65**, 175–184.

HAMILTON, W. (1978) Mesozoic tectonics of the western United States. In: *Mesozoic Paleogeography of the Western United States: Society of Economic Paleontologists and Mineralogists Pacific Section Pacific Coast Paleogeography Symposium* **2**, pp. 33–70.

HAMILTON, W. (1981) Plate-tectonic mechanism of Laramide deformation. *Univ. Wy. Contr. Geol.* **19**, 87–92.

HAMMONS, P.M. (1981) Structural observations along the southern trace of the Tendoy fault, southern Beaverhead County, Montana. In: *Southwest Montana: Montana Geological Society 1981 Field Conference*, pp. 253–260.

HANSEN, W.R. (1984) Post-Laramide tectonic history of the eastern Uinta Mountains, Utah, Colorado, and Wyoming. *Mountain Geol.* **21**, 5–29.

HENDERSON, L.J., GORDON, R.G. & ENGEBRETSON, D.C. (1984) Mesozoic aseismic ridges on the Farallon plate and southward migration of shallow subduction during the Laramide orogeny. *Tectonics*, **3**, 121–132.

HOFFMAN, J., HOWER, J. & ARONSON, J.L. (1976) Radiometric dating of time of thrusting in the disturbed belt of Montana. *Geology*, **4**, 16–20.

HYNDMAN, R.D. (1972) Plate motions relative to the deep mantle and the development of subduction zones. *Nature*, **238**, 263–265.

JEFFERSON, W.S. (1982) Structural and stratigraphic relations of Upper Cretaceous to Lower Tertiary orogenic sediments in the Cedar Hills, Utah. In: *Overthrust Belt of Utah. Publ. Utah geol. Ass.* **10**, 65–80.

JORDAN, T.E. (1981) Thrust loads and foreland basin evolution, Cretaceous, western United States. *Bull. Am. Ass. Petrol. Geol.* **65**, 2506–2520.

JURDY, D.M. (1984) The subduction of the Farallon plate beneath North America as derived from relative plate motions. *Tectonics*, **3**, 107–114.

KEEFER, W.R. (1965a) Stratigraphy and geologic history of the uppermost Cretaceous, Paleocene, and Lower Eocene rocks in the Wind River Basin, Wyoming. *Prof. Pap. U.S. geol. Surv.* **495-A**, 77 pp.

KEEFER, W.R. (1965b) Geologic history of Wind River Basin, central Wyoming. *Bull. Am. Ass. Petrol. Geol.* **49**, 1878–1892.

KEEFER, W.R. (1970) Structural geology of the Wind River Basin, Wyoming. *Prof. Pap. U.S. geol. Surv.* **495-D**, 35 pp.

KEEFER, W.R. & LOVE, J.D. (1963) Laramide vertical movements in central Wyoming. *Univ. Wy. Contr. Geol.* **2**, 47–54.

KEITH, S.B. (1978) Paleosubduction geometries inferred from Cretaceous and Tertiary magmatic patterns in southwestern North America. *Geology*, **6**, 516–521.

KELLEY, V.C. (1955) Monoclines of the Colorado Plateau. *Bull. geol. Soc. Am.* **66**, 789–804.

KNIGHT, S.H. (1953) *Summary of the Cenozoic History of the Medicine Bow Mountains, Wyoming*, pp. 65–76. *Wy. geol. Ass. 8th Annual Field Conf. Guidebook.*

KOPANIA, A.A. (1983) Deformation consequences of the impingement of the foreland and northern thrustbelt, eastern Idaho and western Wyoming. *Abstr. Prog. Geol. Soc. Am.* **15**, 296.

LAWTON, T.F. (1983) Late Cretaceous fluvial systems and the age of foreland uplifts in central Utah. In: *Rocky Mountain Foreland Basins and Uplifts*, pp. 181–199. Rocky Mountain Association of Geologists.

LAWTON, T.F. & MAYER, L. (1982) Thrust load-induced basin subsidence and sedimentation in the Utah foreland: temporal constraints on the Upper Cretaceous Sevier orogeny. *Abstr. Progr. geol. Soc. Am.* **14**, 542.

LIPMAN, P.W. (1980) Cenozoic volcanism in the western United States: implications for continental tectonics. In: *Continental Tectonics*, pp. 161–174. National Academy of Sciences.

LIPMAN, P.W., PROSTKA, H.J. & CHRISTIANSEN, R.L. (1972) Cenozoic volcanism and plate-tectonic evolution of the western United States. 1. Early and middle Cenozoic. *Phil. Trans. R. Soc.* **271**, 217–248.

LISENBEE, A.L. (1978) Laramide structure of the Black Hills uplift, South Dakota–Wyoming–Montana. In: *Laramide Folding Associated with Basement Block Faulting in the Western United States. Mem. geol. Soc. Am.* **151**, 165–196.

LIVACCARI, R.F., BURKE, K. & SENGOR, A.M.C. (1981) Was the Laramide orogeny related to subduction of an oceanic plateau? *Nature*, **289**, 276–278.

LORENZ, J.C. (1982) Lithospheric flexure and the history of the Sweetgrass arch northwestern Montana. In: *Geologic Studies of the Cordilleran Thrust Belt*, pp. 77–89. Rocky Mountain Association of Geologists.

LORING, A.K. & LORING, R.B. (1980) Age of thrust faulting, Little Hatchet Mountains, southwestern New Mexico. *Isochron/West*, **27**, 29–30.

LOVE, J.D. (1978) Cenozoic thrust and normal faulting, and tectonic history of the Badwater area, northeastern margin of Wind River Basin, Wyoming. *Wyoming geol. Ass. 30th Annual Field Conf. Guidebook*, pp. 235–238.

LOVE, J.D. (1971) Relation of Cenozoic geologic events in the Granite Mountains area, central Wyoming, to economic deposits. In: *Wyoming Tectonics Symposium*, pp. 71–80. *Wy. geol. Ass. 23rd Annual Field Conf. Guidebook.*

LOVE, J.D. (1973) Harebell Formation (Upper Cretaceous) and Pinyon Conglomerate (uppermost Cretaceous and Paleocene), northwestern Wyoming. *Prof. Pap. U.S. geol. Surv.* **734-A**, 54 pp.

LOVE, J.D. (1960) Cenozoic sedimentation and crustal movement in Wyoming. *Am. J. Sci.* **258-A**, 204–214.

LOVE, J.D. (1970) Cenozoic geology of the Granite Mountains area, central Wyoming. *Prof. Pap. U.S. Geol. Surv.* **495-C**, 154 p.

LOVE, J.D. LEOPOLD, E.B. & LOVE, D.W. (1978) Eocene rocks, fossils, and geologic history, Teton Range, northwestern Wyoming. *Prof. Pap. U.S. geol. Surv.* **932-B**, 40 pp.

MACKIN, J.H. (1947) Altitude and local relief of the Bighorn area during the Cenozoic. *Wyoming Geological Association Field Conference in the Bighorn Basin*, pp. 103–120.

MUDGE, R.R. (1972a) Pre-Quaternary rocks in the Sun River Canyon area, northwestern Montana. *Prof. Pap. U.S geol. Surv.* **663A**, 142 pp

MUDGE, R.R. (1972b) Structural geology of the Sun River Canyon and adjacent areas, northwestern Montana. *Prof. Pap. U.S. geol. Surv.* **663B**, 152 pp.

MUDGE, R.R. (1980) The Lewis thrust fault and related structures in the Disturbed Belt, northwestern Montana. *Prof. Pap. U.S. geol. Surv.* **1174**, 18 pp.

MUDGE, R.R. (1982) A resume of the structural geology of the Northern Disturbed Belt, northwestern Montana. In: *Geologic Studies of the Cordilleran Thrust*, pp. 91–122. Rocky Mountain Association of Geologists.

NICHOLS, D.J., PERRY, W.J. Jr & HALEY, J.C. (1985) Reinterpretation of the palynology and age of Laramide syntectonic deposits, southwestern Montana, and revision of the Beaverhead Group. *Geology*, 149–153.

PALMER, A.R. (1983) The decade of North American Geology 1983 geologic time scale. *Geology*, **11**, 503–504.

PERRY, W.J. Jr, RYDER, R.T. & MAUGHAN, E.K. (1981) The southern part of the southwest Montana thrust belt: a preliminary re-evaluation of structure, thermal maturation and petroleum potential. In: *Southwest Montana*, pp. 261–273. Montana Geological Society 1981 Field Conference.

PERRY, W.J. Jr & SANDO, W.J. (1982) Sequence of deformation of Cordilleran thrust belt in Lima, Montana region. In: *Geologic Studies of the Cordilleran Thrust Belt*, pp. 137–144. Rocky Mountain Association of Geologists.

REYNOLDS, M.W. (1971) Geologic map of the Bairoil Quadrangle, Sweetwater and Carbon Counties, Wyoming. *U.S. geol. Surv. Quadrangle Map GQ-913.*

REYNOLDS, M.W. (1976) Influence of recurrent Laramide structural growth on sedimentation and petroleum accumulation, Lost Soldier area, Wyoming. *Bull Am. Ass. Petrol. Geol.* **60**, 12–33.

REYNOLDS, M.W. (1978) Late Mesozoic and Cenozoic structural development and its effect on petroleum accumulation, southwest arm of the Wind River Basin, Wyoming. *Wy. geol. Ass. 30th Annual Field Conf. Guidebook*, pp. 77–78.

RICHARDS, P.W. (1957) Geology of the area east and southeast of Livingston, Park County, Montana. *Bull. U.S. geol. Surv.* **1021-L**, 385–438.

RITZMA, H.R. (1955) Late Cretaceous and Early Cenozoic structural pattern, Southern Rock Springs Uplift, Wyoming. *Wy. geol. Ass. 10th Annual Field Conf. Guidebook*, pp. 135–137.

ROBINSON, G.D., KLEPPER, M.R. & OBRADOVICH, J.D. (1968) Overlapping plutonism, volcanism, and tectonism in the Boulder batholith region, western Montana. In: *Studies in volcanology. Mem. geol. Soc. Am.* **116**, 557–608.

ROEHLER, H.W. (1961) The Late Cretaceous-Tertiary boundary in the Rock Springs uplift, Sweetwater County, Wyoming. In: *Symposium on Late Cretaceous Rocks, Wyoming and Adjacent Areas*, pp. 96–100. *Wy. geol. Ass. 16th Annual Field Conf. Guidebook.*

RUPPEL, E.T. (1963) Geology of the Basin Quadrangle, Jefferson, Lewis and Clark, and Powell Counties, Montana. *Bull. U.S. geol. Surv.* **1151**, 121 pp.

RUPPEL, E.T., WALLACE, C.A., SCHMIDT, R.G. & LOPEZ, D.A. (1981) Preliminary interpretation of the thrust belt in southwest and west-central Montana and east central Idaho. In: *Southwest Montana*, pp. 139–159. Montana Geological Society 1981 Field Conference.

RYDER, R.T. & SCHOLTEN, R. (1973) Syntectonic conglomerates in southwestern Montana: their nature, origin, and tectonic significance. *Bull. geol. Soc. Am.* **84**, 773–796.

SCLATER, J.G. & CHRISTIE, P.A.F. (1980) Continental stretching: an explanation of the post-mid-Cretaceous subsidence of the central North Sea basin. *J. geophys. Res.* **85**, 1711–1739.

SCHMIDT, C.J. & GARIHAN, J.M. (1983) Laramide tectonic development of the Rocky Mountain foreland of southwestern Montana. In: *Rocky Mountain Foreland Basins and Uplifts*, pp. 271–294. Rocky Mountain Association of Geologists.

SCHMIDT, R.G. (1978) Rocks and mineral resources of the Wolf Creek area, Lewis and Clark and Cascade Counties, Montana. *Bull. U.S. geol. Surv.* **1441**, 91.

SNYDER, W.S., DICKINSON, W.R. & SILBERMAN, M.J. (1976) Tectonic implications of space-time patterns of Cenozoic magmatism in the western United States. *Earth planet. Sci. Lett.* **32**, 91–106.

STANDLEE, L.A. (1982) Structure and stratigraphy of Jurassic rocks in central Utah: their influence on tectonic development of the Cordilleran foreland thrust belt. In: *Geologic Studies of the Cordilleran Thrust Belt*, pp. 357–382. Rocky Mountain Association of Geologists.

STANLEY, K.O. & COLLINSON, J.W. (1979) Depositional history of Paleocene—Lower Eocene Flagstaff Limestone and coeval rocks, central Utah. *Bull. Am. Ass. Petrol. Geol.* **63**, 311–323.

STEIDTMANN, J.R., McGEE, L.C. & MIDDLETON, L.T. (1983) Laramide Sedimentation, folding, and faulting in the southern Wind River Range, Wyoming. In: *Rocky Mountain Foreland Basins and Uplifts*, pp. 161–167. Rocky Mountain Association of Geologists.

SYKES, L.R. (1978) Intraplate seismicity, reactivation of preexisting zones of weakness, alkaline magmatism, and other tectonism postdating continental fragmentation. *Rev. Geophys. Space Phys.* **16**, 621–688.

THOMAIDIS, N.D. (1973) *Church Buttes arch, Wyoming and Utah*, pp. 35–39. *Wy. geol. Ass. 25th Field Conf. Guidebook.*

THOMAS, J. (1981) Structural geology of the Badger pass area, southwestern Montana. In: *Southwest Montana*, pp. 211–214. Montana Geological Society 1981 Field Conference.

TRIMBLE, D.E. (1980) Cenozoic tectonic history of the Great Plains contrasted with that of the southern Rocky Mountains. A synthesis. *Mountain Geol.* **17**, 59–69.

TWETO, O. (1975) Laramide (Late Cretaceous—Early Tertiary) orogeny in the southern Rocky Mountains. In: *Cenozoic History of the Southern Rocky Mountains. Mem. geol. Soc. Am.* **144**, 1–14.

VAN HINTE, J.E. (1978) Geohistory analysis—application of micropaleontology in exploration geology. *Bull. Am. Ass. Petrol. Geol.* **62**, 201–222.

WACH, P.H. (1977) The Moxa arch, an overthrust model?. In: *Rocky Mountain Thrust Belt Geology and Resources*, pp. 651–664. Wyoming Geological Association 29th Annual Field Conference.

WANLESS, H.R., BELKNAP, R.L. & FOSTER, H. (1955) Paleozoic and Mesozoic rocks of Gros Ventre, Teton, Hoback, and Snake River Ranges, Wyoming. *Mem. geol. Soc. Am.* **63**, 90 pp.

WATTS, A.B. & STECKLER, M.S. (1979) Subsidence and eustacy at the continental margin of eastern North America. In: *Deep Drilling Results in the Atlantic Ocean: Continental Margins and Paleoenvironment: Am. geophys. Un., M. Ewing series*, **3**, 218–234.

WEIMER, R.J. & DAVIS, T.L. (1977) Stratigraphic and seismic evidence for Late Cretaceous growth faulting, Denver Basin, Colorado. In: *Seismic stratigraphy—Applications to Hydrocarbon Exploration: Mem. Am. Ass. Petrol. Geol.* **26**, 277–299.

WILTSCHKO, D.V. & DORR, J.A. Jr (1983) Timing of deformation in overthrust belt and foreland of Idaho, Wyoming, and Utah. *Bull. Am. Ass. Petrol. Geol.* **67**, 1304–1322.

WINTERFELD, G.F. & CONARD, J.B. (1983) Laramide tectonics and deposition, Washakie Range and northwestern Wind River Basin, Wyoming. In: *Rocky Mountain Foreland Basins and Uplifts*, pp. 137–148. Rocky Mountain Association of Geologists.

WOODWARD, L.A., KAUFMAN, W.H. & Anderson, J.B. (1972) Nacimiento fault and related structures, northern New Mexico. *Bull. geol. Soc. Am.* **83**, 2383–2396.

ZOBACK, M.L. & ZOBACK, M. (1980) State of stress in the conterminous United States. *J. geophys. Res.* **85**, 6113–6156.

Spec. Publs int. Ass. Sediment. (1986) **8**, 41–61

The stratigraphic and structural evolution of the central and eastern Magallanes Basin, southern South America

K. T. BIDDLE, M. A. ULIANA, R. M. MITCHUM JR,
M. G. FITZGERALD *and* R. C. WRIGHT

Exxon Production Research Company, P.O. Box 2189, Houston, TX 77001, U.S.A.

ABSTRACT

The Magallanes Basin is located at the southern edge of the South American plate and is underlain by crust of Palaeozoic age. The initial history of the basin is one of extension associated with the breakup of the South American sector of Gondwanaland. Triassic to Late Jurassic extension produced a normal-faulted terrane with numerous grabens and half grabens. This extensional event also resulted in extensive, dominantly silicic volcanism. The basin floor subsided from the Late Jurassic to the Late Cretaceous with decay of the thermal anomaly associated with extension. During the Late Cretaceous and Tertiary, uplift and shortening occurred along the western and southern edges of the basin, forming the Patagonian Andes and the fold and thrust belt of southernmost South America. Subsidence in the basin during this interval of time was the result of lithospheric flexure caused by loading.

The sedimentary fill of the basin is related to three major phases of basin development. The rift-related Triassic to Middle/Upper Jurassic succession consists of mostly non-marine volcanic and volcaniclastic rocks largely restricted to isolated grabens. Upper Jurassic to Upper Cretaceous, largely retrogradational sedimentary units were deposited while the basin passively subsided on the remnant-arc side of a small marginal sea. Uppermost Cretaceous and Tertiary units were derived from the south, west, and north-west, and show a progressive onlap geometry from west to east. These deposits mark the onset of sedimentation from the Andes, although subsidence caused by tectonic loading started somewhat earlier in the Late Cretaceous. Depositional patterns for this interval consist of fanglomerates separated by deep-water shales from an eastern complex of low sedimentation rate glauconitic sandstones which onlap a long-lived basement high. The most impressive feature formed during the foreland basin stage is a regional composite unconformity that separates rocks as old as Palaeocene from the Mesozoic section.

Production or shows of oil and gas occur in many of the stratigraphic sequences defined in the basin. The producing interval, the Springhall Sandstone, and the major source-rock units were deposited while the basin was a westward-facing remnant-arc margin. Burial of these rocks during the foreland basin stage led to the maturation and migration of hydrocarbons. Thus, the Magallanes Basin is a polyphase foreland basin and each phase of evolution has had a role in making the basin a productive one.

INTRODUCTION

The Magallanes Basin is located at the southern tip of South America (Fig. 1). The early history of the basin is one of Triassic and Jurassic extension associated with the opening of a small marginal sea behind a developing magmatic arc. That marginal sea closed in the mid-Cretaceous, and the Late Cretaceous through Cenozoic history of the basin is one of a foreland basin in front of the rising Andes Mountains. The Magallanes Basin produces significant oil and gas in both Argentina and Chile. The maturation, migration and entrapment of hydrocarbons in the basin is a result of the interplay between the various episodes of basin evolution. Here, we summarize the evolution of the basin and discuss how this has controlled the development of oil and gas accumulations.

Our study is based primarily on subsurface data from the central and eastern parts of the basin. We have used reflection seismic, well-log, lithological, and

palaeontological data to divide the sedimentary fill of
the basin into 17 unconformity-bound sequences.
Thickness and generalized depositional settings for
six sequences, or groups of sequences, are presented
here. These sequences and sequence groups were
chosen to document important steps in the evolution
of the basin. The reflection seismic data also show the
types of structures found in the basin, and provide
information on the timing of their formation. Because
our study is dominantly a subsurface one, it comple-
ments previous work on the outcropping basin fill (see

below), and provides a more complete view of the
basin than has been possible in the past.

Previous work

Early geological research in the southern Andes
was related to European scientific expeditions (Wilckens, 1907; Quensell, 1911, 1912) and to exploration
for coal (Bruggen, 1913; Felsch, 1916; Bonarelli,
1917). Work on the area up to the 1930s was
summarized by Kranck (1930a, b, 1932) and Wind-

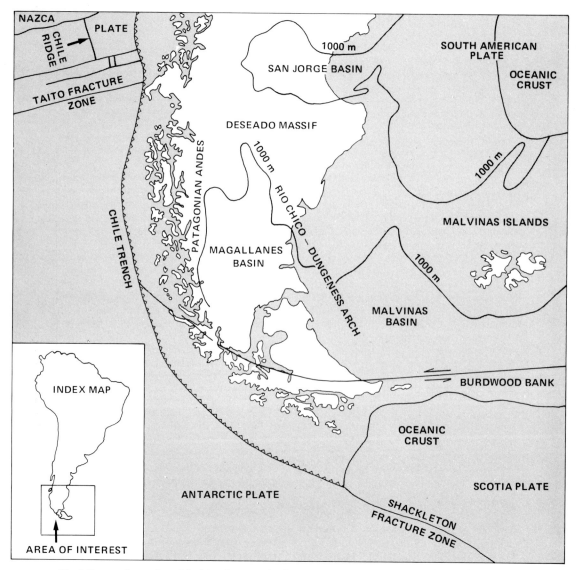

Fig. 1. Present-day setting of the Magallanes Basin. Contour line represents 1,000 m of sedimentary fill.

hausen (1931). In the 1940s, the search for oil and gas in the Magallanes Basin began in earnest, and a series of systematic geological surveys was initiated (Thomas, 1949a, b; Feruglio, 1949–50).

Modern studies of the basin and surrounding area have focused on the stratigraphy or the structural and tectonic development of the region. Information on stratigraphy can be found in Yrigoyen (1962, 1970), Katz (1963), Malumian, Masiuk & Riggi (1971), Flores *et al.* (1973), Natland *et al.* (1974), Riccardi & Rolleri (1980), Russo, Flores & Di Beneletto (1980), Winslow (1980) and Wilson (1983). Early attempts to explain the tectonic evolution of southern South America were based on geosynclinal theory (Borrello, 1969; Auboin *et al.*, 1973). More recently, work along the Chilean coast and in the Andean fold and thrust belt in southern Chile and westernmost Argentina has placed the basin in a plate-tectonic framework (Dalziel, de Wit & Palmer, 1974; Bruhn, 1979; Dalziel & Palmer, 1979; Saunders *et al.*, 1979; Nelson, Dalziel & Milnes, 1980; Dalziel, 1981; Winslow, 1980, 1982; Dott, Winn & Smith, 1982).

Geological setting

The Magallanes Basin is located near the south-western edge of the South American plate (Fig. 1), and today, as in the past, plate interactions along the western edge of South America control the evolution of the region. The southern margin of the basin is a complicated zone of shortening and strike-slip deformation that forms a segment of the boundary between the South American and Scotian plates (Fig. 1). To the west, the southern continuation of the Chile Trench forms the plate boundary between the Antarctic and South American plates, but the basin edge there is defined by the Patagonian batholith and the deformed belt of the southern Andean mountains. To the north and east, the basin is bounded by the Rio Chico–Dungeness arch (Fig. 1), a long-lived basement high. To the south-east, the basin connects with the Malvinas Basin.

The Magallanes Basin itself has a NNW trend, covers roughly 160,000 km² and contains a sedimentary fill that is more than 7,000 m thick (Fig. 2). At its widest point, the basin is about 370 km across. It is approximately 700 km long, but at one time probably connected with the San Jorge Basin to the north (see Fig. 1).

Basement rocks crop out along the present edges of the Magallanes Basin in small areas on the western Deseado massif (Fig. 1), and along a discontinuous belt in the Patagonian Andes and coastal Chile. Basement also has been encountered in a few wells in the basin (Lesta & Ferello, 1972; Natland *et al.*, 1974). Most of the known basement rocks are lower to middle greenschist-grade slates, phyllites, mica schists, and metacherts derived from clayey and sandy sedimentary protoliths (Miller, 1976; de Guisto, Di Persia & Pezzi, 1980; Nelson *et al.*, 1980; Hervé *et al.*, 1981). Rare gneisses and amphibolites also occur (Miller, 1976; Nelson *et al.*, 1980). In coastal Chile, the basement complex contains a limited suite of blueschists and less-metamorphosed, obviously allochthonous elements such as shallow-water limestones and pillow basalts with mid-ocean ridge affinities (Forsythe & Mpodozis, 1979; Hervé *et al.*, 1981; Mpodozis & Forsythe, 1983). Field observations, fossil content, and limited radiometric dating indicate that most of the basement rocks range in age from Early to Late Palaeozoic (Halpern, 1972; Forsythe & Mpodozis, 1979; de Guisto *et al.*, 1980; Hervé *et al.*, 1981; Forsythe, 1982; Ling, Forsythe & Douglass, 1985). Ramos (1983) reported a Precambrian date from the Deseado massif; the distribution of these rocks is limited and their regional extent is unknown. There is also an indication that some of the basement includes rocks as young as Early Triassic (Hervé *et al.*, 1981). Where seen the basement is pervasively deformed into a complex set of folds formed during several deformational events. The last major pre-Andean deformation appears to have been mostly Late Palaeozoic (Forsythe, 1982). In Chile, basement structures trend generally NW (Miller, 1976; Forsythe, 1982).

The above information suggests that the basement rocks found along the present western and southern flanks of the Magallanes Basin represent material accreted to the South American continent by subduction-related processes during the Late Palaeozoic–Early Mesozoic (Forsythe, 1982; Ling *et al.*, 1985). The genesis of basement rocks beneath the axial and eastern parts of the Magallanes Basin is less certain. They may have formed in a Late Palaeozoic forearc or arc setting, as suggested by Forsythe (1982), represent older crust along the edge of Gondwanaland (Ramos, 1983), or result from a combination of events and processes.

The sedimentary fill of the basin is dominated by shaly rocks, and can be divided into three major packages: (1) a syn-rift package deposited in grabens and half grabens created during the extensional event that led to the formation of the Rocas Verdes marginal basin to the SW (see Dalziel, 1981, and the references

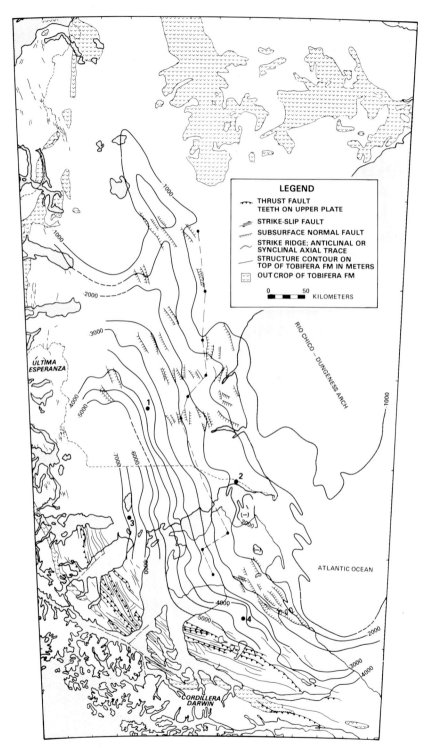

Fig. 2. Structure-contour map of the Magallanes Basin drawn on top of the Tobifera volcanic rocks; values in metres. Structures of the southern South American fold and thrust belt are from Winslow (1980). Dashed line shows location of the chronostratigraphic chart (Fig. 3). Well locations: 1—El Fondo 1; 2—Condor 1; 3—Manzano 7; 4—Evans 1.

therein, for a discussion of this event); (2) a late to post-rift remnant-arc-margin package deposited just before and after oceanic crust was formed in the Rocas Verdes marginal sea as the area of the Magallanes Basin underwent thermally driven subsidence; and (3) a foreland basin package deposited in front of the rising Andes as the basin subsided due to loading along its western and southern edges. Representative elements of each of these packages are discussed below, but we will concentrate on the last two.

SEQUENCE STRATIGRAPHY

We divided the sedimentary fill of the basin into unconformity-bound sequences using the criteria outlined by Vail *et al.* (1977). Briefly, sequence boundaries were defined by seismic reflection terminations, tied to palaeontological and lithological data from wells by synthetic seismograms, and the sequences were mapped around the basin. Reflection configuration, amplitude variations, and bore-hole data were used to define gross depositional settings for each of the sequences. Our data set included 115 wells and about 8,500 km of seismic data.

We defined 17 sequences in the basin, but only six sequences or groups of sequences will be presented here. These six were chosen to document important steps in the history of the basin. For each sequence or group of sequences approximate thickness was taken from wells and calculated from seismic data by using time–depth relationships derived from sonic logs and check-shot surveys.

The stratigraphic terminology applied in the literature to the basin fill is complicated. This is a result of scattered outcrops around the basin, difficulties in dating some of the units, and uncertainties in correlation from place to place. In addition, our study has shown that many of the lithostratigraphic units in the basin are significantly time transgressive. We have chosen to use the terminology applied in the subsurface by the petroleum industry. The stratigraphic terms we have used and our interpretation of their time significance are shown on a chronostratigraphic chart (Fig. 3) that runs north–south through the basin. The physical relationships between stratigraphic units are shown on a companion well-log cross-section (Fig. 4), and in sketches from seismic lines (Fig. 5).

Biostratigraphic control

A key element of our approach is the dating of the identified sequences. Available well samples were zoned using foraminifera, dinoflagellates, and calcareous nannofossils, and the results were compiled into an integrated zonation for each well. These, plus biostratigraphic data from the adjacent Malvinas Basin and Falkland Plateau, were then used to reinterpret published faunal lists (Malumian *et al.*, 1971; Malumian, Masiuk & Rossi de Garcia, 1972; Malumian & Masiuk, 1975, 1976a, b; Wright, 1984).

Age control in the Tertiary section was hampered by several factors. The upper Neogene section in the basin is commonly non-marine and lacks the microfossils used in dating. Except for the lower Palaeogene section, Tertiary sedimentary rocks did not contain abundant calcareous nannoplankton, nor did they yield biostratigraphically useful dinoflagellates. Consequently, foraminifera provided most of the age control. Cretaceous sections yielded rich microfossil assemblages and were relatively easy to date. Age determinations using dinoflagelates from these rocks were based on Australian zonation schemes; those using calcareous nannoplankton were based on standard zonation schemes. Jurassic sections are notoriously difficult to date using microfossils; we relied on K/Ar dates on Jurassic volcanic rocks and limited, published faunal control to date these rocks.

The Tobífera Formation and older sedimentary rocks (Triassic ? to Upper Jurassic)

Over large areas of the Magallanes Basin, the basement is overlain by a heterogeneous suite of silicic volcanic and non-marine volcaniclastic rocks. These rocks are known as the Tobífera Formation (Thomas, 1949a), and are associated with the last stages of a Triassic to Jurassic extensional event (Bruhn, Stern & De Wit, 1978; Gust *et al.*, 1985). The Tobífera volcanic rocks and their equivalents cover more than 1 M km^2 in southern South America, and represent a key event in the evolution of Patagonia and Tierra del Fuego (Lesta & Ferello, 1972; Gust *et al.*, 1985). In the Magallanes Basin, these rocks are absent on some basement highs and can be over 2,000 m thick in intervening lows.

Radiometric dates from outcrop samples taken mostly to the north of the Magallanes Basin indicate a main period of volcanic activity around 160–150 Ma (Cazeneuve, 1965; Halpern, 1973; Nullo, Proserpio &

Ramos, 1978; Gust *et al.*, 1985). We have used these data and a previously unpublished date of $168 + 3$ Ma from the adjacent Malvinas Basin to constrain the age of the top of the Tobífera volcanic rocks in the Magallanes Basin. We have assigned an age of 151 Ma to the sequence boundary that we carry as the top of the Tobífera volcanic rocks (Fig. 3). Additional constraints are supplied by interbedded and overlying rocks that crop out to the west of the basin and contain Oxfordian and Kimmeridgian fossils (Sigal *et al.*, 1970; Cecioni & Charrier, 1974; Natland *et al.*, 1974).

A number of isolated pods of layered rocks occur in grabens beneath the Tobífera volcanic rocks (shown schematically on Fig. 3). These are clearly visible on seismic data from the basin, but rarely have been penetrated by drilling. The age of the oldest graben fill is unknown, but based on analogy with a few outcrops in the Andes (at Lago Belgrano, Lago Pueyrredon, and Peninsula Brunswick) and on the Deseado massif (Leanza, 1972; Riccardi & Rolleri, 1980), we believe that they are of Late Triassic and Early Jurassic age.

We did not map either the Tobífera Formation or the underlying graben fill because of a lack of adequate seismic coverage. We point out their existence, however, because they represent important events in the history of the basin.

The Springhill and Lower Inoceramus Formations (Upper Jurassic to Lower Cretaceous)

This sequence group includes the Springhill and Lower Inoceramus formations (Fig. 3). The interval ranges in age from Late Jurassic to mid-Aptian. The age of the Springhill sandstones is controlled by three reliable biostratigraphic dates of Tithonian–Oxfordian, Berriasian, and Neocomian (Sigal *et al.*, 1970; Riccardi, 1976). The Lower Inoceramus shales are easily datable using microfossils as pre-mid-Aptian Early Cretaceous.

The Springhill sandstones are a succession of retrogradational fluvial, shoreline, and shallow-marine sandstones (Hinterwimmer, Messinger & Soave, 1984; Kielbowicz, Ronchi & Stach, 1984), and are the major hydrocarbon reservoirs in the basin. The equivalent offshore marine shales are called locally 'Estratos con Favrella'. These rocks are poor to fair source rocks in some areas.

The Springhill sandstones are generally thin and acoustically unresolvable. They are absent over some basement highs, and locally thicken over the pre-Tobífera grabens. As a unit, these sandstones form a classic basal transgressive sandstone sheet (Riccardi, 1976; Marinelli, 1982). Our work suggests, however, that the formation consists of at least three backstepping, but individually prograding sandstone intervals (Fig. 3). These may or may not in contact with each other.

The overlying shales of the Lower Inoceramus Formation form a broad, prograding wedge that onlaps the Dungeness Arch to the east and thickens basinward to the west and SW (Fig. 6). The unit consists of dark grey to dark brown claystones and shales with some glauconite (Flores *et al.*, 1973) deposited under disaerobic to anaerobic conditions. Older portions of these shales are the best source rocks in the basin. The parafinic composition of oil indicates an important amount of terrestrially derived organic matter. This composite interval is as thick as 800 m in south-western Argentina (Fig. 6). In the northernmost part of the basin, these rocks have been partially eroded (Fig. 3).

Seismic facies interpretation, lithology, and interpreted depositional settings from log patterns suggest a generally low-energy depositional environment with a sediment source to the NE for much of this interval. On the eastern margin of the basin, high-amplitude, parallel reflections are interpreted as representing shelfal rocks. These reflections grade basinward to low-angle, sigmoidal, downlapping reflections indicative of a low-energy (shale-prone) shelf and slope. The NW-trending shelf margin is a subtle feature on seismic data, because of the gradual slope of the basin floor and the shale-prone sedimentary fill. To the west of the shelf margin, a 'shallow' slope is defined by discontinuous, low-amplitude, parallel-to-sigmoidal reflections with some downlapping terminations. This slope passes westward into a thin, basin-plain interval characterized by parallel, low-amplitude seismic cycles. A structural break divides the basin plain into a shallow and deep basin. The deep basin corresponds to an area of thick sedimentary rocks in south-western Argentina and adjacent Chile (Fig. 6).

The top of this sequence group is a regional mid-Aptian unconformity. This corresponds to the C_5 marker horizon of Harambour Giner (1965).

Margas Verdes and Middle Inoceramus Formations (mid-Cretaceous to mid–Late Cretaceous)

This interval includes the Marges Verdes marls of mid-Aptian to mid-Cenomanian age, and the Middle

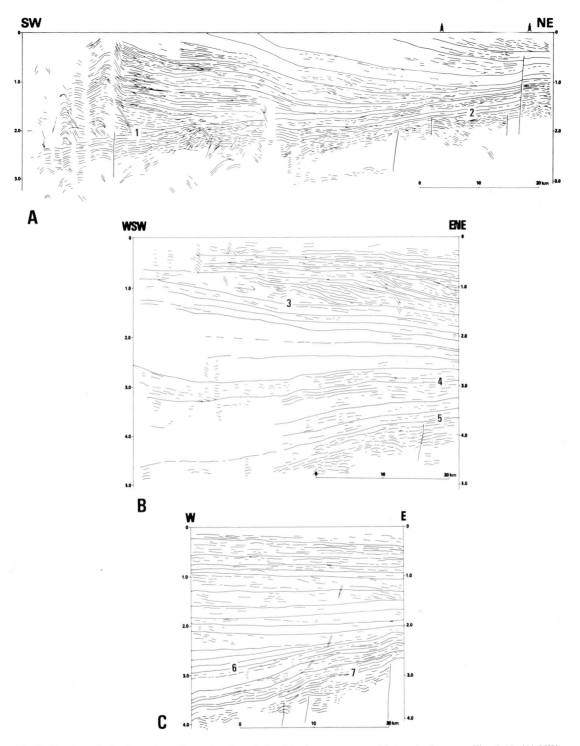

Fig. 5. Sketches of seismic sections illustrating the relationships between mapped intervals shown on Figs 6–11. (A) NW-trending line from the frontal fold of the Chilean fold and thrust belt to the edge of the Dungeness Arch. Note the pronounced onlap pattern on the near-top-of-Cretaceous horizon and the Tertiary wedge-shape units interpreted as a progradational set of fan-deltas (loc. 1; also see Figs 3 and 10). Location 2 shows the approximate top of the Tobífera volcanic rocks. (B) ENE-trending line near the western basin edge just north of the Straits of Magellan. Note the apparent east-directed progradation at location 3. Seismic stratigraphic correlation shows that the progradational units range in age from early Oligocene to early Miocene, and correspond in part to the Loreto Formation (see Figs 3 and 11). Top of the Cretaceous is indicated at location 4. Top of the Tobífera volcanic rocks is shown at location 5. (C) East-trending line just to the east of line B. Note pronounced onlap unconformity that separates uppermost Cretaceous rocks from Palaeocene rocks in the west (loc. 6). The time significance of this onlap unconformity is shown on Fig. 3. Location 7 represents the top of the Tobífera volcanic rocks.

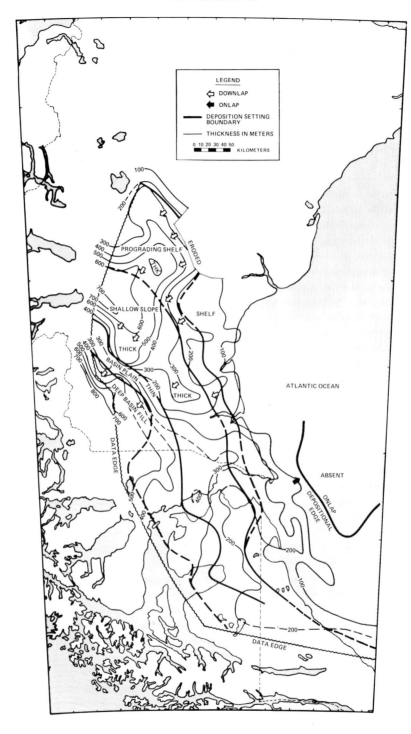

Fig. 6. Thickness and inferred, generalized depositional settings, Upper Jurassic to Lower Cretaceous Springhill sandstone and Lower Inoceramus Formations. Note the general NNW trend of depositional settings and the SW-ward progradation and thickening.

Inoceramus shales of mid-Cenomanian to mid-Coniacian age (Fig. 3). The Margas Verdes is a succession of light grey to grey-green calcareous claystones, shaly limestones, and glauconitic sandstones with abundant analcime and heulandite (Flores *et al.*, 1973). The zeolites represent the diagenetic products of altered pyroclastic debris derived from the west. The Middle Inoceramus Formation consists of monotonous grey-green claystone and shale with rare glauconite. Both units were deposited in open-marine settings under aerobic to disaerobic conditions. The distribution of these two sequences is similar to that of the underlying Springhill–Lower Inoceramus unit. A NW-tending sediment wedge onlaps the Dungeness Arch to the east and thickens to the west and SW (Fig. 7). Thicknesses in excess of 1,500 m are observed in south-western Argentina. A second thick accumulation is located in north-eastern Tierra del Fuego, and appears to be related to an underlying graben. Post-Cretaceous erosion in the northernmost part of the basin removed most or all of this interval (Figs 3 and 7).

Seismic facies suggest a gently sloping basin floor with low-energy shale-prone depositional environments and a north-easterly sediment source (Fig. 7). Continuous, parallel, medium-amplitude reflections of the shelf grade westward into downlapping cycles with some toplap but mostly a sigmoidal geometry. The shelf margins for the sequences are subtle features because of the gentle slopes and shale-prone sediments. To the west of the shelf margins are shallow slope and basin settings (Fig. 7). These are characterized by low-amplitude, wavy to subparallel reflections. The greatest sedimentary thickness is associated with the basinal setting.

Although the top of the Margas Verdes marl (C_1 marker horizon; Harambour Giner, 1965; Flores *et al.*, 1973) commonly exhibits a high impedence contrast, perhaps suggesting a higher carbonate content than the overlying unit, there is no seismic evidence of shelf-edge carbonate buildups. The top of the Middle Inoceramus shales is a mid-Coniacian unconformity (Fig. 3).

Upper Inoceramus and Arcillas Fragmentosas Formations (mid-Coniacian to mid-Maastrichtian)

This interval is a composite of two sequences of similar facies and distribution: a mid-Coniacian to mid-Campanian sequence known as the Upper Inoceramus Formation, and a mid-Campanian to mid-Maastrichtian sequence called the Arcillas Fragmen-tosas shale (Fig. 3). Both units are silty to micritic claystone with some interbedded very fine grained sandstone.

These two sequences are less widespread than the previous sequences because of extensive post-Cretaceous erosional truncation in the northern part of the basin. The distribution of isopachs, however, still shows the presence of a NW-trending sedimentary wedge (Fig. 8). Maximum thickness within the area of our data coverage exceeds 1,400 m in western Argentina (Fig. 8). A second thick accumulation in excess of 300 m is present in north-eastern Tierra del Fuego. These two sequences thin dramatically eastward toward the Dungeness Arch, but do not pinch out depositionally as do the underlying units.

A gently sloping depositional surface and low-energy environments are suggested by the seismic facies of this interval. Directions of progration indicate a sediment source to the NE. A thin shelfal area in the east is characterized by high-amplitude, concordant reflections. This shelfal area gives way westward to a prograding shelf with downlapping, medium-amplitude, discontinuous seismic reflections. The shallow slope widens toward the SE and is defined by medium-amplitude, discontinuous reflections with low-angle sigmoidal geometry. The shallow basin area corresponds to the greatest thickness of this interval, and has discontinuous, low-amplitude reflections with a parallel to sigmoidal geometry.

The upper boundary of this interval corresponds to the prominent G_7 marker horizon of Harambour (1965) and Flores *et al.* (1973). Roughly equivalent Upper Cretaceous rocks crop out in the westernmost Magallanes Basin outside the area of our data coverage. There deep-water units, known as the Cerro Toro and Lago Sofia formations, show transport directions from the north down the axis of the basin (Scott, 1966; Winn & Dott, 1979).

Chorillo Chico and Dorotea Formations (mid-Maastrichtian to mid-Thanetian)

Two sequences compose this interval, and are approximately equivalent to the shallow-water Dorotea Formation in Argentina and Chile (Riccardi & Rolleri, 1980) and the deeper water Chorrillo Chico Formation in Chile (Thomas, 1949a; Charrier & Lahsen, 1969, fig. 3). Within the area covered by our data, this interval is restricted to south-central Chile and south-western Argentina (Fig. 9). It forms a NW-trending sedimentary wedge which narrows to the SE and attains a maximum thickness in excess of 700 m

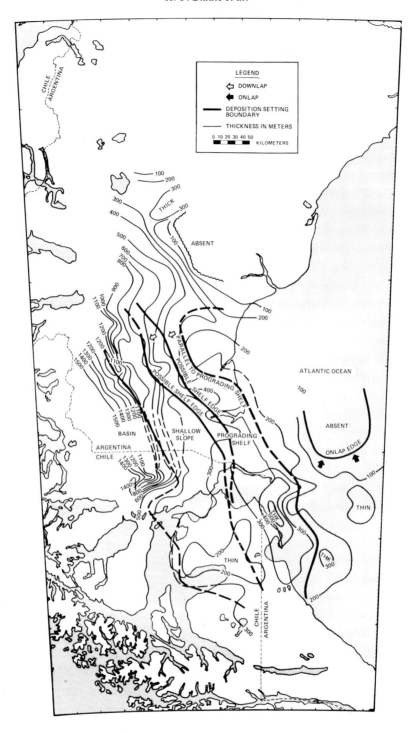

Fig. 7. Thickness and inferred, generalized depositional settings, Lower to Upper Cretaceous Margas Verdes and Middle Inoceramus Formations. Note the NNW trend of depositional settings and the SW-ward progradation and thickening.

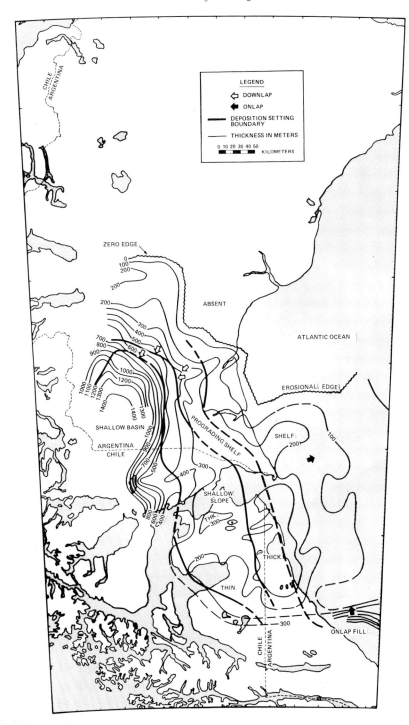

Fig. 8. Thickness and inferred, generalized depositional settings, Upper Cretaceous Upper Inoceramus and Arcillas Fragmentosas Formations. Note the NNW to NW trend of depositional settings and the SW-ward progradation and thickening.

(Fig. 9). This wedge thins by erosion and onlap to the east. Thin, erosional remnants are present beyond the general limit of truncation shown on Fig. 9.

Seismic reflections are parallel in this unit, and show limited to pronounced onlap at the base of the interval and erosional truncation at the top. Reflections are of medium to high amplitude within the older sequence, but of low to medium amplitude within the younger sequence. The higher amplitudes of the older sequence (Danian) indicates more interbedding and perhaps a higher sand content. In the west the Danian sandstones are glauconitic, particularly at the base of the unit. These basal glauconitic sandstones appear to become progressively younger to the east.

The top of this interval is a mid-Thanetian (late Palaeocene) unconformity (Fig. 3). This unconformity is of regional extent and can be seen in outcrop in southern South America and in the subsurface of both the Magallanes and Malvinas Basins. The unconformity shows evidence of erosion of the rocks below it, particularly along the flanks of the Dungeness Arch, and progressively onlap above it (Fig. 3). The cause (or causes) of this major unconformity are unknown.

Ballena, Tres Brazos, and Leña Dura Formations and the Zona Glauconitíca (mid-Eocene to Lower Oligocene)

This composite interval consists of four sequences and is roughly equivalent to the coarse-grained Ballena Formation in southern Tierra del Fuego (Natland et al., 1974), the deeper-water Tres Brazos and Leña Dura formations (Natland et al., 1974), and the Zona Glauconitíca in the Chilean portion of the basin (Hauser, 1964; Fig. 3). The mapped interval ranges in age from mid-Lutetian to mid-Rupelian. It has a NW depositional strike through central and western Chile and south-eastern Tierra del Fuego, but is also of limited areal extent (Fig. 10). Thickness increases gradually from east to west and rapidly to the south, where we have measured over 3,600 m of this interval (Fig. 10).

Seismic facies in western Chile show a north-eastward-prograding shelfal area (Fig. 10). Medium- to high-amplitude relections display a parallel to sigmoidal geometry and a distinct shelf edge. Directions of progradation indicate that the sediment source for these deposits in western Chile was to the SSW.

The area of very thick sedimentary rocks in southern Tierra del Fuego (Fig. 10) appears to correspond with the coarse-grained clastic rocks of the Ballena Formation. We interpret this unit as a fan-delta derived from an area of high relief to the south, and deposited in fairly deep water along a steep sedimentary gradient. This unit is characterized by several sequences of high-amplitude, parallel to convergent reflections that downlap at the base of the sequence. Poorly sorted, conglomeratic sandstones of this unit were encountered in the Evans 1 well in southern Tierra del Fuego (Natland et al., 1974). The composition of the coarse fraction of these sandstones indicates that they were derived from the Palaeozoic and Mesozoic rocks now exposed in the Andes.

To the east and north of the prograding shelfal area and the Ballena fan-delta, the seismic reflection pattern of this interval is one of simple onlap fill (Fig. 10). Discontinuous reflections are conformable at the top of the interval and onlapping at the base. In areas of distal onlap, time-transgressive glauconitic sandstones, known as the Zona Glauconitíca in Argentina, dominate the section (Fig. 3). These sandstones reflect a low sedimentation rate and long periods of near starvation of sediment. They represent the distal toes of the sedimentary units derived from the Andes. The glauconitic sandstones become younger to the east as the interval progressively onlaps the Dungeness Arch (Fig. 3).

Lower Loreto Formation (Lower to Upper Oligocene)

This composite sequence is roughly equivalent to the lower Loreto Formation, and is of mid-Rupelian to mid-Chattian in age (Fig. 3). It forms a silty and sandy, shallowing-upward succession. This interval extends farther to the east than the older Tertiary units (Fig. 11). Isopachs define a NW-trending wedge that thickens rapidly to the west and SW (Fig. 11). Interval thickness exceeds 1,600 m in a restricted area north of Bahía Inútil. To the west and south this composite sequence has been extensively eroded. Difficulty in well-log correlations and gaps in our seismic data precluded mapping the eastern limit of this interval.

Seismic data from western Chile reveal a progression of eastward-prograding shelf margins. Downlap patterns suggest that the sediment supply was located to the west and SW. The prominent shelf edges display a complex, oblique geometry and a pronounced toplap pattern. This gives way to the east to a shallow-slope

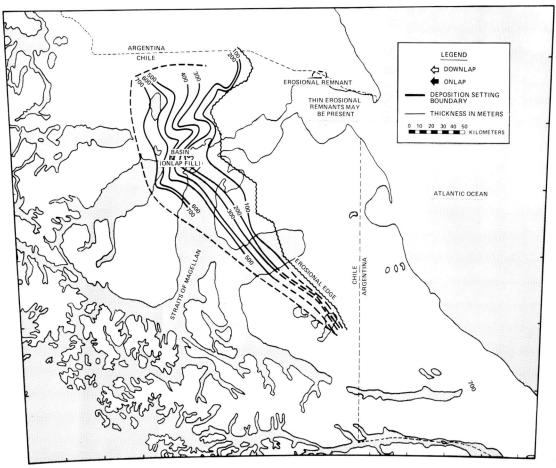

Fig. 9. Thickness and inferred, generalized depositional settings, Maastrichtian to Danian Dorotea and Chorillo Chico Formations. Note the NW trend of thickness contours and the pattern of onlap fill. This is the first interval in our study area that may have an Andean source.

area defined by low-angle prograding, discontinuous, low-amplitude reflections. The onlap fill area shown on Fig. 11 is characterized by onlapping reflection patterns that are formed by glauconitic sandstones (Fig. 3).

This is the last interval for which we present a thickness and depositional-settings map. By the end of the Oligocene, deposition was once again basin-wide (Fig. 3). The post-upper Oligocene rocks include the upper Loreto, Pampa Larga, and Magallanes Inferior Formations in Chile and the Patagoniano Formation in Argentina (Fig. 3). In general, these units were derived from the Andes to the west and south of the basin, but some material also came from the Deseado area to the north-east.

STRUCTURAL GEOLOGY

The Magallanes Basin can be divided into two structural provinces, an eastern one dominated by normal-fault-related structures, and a western province formed by the southern Andes fold and thrust belt. Strike-slip deformation is confined to the southernmost edge of the basin along the South America–Scotia plate boundary or to tear faults in the fold and thrust belt. Elsewhere in the basin, strike-slip is insignificant. The normal-fault dominated province covers over two-thirds of the basin; the fold and thrust belt affects less than one-third of the total area of the basin (Fig. 2). Both areas contain viable structural traps, although most of the basin's production comes

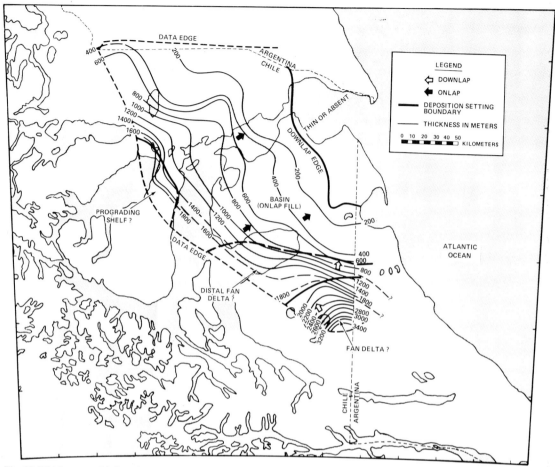

Fig. 10. Thickness and inferred, generalized depositional settings, mid-Eocene to lower Oligocene Ballena, Tres Brazos, and Leña Dura Formations, and the Zona Glauconítica. Note the general NW strike of thickness contours and the extremely thick fan-delta (?) in central Tierra del Fuego. Also note the NW to north progradation in central Tierra del Fuego and the pattern of onlap in the rest of the study area.

from subtle normal-fault traps or combination normal fault-stratigraphic traps in the eastern part of the basin.

Two major types of normal faults exist in the basin, those that involve the Tobífera and pre-Tobífera rocks (Fig. 5), and those that flatten downward and die out in the Mesozoic sedimentary section. The former type is by far the most important. These faults were formed during the Triassic to Late Jurassic rift to remnant-arc margin stage of basin evolution. Most of these faults either die out in or below the Tobífera volcanic rocks, but a few cut somewhat higher in the section, although with less displacement than at lower stratigraphic levels. The majority of these faults strike to the NW or NNW (Fig. 2). They commonly bound, or

occur within, NW- to NNW-trending grabens and half grabens with half grabens being the most prevalent. The detached normal faults die out in the Jurassic and Lower Cretaceous shales. These detached normal faults are generally unimportant from an exploration point of view.

The structures of the western third of the basin are dominated by folds and thrust faults of the southern Andes. The exposed fold and thrust belt has been well described by Winslow (1980, 1981, 1982). We will restrict our discussion to areas to the east of Winslow's studies where we have data over individual structures. The most obvious structures associated with the fold and thrust belt in the Chilean subsurface are long, sinuous folds. These folds overlie areas where thrust

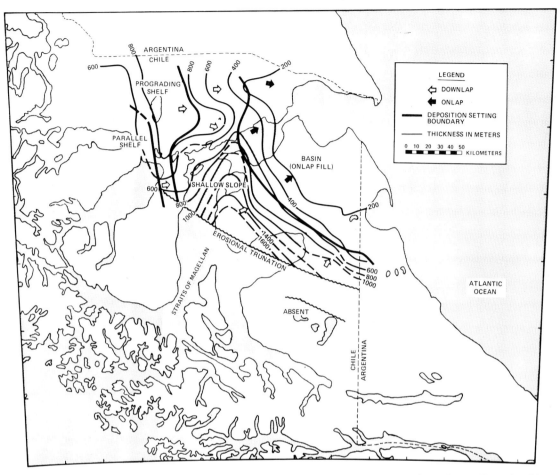

Fig. 11. Thickness and inferred, generalized depositional settings, lower Oligocene to lower Miocene Loreto Formation. Note the east to NE patterns of progradation and onlap to the east against the Dungeness Arch.

faults cut upward to higher stratigraphic levels (Fig. 5). The folds are open and symmetrical to asymmetrical with eastward vergence. They are separated by long, nearly horizontal segments above nearly flat décollement surfaces in the Lower Cretaceous shales. We have no direct evidence on the timing of deformation at the leading edge of the fold and thrust belt, but the folds deform rocks that are as young as early Miocene in age in Tierra del Fuego.

Although the folds and thrust faults are underlain by older normal faults, we do not see any evidence of reactivation of those normal faults as reverse faults in the area of our data coverage. Winslow (1981) has described such features (based on ENAP subsurface data) to the west of the area covered by our study.

DISCUSSION

Timing

Most of the data that bear on the timing of development of the Magallanes Basin come from the outcrop belt that fringes the basin, or from the crystalline terranes of coastal Chile. These data show that the basin had two major tectonic stages of development; one associated with the breakup of Gondwanaland and the formation of the Rocas Verdes back-arc marginal basin, and one related to the destruction of that basin and deformation and uplift in the Andes. Our study provides two additional data sets that can be used to constrain the history of the

Magallanes Basin. These are, first, directions of progradation and inferences on locations of sedimentary source areas for different stratigraphic units, and, second, subsidence data that reflect the mechanisms responsible for forming the basin.

Briefly, stratigraphic and structural data show that the ophiolitic floor of the Rocas Verdes marginal basin had begun to form by 140 Ma (Dalziel, 1981). The fill of this basin can be dated at Isla Hoste where the Yahgan Flysch and Tekenika beds crop out. The Yahgan Flysch is of Neocomian age and the Tekenika beds are possibly Albian or Aptian in age (post-Neocomian–pre-81 to 77 Ma; Dott *et al.*, 1977). Both of these units were deformed by the event that closed the Rocas Verdes basin (Dott *et al.*, 1977). On Isla Navarino, the marginal-basin fill is intruded by a 93 Ma post-tectonic grandiorite dike of the Santa Rosa pluton (Hervé, Suárez & Puig, 1984). This, and supporting information, shows that oceanic crust existed in the marginal basin before the beginning of the Cretaceous and that the basin was deformed and closed between about 100 and 80 Ma (Dalziel, 1981; Hervé *et al.*, 1984).

Part of the cratonward edge of the Rocas Verdes basin is represented by the Cordillera Darwin of southernmost Chile (Nelson *et al.*, 1980). Here, three phases of penetrative ductile deformation occurred before 90–80 Ma (Nelson *et al.*, 1980). Uplift of the Cordillera Darwin was rapid between about 95 and 65 Ma, and slower after 60 Ma (Nelson, 1982).

In the Última Esperanza area of Chile, deposition of the deep-water Punta Barrosa Formation has been cited as the beginning of the foreland basin stage of development of the Magallanes Basin (Wilson, 1983). This is based on a western source area for the upper Albian–Cenomanian unit. The overlying Cenomanian to Santonian Cerro Toro Formation has also been interpreted as being derived from the west (Scott, 1966; Winn & Dott, 1979; also see Arbe & Hechem, 1984). In western Chile, benthonic foraminifera suggest that there was a water-depth change near the Albian–Cenomanian boundary from less that 300 m to about 2,000 m (Natland *et al.*, 1974). This deepening has also been interpreted as the start of foreland basin subsidence.

The timing of initiation of shortening in the Patagonian fold and thrust belt is difficult to document. Winslow (1982) stated that Lower Cretaceous to Palaeogene rocks in the deepest part of the basin were folded in the middle Tertiary, and that this deformation started earlier in the SE than it did in the NW. The cessation of folding and thrusting was also diachronous; in the outcropping basin fill in Tierra del Fuego middle Eocene strata are the youngest units deformed; to the NW post-middle Miocene beds are involved, but not Pliocene–Pleistocene units; farther north Pliocene rocks are warped and tilted (Winslow, 1982).

The above information is summarized in Fig. 12, and shows a progression of deformation and uplift from the margin of the South American continent toward the craton. Our data come from an area that is generally to the east of the other localities shown on Fig. 12, and reinforce the cratonward migration with time of events associated with the foreland basin stage of evolution of the Magallanes Basin.

First, the thickness patterns and directions of progradation seen on seismic data show that, in areas where we have data, the Upper Jurassic to Campanian units were derived from the north, NE and east. The first evidence of a western or southern source may be the Maastrichtian–Danian Chorillo Chico Formation and its equivalents. We do not seen progradation within this unit, only onlap to the east (Fig. 9). These rocks could have been derived from a western source, but could have also come from the north; the onlap pattern is not diagnostic. The first unequivocal evidence in our data of a source from the Andes comes from the overlying Eocene and younger units (Figs 10 and 11). These rocks are separated from the underlying units by a major composite unconformity (Figs 3, 4 and 5). In the deep part of the basin, this unconformity encompasses a relatively small amount of time, but along the Rio Chico–Dungeness Arch it represents a major, combination erosional/onlap surface with Oligocene rocks sitting directly on Upper Jurassic rocks (Fig. 3).

We have also produced subsidence curves for four wells in the basin, the El Fondo 1, Condor 1, Evans 1, and Manzano 7 wells (Fig. 13). These curves are corrected for the effects of loading by sediment and for changes in water depths with time. They reflect the subsidence of the basin floor due to thermal or tectonic causes. Although the curves are difficult to interpret rigorously because of possible errors in palaeowater-depth estimates, they do provide qualitative information on basin evolution.

The early part of these curves all have a similar shape (Fig. 13). We interpret this to represent decay of the thermal anomaly associated with Triassic to Late Jurassic rifting. The first part of each curve is generally concave upward and shows greater subsidence during the Early Cretaceous and less in the mid-Cretaceous. The early part of the curves suggest only

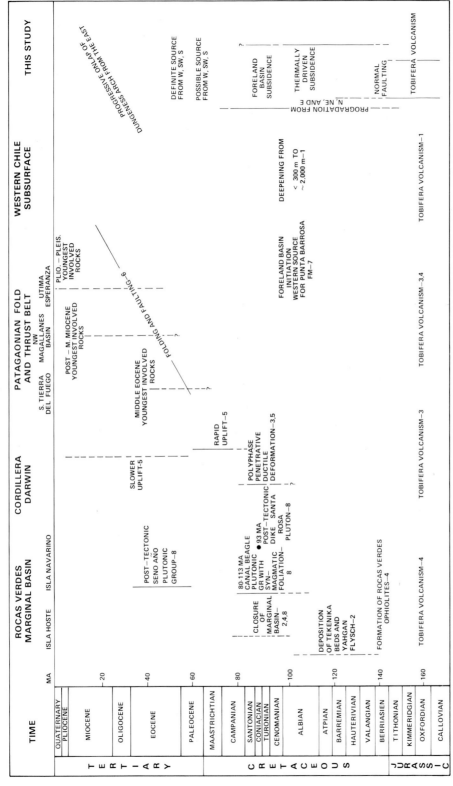

Fig. 12. Timing of important Mesozoic and Cenozoic events in southern South America compared with the results of this study. Numbers 1 to 8 represent key references. 1—Natland *et al.* (1974); 2—Dott *et al.* (1977); 3—Nelson *et al.* (1980); 4—Dalziel (1981); 5—Nelson (1982); 6—Winslow (1982); 7—Wilson (1983); 8—Hervé *et al.* (1984).

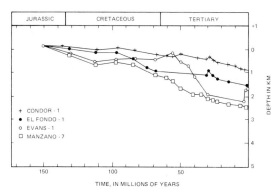

Fig. 13. Subsidence curves for the Condor 1, El Fondo 1, Evans 1, and Manzano 7 wells, Magallanes Basin (see Fig. 1 for location). These curves are corrected for the effects of water-depth changes and sediment loading. The early parts of the curves represent subsidence caused by thermal contraction as the heat-flow anomaly associated with Triassic to Late Jurassic rifting decayed. The later parts of the curves represent subsidence caused by tectonic loading along the Andean edge of the basin. See text for discussion.

limited lithospheric thinning, but it should be noted that most of these wells were drilled on structural highs. We interpret the subsidence represented by the last half of the curves to be the result of flexure caused by loading around the western and southern perimeter of the basin. The time of inception of this subsidence event is subject to interpretation, but we believe that it started at about 90 Ma in the Manzano 7 and El Fondo wells, and at approximately 50 Ma in the Evans well. The Condor 1 well does not seem to be much affected by this event. This well is located high on the Dungeness Arch, however. The Manzano 7 well is closest to the Andean deformational front and the El Fondo 1 is from the centre of the basin in Argentina. The Evans 1 is in an intermediate position in Tierra del Fuego.

The subsidence curves and directions of progradation indicate that the foreland basin stage evolution of the central and eastern Magallanes basin did begin at about 90 Ma, which is similar to previously published estimates, but that the first stages of this evolution were characterized by deepening only in the west and south. In the area covered by this study, sedimentation was still from the north and east at this time. It was not until perhaps the latest Cretaceous or earliest Tertiary that significant sediment began to arrive in our study area from the Andes. Once this took place most of the rest of the basin experienced either erosion or very slow sedimentation (Fig. 3).

Hydrocarbon occurrence

The hydrocarbons that occur in the basin are the result of the interplay between the rifting and foreland stages of basin development. The major reservoir in the basin, the Springhill sandstone, and the major source-rock intervals, the Springhill-equivalent shales and Lower Inoceramus shale, were deposited when the basin was a westward-facing remnant-arc margin to the Rocas Verdes marginal basin. The Springhill sandstone is composed of at least three backstepping, but prograding, sand-rich intervals deposited while the basin floor subsided because of thermal contraction. Each of these intervals was derived from, and progressively onlapped, the Dungeness Arch. The best hydrocarbon source interval, the lowermost Lower Cretaceous Lower Inoceramus shale, was deposited in relatively deep water under anoxic conditions in the centre of the basin. These source rocks did not mature, however, until well into the foreland basin stage of evolution. The thermal anomaly associated with the Triassic to Late Jurassic rifting had nearly disappeared by 80 Ma. The depth of burial produced by post-80 Ma sedimentation led to the maturation. By this time, heat flow in the basin had returned to near background levels. The loading created by the evolving Andes and the sediment shed from them also created an enhanced westward tilt to the earlier-deposited basin fill. This facilitated migration from the maturing source rocks toward the Dungeness Arch. Limited faulting and fracturing in the Early Cretaceous and younger units led to long-distance, stratigraphically controlled lateral migration along the Springhill sandstones at the expense of significant vertical migration. Without the interplay between the rifting and foreland basin stages, the pattern of hydrocarbon occurrence in the Magallanes Basin would be much different than it is today.

CONCLUSIONS

The Magallanes Basin is currently a foreland basin in front of the Patagonian fold and thrust belt and southern Andes mountains. Like many other foreland basins, it has had a multi-stage history of evolution, and each stage has contributed to making the basin a productive one. A syn-rift stage from the Triassic to Late Jurassic formed the normal faults that control many of the producing traps along the Dungeness Arch, and modified the early heat-flow history of the basin. The major reservoir and source-rock intervals were deposited while the basin was a westward-facing remnant-arc margin to a small marginal sea from the latest Jurassic to the mid-Cretaceous. From the Late Cretaceous to the present, the basin has subsided in response to tectonic and sedimentary loads supplied by the orogenic belt that fringes the basin on the west and south. Burial of the earlier-deposited source rocks during this stage in the basin's history led to the maturation of hydrocarbons and enhanced tilt of conduit beds away from the Dungeness Arch, thus facilitating migration.

Our work covers the central and eastern parts of the basin, and it provides a slightly different picture of basin evolution than previous studies that have concentrated on the outcrop belt to the west and south. First, all of the Upper Jurassic to mid-Maastrichtian basin fill in our study area appears to have been derived from the north, NE and east, and prograded into the basin from the Deseado massif and the Rio Chico–Dungeness Arch. The first evidence here of a 'western' or Andean source comes from the latest Cretaceous (?) and earliest Tertiary sedimentary units. Second, the first evidence of subsidence caused by a flexural response to Andean tectonic loads appears in the Late Cretaceous section in our study area at about 90 Ma, but may be as young as 50 Ma in the eastern part of the basin. This is slightly younger than the inception of foreland basin subsidence suggested from the Última Esperanza area of Chile. The initial effect of foreland basin subsidence was to create greater water depths in the western part of the basin; almost no effect of this event is seen in the eastern parts of the basin until well into the Tertiary.

The Magallanes Basin may well be unique among foreland basins in that it is filled almost completely with shaly rocks. This is due to three factors; the palaeogeographic position of the basin at the isolated end of the South American plate for much of its history, the apparently limited topographic relief in the area from the Late Palaeozoic until the Late Cretaceous, and the short transport distances into the basin of coarse material derived from the Andes from the Late Cretaceous to the present. Coarse-grained material from the Andes was deposited only along the Andean edge of the basin, localized by the rapid subsidence and a narrow zone of deep water there. Large areas of the basin were sediment starved for much of the foreland basin stage of evolution, and were the site of time-transgressive shale or glauconitic sandstone deposition. The most pronounced feature of the Tertiary foreland basin fill is the composite mid-Palaeocene unconformity on which the glauconitic sandstones rest. The tectonic significance of this unconformity is unknown at present, but it clearly marks an important event in the evolution of the Magallanes Basin.

ACKNOWLEDGMENTS

Early versions of this paper were reviewed by I. W. D. Dalziel and R. H. Dott Jr, and we thank them for their constructive comments. We also thank YPF— the national oil company of Argentina, Esso Exploration, Inc., and Exxon Production Research Company for approval to publish this work.

REFERENCES

ARBE, H.A. & HECHEM, J.J. (1984) Estratigrafía y facies de depósitos marinos profundos del Cretácico superior, Lago Argentino, Provincia de Santa Cruz. *Noveno Cong. Geol. Argent. Actas*, **5**, 7–41.

AUBOUIN, J., BORELLO, G., CECIONI, R., CHOTIN, P., FRUTOS, T., THIELE, R. & VICENTE, J.C. (1973) Esquisse Paleogeographique et Structurale des Andes Meridionales. *Rev. Geogr. Phys. Geol. dyn.* **15**, 11–72.

BONARELLI, G. (1917) Tierra del Fuego y sus Turberas. *Anal. del Ministerio de Agricultura de la Nación. Sec Geol. Min. Minera*, **12**, 119 pp.

BORRELLO, A.V. (1969) Los Geosinclinales de la Argentina. *An. Dir. Nac. Geol. Min.* **14**, 1–188.

BRUGGEN, E. (1913) Informe sobre las exploraciones geológicas de la región carbonífera del Sud de Chile. *An. Soc. Nac. Minería*, 15–19.

BRUHN, R.L. (1979) Rock structures formed during back-arc basin deformation in the Andes of Tierra del Fuego. *Bull. geol. Soc. Am.* **90**, 998–1012.

BRUHN, R.L., STERN, C.R. & DE WIT, M.J. (1978) Field and geochemical data bearing on the development of a Mesozoic volcano-tectonic rift zone and back-arc basin in southernmost South America. *Earth planet. Sci. Lett.* **41**, 32–46.

CAZENEUVE, H. (1965) Datación de una toba de la Formación Chon Aike (Jurásico de Santa Cruz, Patagonia) por el método de Potasio-Argon. *Ameghiniana*, **5**, 156–158.

CECIONI, G. & CHARRIER, R. (1974) Relaciones entre la Cuenca Patagónica, la Cuenca Andina y el Canal de Mozambique. *Ameghiniana*, **11**, 1–38.

CHARRIER, R. & LAHSEN, A. (1969) Stratigraphy of Late Cretaceous and early Eocene, Seno Skyring—Straights of Magallan area, Magallanes Province, Chile. *Bull. Am. Ass. Petrol. Geol.* **53**, 568–590.

DALZIEL, I.W.D. (1981) Back-arc extension in the southern Andes: a review and critical reappraisal. *Phil. Trans. R. Soc. A*, **300**, 319–335.

DALZIEL, I.W.D. & PALMER, K.F. (1979) Progressive deformation and orogenic uplift at the southern extremity of the Andes. *Bull. geol. Soc. Am.* **90**, 259–280.

DALZIEL, I.W.D., DE WIT, M.J. & PALMER, K.F. (1974) A fossil marginal basin in in the Southern Andes. *Nature*, **250**, 291–294.

DEGUISTO, J.M., DI PERSIA, C.A. & PEZZI, E. (1980) Nesocratón del Deseado. *Geol. Reg. Argent.* **2**, 1389–1430. Academia National de Ciencias, Córdoba.

DOTT, R.H., WINN, R.D., DE WIT, M.J. & BRUHN, R.L. (1977) Tectonic and sedimentary significance of Cretaceous Tekenika beds of Tierra del Fuego. *Nature*, **266**, 620–622.

DOTT, D.R., WINN, R.D. & SMITH, C.H.L. (1982) Relationship of Late Mesozoic and Early Cenozoic sedimentation to the tectonic evolution of the southernmost Andes and Scotia arc. In: *Antarctic Geoscience* (Ed. by C. Craddock), pp. 193–202. University of Wisconsin, Madison.

FELSCH, J. (1916) Reconocimiento geológico de los terrenos petrolíferos de Magallanes del Sur. *Bol. Soc. Nacional Minería, Chile*, **1**, 214–223, 309–315.

FERUGLIO, E. (1949–50) Descripción Geológica de la Patagonia. *Yacimientos Petrolíferos Fiscales, Buenos Aires*, 1, 2, 3.

FLORES, M.A., MALUMIÁN, N., MASIUK, V. & RIGGI, J.C. (1973) Estratigrafía cretácica del subsuelo de Tierra del Fuego. *Rev. As. Geol. Argent.* **28**, 407–437.

FORSYTHE, R.D. (1982) The Late Paleozoic to Early Mesozoic evolution of southern South America: a plate tectonic interpretation. *J. geol. Soc. London*, **139**, 671–682.

FORSYTHE, R.D. & MPODOZIS, C. (1979) El Archipiélago Madre de Dios, Patagonia occidental, Magallanes: Rasgos generales de la estratigrafía y estructura del 'basamento' pre-Jurásico superior. *Rev. Geol. Chile*, **7**, 13–29.

GUST, D.A., BIDDLE, K.T., PHELPS, D.W. & ULIANA, M.A. (1985) Associated Middle to Late Jurassic volcanism and extension in southern South America. *Tectonophys.* **116**, 223–253.

HALPERN, M. (1972) Rb-Sr and K-Ar dating of rocks from southern Chile and west Antarctica. *Antarctic J. U.S.* **7**, 149–150.

HALPERN, M. (1973) Regional geochronology of Chile south of 50 latitude. *Bull. geol. Soc. Am.* **84**, 2407–2422.

HARAMBOUR GINER, S. (1965) *Geología de los yacimientos petrolíferos del sector Cóndor-Dungeness, provincia de Magallanes*. Unpublished Thesis. Universidad de Chile, Escuela Geología, Santiago.

HAUSER, A. (1964) *La "Zona Glauconítica" en la plataforma Springhill, Magallanes*. Unpublished Thesis. Universidad de Chile, Escuela Geología, Santiago.

HERVÉ, F., DAVIDSON, J., GODOY, E., MPODOZIS, C.M. & COVACEVICH, V. (1981) The Late Paleozoic in Chile: stratigraphy, structure, and possible tectonic framework. *Acad. Bras. Cienc. Ann.* **53**, 361–373.

HERVÉ, F., SUÁREZ, M. & PUIG, A. (1984) The Patagonian batholith south of Tierra del Fuego: timing and tectonic implications. *J. geol. Soc. London*, **141**, 909–917.

HINTERWIMMER, G.A., MESSINGER, V.E. & SOAVE, L.A. (1984) Análisis de facies, porosidad y diagénesis de una secuencia de playa—Formación Springhill—en el sondeo Puesto Barros, Provincia de Santa Cruz. *Noveno Congr. Geol. Argent. Actas*, **5**, 136–145.

KATZ, H.R. (1963) Revision of Cretaceous stratigraphy in Patagonian Cordillera of Ultima Esperanza, Magallanes province, Chile. *Bull. Am. Ass. Petrol. Geol.* **47**, 506–524.

KIELBOWICZ, A.A., RONCHI, D.I. & STACH, N.H. (1984)

Foraminíferos y ostrácodos valanginianos de la Formación Springhill, Patagonia Austral. *Rev. As. geol. Argent.* **38**, 313–339.

KRANCK, E.H. (1930a) Sur la tectonique Cordillere de la Terra de Feu. *Extrait C.R.S. Soc. Geol. Fr.* **7**, 66–67.

KRANCK, E.H. (1930b) Sur le profil longitudinal de la Cordillere de la Terre de Feu. *Extrait C.R.S. Soc. Geol. Fr.* **10**, 102–103.

KRANCK, E.H. (1932) Geological investigations in the Cordillera of Tierra del Fuego. *Acta Geogr., Soc. Geogr. Fenn.* **4**, 1–231.

LEANZA, A.F. (1972) Andes Patagónicos Australes. In: *Geología Regional Argentina* (Ed. by A. F. Leanza), pp. 689–706. Academia Nacional de Ciencias, Cordoba.

LESTA, P.J. & FERELLO, R. (1972) Región Extraandina de Chubut y Norte de Santa Cruz. In: *Geología Regional Argentina* (Ed. by A. F. Leanza), pp. 601–653. Academia Nacional de Ciencias, Córdoba.

LING, H.Y., FORSYTHE, R.D. & DOUGLASS, R.C. (1985) Late Paleozoic microfaunas from southernmost Chile and their relation to Gondwanaland and forearc development. *Geology*, **13**, 357–360.

MALUMIÁN, N. & MASIUK, V. (1975) Foraminíferos de la Formación Pampa Rincón (Creatácico inferior) Tierra del Fuego. *Rev. Española Micropaleont.* **7**, 579–600.

MALUMIÁN, N. & MASIUK, V. (1976a) Foraminíferos caraterísticos de las Formaciones Nueva Argentina y Arroyo Alfa, Cretácico inferior, Tierra del Fuego, Argentina. *VI Congr. Geol. Argent. Actas*, **1**, 393–411.

MALUMIÁN, N. & MASIUK, V. (1976b) Foraminíferos de la Formacion Cabeza de León (Cretácico inferior) Tierra del Fuego, República Argentina. *Rev. As. Geol. Argent.* **31**, 180–202.

MALUMIÁN, N., MASIUK, V. & RIGGI, J.C. (1971) Micropaleontología y sedimentología de la perforación SC-1, Provincia Santa Cruz, República Argentina. *Rev. As. Geol. Argent.* **26**, 175–208.

MALUMIÁN, N., MASIUK, V. & ROSSI DE GARCIA, E. (1972) Microfósiles del Cretácico superior del la perforación SC-1, Provincia Santa Cruz, Argentina. *Rev. As. Geol. Argent.* **23**, 265–272.

MARINELLI, R.V. (1982) Distribución de campos productores de hidrocarburos en el área de plataforma de cuenca Austral: su relación con antiguas líneas de costa. *Primer. Congr. Nac. de hidrocarburos, petróleo y gas*, Buenos Aires, pp. 209–216.

MILLER, H. (1976) El basamento de la provincia de Aysén (Chile) y sus correlaciones con las rocas premesozoicas de la Patagonia Argentina. *Actas 6 Geol. Congr. Argent.* **1**, 125–141.

MPODOZIS, C. & FORSYTHE, R. (1983) Stratigraphy and geochemistry of accreted fragments of the ancestral Pacific floor in southern South America. *Palaeogeogr. Palaeoclim. Palaeoecol.* **41**, 103–124.

NATLAND, M.L., GONZÁLEZ, E., CAÑÓN, A. & ERNST, M. (1974) A system of stages for correlation of Magallanes Basin sediments. *Mem. geol. Soc. Am.* **139**, 126 pp.

NELSON, E.P. (1982) Post-tectonic uplift of the Cordillera Darwin orogenic complex: evidence from fission track geochronology and closing temperature-time relationships. *J. geol. Soc. London*, **139**, 755–761.

NELSON, E.P., DALZIEL, I.W.D. & MILNES, A.G. (1980) Structural geology of the Cordillera Darwin—collisional-style orogenesis in the southernmost Chilean Andes. *Eclog. geol. Helv.* **73**, 727–751.

NULLO, F.E., PROSPERPIO, C. & RAMOS, V.A. (1978) Estratigrafía y tectónica de la vertiente este del Hielo Continental Patagónico, Argentina-Chile. *Actas Séptimo Congr. Geol. Arg.* pp. 455–470.

QUENSELL, P. (1911) Geologisch-Petrographische Studien in der Patagonischen Cordillera. *Bull. Geol. Inst. Upsala*, **2**, 114 pp.

QUENSELL, P. (1912) Die quarzporphyr- und Porphyroidformation in Sudpatagonian un Feuerland. *Bull. Geol. Inst. Upsala*, **2**, 1–40.

RAMOS, V.A. (1983) Evolución tectónica y metalogénesis de la cordillera Patagónica. *Segundo Congr. Nac. Geol. Econ. Actas*, San Juan, **1**, 107–124.

RICCARDI, A.C. (1976) Paleontología y edad de la Formación Springhill. *Primer. Congr. Geol. Chileno, Actas*, **1**, C41–C56.

RICCARDI, A.C. & ROLLERI, E.O. (1980) Cordillera Patagónia Austral. *Geol. Reg. Argent.* **2**, 1173–1306. Academia Nacional de Ciencias, Córdoba.

RUSSO, A., FLORES, M.A. & DI BENEDETTO, H. (1980) Patagonia Austral Extraandina. *Geol. Reg. Argent.* **2**, 1431–1462. Academia Nacional de Ciencias, Córdoba.

SANDERS, A.D., TARNEY, J., STERN, C.R. & DALZIEL, I.W.D. (1979) Geochemistry of Mesozoic marginal basin floor igneous rocks from southern Chile. *Bull. geol. Soc. Am.* **90**, 237–258.

SCOTT, K.M. (1966) Sedimentology and dispersal pattern of Cretaceous flysch sequence. Patagonian Andes, southern Chile. *Bull. Am. Ass. Petrol. Geol.* **50**, 72–107.

SIGAL, J., GREKOFF, N., SINGH, N.P., CAÑÓN, A. & ERNST, M. (1970) Sur l'age et les affinities "gondwaniennes" de microfaunes (foraminiferes et ostracodes) malgaches, indiennes, et chiliennes au sommet du jurassique et a la du cretace. *C. r. Acad. Sci., Paris*, **271**, 24–27.

THOMAS, C.R. (1949a) Geology and petroleum exploration in Magallanes province, Chile. *Bull. Am. Ass. Petrol. Geol.* **33**, 1553–1578.

THOMAS, C.R. (1949b) Manantiales field, Magallanes province, Chile. *Bull. Am. Ass. Petrol. Geol.* **33**, 1579–1589.

VAIL, P.R., MITCHUM, R.M., JR, TODD, R.G., WIDIMIER, J.M., THOMPSON, S., JR, SANGREE, J.B., BUBB, J.N. & HATELID, W.G. (1977) Seismic stratigraphy and global changes of sea level. In: *Seismic Stratigraphy—Applications to Hydrocarbon Exploration* (Ed. by C. E. Payton). *Mem. Am. Ass. Petrol. Geol., Tulsa*, **26**, 49–212.

WILCKENS, O. (1907) Erlauterung zu R. Hathals geologischer Skizze des Gebietes zwischen dem Lago Argentino und dem Seno de la Ultime Esperanza (Sudpatagonien). *Naturf. Gesell. Freiburg*, **15**, 75–96.

WILSON, T.J. (1983) *Stratigraphic and structural evolution of the Ultima Esperanza foreland fold-thrust belt, Patagonian Andes, southern Chile.* Unpublished Ph.D. dissertation. Colombia University, New York, 360 pp.

WINDHAUSEN, A. (1931) *Geología Argentina. Segunda Parte.* Jacobo Peuser, Buenos Aires, 645 pp.

WINN, R.D., JR & DOTT, R.H., JR (1979) Deep-water fan-channel conglomerates of Late Cretaceous age, southern Chile. *Sedimentology*, **26**, 203–228.

WINSLOW, M.A. (1980) *Mesozoic and Cenozoic tectonics of the fold and thrust belt in southernmost South America and stratigraphic history of the Cordilleran margin of the Magallanes Basin.* Unpublished Ph.D. dissertation. Columbia University, New York.

WINSLOW, M.A. (1981) Mechanisms for basement shortening in the Andean foreland fold and thrust belt of southern South America. In: *Thrust and Nappe Tectonics* (Ed. by K. R. McClay & N. J. Price). *Spec. Publ. geol. Soc. London*, **9**, 513–528. Blackwell Scientific Publications, Oxford.

WINSLOW, M.A. (1982) The structural evolution of the Magallanes Basin and neotectonics in the southernmost Andes. In: *Antarctic Geoscience* (Ed. by C. Craddock), pp. 143–154. University of Wisconsin, Madison.

WRIGHT, R. (1984) Cretaceous benthic foraminiferal biostratigraphy, Malvinas Basin, Argentina. *Benthos '83, 2nd. int. Symp. Benthic Foraminifera, Pau, April 1983*, p. 619.

YRIGOYEN, M. (1962) Evolución de la Exploración Petrolera en Tierra del Fuego. *Petrotecnia*, Buenos Aires, **12**, 28–38.

YRIGOYEN, M. (1970) Problemas Estratigráficos del Terciario de Argentina. *Rev. As. Paleont. Argent.* **6**, 315–329.

Spec. Publs int. Ass. Sediment. (1986) **8**, 63–75

Magnetic polarity stratigraphy, age and tectonic setting of fluvial sediments in an eastern Andean foreland basin, San Juan Province, Argentina

NOYE M. JOHNSON*, TERESA E. JORDAN†, PATRICIA A. JOHNSSON* *and* CHARLES W. NAESER‡

*Earth Sciences Department, Dartmouth College, Hanover, New Hampshire 03755, U.S.A.;
†INSTOC, Cornell University, Ithaca, New York 14853, U.S.A.;
‡Branch of Isotope Geology, U.S. Geological Survey, Denver, Colorado 80225, U.S.A.

ABSTRACT

A 5·4 km thick sequence of tectonically derived sediment has been dated by means of magnetic polarity stratigraphy and fission track methods. The sequence is located at Sierra de Huaco in the Precordillera of western Argentina. Sedimentation occurred over some 12 Myr from 14 to 2 Myr. Throughout this period the climate was arid as indicated by the dominance of alluvial fan and playa deposits. Sedimentation rates increased radically at about 9 Myr going from 0·17 to 0·92 mm yr^{-1}, a five-fold change. At about this time sedimentary style also changes and by 8·6 Myr new detrital minerals appear in the sequence. We attribute these transitions to the initiation of uplift in the Central Precordillera, uplift activity which continues to the present day. Locally derived conglomerate first appears in the section at 4·8 Myr and subsequently this conglomerate facies becomes pervasive. These conglomerates suggest that the Sierra de Huaco structure itself was being formed and eroded perhaps as early as 4·8 Myr. The pediments and gravel straths flanking the Sierra de Huaco today are the modern extensions of this process.

INTRODUCTION AND TECTONIC SETTING

Our purpose here is to report on the age and rate processes which characterize the fluvial sediments of a portion of the eastern Andean foreland basin. Our age data have been obtained from magnetic polarity stratigraphy and by isotopic methods. The stratigraphic section we have studied is 5·4 km in thickness and exemplifies the Tertiary molasse deposits occurring in San Juan Province, Argentina (Fig. 1). Tectonically, this area is situated above a subducting plate which is plunging at a relatively shallow angle and is nearly horizontal beneath San Juan Province (Jordan *et al.*, 1983a). The fluvial sequence is composed almost entirely of alluvial fan and playa deposits, indicating an arid environment with internal drainage. Mammalian fossils are conspicuously absent. Some 300 km farther to the north, however, stratigraphically equivalent sediments are quite fossiliferous, indicating that hostile conditions were not universal in the eastern Andean Foreland Basin (Butler *et al.*, 1984).

Our study section is a clastic sequence exposed in a series of north-trending synclines and anticlines (Fig. 2) that are part of the west-verging eastern Precordillera structural province (Ortiz & Zambrano, 1981). The eastern Precordillera is the eastern belt of the largely east-verging Precordillera thin-skinned fold and thrust belt (Ortiz & Zambrano, 1981) which are the eastern 'foothills' of the Andes Mountain chain (Figs 1 and 3). To the east, the Sierra de Huaco grades into the Bermejo valley, which is about 30 km wide at the latitude of our study.

The Sierra de Huaco forms the northern half of a 120 km long belt of continuous exposure of Cenozoic clastic strata. No detailed geological map of the Sierra de Huaco has been previously published, although the region is included in a 1:500,000 scale map of the province (Zambrano, unpublished; Ortiz & Zambrano, 1981). The south-western part of the Sierra de Huaco has been mapped at a 1:200,000 scale (Furque,

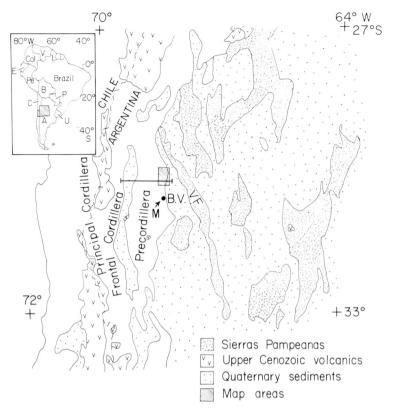

Fig. 1. Geographical setting of the Sierra de Huaco study area. Bermejo Valley is labelled BV, the Valle Fertil Range by VF, and Mogna by M. The area represented in Fig. 2 is indicated by the highlighted rectangle. The east-trending line across the Precordillera and study area indicates the position of the structural cross-section shown in Fig. 3.

1979) as has the southern half of the outcrop belt of Cenozoic strata at Sierra de Mogna (Furque, 1983; Cuerda *et al.*, 1984). We used low altitude air photos of the Sierra de Huaco to map structures and to establish lithostratigraphic units. Because of the excellent exposure, the air photos allowed precision in tracing units laterally, although lateral facies changes caused some difficulties in deciding where to place formation boundaries.

The stratigraphic framework of the Sierra de Huaco–Sierra de Mogna outcrop belt was first established by J. G. Kelly (1962) in an unpublished report for Yacimientos Petroliferos Fiscales (the Argentina state oil company). Cuerda *et al.* (1981) utilized Kelly's units in mapping the Sierra de Mogna. However, Furque (1979, 1983) and Cuerda *et al.* (1984) revised the stratigraphic nomenclature. A third set of stratigraphic terms has been employed in regional and subsurface petroleum resource studies (e.g. Ortiz & Zambrano, 1981). Thus, stratigraphic terminology is at present rather confused, with at least three systems

in use, and some of the formation names are used in different positions in the different nomenclature systems. We have used the terminology employed by Cuerda *et al.* (1981)—future lithostratigraphic and chronostratigraphic work will ultimately demonstrate the degree to which lithostratigraphic units are of regional utility and lead to a lasting resolution of the nomenclature problem.

The Huaco section and the basin of which it is a part (the Bermejo basin) is a foreland basin, analogous in geography and tectonic setting to the Hoback-Green River basin of the western United States (Jordan *et al.*, 1983a; Dorr, Spearing & Steidtmann, 1977). It overlies a region of subducted Nazca plate that is now sub-horizontal at about 120 km depth. Several anomalous features of the Andes orogenic system appear to correlate with the sub-horizontal segment of the subducted plate: the magmatic arc has been inactive for approximately 10 Ma, the Andes Cordillera is particularly narrow (with the Precordillera along its eastern margin), and a broad part of the

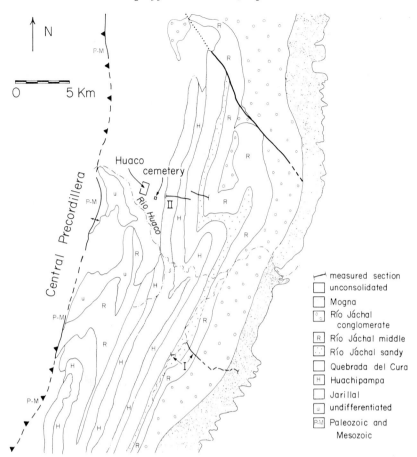

Fig. 2. Geological map of the Sierra de Huaco study area. The spatial and structural orientation of sample sections I and II are indicated. Lower and middle members of the Río Jáchal Formation as mapped correspond to the finer grained facies discussed in the text. In the western anticline, the members of the Río Jáchal Formation are not subdivided and it is labelled 'R'.

foreland is broken by reverse faults that uplift crystalline basement (Sierras Pampeanas) (Jordan *et al.*, 1983a, b) (Fig. 1). The main Andean Cordillera, Precordillera thrust belt, and Sierras Pampeanas structural belts expose significantly different bedrock lithologies. Of the different lithologies available in each source area, the following types are diagnostic: volcanic, plutonic and low grade metamorphic rocks come from the Andean Cordillera; clastic sedimentary rocks and limestones come from the thrust sheets of the Precordillera (Fig. 3); and medium grade metamorphic rocks come from the Sierras Pampeanas. These associations, which will be reported in detail in a later publication, enable recognition of source areas for the synorogenic deposits of the Sierras de Huaco sequence. The chronostratigraphy presented in this paper is a significant first step toward utilizing the synorogenic strata of the Bermejo basin to date the

deformation history in the Precordillera and Sierras Pampeanas, and to examine vertical motions in a foreland basin at the boundary between these two structural provinces.

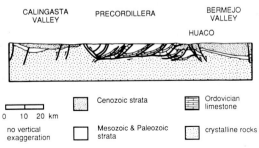

Fig. 3. Simplified structural cross-section of the Precordillera thrust belt, including the Sierra de Huaco at its eastern edge. The Frontal Cordillera rises immediately to the west of the Calingasta Valley. After Ortiz & Zambrano (1981).

LITHOSTRATIGRAPHY AND DEPOSITIONAL ENVIRONMENTS

Five formations are recognized in the Sierra de Huaco (Fig. 4). The names applied to these units follow Cuerda *et al.* (1981). All of the units are non-marine, and at least the upper four represent deposition in an arid environment with ephemeral streams, much like the climate today. Their sedimentology and depositional environments will be elaborated in subsequent publications. The interpretations presented here are drawn in part from comparing strata to the depositional landforms of modern San Juan Province. We have not systematically studied the cements, but gypsum cement and veins are common. The units will be described in ascending order.

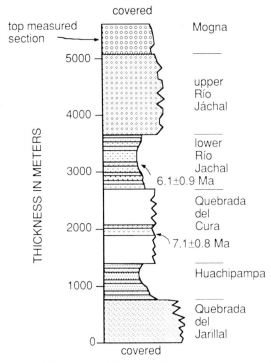

Fig. 4. Lithostratigraphic units of Huaco, showing ages of bentonites collected in the sequence. The right side of the column represents relative resistance of the units to erosion.

Quebrada del Jarillal Formation

The Quebrada del Jarillal Formation is a series of dark, reddish-brown weathering, resistant sandstones and minor siltstones. The base of the formation is not exposed in the Sierra de Huaco section, but it is reported to be concordant on strata cut by middle Cretaceous dykes in the Sierra de Mogna, about 60 km along strike of our line of section (Cuerda *et al.*, 1981). We measured 760 m of Jarillal in our section II; 800 m are reported in a complete section in the Sierra de Mogna (Cuerda *et al.*, 1981).

The Jarillal Formation consists of fine to medium grained sandstones and minor siltstone, with local very coarse sandstone near the top of the unit. It is thin to medium bedded, gradually thickening upsection to relatively thick bedded sandstone. The medium grained sandstones are parallel laminated and trough and planar cross-bedded, with sharp lower boundaries. The beds are laterally discontinuous but only very locally channelized. Desiccation features, rip-up clasts, and burrows are common.

The Quebrada del Jarillal Formation was apparently deposited in a low-relief, braided, fluvial system. The low proportion of mudstone suggests that most bar tops and interchannel areas were frequently washed by succeeding floods, transporting the muds to a more distal setting.

Huachipampa Formation

The Huachipampa Formation is conformable on the Jarillal Formation; the contact is gradual (Fig. 4). The unit is significantly less resistant to erosion than the Quebrada del Jarillal Formation in the lower half, and even less resistant in the upper half, forming a broad topographic low in the range. The lower contact is marked by a significant reduction in abundance of medium grained, thick bedded sandstones and an increase in mudstone. There is a notable cyclicity to the fine to medium grained sandstones, siltstones, and shales: 1–10 m thick sequences gradationally fine and thin upward. These cycles are occasionally symmetrical. Medium grained, parallel laminated sandstones that are not preserved as part of the cycles form occasional thick beds. Cross-bedding and channelling are virtually absent.

Locally the sandstones are tuffaceous, and on the basis of thin section studies the first appearance of limestone clasts among the sand grains occurs in the upper half of the formation. Furque (1979, 1983) reported conglomerate in the equivalent strata of the Huachipampa Formation at the south end of the Sierra de Huaco. We measured 610 m of the Huachipampa Formation in section II. To the south, 400–500 m of the unit are reported (Furque, 1983; Cuerda *et al.*, 1981).

The cycles and sedimentary structures are suggestive of deposition on a distal alluvial fan that grades into a playa or an alluvial plain environment, with highly ephemeral discharge.

Quebrada del Cura Formation

The Quebrada del Cura Formation is conformable on the Huachipampa Formation. The boundary is marked by a rapid increase in erosional resistance accompanying a decrease in the percentage of siltstone (Fig. 4). Cycles like those of the Huachipampa continue, but there are also common medium to coarse sandstone beds with tool and scour marks on the base, heavy mineral laminae, and cross beds. The finer grained facies include repeated 10–20 cm thick cycles of fining-upward, medium to fine grained sandstone and siltstone, representing deposition from declining flows currents. The sandstone beds and finer members thin laterally over short distances (10^2 m); there is local channelization with relief of about 1 m.

The middle of the Quebrada del Cura includes minor conglomerate and a bentonite bed that was dated for this study. Furque (1983) also reported minor conglomerate in the upper part of the equivalent unit, along strike to the south.

There are 1315 m of Quebrada del Cura Formation along section II. Thicknesses of 850–925 m are reported for exposures to the south (Furque, 1983; Cuerda *et al.*, 1981).

The Quebrada del Cura Formation was probably deposited in a more proximal part of an ephemeral, alluvial fan/braid plain system than was the Huachipampa Formation. The sandstone beds represent bars and migrating bedforms in low relief channels, and the thin fining-upward sandstones represent pulses in floods in overbank settings.

Río Jáchal Formation

The Río Jáchal Formation is conformable on the Quebrada del Cura Formation. It is characterized by decreasing resistance to erosion and, at least locally, renewed appearance of minor pebble conglomerate. We studied and sampled across this contact on both sections I and II (Fig. 2); the character is grossly similar at the two sites, but highlights the fact that the facies vary along strike. The upper part of the unit was not studied on section II, because a large-scale syncline repeats the sequence. A bentonite bed was found in the lower Río Jáchal Formation and was dated for this study (Fig. 4).

The lower member of the Río Jáchal of section I (about 940 m thick) and the part exposed on section II consist of highly gypsiferous, medium to thick bedded, medium grained sandstones, interbedded with thin to medium bedded shale, siltstone and fine sandstone. In section II the sandstones are commonly coarse grained with local pebbles conglomerates; in section I they coarsen gradually up-section into an upper conglomeratic member. Beds of both the sandstone facies and the mudstone facies pinch out rapidly laterally. The sandstones are variably parallel laminated to cross-bedded, locally with clay drapes on the cross-bed laminae and heavy mineral laminations. The bases of sandstone beds are sharp and locally channelized; the upper boundaries are variably sharp or pass upward through finer and thinner sandstone beds. The fine grained members are laminated, with common desiccation cracks. Locally there are thin, distinctive bluish-green shales that can be traced along strike over hundreds of metres distance.

The conglomeratic upper member of the Río Jáchal Formation was studied in section I, where it is about 1435 m thick. This unit is resistant because of a calcite cement in the conglomerates. Facies similar to those of the lower member continue, with the addition of medium to very thick beds of pebble conglomerate, with local cobbles and boulders. The dominant type of conglomerate consists of thick bed-sets of repeated thin to medium, horizontally bedded and planar cross-bedded, pebble conglomerate grading up to coarse sandstone. Upsection these are commonly trough cross-bedded and better sorted, with greater distinction between thick beds of gravel and coarse sandstone interbeds. Clast imbrication and convex-up laminae occur locally. Less common cobble conglomerate fills channels with up to 1 m relief, and occurs in lenses in the pebble conglomerates. Both fine grained and coarse grained units typically pinch out laterally over distances of tens of metres. Four to five centimetre diameter vertical burrows are common in the sandstone beds, but generally bioturbation is not sufficient to destroy primary depositional structures.

Clasts in the Río Jáchal conglomerates are subangular to sub-rounded. In order of diminishing abundance, they are composed of limestone, very low grade metasandstone, metapelite, red sandstone, volcanic rocks, quartz, and granitic rock fragments.

We measured a total thickness of 2375 m of Río Jáchal Formation in section I. Sixty kilometres to the south, Cuerda *et al.* (1981) reported that the unit is not conglomeratic and is 1300 m thick.

The Río Jáchal Formation was deposited by ephemeral braided streams, in an alluvial fan setting. Low relief of the channels and bars ($\sim 1 \cdot 0$ m) is indicated by the paucity of channelling and the thickness of the beds. The lower member is suggestive of an environment much like the Quebrada del Cura, with the bluish-green shales accumulating in broad ponds that trapped flood waters between the more active channels. The gradient of the fan and/or the volume of water gradually increased upsection, such that both the evidence of migrating bedforms and competence to transport coarse material increased, without significantly increasing the depositional relief of the channels and bars. The coarse sandstones and gravels represent bar deposits, whereas finer grained units were deposited on bar tops and interchannel areas.

Mogna Formation

The Mogna Formation overlies the Río Jáchal Formation. The contact is reported to be an erosional or angular unconformity at other locations in the Eastern Precordillera (Ortiz & Zambrano, 1981; Furque, 1983). However, along the line of section I the exposed sequence is located on the flank of a north-trending anticline on the far eastern margin of the fold belt (Fig. 2), and at that location the contact appears to be structurally concordant. That may not be true at other structural sites, such as on the noses of folds or farther to the west. The upper contact of the Mogna Formation is covered by younger unconsolidated gravels on modern pediment surfaces that form the western slope of the Bermejo Valley.

The lower part of the Mogna Formation is characterized by thick to very thick, massive conglomerate beds of well sorted, rounded, and imbricated pebbles and cobbles. Higher in the unit the conglomerates tend to be planar bedded, but more poorly sorted, similar to those of the Río Jáchal Formation. Thin sandstone and siltstone beds are interbedded, but become rare upsection. Erosional bases on the coarse beds are common for the first time in the section, with about 1 m of relief. Successive beds are as deeply eroded into one another as they are into the underlying Río Jáchal Formation.

The Mogna conglomerates are compositionally distinct from the underlying Río Jáchal conglomerates. There are few limestone clasts, and the clasts are generally better rounded. Because of the increasing frequency of conglomerates and diminution of silt in the upper part of the Mogna Formation, we were not able to sample palaeomagnetically the sequence to its top. We did, however, sample and measure 270 m of Mogna Formation, which compares with 600–800 m

of this unit reported to the south (Cuerda *et al.*, 1981; Furque, 1979). Our 270 m value is a minimum value, because steeply dipping Mogna conglomerate overlies our highest palaeomagnetic site to the east (Fig. 4). The Mogna Formation is probably much thicker farther to the east under the floor of the Bermejo Valley.

The Mogna Formation was formed in an ephemeral braided stream environment, where there was considerable relief, perhaps 1–2 m, between the channels and interchannel areas. The clast imbrications and channel incision suggests that flood waters were more confined to the channels than was true of the underlying units. The Mogna is the most proximal part of an alluvial fan environment represented in the Huaco sequence.

STRATIGRAPHIC OVERVIEW

The formations in the Huaco section represent fluvial deposition in an arid, ephemeral stream setting, as was established in adjacent areas by Cuerda *et al.* (1981) and Furque (1983). The Quebrada del Jarillal Formation formed in a system of braided streams that winnowed the muds out of the deposits. A new cycle of depositional systems is represented by the overlying 4600 m of section. There was a short interval of transition at the top of the Jarillal Formation, followed by a long-term succession from distal to proximal alluvial fan/braid plain environments. Deposits in the Huachipampa, Quebrada del Cura, and lower Río Jáchal formations indicate low energy conditions in low relief distal fan to playa environments, to which sediment was fed during floods but which underwent little erosion between or during floods. The conglomeratic member of the Río Jáchal Formation is a more proximal deposit, but still suggests low relief bars and channels with little erosional migration of stream channels. The uppermost unit, the Mogna Formation, represents a significantly higher energy depositional system, still probably comprising braided stream channels.

EXPERIMENTAL METHODS

The sedimentary sequence was studied in two sections, the first (section I, Fig. 2) spanning the interval from the lower Mogna Formation through the entire Río Jáchal Formation; the second (section II, Fig. 2) spanning the interval from the lower Río

Jáchal Formation to the Jarillal Formation. The two sections share a 500 m overlap of the lower Río Jáchal Formation which secures their physical correlation.

Stratigraphic thicknesses were measured by means of a tape and compass survey, traversing the sections essentially at right angles. Sandstones were sampled throughout the sequence for clast provenance studies, which will be reported in detail in another report. The sections were selected for their fresh, clear exposures and access by means of actively dissecting streams.

Two hundred and sixty-five palaeomagnetic sites were sampled throughout the sequence. Triplicate block samples were taken from each site by standard methods (Johnson, Opdyke & Lindsay, 1975). The magnetic mineralogy of the Huaco deposits is simple, consisting for the most part of magnetite as the principal carrier of a depositional remanence, with accessory maghemite and/or titanohematite (Johnsson, 1984). Accordingly, a thermal demagnetization treatment of 300–500°C was, with few exceptions, sufficient to isolate the primary remanence. The exceptions were confined invariably to the Jarillal Formation where hematite is more prevalent and required thermal demagnetization at higher temperatures, occasionally up to 640°C. In every case, thermal demagnetization was continued until the primary component revealed itself. Of the 265 paleomagnetic sites samples, 212 (80%) were statistically significant (according to the Watson criteria; Watson, 1956), 49 (18%) were not statistically significant but their polarity was not in doubt, and four (2%) were lost, destroyed during treatment, or had ambiguous polarity results. In those samples where statistical significance was not obtained, the cause was either a lack of sample size ($N \le 2$) or the presence of a sample whose magnetization direction was at variance with the others. Significantly, a disproportionate percentage of the statistically non-significant sites, 30% versus 15%, came from the Jarillal Formation, whose magnetic properties were distinctly more complicated than those of the younger formations. The population of statistically significant sites from section I satisfy the requirements of a reversal test (Johnsson, 1984; Johnsson *et al.*, 1984), as do the equivalent data from section II (Fig. 5). The palaeomagnetic data from the Huaco sequence therefore are accurately reflecting changes of the earth's magnetic field in times past. Both sections I (Johnsson *et al.*, 1984) and II (Fig. 5) show a 10° clockwise rotation, which further testifies to the reliability of the palaeomagnetic data. The magnetic properties of the rocks at Sierra de Huaco are quite similar to those reported by Butler *et al.*

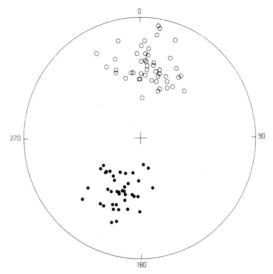

Fig. 5. Stereographic plot of all the statistically significant palaeomagnetic data from the Huachipampa to Río Jáchal Formations in section II. The Quebrada del Jarillal data are omitted from this diagram for reasons discussed in the text. Open symbols are projections from the lower hemisphere, closed symbols from the upper hemisphere. The normal sample has a mean declination of 002°, mean inclination of −38·6° with an average deviation from the mean of 17·3°. The reversed sample has a mean declination of 198°, mean inclination of +51·0° with an average deviation from the mean of 14·6°. With respect to their means and average deviations, the normal sample is antipodal to the reversed sample, thus satisfying the requirements of a positive reversal test.

(1984) for Cenozoic rocks farther north in western Argentina.

Bentonized volcanic tuffs were found in both section I (Johnsson *et al.*, 1984) and section II. Zircons were isolated from these air-fall deposits and fission track dated by the external detector method (Naeser, 1979).

RESULTS

We show in Fig. 6 the stratigraphic distribution of our palaeomagnetic data for section II, which defines its magnetic polarity stratigraphy. The comparable magnetic polarity stratigraphy for section I has been published previously (Johnsson, 1984; Johnsson *et al.*, 1984). In Fig. 7 we show our correlation between sections I and II, and also the correlation of our total magnetic stratigraphy with the polarity time-scale of Berggren *et al.* (1985).

The criteria we used to correlate sections I and II are two-fold. First, we used the lithostratigraphy of

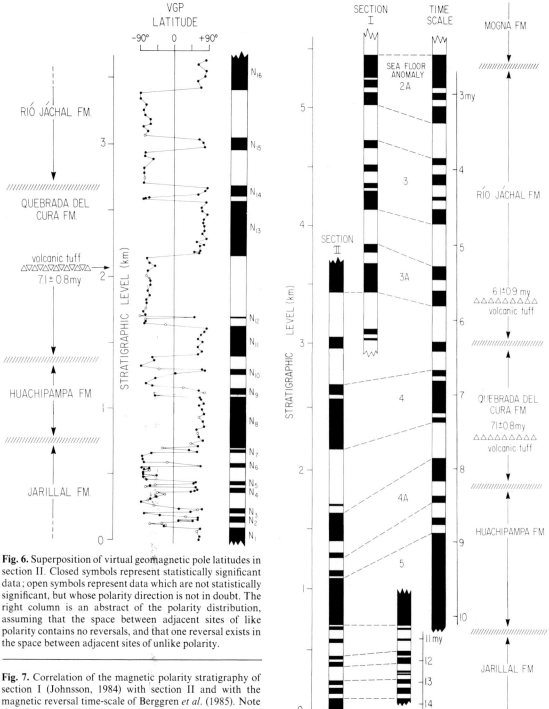

Fig. 6. Superposition of virtual geomagnetic pole latitudes in section II. Closed symbols represent statistically significant data; open symbols represent data which are not statistically significant, but whose polarity direction is not in doubt. The right column is an abstract of the polarity distribution, assuming that the space between adjacent sites of like polarity contains no reversals, and that one reversal exists in the space between adjacent sites of unlike polarity.

Fig. 7. Correlation of the magnetic polarity stratigraphy of section I (Johnsson, 1984) with section II and with the magnetic reversal time-scale of Berggren *et al.* (1985). Note that the time-scale is portrayed in two segments that are scaled differently.

the lower Río Jáchal Formation and traced beds on the air photos to establish the basic register of the two sections. Secondly, we used a magnetic time-line from each section and made them equivalent. In this regard the lower Río Jáchal Formation is largely of reversed polarity in both sections. We used the top of this reversed polarity zone, that is the base of N3 and N16 in sections I and II respectively, as the effective correlation tie-line (Fig. 7).

Our correlation of the entire Huaco magnetic stratigraphy to the reversal time scale was constrained by two independent lines of evidence. The isotopic dates in section I (6·1 ± 0·9 Myr) and section II (7·1 ± 0·8 Myr) provide strong indicators as to what part of the time-scale concerns us (Table 1). Also, knowing the stationary, stochastic properties of magnetic reversals in the past (Lowrie & Kent, 1983) and the efficiency of our palaeomagnetic sampling array (Johnson & McGee, 1983) we were able to estimate the length of time sampled in our section (Table 2). We uncovered 49 magnetic reversals with an array of 233 palaeomagnetic sites (allowing for the replication and overlap of samples in the lower Río Jáchal Formation). If we assume that our palaeomagnetic sample was exponentially distributed over the section, the time spanned by our section should be more than 9·7 Myr but less than 13·3 Myr (at the 68% confidence level). This compares with the 11·5 Myr time span we

assigned to the section by visually matching our observed magnetic stratigraphy to the time-scale (Fig. 7). In other words the time span estimated for the section ($\Delta t = 11·5 \pm 1·8$ Myr, Table 2) and the isotopic ages obtained from within the section (6·1 and 7·1 Myr, Table 1) constrain rather tightly our correlation.

A visual comparison of the reversal time-scale and the observed magnetic polarity pattern makes a reasonable one-to-one match (Fig. 7). Several small polarity zones are missing in the observed record compared to the time-scale, and there is an extra one that is not included in the time-scale. These discrepancies are readily explained, however, by sampling fluctuations and the uncertainties in the fine-structure of the time-scale (Johnson & McGee, 1983; Lowrie & Kent, 1983).

A closer inspection reveals that the distribution and length of magnetic polarity zones is uneven in the Huaco section (Fig. 7). For convenience in the visual identification and correlation of polarity zones, we have used two scaling factors to portray the reversal time-scale: an expanded scale for the Huachipampa Formation and younger strata, and a compressed scale for the Jarillal Formation. The implications of this change-in-scale will be discussed later.

Assuming that the correlations shown in Fig. 7 are accurate, we are then in a position to assign ages to

Table 1. Fission track data on air-fall zircon and estimated age

Sample	Formation	ρ_s tracks cm^{-2}	ρ_i tracks cm^{-2}	ϕ n cm^{-2}	No of grains	U ppm	r	Age Myr	$\pm 2\sigma$* Myr
HH-15	Río Jáchal	$9·40 \times 10^5$ (253)	$4·85 \times 10^6$ (1306)	$1·06 \pm 0·04$ $\times 10^{15}$	6	290	0·999	6·1	0·9
HH-179	Quebrada del Cura	$1·42 \times 10^6$ (392)	$6·36 \times 10^6$ (1756)	$1·06 \pm 0·04$ $\times 10^{15}$	6	381	0·950	7·1	0·8

† $\lambda_F = 7·03 \times 10^{-17}$, $\lambda_T = 1·55 \times 10^{-10}$, $\sigma = 580$, $I = 7·25 \times 10^{-3}$.
* Poisson error.

Table 2. Estimate of the time interval (ΔT) in the Huaco Sequence

Number of palaeomagnetic sites (N)	Number of reversals found (R)	Probability $P = R/(N-1)$	Mean spacing of palaeomagnetic sites[a] $S = f(p)$	Expected time interval[b] $\Delta t = \bar{S}\tau N$ (Myr)
233	49	0·21	0·40	$11·2 \pm 1·8$ (1σ)

[a] From Johnson & McGee (1983, equation 7 ($P = \bar{S}/(2\bar{S}+1)$). Assuming that samples are distributed with an exponential randomness over the stratigraphic interval.
[b] An estimate for $\bar{\tau}$ for the late Neogene is $1·2 \times 10^5$ yr (Johnson & McGee, 1983).

any part of the Huaco sequence. On this basis we have assigned ages to the formational contacts in the section (Table 3). From these data we can also say that the first appearance of limestone clasts occurs at 8·6 Myr, and that the first appearance of systematic conglomerates in the sequence is at 4·8 Myr.

Table 3. Ages of formational boundaries in the Huaco Sequence (Myr)

Jarillal/Huachipampa	10·3
Huachipampa/Quebrada del Cura	8·4
Quebrada del Cura/Río Jáchal	6·7
Río Jáchal/Mogna	2·6

Knowing the age of the strata level-by-level also allows us to calculate the sediment accumulation rates (Fig. 8). In section II sediment accumulation is characterized by two distinct linear phases, separated by a relatively short transition period (Fig. 8). Section I shows a single, linear sedimentation phase which is substantially different from that in the underlying strata of section II (Fig. 8). The fact that a break in mean sediment accumulation rate coincides exactly with the point where we joined sections I and II leads us to believe it is a measure of local geographic variation rather than a variation in the sediment accumulation rate over time.

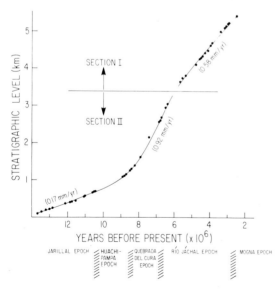

Fig. 8. Plot of stratigraphic level of magnetic time lines as a function of age of the time line. Data taken from Fig. 7. Note that slope of the line is equivalent to sediment accumulation rate.

DISCUSSION AND CONCLUSIONS

Perhaps the most conspicuous finding emerging from our chronology is the quantitative change in sediment accumulation that occurs at about 9 Myr. Note also that about this time, 8·6 Myr, a subtle but important change also occurs in sediment provenance, the first appearance of limestone clasts in the sand assemblage. Consider further that at 10·3 Myr, the Jarilla–Huachipampa boundary, a rather significant transition in fluvial environment takes place; the conversion of a throughgoing, braided stream system into a distributary channel system. The time interval 10·3–8·6 Myr in the Huaco sequence thus represents a period of profound stratigraphic change.

We interpret these changes in the Huaco sequence to result from uplift and erosion in the Central Precordillera. The five-fold increase in sediment accumulation rate at 9 Myr implies that a new or closer source of sediment supply became available at this time, and/or that the rate of subsidence increased. The concurrent appearance of limestone clasts (8·6 Myr) in this flood of sediment indicates also that lower Palaeozoic strata were being unroofed and eroded. These lower Palaeozoic limestones are presently exposed at the base of the thrust sheets of the Precordillera immediately to the west of Sierra de Huaco (Fig. 3). In this context we view the Jarillal–Huachipampa lithologic transition at 10·3 Myr as a precursory stage in the formation of a major foreland basin at the eastern margin of the Precordillera thrust slices.

As reported previously by Jordan et al. (1983a, b), volcanic activity essentially ceased about 10 Myr ago in the flat-subduction segment of the Nazca plate, and modern tectonic segmentation also came into existence at about this time. The synchroneity between these plate tectonic events (~ 10 Myr, Jordan et al., 1983b) and the sedimentary events taking place at Sierra de Huaco (10·3–8·6 Myr, see above) is obvious and striking. A causal relationship between plate tectonics and sedimentary response is suggested. Perhaps Precordillera thrusting and uplift were due to the onset and progression of flat-subduction, and in turn the sediments derived from these Precordillera structures were preserved in part in the Huaco foreland basin.

The end of sediment accumulation over the Sierra de Huaco area also has a tectonic and structural cause. Basically, sediment accumulation ceased because the foreland basin has been progressively destroyed on the side adjacent to the thrust belt. The original depositional site at Sierra de Huaco was systematically transformed into breached and faulted anticlines and synclines and became a source of sediment for the surrounding basins. In a broad sense then sedimentation has not really ceased, but has just migrated laterally. The uppermost units in our Huaco sequence are, in fact, proximal extensions of sediments which grade laterally into and under the present Bermejo Valley.

The youngest beds we observed folded in the Sierra de Huaco structure are about 2·4 Myr in age (Johnsson *et al.*, 1984), so deformation certainly post-dates 2·4 Myr. On the other hand, extensive conglomerates first appear in the Huaco sequence at 4·8 Myr (Johnsson *et al.*, 1984). These early conglomerates may signal the incipient stages of local uplift. They persist thereafter and eventually attain regional distribution exemplified in the Mogna Formation. In the Sierra de Huaco area the younger conglomerates of the Mogna Formation and successive straths and pediments perhaps are from the cannibalization of the Sierra de Huaco structure itself. Extensive conglomerates also are present in the fluvial sediments near the city of San Juan, some 150 km to the south of Sierra de Huaco. These conglomerates first appear in the San Juan sedimentary record at 8·0 Myr (Bercowski, Ruzycki de Berenstein & Johnson, 1985). They subsequently increase in frequency and extent until they become the dominant lithology at the top of the San Juan sequence. Evidently, the capping conglomerates, such as those at San Juan and the Mogna Formation of Sierra de Huaco, are diachronous from place-to-place in the eastern Precordillera.

A distinctive feature of the sediment accumulation curve for the Huaco sequence is the long periods (10^6 yr) over which mean sedimentation rate was maintained at a constant value (Fig. 8). These episodes of linear sedimentation continue despite considerable changes in lithology and sedimentary environment. In particular note that linear sediment accumulation continues throughout the period when systematic conglomerates first appear and eventually dominate the system (4·8–2·5 Myr, Fig. 8). The conditions necessary to maintain constant sedimentation rate are quite stringent; there must exist a tight balance among the various geological processes involved, sediment supply rate, basin subsidence rate, sediment compaction rate, isostasy and river grade. That such a delicate balance would obtain for millions of years might not seem a likely circumstance for a tectonically active terrane. Yet, our data show that mean sedimentation rates were indeed constant most of the time during the accumulation of the Huaco sequence. The existence of this steady-state condition, although perhaps unexpected, is not unique. Similar, steady-state sedimentation has been reported for the fluvial sediments in a foreland basin of Pakistan (Johnson *et al.*, 1982, 1985).

Lastly, we take note of the ubiquitous evidence for arid conditions during deposition of the Huaco sequence. A desert environment has imposed a dominant imprint on the stratigraphy of the Huaco sequence. What is visible first and foremost in these sediments is their desert origin. Their tectonic derivation is seen only after the effects of the desert environment are accounted for. That is, the effects of tectonism on the Huaco sequence are manifested only through such subtle parameters as mean sedimentation rates, and the changes thereof, and clast compositions. At face value the Huaco sediments are otherwise just thick desert deposits.

COMPARISONS WITH THE SIWALIK MOLASSE OF PAKISTAN

To gauge and evaluate better the Huaco data presented above, it might be useful to compare these data with a similar foreland basin of fluvial deposition. For this purpose the Siwalik sediments from the Himalayan foredeep will provide a convenient and pertinent contrast (Johnson *et al.*, 1985; Burbank & Reynolds, 1984). We present these comparisons in Table 4. Some striking similarities and differences may be seen. Most conspicuous among these is the fact that the Siwalik deposits and the Huaco deposits are exact contemporaries, both ranging in age from the Miocene to the present. Mountain building characterizes both areas and continues to the present day.

Rather large differences in tectonic setting exist between the Siwalik and the Huaco foreland basins. The Siwalik basins are situated in front of a continent-to-continent collision zone; while the Huaco basin lies behind an ocean–continent subduction zone. Despite these differences, the sediment accumulation rates and sedimentary thicknesses obtained for both basins are quite similar (Table 4). Both areas also are characterized by long intervals of constant sediment accumulation, interspersed with relatively brief periods of change. In both the Siwalik system and the Huaco system the sedimentary response to accelerated uplift in the source area is an increase in mean sediment accumulation rate. Accompanying the in-

Table 4. Comparison of the Huaco sequence with the Siwalik Group of Pakistan

	Siwalik sequence	Huaco sequence
Chronology and rates		
Age range (Myr)	1·6–18·3	2·2–14·0
Strata thickness	1–6	5
Sedimentation rates	0·1–0·5	0·1–0·9
Sedimentary episodes	Long intervals (10^6 yr) of linear sediment accumulation	Long intervals (10^6 yr) of linear sediment accumulation
Sedimentary transition	Sudden (10^5 yr) shifts between sedimentary phases	Sudden (10^5 yr) shifts between sedimentary phases
Tectonic setting		
Source	Himalaya	Andes
Mechanics	Continent–continent collision	Ocean–continent subduction
Orientation	Basin ahead of collision zone	Basin behind subduction zone
Initiation of sedimentation	Closure between basin and collision zone	Migration of uplift toward basin
Termination of sedimentation	Basin itself uplifted	Basin itself uplifted
Volcanism	Rare air fall deposits	Rare air-fall deposits
Stratigraphy setting		
Lithology	Fluvial clastics	Fluvial clastics
Climate	Savanna, open woodland	Desert
Drainage	Throughgoing	Dominantly internal with occasional throughgoing
River	Major trunk stream (ancestral Indus)	Multiple distributaries
Environment	Floodplain	Pediplane, playa
Style	Fluvial cycles	Alluvial fan, playa beds
Fossils	Abundant and varied	Absent or rare

creased sediment accumulation rate is the appearance of new detrital minerals or lithologies. In both the Siwalik and the Huaco sequences sedimentation was terminated by the destruction of the foreland basin itself.

Also noteworthy are the profound differences in sedimentary style between the Siwalik and Huaco sequences. The stratigraphic architecture of the Siwaliks bears little resemblance to that of the Huaco region. The Siwaliks represent throughgoing stream deposits characterized by the classic fluvial cycles of Allen (1965). In contrast the Huaco sequence for the most part shows an internal drainage system in the form of alluvial fan and playa deposits. It is clear from these data then that tectonic sediments in a non-marine environment can assume a variety of guises. It would seem also that the deciding factors in determining exact stratigraphic form are climate and hydrology and not tectonism. The sequence at Sierra de Huaco nicely illustrates the case of desert conditions imposed on tectonically derived sediments.

ACKNOWLEDGMENTS

We gratefully thank V. Ramos, D. Valencio, F. Bercowski, R. Allmendinger and S. Reynolds for their contributions to this project, and especially E. Fielding and J. Reynolds whose efforts in the field were vital. Much of the motivation and groundwork for this study can be attributed to A. Ortiz, to whom we are deeply indebted. Financial support was provided by NSF grants EAR-8206787, EAR-8218067, EAR-8418131, the donors to the Petroleum Research Fund of the American Chemical Society, and the Cornell Program for the study of the Continents (COPSTOC), Institute for Study of the Continents (INSTOC) contribution number 31.

REFERENCES

ALLEN, J.R.L. (1965) Fining-upwards cycles in alluvial successions. *Geol. J*, **4**, 229–246.

BERCOWSKI, F., RUZYCKI DE BERENSTEIN, L. & JOHNSON, N.M. (1985) Magnetic polarity stratigraphy and age of the Ullum Formation, San Juan, Argentina. *Primeras Hornadas sobre Geologia de Precordillera*, San Juan, Argentina, 9–11 October 1985 (abstract).

BERGGREN, W.A., KENT, D.V., FLYNN, J.J. & VAN COUVERING, J.A. (1985) Cenozoic geochronology. *Bull. geol. Soc. Am.* **96**, 1407–1418.

BURBANK, D.W. & RAYNOLDS, R.G.H. (1984) Sequential late Cenozoic disruption of the northern Himalayan foredeep. *Nature*, **311**, 114–118.

BUTLER, R.F., MARSHALL, L.G., DRAKE, R.E. & CURTIS, G.N. (1984) Magnetic polarity stratigraphy and ^{40}K-^{40}Ar dating of Late Miocene and Early Pliocene continental deposits, Catamarca Province, NW Argentina. *J. Geol.* **92**, 623–636.

CUERDA, A.J., CINGOLANI, C.A., VARELA, R. & SCHAUER, O.C. (1981) Geologia de la Sierra de Mogna, Provincia de San Juan. *VIII Congreso Geologico Argentino*, San Luis, **3**, 139–158.

CUERDA, A.J., CINGOLANI, C.A., VARELA, R. & SCHAUER, O.C. (1984) Descripcion Geologica de la Hoja 19d, Mogna. *Servicio geol. Nacional (Argentina), Bol.* **192**, 86 pp.

DORR, J.A., JR, SPEARING, D.R. & STEIDTMANN, J.R. (1977) Deformation and deposition between a foreland uplift and an impinging thrust belt: Hoback basin, Wyoming. *Spec. Pap. geol. Soc. Am.* **177**, 82 pp.

FURQUE, G. (1979) Descripcion geologica de la Hojo 18c, Jachal. *Servicio geol. Nacional (Argentina), Bol.* **164**, 79 pp.

FURQUE, G. (1983) Descripcion geologica de la Hoja 19c, Cienaga de Gualilan. *Servicio geol. Nacional (Argentina), Bol.* **193**, 111 pp.

JOHNSON, N.M. & MCGEE, V.E. (1983) Magnetic polarity stratigraphy: stochastic properties of data, sampling problems and the evaluation of interpretation. *J. geophys. Res.* **88**, B2, 1213–1221.

JOHNSON, N.M., OPDYKE, N.D. & LINDSAY, E.H. (1975) Magnetic polarity stratigraphy of Pliocene/Pleistocene terrestrial deposits and vertebrate faunas, San Pedro Valley, Arizona. *Bull. geol. Soc. Am.* **86**, 5–12.

JOHNSON, N.M., OPDYKE, N.D., JOHNSON, G.D., LINDSAY, E.H. & TAHIRKHELI, R.A.K. (1982) Magnetic polarity stratigraphy and ages of Siwalik group rocks of the Potwar Plateau, Pakistan. *Palaeogeogr. Palaeoclim. Palaeoecol.* **37**, 17–42.

JOHNSON, N.M., STIX, J., TAUXE, L., CERVENY, P.F. & TAHIRKHELI, R.A.K. (1985) Paleomagnetic chronology, fluvial processes, and tectonic implications of the Siwalik deposits near Chinji Village, Pakistan. *J. Geol.* **93**, 27–40.

JOHNSSON, P.A. (1984) *Magnetic polarity stratigraphy and age of the Río Jáchal and Mogna Formations, San Juan Province, Argentina.* M.A. Thesis, Dartmouth College, Hanover.

JOHNSSON, P.A., JOHNSON, N.M., JORDAN, T.E. & NAESER, C.W. (1984) Magnetic polarity stratigraphy and age of the Río Jáchal and Mogna Formations, San Juan Province, Argentina. *Proc. IX Congr. Geol. Argentino,* **3**, 81–96.

JORDAN, T.E., ISACKS, B.L., ALLENDINGER, R.W., BREWER, J.A., RAMOS, V.A. & ANDO, C.J. (1983a) Andean tectonics related to geometry of subducted Nazca plate. *Bull. geol. Soc. Am.* **93**, 341–361.

JORDAN, T.E., ISACKS, B.L., RAMOS, V.A. & ALLMENDINGER, R.W. (1983b) Mountain building in the Central Andes. *Episodes,* 1983, no. 3, 20–26.

KELLY, J.G. (1962) Geología de la Sierra de Móquina y perspectivas petrolíferas, Departamento de Jáchal, Provincia de San Juan. *Yacimientos Petroliferos Fiscales, Exploration Division,* unpublished report.

LOWRIE, W. & KENT, D.V. (1983) Geomagnetic reversal frequency since the Late Cretaceous. *Earth planet. Sci. Lett.* **62**, 305–313.

NAESER, C.W. (1979) Fission track dating and geologic annealing of fission tracks. In: *Lectures in Isotope Geology* (Ed. by E. Jager & J. L. Hunziker), pp. 154–169. Springer-Verlag, Berlin.

ORTIZ, A. & ZAMBRANO, J.J. (1981) La provincia geologica Precordillera oriental. *VIII Congreso Geologico Argentino,* San Luis, **3**, 59–74.

WATSON, C.S. (1956) A test for randomness of directions. *Mon. Not. R. astr. Soc. geophys. Suppl.* **7**, 160–161.

ZAMBRANO, J.J. (unpublished). *Mapa geologico de la provincia de San Juan* (1:500,000). Prepared for Instituto Nacional de Prevencion Sismica.

Spec. Publs int. Ass. Sediment. (1986) **8**, 77–90

The evolution of foreland basins to steady state: evidence from the western Taiwan foreland basin

MICHAEL COVEY*

Department of Geological and Geophysical Sciences, Princeton University, Princeton, New Jersey 08544, U.S.A.

ABSTRACT

The three-dimensional distribution of lithofacies within the Pliocene–Pleistocene Taiwan foreland basin shows that the basin evolved from an early deep-water stage to a later shallow-water stage. The early stage occurred while the adjacent orogenic belt grew in size from a submarine ridge to a rugged mountain range. Crustal flexure adjacent to the growing orogenic load created a subsiding foreland basin, but low relief in the emerging mountains produced relatively little sediment, resulting in deep-water conditions in the foreland basin. The later stage occurred after the orogenic belt reached a maximum, steady-state size, where erosion balanced uplift. As the steady-state orogen migrated further on to the continental margin, crustal flexure caused by the mountain load remained constant while erosion and sedimentation reached a maximum, resulting in shallowing conditions as the foreland basin filled with sediment.

Once the basin filled, high-energy processes of fluvial, marginal-marine, and shallow-marine environments removed some sediment longitudinally out of the basin, around the orogenic belt and into the open ocean. The resulting decrease in sediment accumulation rates created a balance between sedimentation and subsidence and led to a steady state within the foreland basin. Thus tectonic and sedimentary processes interacted to give the foreland basin a constant cross-sectional area as both it and the mountain belt migrated farther towards the craton.

INTRODUCTION

Foreland basins lie adjacent to orogenic belts, between the orogen and the nearby continent. They develop by flexure of the lithosphere under the load of the orogenic belt (Price, 1973; Beaumont, 1981; Jordan, 1981). The sedimentary fill of these basins consists primarily of detritus eroded from the orogenic belt, and as such records the unroofing history of the orogen, the patterns of subsidence, and the nature of sediment dispersal across the basin. Furthermore, because the foreland basin is located on the craton, the deposits record the encroachment of the orogenic belt on to the craton. Clearly, foreland basin deposits constitute an important record of orogenic history and processes.

Most foreland basins that have been studied are adjacent to inactive orogenic belts, so that the history of the basins is known in more detail than that of the

* Present address: Exxon Production Research Center, P.O. Box 2189, Houston, Texas 77001, U.S.A.

associated orogens. Indeed, the history of many ancient orogenic belts is inferred in large part from the sedimentary records within the associated foreland basins. In order to understand the detailed relationships between the sedimentary record of a foreland basin and the tectonic history of the adjacent orogen, an active orogenic belt and its foreland basin must be studied.

TECTONIC SETTING OF THE WESTERN TAIWAN FORELAND BASIN

The island of Taiwan is an active orogenic belt located at the site of collision of the Asian continent with the Luzon volcanic arc (Fig. 1A; Chi, Namson & Suppe, 1981). The development of the orogen has flexed the surrounding crust, creating a foreland basin on its western, cratonic margin (Fig. 1). The collision has deformed, uplifted, and exposed the part of the

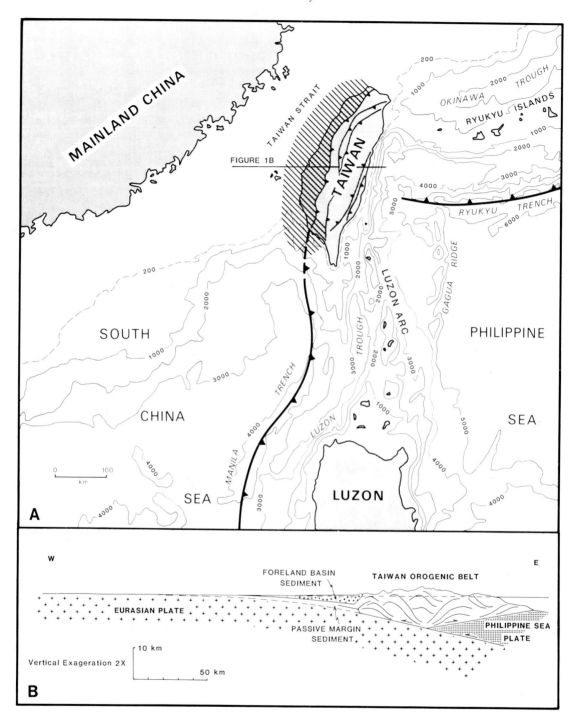

Fig. 1. Tectonic setting of the western Taiwan foreland basin. (A) Location of Taiwan where the passive margin of Asia encounters the Manila Trench subduction zone. The western Taiwan foreland basin (hachured) is located just west of the collision zone. (B) Schematic west–east cross-section of the foreland basin and orogen across the central part of the island.

foreland basin nearest to the orogen. Thus Taiwan presents an exceptional opportunity for studies of orogenic sedimentation: it displays both a sedimentary record of orogenic events and the orogenic and sedimentary processes responsible for that record. Here we can compare orogeny and its sedimentary record directly.

Orogenesis in the Taiwan mountain belt is well understood, facilitating comparison with the sedimentary record in the adjacent foreland basin. Plate convergence in the Taiwan area is approximately 70 km Myr^{-1} (Seno, 1977). Deformation across Taiwan has resulted in at least 160 km of shortening (Suppe, 1980), and the high level of seismic activity there attests to continuing deformation (Wu, 1978). Tectonic uplift rates have been measured by dating raised coral reefs at the north and south ends of Taiwan, they average $5\cdot0\pm0\cdot7$ mm yr^{-1} (Peng, Li & Wu, 1977). The measured denudation rate for the entire island is approximately 5·5 mm yr^{-1} (Li, 1976).

The arc–continent collision in Taiwan is occurring obliquely (Suppe, 1981). The NE–SW trend of the mainland China passive margin makes a high angle with the north–south trending Luzon volcanic arc (Fig. 1A). Thus orogenesis is beginning just south of Taiwan but began 4 Myr ago in northern Taiwan (Suppe, 1981).

Pre-collision sedimentary rocks exposed in the southern part of the Taiwan orogen were deposited in deep water of the distal part of the mainland China passive margin, while those exposed in the northern part of the orogen were deposited in shallow-marine and fluvial environments in a more landward position on the passive margin, showing that orogenesis has progressed farther on to the continent in the northern, older part of the collision zone. Thus we can examine the orogen, and the adjacent foreland basin, at advancing stages of development by moving northward along the island. This time–space equivalence is a powerful tool in determining the history of the orogen (Suppe, 1981) and of the foreland basin.

EVOLUTION OF THE WESTERN TAIWAN FORELAND BASIN TO STEADY STATE

The sedimentary record

The western Taiwan foreland basin is approximately 400 km long and 100 km wide (Fig. 1A). Sediment within the basin lies on deposits of the mainland China passive margin (Fig. 1B). The transition from passive margin to foreland basin deposits is marked by a change from eastward-directed to westward-directed palaeocurrents (Chou, 1973), the first occurrence of slate clasts derived from the Taiwan orogenic belt (Lee, 1963), and the first occurrence of reworked calcareous nannoplankton, also derived from the orogenic belt (Chi & Huang, 1981). This transition occurred approximately 4 Myr ago in the northern part of the basin (Chi & Huang, 1981). Although only 4 Myr old, 4–5 km of sediment has accumulated along the deeper, eastern edge of the basin. This edge has been uplifted by the westward-migrating orogen, so that the entire Pliocene–Pleistocene foreland basin sequence crops out along the western edge of the mountain belt.

Continuing orogeny in Taiwan means that erosion, transport and deposition of orogenic sediment is also continuing, and observable. At the same time, Pliocene–Pleistocene deposits exposed and drilled in western Taiwan constitute a record of these processes throughout the history of the foreland basin. Thus evidence from both the modern and ancient can be used to understand the sedimentary processes and depositional environments that have been active in the foreland basin.

The foreland basin depositional environments are briefly summarized in this paper. The summary is based on:

(1) Lithofacies analysis of foreland basin deposits exposed along the uplifted eastern edge of the basin (Covey, 1984a, b). Eleven sections at eight localities distributed along the length of Taiwan were analysed during six months of field work. In general, the entire foreland basin sequence is exposed at each locality.

(2) Lithofacies analysis of wireline records of wells drilled by the Chinese Petroleum Corporation (Covey, 1984a, b). Seventeen wells drilled in the structurally undeformed part of the basin, where there are no surface exposures, were examined.

(3) Published descriptions of processes and deposits of the contemporary foreland basin (Boggs, 1974; Boggs *et al.*, 1979; M.-P. Chen, 1981, 1986; R.-L. Chen, 1982; Chen & Covey, 1983; Chu, 1971; Covey, 1984a; Fan & Yu, 1981; Hsu, 1962; Liao, 1982; Shih, 1980; Water Resources Planning Commission, 1980). A summary of this published literature was presented in Covey (1984a).

Depositional environments in the western Taiwan foreland basin were quite diverse, ranging from deep marine to fluvial (Fig. 2). Five depositional environments have been delineated from lithofacies analysis, modern analogues of each are present within the contemporary basin.

The deepest environment, termed the offshore marine environment, was characterized by deposition of mud and silt from suspension. Sediment deposited in this environment reaches thicknesses as great as 4000 m. Burrowing organisms reworked much of the deposits as they accumulated, so that the dominant lithofacies is mottled silty mudstone. Rare silt-laminated mudstone and mudstone with cross-laminated siltstone indicate the intermittent presence of water currents. No deposits of turbidity currents have been recognized. Deposition was below storm-wave base, probably deeper than 200 m. Similar lithofacies have been piston cored in the southern part of the contemporary foreland basin in depths ranging from 500 to 2000 m (M. P. Chen, personal communication, 1984). Local slump features indicate that some deposition occurred in a slope environment.

Shallow-marine deposits are heterogeneous successions of fossiliferous mud, silt and fine-grained sand with varied lithofacies (Fig. 3). Successions from 100 to 2000 m thick, lacking associated non-marine deposits, suggest an environment similar to the present Taiwan Strait, the body of water between Taiwan and mainland China that is characterized by water depths of less than 80 m. Locally, associations of herringbone cross-bedding, trough cross-bedding, flaser bedding, and wavy bedding indicate that tidal processes

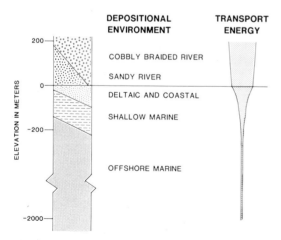

Fig. 2. Approximate water depths and relative transport energy of depositional environments within the western Taiwan foreland basin.

dominated during deposition (Fig. 3A). Elsewhere, sandstone with horizontal lamination, low-angle cross-bedding, and hummocky cross-bedding suggest storm currents were dominant (Fig. 3B). Still other successions, characterized by tens of metres of muddy sandstone completely homogenized by bioturbation, show that biological processes were important. In many localities tidal, storm, and biologic processes shaped the deposits concurrently.

Shoreline deposits differ from shallow-marine deposits by virtue of coarser-grained sand, less mud within sequences, and the presence of interbedded non-marine strata. Shoreline environments comprised both deltaic and non-deltaic settings. Deltaic deposits are thicker and generally coarser grained than non-deltaic deposits, and are capped by thick fluvial channel sequences. Non-deltaic deposits are generally less than 40 m thick, contain maximum grain sizes of fine- to medium-grained sand, and are capped by floodplain or shallow-marine deposits. Like the shallow-marine deposits, lithofacies of the shoreline deposits document the influence of tidal and storm processes (Fig. 4). Analogous environments in the contemporary foreland basin consist of tide-dominated, wave-dominated, and tide- and wave-dominated shorelines (Hsu, 1962).

Two types of fluvial systems were active in the foreland basin: sand-dominated and cobble-dominated. Sandy fluvial deposits are characterized by poor outcrops, so that detailed lithofacies analysis was not possible. Modern rivers with sand-dominated bedloads in the contemporary foreland basin are either braided or low sinuosity, it is likely that both types were present during the Pliocene and Pleistocene as well. Cobble-dominated rivers were braided and deposited clast-supported conglomerates up to 600 m thick. Clast size in the conglomerates average 15–20 cm, maximum clast size is 1 m. In the contemporary foreland basin rivers with cobble-dominated bedloads are up to 3 km wide, with bedload transport usually occurring during typhoons. Actively depositing fluvial environments occur at elevations between 0 and 200 m above sea-level in the present-day foreland basin.

The distribution of the above depositional environments in time and space is shown in Fig. 5, and provides a record of the evolution of the western Taiwan foreland basin. Figure 5 shows that in general the foreland basin fill consists of a westward-thinning clastic wedge, and that deposits representing progressively shallower environments prograde westward, away from the orogenic belt. The oblique collision in Taiwan means that the southern part of the foreland

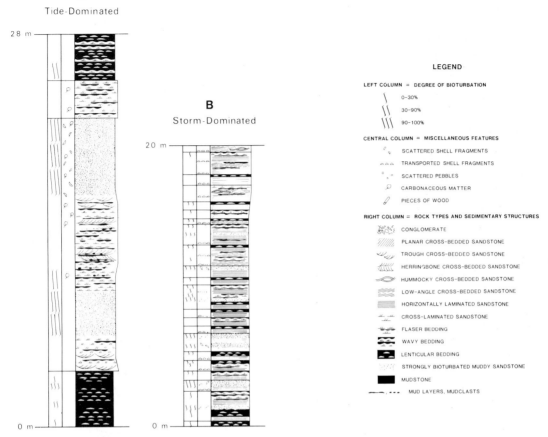

Fig. 3. Column A—example of Pliocene tide-dominated shallow-marine lithofacies assemblage, Miaoli section. Column B— example of Pliocene storm-dominated shallow-marine lithofacies assemblage, SW Taiwan. Legend to columns on right. Location of column A given in Fig. 5.

basin is less developed than the northern part, so that Fig. 5 also shows that the younger part of the basin is dominated by deep-water deposits, while the older part contains only shallow-water and fluvial deposits.

Serial cross-sections detail the changes from south to north in the basin (Fig. 6). The eastern end of each of these cross-sections are surface exposures that lie within the western foothills of the Taiwan orogenic belt, the western ends lie within the undeformed part of the foreland basin. Thus the sections presently lie at a similar distance from the orogenic belt, and probably did throughout the history of the basin, since the orogenic belt appears to have migrated westward cylindrically through time (Chi *et al.*, 1981). The northern two cross-sections consist entirely of fore-land basin deposits, as evidenced by the first

occurrence of reworked calcareous nannoplankton derived from the Taiwan orogen well below the nn15/ nn16 nannoplankton datum (Chi & Huang, 1981). The base of the southern cross-sections probably consist of slope deposits of the mainland China passive margin sequence. The transition from the passive margin sequence to the foreland basin se-quence is not known precisely in the two southern cross-sections, however all the deposits above the nn18/nn19 datum are assigned to the foreland basin sequence because of their high accumulation rates.

The southernmost cross-section (Fig. 6A) shows that this youthful part of the foredeep is dominated by offshore-marine deposits. Sediment deposited in shal-lower environments is restricted to a thin wedge at the top of the section. The cross-section located slightly to

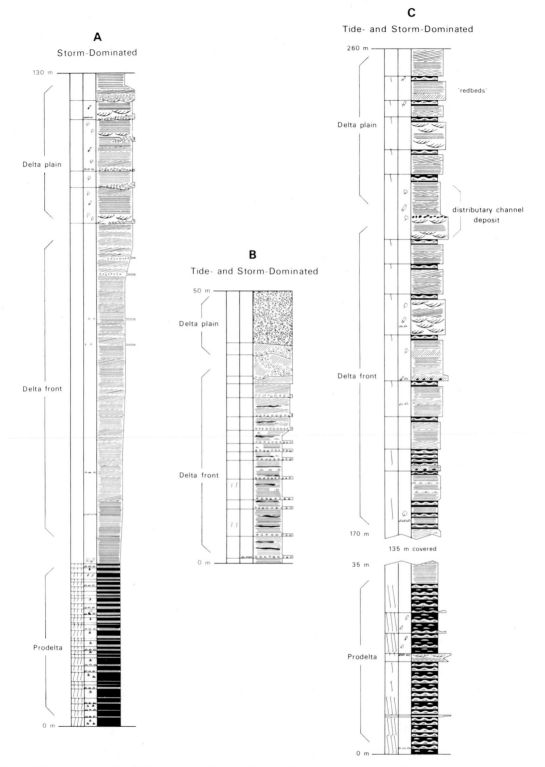

Fig. 4. Column A—example of Pliocene storm-dominated deltaic lithofacies assemblage, Chinanshan section. Column B—example of Pleistocene tide- and storm-dominated deltaic lithofacies assemblage, Miaoli section. Column C—example of Pleistocene tide- and storm-dominated deltaic lithofacies assemblage, Chunkungliao-chi section. Legend to columns in Fig. 3. Location of all three columns given in Fig. 5.

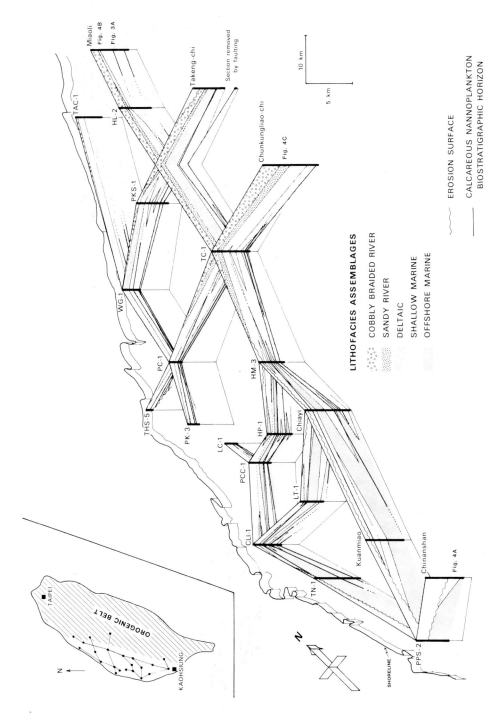

Fig. 5. Simplified fence diagram of studied part of Taiwan foreland basin. The top of each panel represents total sedimentary column prior to erosion. Each panel extends to 5 km depth, except that between the TC-1 well and the Chunkungliao-chi section, which extends to 6 km at its eastern end. Nannoplankton time lines provided by W.-R. Chi of the Chinese Petroleum Corporation. Locations of panels indicated on inset map, hachured area on inset map indicates location of Taiwan orogen. Locations of Figs 3(A), 4(A, B, C) indicated in fence diagram.

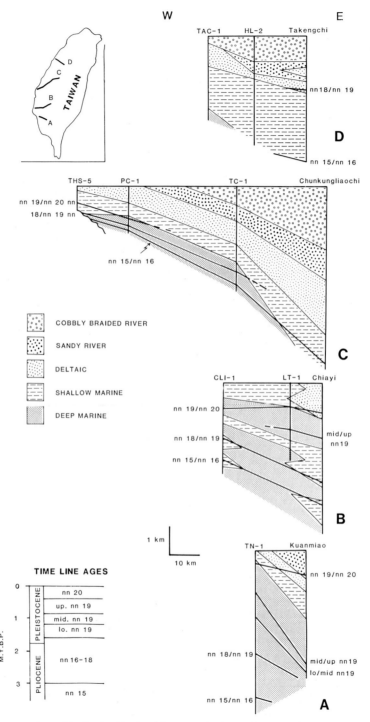

Fig. 6. Stratigraphic cross-sections of the foreland basin sedimentary fill showing the depositional environment of the deposits. Locations of sections indicated on inset map. Boundaries of calcareous nannoplankton zones used to define time lines provided by W. R. Chi of the Chinese Petroleum Corporation.

the north (Fig. 6B) is also dominated by offshore marine deposits, but interbedded shallow-marine deposits show that sediment intermittently filled the basin to near sea level. Still farther north, the next cross-section (Fig. 6C) is dominated by shallow-marine, deltaic, and fluvial deposits, with only a thin accumulation of deep-marine sediment at the base of the cross-section. The northernmost cross-section (Fig. 6D) is almost completely lacking in deep-marine deposits. Instead it consists of a progradational wedge of shallow-marine, deltaic, and fluvial sediment.

The four cross-sections of Fig. 6 shows that the western Taiwan foreland basin evolved from an early, deep-water stage to a later, shallow-water stage. In addition, the northern cross-sections, especially Fig. 6(C), shows that once the basin filled with sediment to depths near sea-level, the configuration of shallow-marine sedimentation far from the orogen and fluvial sedimentation near to the orogen persisted throughout the late Pliocene and Pleistocene, prograding toward the craton in front of the orogen. This indicates a balance between sedimentation and subsidence during the late, shallow-water stage, and that the foreland basin reached a steady state.

Tectonic and sedimentary controls of the sedimentary record

The evolution of the western Taiwan foredeep to steady state resulted from the interaction of tectonic and sedimentary processes. Tectonic loading of the craton by the Taiwan orogen initiated and sustained the development of the adjacent foreland basin. The fact that the orogen matures northward due to the oblique collision in Taiwan can be used to show how the tectonic load changed during its history (Suppe, 1981). When the Asian passive margin first encountered the Manila Trench subduction zone, the increased sediment thickness entering the subduction zone caused the accretionary wedge to enlarge. With continued collision, the accretionary wedge eventually emerged above sea-level, to become the Central Range of Taiwan. Once the mountain belt reached 3·5–4·0 km in height, the cross-sectional width and area of the mountain belt no longer increased, indicating that erosion balanced uplift, and that the mountain belt reached a steady-state size (Suppe, 1981).

Because early growth of the orogenic belt took place while it was submerged, tectonic loading and resultant

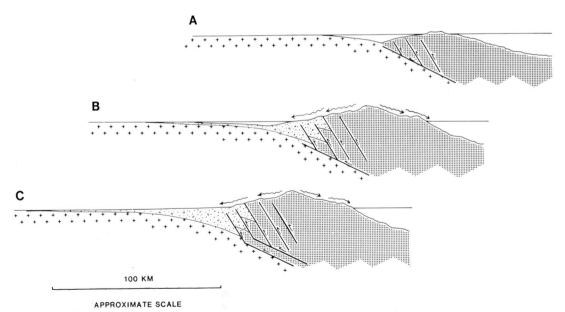

Fig. 7. Diagrammatic portrayal of the development of the foreland basin as the orogen migrates toward the continent. Faults are schematic, wavy arrows indicate transport of eroded debris. (A) Early development of the orogen occurs while it is submarine. Loading causes flexure of the crust prior to substantial sediment production, so that the foreland basin is characterized by deep water. (B) The subaerially exposed orogenic belt reaches steady state, and erosion is at a maximum. The foreland basin grows cross-sectionally because of sediment loading, and shoals through time as sediment displaces water. (C) The foreland basin is filled with sediment.

basin subsidence occurred before sediment was shed from the orogen (Fig. 7A). Even after the orogenic belt emerged above sea-level, erosion of the orogen and resultant sediment infilling of the foreland basin did not reach its maximum rate until the mountain belt reached its maximum size, when steady state was established. The delay in sedimentation meant that initially subsidence was greater than sediment accumulation. The result was the deposition of deep-water foreland basin sediment upon shallow-water passive margin deposits.

Once the orogenic belt reached a steady-state size, the magnitude of the tectonic load remained constant through time as it moved toward the craton (Fig. 7B, C). Further cross-sectional expansion of the sedimentary basin was caused by sediment influx only (Fig. 7B). Because subsidence induced by the sedimentary load was less than sediment thickness, depositional environments became progressively shallower.

Eventually sediment filled the foreland basin to near sea-level. At this point the high-energy processes of the shallow-marine system (tide and storm currents)

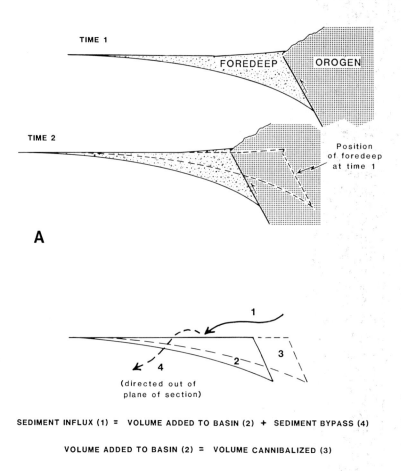

SEDIMENT INFLUX (1) = VOLUME ADDED TO BASIN (2) + SEDIMENT BYPASS (4)

VOLUME ADDED TO BASIN (2) = VOLUME CANNIBALIZED (3)

Fig. 8. Foreland basin steady state and mass balance. (A) Schematic representation of foreland basin steady state. Fault is symbolic, separating region of uplift from region of subsidence. Through time the orogen migrates continentward, uplifting and destroying earlier foreland basin deposits. This is offset by basin growth as the flexure associated with the orogenic load migrates continentward. The sediment surface remains at the same elevation through time because sediment bypassing produces a balance between sedimentation and subsidence. (B) Mass balance of the steady-state foreland basin. Sediment influx (1) comprises both cannibalized foreland basin material (3) and underlying passive margin deposits (not shown). Volume added to basin at its upper surface (2) equals the amount cannibalized (3). Excess material bypasses the basin (4).

moved finer-grained sediment out of the basin, through the open north and south ends. Although sediment influx into the basin was greater than could be accommodated by subsidence, sediment bypassing produced a balance between sediment accumulation and subsidence.

The balance between subsidence and sedimentation is possible because the orogen and foredeep pair is a dynamic system. As the Taiwan orogen migrated toward the craton, foreland basin deposits were uplifted and eroded along the eastern edge of the basin (Fig. 8). This cannibalism of the foreland basin was balanced by growth along its western edge, as the crust flexed in front of the migrating load. Sediment bypassing constrained the sediment surface to a position close to sea-level, so that the cross-sectional width and area of the basin remained constant through time (Fig. 8). Like the orogenic belt, the foreland basin attained a steady-state cross-sectional size as it moved continentward.

The evolution of the sedimentary fill as the western Taiwan foreland basin evolved to a steady state is summarized in Fig. 9(A–C). Early in the basin's history (Fig. 9A) deep-water sediments were deposited because of the initial delay in sedimentation relative to subsidence. Later, as the orogen attained its maximum elevation (steady state within the orogen) high sedimentation rates caused filling of the basin and produced a shallowing-upward sequence (Fig. 9B). Once the basin filled with deposits and sediment bypassing became effective, sedimentation and subsidence became balanced, and steady state was attained in the foreland basin (Fig. 9C). In the western Taiwan foreland basin, shallow-marine processes produce the necessary sediment bypassing for steady state; thus the steady-state sedimentary sequence consists of prograding shallow-marine, deltaic and fluvial deposits.

Since deep-water sediment no longer accumulated once steady state was attained, uplift and erosion of

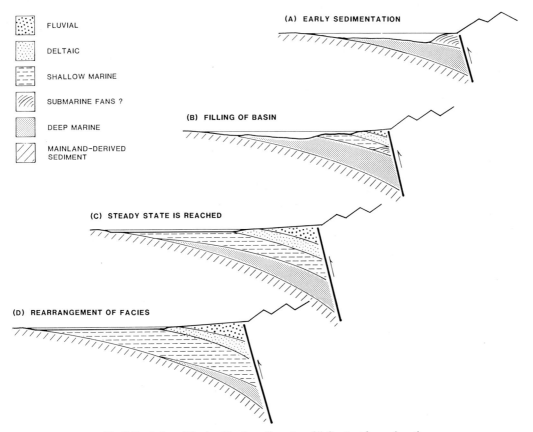

FLUVIAL

DELTAIC

SHALLOW MARINE

SUBMARINE FANS ?

DEEP MARINE

MAINLAND–DERIVED SEDIMENT

(A) EARLY SEDIMENTATION

(B) FILLING OF BASIN

(C) STEADY STATE IS REACHED

(D) REARRANGEMENT OF FACIES

Fig. 9. Evolution of foreland basin sedimentary fill. See text for explanation.

the deposits by the encroaching mountain belt continued to alter the sedimentary fill of the basin, so that eventually the deep-water deposits were partially or totally removed (Fig. 9D). Thus I hypothesize that sediment deposited prior to and to the east of the mature part of the basin were similar to those of the youthful part of the basin, but have since been uplifted and eroded away.

SUMMARY AND DISCUSSION

An understanding of the role that tectonism played in the development of the western Taiwan foreland basin is possible only because the associated orogenic belt is active at present and is well understood. In particular, the oblique collision in Taiwan allows observation of the orogen at various stages of development (Suppe, 1981). We see that early growth of the mountain belt, and resultant subsidence in the foreland basin, took place before a large amount of sediment was shed into the basin. Thus subsidence in the foreland basin initially outpaced sediment accumulation, and deep-water deposits were laid down upon shallower-water passive margin deposits that pre-date orogeny. The oblique collision also shows that erosion limited the orogenic belt to a maximum size, which the belt maintained as it migrated farther on to the continental margin. Once the orogen attained this steady-state size, the tectonic load remained constant, and the tectonic impetus for growth of the basin ceased. Maximum erosion rates associated with steady state in the orogen meant that sediment accumulation rates became greater than subsidence rates, and the basin filled with sediment. The evolution of the basin through the early, deep-water stage to the later, shallow-water stage directly reflects the evolution of the orogenic belt through its growth stage to its stage of steady state.

Shallow-marine deposits form sequences up to 2000 m thick in the northern part of the Taiwan foreland basin, and the shallow-marine environment makes up a large part of the contemporary basin. The persistence of the shallow-marine environment through time indicates that once the basin filled close to sea-level, a balance between sedimentation rates and subsidence rates was achieved, caused by a reduction in sedimentation rates. This reduction was brought about by transport of some of the sediment load longitudinally, out of the basin, by tide and storm currents. The balance between subsidence and sediment accumulation meant that the cross-sectional size

of the foreland basin no longer increased. Uplift and erosion of foreland basin deposits by the migrating orogenic belt was matched by subsidence at the distal end of the basin (Fig. 8). Thus both the orogen and the foreland basin remained a steady size as they migrated further toward the craton.

Deep-water deposits no longer accumulated once the Taiwan foreland basin reached steady state. Instead, deep-water deposits were removed from the foreland basin sequence as the encroaching orogenic belt destroyed earlier basin deposits. In the mature parts of the basin the orogenic belt has migrated to the point that all pre-steady-state deposits have been removed, and the basin record consist of shallow-water and fluvial deposits only.

The pattern of initial deep-water sedimentation followed by subsequent shallow-water sedimentation is common to several foreland basins (Miall, 1978). In some basins the early, deep-water stage is marked by the deposition of turbidites to form a distinctive flysch sequence. Although the western Taiwan foreland basin lacks known turbidite sands, the offshore muds of Taiwan are in a similar tectono-stratigraphic position as flysch deposits of other foreland basins; both are associated with the early stages of orogenesis and record deposition before the basin filled with sediment.

The shallow-water stage is achieved by almost all foreland basins. I suggest that thick shallow-marine and fluvial deposits in these basins accumulated after the basins reached steady state. Like the Taiwan foreland basin, some of these basins display thick shallow-marine deposits, implying that the shallow-marine environment was dominated by processes able to carry enough sediment out of the basin to produce a balance between subsidence and sediment accumulation. The Upper Marine Molasse of the Alpine foreland basin, for example, shows that a palaeogeography of fluvial environments proximal to the orogen and shallow-marine environments distal to the orogen remained relatively stable (Homewood & Allen, 1981). As in Taiwan, the high tidal and wave energy of the shallow-marine system (Homewood & Allen, 1981) may have provided the mechanism for sediment bypassing necessary for maintenance of steady state.

Conversely, steady-state sequences in many foreland basins lack shallow-marine deposits, and consist of fluvial deposits only. During deposition of the Lower and Upper Freshwater Molasse, fluvial environments extended across the entire Alpine foreland basin (Matter & Homewood, 1980). Perhaps the currents that previously existed in the shallow-marine realm became too weak to transport enough sediment out of the basin. Instead sediment remained in the basin, filling it above sea-level, so that axial fluvial systems were created (Fuchtbauer, 1967). I suggest that these fluvial systems served to bypass enough sediment to establish steady state in the basin. A similar situation exists in the Himalayan foreland basin (Indus–Ganges alluvial plain), an active basin receiving sediment from the adjacent Himalayan Mountains to the north. Rivers flow out of the orogen and across the basin until they reach the Indus and Ganges rivers, which flow parallel to the axis of the basin along its southern margin. The sedimentary fill within this basin consists almost entirely of non-marine deposits of the Miocene to Pleistocene Siwalik Group, which record a similar palaeoflow pattern to that of today (Graham, Dickinson & Ingersoll, 1975; Parkash, Sharma & Roy 1980; Johnson *et al.*, 1985). Thus major rivers flowing longitudinally to the basin provided, and still provide, the sediment bypassing necessary for steady state.

Some foreland basins lack early stage deep-water deposits, instead they contain later stage shallow-water and fluvial deposits only. I hypothesize that, like the northern part of the Taiwan foreland basin, these basins at one time contained deep-water deposits, but these deposits were uplifted and eroded as the orogen migrated toward the continent. The Himalayan foreland basin, for example, contains fluvial deposits exclusively. These overlie marine carbonates of Eocene age (Gansser, 1964; Johnson *et al.*, 1985), that were not related to the India–Asia collision. I suggest that during the early stages of collision and orogenesis, subsidence outpaced sediment accumulation, and deep-water deposits were laid down on the northern Indian margin in a foreland basin position. It has been estimated between 900 and 1500 km of the northern margin of India has been overridden by the Himalayan orogen (reconstructions of Zhao & Morgan, 1985), indicating that any foreland basin deposits originally deposited on the margin have been totally removed. Although it is not possible to prove that some of the now-eroded deposits accumulated in a deep-water environment, the evidence from Taiwan presented here suggests that they were.

To conclude, sedimentary fill of the western Taiwan foreland basin evolved through stages that reflected the maturity of the associated orogenic belt. Early on, deep-water deposits accumulated in the basin when the youthful orogenic belt increased in size but shed a limited amount of clastic debris. Subsequently, a shallowing-upward sequence was laid down as the

orogen attained its steady-state size, and erosion was maximized. The basin then filled until processes on the sediment surface were efficient at transporting some of the debris out of the foreland basin, creating a basin of steady-state size. Finally, the amount of deep-water deposits in the basin sedimentary fill volumetrically decreased through time, evidence that the steady-state orogenic belt had migrated far on to the continent.

ACKNOWLEDGMENTS

This paper stemmed from my PhD thesis work, supervised by Drs Franklin Van Houten and John Suppe of Princeton University. My thanks are extended to both of them for their guidance and support. Logistical support for work in Taiwan was provided by the Chinese Petroleum Corporation. Thanks to Dr Stanley S. L. Chang, Dr Jenn-Wei Yuan and Mr Wen-Rong Chi, all of the CPC, for support and friendship during my stay in Taiwan. Mr Wen-Rong Chi kindly provided the nannoplankton biostratigraphic data used in this study. Tim Cross, Neil Lundberg and John Suppe thoroughly reviewed the manuscript, their criticisms and suggestions improved it immensely. This work was supported by National Science Foundation grant EAR 8121196.

REFERENCES

BEAUMONT, C. (1981) Foreland basins. *Geophys. J. R. astr. Soc.* **65**, 291–329.
BOGGS, S., JR (1974) Sand-wave fields in Taiwan Strait. *Geology*, **2**, 251–253.
BOGGS, S., Jr, WANG, W.-C., LEWIS, F.S. & CHEN, J.-C. (1979) Sediment properties and water characteristics of the Taiwan shelf and slope. *Acta Ocean. Taiwanica*, **10**, 10–49.
CHEN, M.-P. (1981) Geotechnical properties of sediments of Hsinchu–northwest Taiwan related to sedimentary environment. *Acta Ocean. Taiwanica*, **12**, 28–53.
CHEN, M.-P. (1986) Physical properties of the sediments from the continental slope off Kaohsiung. *Acta Ocean. Taiwanica*, in press.
CHEN, M.-P. & COVEY, M. (1983) Radiocarbon dating of piston cored sediments from the Taiwan Strait: preliminary results. *Acta Ocean. Taiwanica*, **14**, 9–15.
CHEN, R.-L. (1982) *Sediments from Hsinchu offshore area.* Unpublished Masters Thesis, National Taiwan University, 158 pp. (in Chinese).
CHI, W.-R. & HUANG, H.-M. (1981) Nannobiostratigraphy and paleoenvironments of the Late Neogene sediments and their implications in the Miaoli area, Taiwan. *Petrol. Geol. Taiwan*, **18**, 111–129.

CHI, W.-R., NAMSON, J. & SUPPE, J. (1981) Stratigraphic record of plate interactions in the Coastal Range of eastern Taiwan. *Mem. geol. Soc. China.* **4**, 491–530.

CHOU, J.-T. (1973) Sedimentology and paleogeography of the Upper Cenozoic System of western Taiwan. *Proc. geol. Soc. China*, **16**, 111–143.

CHU, T.-Y. (1971) Environmental study of the surrounding waters of Taiwan. *Acta Ocean. Taiwanica*, **1**, 15–32.

COVEY, M. (1984a) *Sedimentary and tectonic evolution of the western Taiwan foredeep.* Unpublished Ph.D. thesis, Princeton University, 152 pp.

COVEY, M. (1984b) Lithofacies analysis and basin reconstruction, Plio–Pleistocene western Taiwan foredeep. *Petrol. Geol. Taiwan*, **20**, 53–83.

FAN, K.-L. & YU, C.-Y. (1981) A study of water masses in the seas of southernmost Taiwan. *Acta Ocean. Taiwanica*, **12**, 94–111.

FUCHTBAUER, H. (1967) Die Sandsteine in der Molasse nordlich der Alpen. *Geol. Rdsch.* **56**, 226–300.

GANSSER, A. (1964) *Geology of the Himalayas.* Wiley (Interscience), London, 289 pp.

GRAHAM, S.A., DICKINSON, W.R. & INGERSOLL, R.V. (1975) Himalayan-Bengal model for flysch dispersal in the Appalachian-Ouachita system. *Bull. geol. Soc. Am.* **86**, 273–286.

HOMEWOOD, P. & ALLEN, P. (1981) Wave-, tide-, and current-controlled sandbodies of Miocene Molasse, western Switzerland. *Bull. Am. Ass. Petrol. Geol.* **65**, 2534–2545.

HSU, T.-L. (1962) A study on the coastal geomorphology of Taiwan. *Proc. geol. Soc. China*, **5**, 29–46.

JOHNSON, N.M., STIX, J., TAUXE, L., CERVENY, P.F. & TAHIRKHELI, R.A.K. (1985) Paleomagnetic chronology, fluvial processes, and tectonic implications of the Siwalik deposits near Chinji village, Pakistan. *J. Geol.* **93**, 27–40.

JORDAN, T.E. (1981) Thrust loads and foreland basin evolution, Cretaceous, western United States. *Bull. Am. Ass. Petrol. Geol.* **65**, 2506–2520.

LEE, P.-J. (1963) Lithofacies of the Toukoshan-Cholan Formation of western Taiwan. *Proc. geol. Soc. China*, **6**, 41–50.

LI, Y.-H. (1976) Denudation of Taiwan island since the Pliocene epoch. *Geology*, **4**, 105–107.

LIAO, Y.-W. (1982) Effects of marine agency on the coastal geomorphology of Taiwan. *Petrol. drill. Product. Engng.* **23**, 254–277.

MATTER, A. & HOMEWOOD, P. (1980) Flysch and Molasse of central and western Switzerland: introduction. In: *Geology of Switzerland, a Guide Book* (Ed. by R. Trumpy). *26th int. Geol. Cong., Paris, Schweiz. Geol. Komm.* pp. 261–293.

MIALL, A.D. (1978) Tectonic setting and syndepositional deformation of molasse and other nonmarine-paralic sedimentary basins. *Can. J. Earth Sci.* **15**, 1613–1632.

PARKASH, B., SHARMA, R.P. & ROY, A.K. (1980) The Siwalik Group (molasse)—sediments shed by collision of continental plates. *Sedim. Geol.* **25**, 127–159.

PENG, T.-H., LI, Y.-H. & WU, F.T. (1977) Tectonic uplift rates of the Taiwan Island since the Early Holocene. *Mem. geol. Soc. China*, **2**, 57–69.

PRICE, R.A. (1973) Large-scale gravitational flow of supracrustal rocks, Southern Canadian Rockies. In: *Gravity and Tectonics* (Ed. by K. A. DeJong & R. Sholten), pp. 491–502. Wiley, New York.

SENO, T. (1977) The instantaneous rotation vector of the Philippine Sea plate relative to the Eurasian plate. *Tectonophys.* **42**, 209–226.

SHIH, T.-T. (1980) The evolution of coastlines and the development of tidal flats in western Taiwan. *Geogr. Res. Inst. Geog. Taiwan Normal Univ.* **6**, 1–28.

SUPPE, J. (1980) A retrodeformable cross-section of northern Taiwan. *Proc. geol. Soc. China*, **23**, 46–55.

SUPPE, J. (1981) Mechanics of mountain building and metamorphism in Taiwan. *Mem. geol. Soc. China*, **4**, 67–89.

WATER RESOURCES PLANNING COMMISSION (1980) *Hydrological Yearbook of Taiwan, Republic of China, for the year 1979.* Ministry of Public Affairs, Republic of China.

WU, F.T. (1978) Recent tectonics of Taiwan. In: *Geodynamics of the Western Pacific* (Ed. by S. Uyeda, R. W. Murphy & K. Kobayshi). *J. Phys. Earth Suppl.* **26**, S265–S299.

ZHAO, W.L. & MORGAN, W.J. (1985) Uplift of Tibetan plateau. *Tectonics*, **4**, 359–369.

Spec. Publs int. Ass. Sediment. (1986) **8**, 91–102

Timor–Tanimbar Trough: the foreland basin of the evolving Banda orogen

M. G. AUDLEY-CHARLES

Department of Geological Sciences, University College London, London WC1E 6BT, U.K.

ABSTRACT

Seismic refraction surveys indicate the 2–3 km deep, 1200 km long and locally up to 70 km wide Timor–Tanimbar Trough is underlain by continental crust varying (west to east) between 31 and 40 km thick. Seismic reflection surveys indicate this Trough is underlain by downbowed Australian continental shelf overlain on the Banda Arc (northern) side by a series of reflectors that dip northwards away from the Australian margin. These reflectors appear to represent a pile of thrust sheets composed of Australian continental margin sedimentary rocks that have moved towards the Australian continent. Immediately south of the Outer Banda Arc islands these reflectors are overlain by seismically chaotic material which is itself overlain locally by a reflective, well bedded section interpreted as Plio–Pleistocene turbidites derived from the rising Banda orogen.

The stratigraphy and structure exposed in the islands of Timor and Tanimbar and the stratigraphy in wells on the Australian shelf are compared with the seismic survey data for the Trough separating these islands from the Australian shelf. These data, and the drilling results from DSDP site 262 are used as a basis for discussion of the Neogene–Quaternary evolution of this foredeep of the Banda collisional orogen. The usual tectonic interpretation is that this foreland basin evolved by the subduction trench migrating into the northward converging Australian continental margin. This concept regards the Outer Banda Arc islands as tectonic melange scraped from the downgoing Australian plate and the Trough, not as a foreland basin or foredeep but, as part of this forearc accretionary wedge melange. Another view considers this continental margin has in the Timor region overridden the trench and Benioff zone by about 240 km. This implies the foreland basin developed partly by the loading effect of the southward migrating pile of thrust sheets downbowing the continental margin after the continental margin had overridden the trench and Benioff zone.

The Timor Trough was initiated immediately after the nappes were emplaced in Timor in the mid-Pliocene. The subsidence history of DSDP site 262 indicates that the axis of the Trough, originally located close to the nappes and rising Banda orogen, migrated away from the collision zone towards the Australian craton at 7·5 cm yr^{-1} from mid- to Late Pliocene. The present axis of the Timor–Tanimbar Trough has subsided 2·3 km at 1 mm yr^{-1} while the nappes have been uplifted nearly 5 km initially at a rate of 3 mm yr^{-1} for about 700,000 yr then at 1 mm yr^{-1} for 2·7 Myr.

INTRODUCTION

The Timor–Tanimbar Trough is a depression filled with sea water to a depth ranging between 2 and 3 km overlying a sedimentary cover 4–7 km thick that sits on crystalline basement varying between 21 and 30 km thick amounting in all to a crustal thickness of between 31 and 40 km (Jacobson *et al.*, 1978).

The thickness, *P*-wave velocity, magnetic and gravity characteristics of this crystalline basement all indicate it is part of the continental crust of Australia that extends, not only below the Timor–Tanimbar Trough but beyond to the northern edge of the islands of the Outer Banda Arc, Timor, Leti, Moa, Sermata, Babar, Tanimbar and Kai (Chamalaun, Lockwood & White, 1976; Jacobson *et al.*, 1978; Bowin *et al.*, 1980; Schluter & Fritsch, 1985; Milsom & Audley-Charles, 1985).

An important question that arises is whether the Timor–Tanimbar Trough (Fig. 1) is the foreland basin (foredeep) of the Banda orogen (Veevers, 1984) or part of the forearc accretionary prism to the volcanic

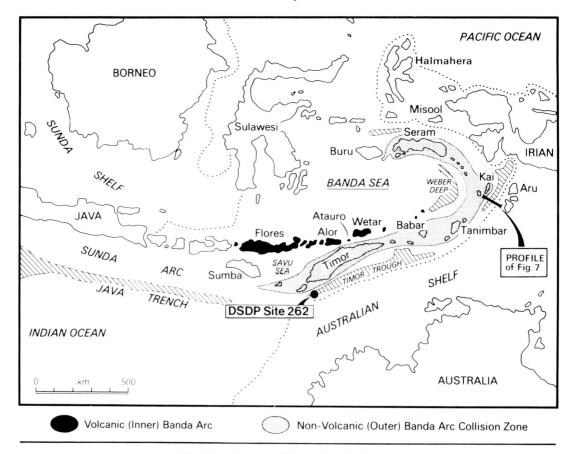

Fig. 1. Location map of Timor–Tanimbar Trough.

Inner Banda Arc (Hamilton, 1979; Silver *et al.*, 1983). Figure 2 indicates that this is not just a semantic matter but a fundamental issue concerning the present structure and mode of evolution of the Timor–Tanimbar Trough and the islands of the Outer Banda Arc. This question has much wider significance because it is directly related to continent–arc collision processes, which can lead to fold and thrust mountain belts with associated foreland basins, as in the Banda Arc. It is perhaps not widely recognized just how large is the Banda Arc. Brunnschweiler (1978) pointed out the similarity in area and relief of Timor (the largest island of the Banda Arc) to the Swiss Alps, both highly complex parts of fold and thrust mountain belts. Figure 3 shows that the Banda Arc is about the same size and shape as the whole of the Alpine mountain chain.

Foredeeps are located (Fig. 4) between the strongly deformed cover rocks of the orogenic zone and the relatively undeformed cover rocks of the craton. Foredeeps are underlain by continental crust and characterized by a sedimentary cover of detritus derived from the erosion of the uplifted orogenic zone. Foredeeps were described by Bally & Snelson (1980) as moats on continental crust to distinguish them from subduction trenches which they described as moats on oceanic crust. There is a tendency for the tectonic deformation to encroach on the foredeeps so that the foothills fold and thrust belt of the orogenic zone incorporates some of the post-orogenic detritus filling the foredeep. This may be clearly demonstrated in the front of the Alps (Bernoulli *et al.*, 1974), in the foothills of the Himalayas (Gansser, 1974) and the foothills of the Rocky Mountains (Wheeler *et al.*, 1974). In contrast, the forearc accretionary wedge is found on oceanic crust and involves the active underthrusting of oceanic crust below the evolving forearc accretionary wedge associated with the volcanic activity of the arc.

The first part of this paper discusses the mode of

A. TIMOR TROUGH AS PART OF FOREARC ACCRETIONARY PRISM

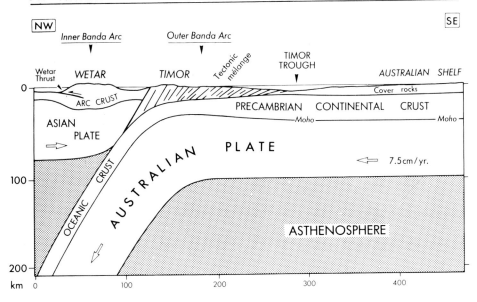

B. TIMOR TROUGH AS FOREDEEP

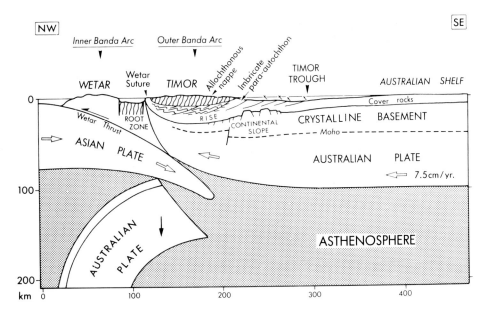

Fig. 2. (A) Tectonic model based mainly on an interpretation of marine seismic reflection data (from Hamilton, 1979, fig. 78 and from von der Borch, 1979, fig. 9) redrawn to locate on the same profile as model (B) below. (B) Model based mainly on surface geological mapping onshore Timor (Rosidi *et al.*, 1979; Audley-Charles, 1968). Development of Wetar Suture plate rupture after Price & Audley-Charles (1983).

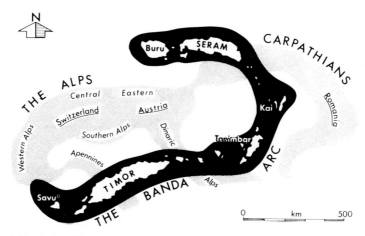

Fig. 3. Banda Arc–Alps comparative shape and area. Only the Outer Banda Arc is shown.

evolution of the Timor–Tanimbar Trough in terms of the stratigraphy and structure revealed in the islands of the Outer Banda Arc. The information from the Trough and Australian shelf is derived from seismic reflection and oil company drilling. Indications are that the orogenic deformation began to migrate into the Timor–Tanimbar Trough away from the uplifting Banda orogen and towards the Australian craton shortly after the main orogenic phase that emplaced the allochthonous nappes in the Banda Arc. The last section of the paper offers some comparison of this foredeep with others from older mountain belts.

TIMOR–TANIMBAR TROUGH: FOREARC ACCRETIONARY WEDGE OR FOREDEEP?

Aspects of stratigraphy and structure bearing on tectonic controversy

It might be argued that, in the case of a passive continental margin converging on an arc-trench system, the position will arise as continental crust begins to enter the trench and be subducted below the forearc accretionary wedge, that the former oceanic trench will become located on the continental crust as the continental margin continues to underthrust the forearc. This is the view generally taken of the present structure (Fig. 2A) and mode of evolution of the Timor–Tanimbar Trough (Jacobson *et al.*, 1978; Hamilton, 1979; von der Borch, 1979; Silver *et al.*, 1983). However, it seems that there are several major

stratigraphical and structural features observed in the islands of the Outer Banda Arc that cannot be reconciled with that model:

(a) It has been shown that the stratigraphical succession of Timor indicates an absence of tectonic compressive deformation, during the time interval early Eocene to late Miocene inclusive, in any of the rocks exposed in the island. In particular, Barber, Audley-Charles & Carter (1977); Audley-Charles *et al.* (1979) and Rosidi, Suwitodirdjo & Tjokrosapoetro (1979) showed that the para-autochthonous succession well exposed in the Kolbano region of southern Timor is a conformable sequence of deep water lutites ranging from the Late Cretaceous to early Pliocene that was imbricated during the middle Pliocene. They also showed that in the allochthon, although the Early Miocene Cablac Limestone sits with an erosional base on the Lolotoi Complex metamorphics and locally on the Eocene Wiluba limestone facies, these observations can be accounted for by tilting and erosion.

(b) Kenyon (1974) and Audley-Charles (1986) have shown that the exposed rocks in Timor indicate that Cenozoic compressive deformation began in the middle Pliocene and stopped in Pleistocene times. The key piece of evidence is the widespread flat-lying Pleistocene coral-algal reef limestones that sit with an erosional base and local angular unconformity on the folded Plio–Pleistocene Noele Marl of the Viqueque Group (see maps of Timor in Audley-Charles, 1968 and Rosidi *et al.*, 1979).

(c) All the post-Middle Jurassic para-autochthonous strata in Timor (i.e. those rocks that underlie the

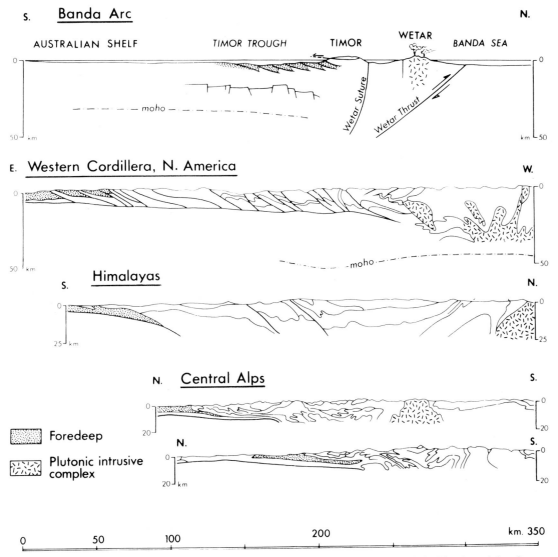

Fig. 4. Comparative cross-sections across four very different types of collisional mountain belt: Himalayas (after Gansser, 1974), Alps (after Bernoulli *et al.*, 1974), Western Cordillera of North America (after Wheeler *et al.*, 1974) and the Banda Arc. Despite their very different modes of origin (e.g. continent–continent collision of Himalayas, continent–continent collision with shear but little subduction of Alps, multiple allochthonous terrane collision with strike-slip of Western Cordillera and arc–continent collision of Banda Arc) these foreland basins are similar in size suggesting the common factor of nappe loading may be influential.

flat lying nappes of the allochthon) are deep water sediments (Fig. 5) that accumulated at the rifted continental margin of Australia (Barber *et al.*, 1977; Audley-Charles *et al.*, 1979).

(d) The rocks of the allochthonous nappes in Timor overlap in age with those of the para-autochthon but are of strikingly different facies. It has been argued by

Earle (1983), and by Brown & Earle (1983) that the allochthonous metamorphic rocks of the Lolotoi–Mutis Complex and the Palelo Group cover of volcanics and turbidites on Timor were overthrust towards Australia from the basement of the forearc accretionary wedge of the Banda volcanic arc during the late Tertiary collision. It was also argued by

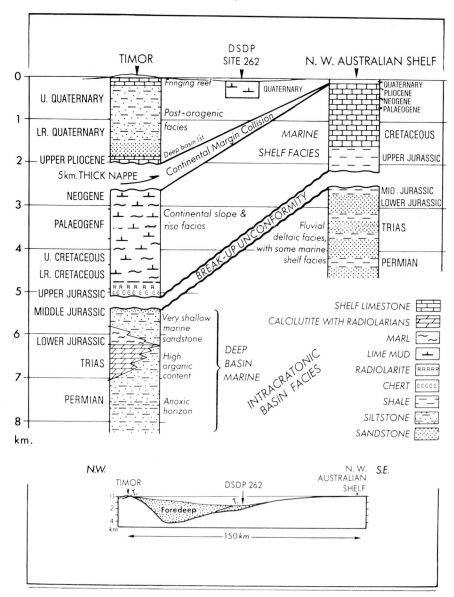

Fig. 5. Stratigraphical column at the margin of the foredeep in Timor, in the Trough and on the Australian shelf. Australian shelf data after Powell (1976). DSDP site 262 data from Veevers *et al.* (1978) and from Johnson & Bowin (1981) who identified 0–366 m as Quaternary and 366–342 m TD as Late Pliocene sediments (not shown here).

Audley-Charles (1985) that the allochthonous nappes of Timor may be closely compared with the exposed pre-Middle Miocene section in Sumba and that these Timor nappes were derived from the basement of the forearc accretionary wedge to the volcanic Banda arc as Carter, Audley-Charles & Barber (1976) and Earle (1983) suggested.

Key arguments relating to tectonic controversy

(a) Von der Borch (1979) and Silver *et al.* (1983) have shown that the seismic reflection profiles across the north slope of the Timor–Tanimbar Trough strongly resemble those of forearc accretionary wedges such as that associated with the Sunda Trench (Bally, 1983).

(b) However, there is an absence in the para-autochthon of sedimentary rocks typical of either oceanic basins or of volcanic forearcs. Instead these rocks, that make up the exposed basement of Timor, typify passive continental margins. They began to be folded and thrust by compressive deformation in the middle Pliocene when this passive continental margin collided with the eastern end of the Java trench. If, as has been so widely argued, the islands of the Outer Banda Arc are part of a forearc accretionary prism, why do they not show indications of compressive deformation and imbrication in post-Eocene to pre-Pliocene times? And, if these islands are part of the forearc accretionary prism, why are they not composed of oceanic and volcanic forearc sediments of post-Eocene to Pliocene age? The only forearc sediments in Timor are of Late Cretaceous–Eocene age forming part of the Palelo Group sequence of the allochthonous nappes that, Earle (1983) and Audley-Charles (1985) have argued, were indeed derived from the Banda volcanic forearc basement by being thrust over the underthrusting Australian continental margin.

There are some tuffs in the Plio–Pleistocene Viqueque Group of Timor, which is a marine molasse facies. These tuffs make up a tiny proportion of the Group and would seem most likely to be the product of the last stages of activity of the volcanic arc opposite Timor, where eruptions appear to have ceased within about 2·5 Myr after the collision.

(c) Thus there is a dual argument against the Outer Banda Arc islands forming part of the forearc accretionary wedge. Firstly, there is an absence of any pre-Pliocene rocks and structures that characterize forearc accretionary wedges from all elements in Timor except the allochthonous nappes and secondly, that in these nappes they are only of late Cretaceous to Eocene age. The strength of this dual argument is reinforced by the fact that the nappes, which carry these only examples of forearc sediments and volcanics to be found in Timor, originated in the Banda volcanic forearc and were thrust over the Australian continental margin rocks from the north during the mid-Pliocene orogenic phase. This nappe emplacement is correlated with the collision between the passive Australian margin and the eastern extension of the Java trench. It also needs to be borne in mind that the Palelo Group deposits are truly allochthonous in having originated in a Late Cretaceous–Eocene forearc that was about 3000 km north of the Timor region of the passive Australian margin at that time (Audley-Charles *et al.*, 1979).

(d) Despite the many papers that employ the accretionary wedge model for the Outer Banda Arc islands none include scale cross-sections that show the position of the flat lying nappes. None have indicated how these nappes were emplaced in their present position by the forearc accretionary wedge processes, and none have indicated from where these very different facies were derived by forearc accretionary wedge mechanisms. Figure 2(A) attempts to illustrate the difficulties faced by the forearc accretionary wedge model applied to the Outer Banda Arc and the Timor–Tanimbar Trough.

WETAR SUTURE AND WETAR THRUST

The Wetar Suture (Fig. 2B) was postulated by Price & Audley-Charles (1983) in order to explain the main structural features mapped in the island of Timor. They attributed the rupture of the lithospheric plate to hydraulic fracture mechanisms by mobilized Australian thick continental rise deposits. It was argued that the Suture was initiated in the Early Pliocene and that it led to the emplacement of the allochthonous nappes in Timor by flake tectonic processes in the mid-Pliocene. The existence of this now steeply dipping rupture just north of Timor, cutting through the downgoing Australian continental plate appears to have been confirmed by the results of a microearthquake survey (McCaffrey, Molnar & Roecker, 1985).

The Wetar Thrust (Fig. 2B) first reported by Hamilton (1979) subsequently mapped by marine seismic surveys (Silver *et al.*, 1983) was regarded on the basis of seismic reflection and gravity surveys as a deep seated structure (McCaffrey & Nabelek, 1984). Price & Audley-Charles (1986) have argued that this fault must (because of spatial and mechanical considerations) join at depth with the Wetar Suture (Fig. 2B). They also argued that whereas the Wetar Suture allowed crustal shortening of about 240 km during the Pliocene, it locked-up in Late Pliocene times as its dip became much steeper. The Wetar Thrust was initiated in the Late Pliocene and accommodated crustal shortening of about 140 km during the Quaternary. All this shortening is associated with the continuing Plio–Pleistocene plate convergence between Asia and Australia (Minster & Jordan, 1978).

STRATIGRAPHICAL-STRUCTURAL EVOLUTION OF THE TIMOR–TANIMBAR TROUGH

The best sources of information about the evolution of the Timor–Tanimbar Trough are the DSDP site

262 (Veevers, Falvey & Robins, 1978) at the western end of the present axis of the Timor Trough (Fig. 6) and the seismic reflection profiles across the Tanimbar Trough (Schluter & Fritsch, 1985) at the eastern end of this foredeep (Fig. 7).

In their account of the drilling results of DSDP site 262 Veevers *et al.* argued that, as the Pliocene sediments at the bottom of the hole are shallow marine shelf-type deposits, overlain by deeper water carbonates which in turn are overlain by deeper nanno-oozes and clays that typify the present axis of the Timor Trough, this sequence demonstrates site 262 has subsided from shelf depths to 2300 m at the present day. Johnston & Bowin (1981) reported micropalaeontological determinations that confirmed this interpretation but indicated that the oldest dated deposits near the bottom of DSDP site 262 are Late Pliocene, probably 2·3 Ma. The date of the orogenic phase in Timor that strongly compressed the para-autochthonous rocks and emplaced the flat lying nappes above has been well established as mid-Pliocene (Carter *et al.*, 1976; Audley-Charles, 1986). By mid-Pliocene times, certainly by 2·3 Ma, the present site of Timor had emerged above sea-level, probably for the first time since the early Permian (with the possible

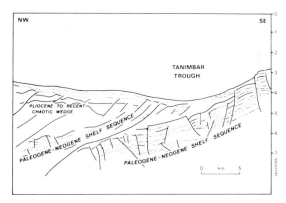

Fig. 7. Late Pliocene and Quaternary thrusting in the Tanimbar Trough (after Schluter & Fritsch, 1985). Australian Tertiary shelf deposits continue north-westwards to crop out in the Kai islands (Achdan & Turkandi, 1982). For location of profile see Fig. 1.

exception of a short-lived phase of emergence in the Middle Jurassic). All available evidence supports the opinion of Veevers *et al.* (1978) that in the Early Pliocene the area of present-day central Timor lay near the axis of a bathyal depression, and from the mid-Pliocene to the present day the area of Timor emerged and that of the Timor Trough subsided (Fig. 6).

Veevers *et al.* (1978) argued that the subsidence of the DSDP site 262, now in the axis of the Timor Trough, occurred as a consequence of the Trough axis migrating away from Timor and towards the Australian continent. The main orogenic phase in Timor was complete about 3·4 Ma (Audley-Charles, 1986) which was about 1 Myr before site 262 showed evidence of subsidence. This means that the onset of uplift of the Timor nappes and their erosion, that produced the Viqueque Group post-orogenic clastic detritus (Kenyon, 1974) which accumulated in central and southern Timor, began about 1 Myr before site 262 showed evidence of subsidence from shelf depths.

One of the major problems in analysing this foredeep, which is now flooded by the sea to depths of 3 km, is how to locate the present position of the pre-collision northern edge of the Australian shelf. DSDP site 262 has shown that the shelf extended at least as far north as the present axis of the western end of the Timor Trough. The indications from many years of geological mapping on the island of Timor are that no Australian shelf facies are exposed there. It is unfortunate that Barber *et al.* (1977) used the term Continental Shelf Unit for the para-autochthon because no Australian shelf deposits have been found in Timor. Instead the rocks involved are of two

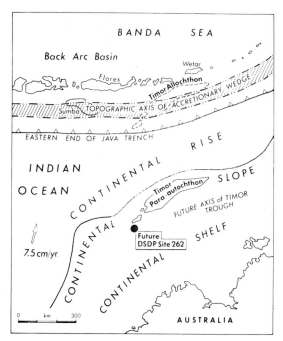

Fig. 6. Palaeogeographical sketch map of the Australian continental margin approaching the eastern end of the Sunda trench at about 7·5 cm yr^{-1} during the late Miocene.

distinct kinds. The pre-mid-Jurassic are various marine facies belonging to an intracratonic basin (Audley-Charles, 1983) while the post-mid-Jurassic sediments are continental slope and rise deposits belonging to the rifted continental margin of NW Australia. The pre-collision shelf edge must therefore be located between these two marker points of the Australian slope and rise deposits now exposed throughout Timor and the Australian shelf deposits drilled at site 262.

No direct evidence of this buried shelf edge is available but one indirect approach to its location is to recognize the potential of the normal faults, developed in the evolving continental margin during the continental rifting in the mid-Jurassic (Powell, 1976), to be rejuvenated as reverse faults when the region opposite the Timor–Tanimbar Trough experienced compressional stresses. Furthermore, the shelf edge would likely be marked by lithological change and thickness of cover rocks. The presence of the normal faults in a region of such changes in lithology and thickness could, under the compressive stresses associated with the collision of the Australian margin with the eastern end of the Java Trench, guide the development of a natural ramp up which a flat thrust surface would turn. Another method is to try to recognize shelf facies in the seismic reflection profiles shot across the Timor–Tanimbar Trough. Schluter & Fritsch have made a good case for recognizing thrust slices of Australian shelf below the northern wall of the Tanimbar Trough (Fig. 7). These two approaches are combined in the interpretation of the palaeogeography of the region before the collision of the continental slope with the trench (Figs 6 and 8).

The seismic reflection indications of thrusting in the Tanimbar end of the Trough (Fig. 7) suggest that crustal shortening may approach 50% in the Trough and that the amount of thrusting diminished significantly in the later part of the Quaternary. As the plate convergence of Australia and Asia has continued (Minster & Jordan, 1978) at about $7.5 \, \text{cmyr}^{-1}$ the convergent strain must have been occurring elsewhere. The most obvious candidate is the Wetar Thrust, which Price & Audley-Charles (1986) estimate accommodated about 220 km of convergence in the last 3 Myr.

Summary of evolution of Timor–Tanimbar Trough (Fig. 8)

Mid-Pliocene—(*c*. 3·8–3·4 Ma) Australian continental slope collides with eastern end Sunda Trench in Timor region. Most of the present Timor–Tanimbar Trough formed part of the pre-collision Australian shelf.

Late Pliocene—(*c*. 3·4 Ma) initiation of Timor–Tanimbar Trough. Subsidence of region between present south coast of Timor and present axis of Trough coincident with onset of uplift of Timor involving its emergence as island.

Deposition of post-orogenic clastics begins in central and southern Timor and built out towards Australia.

Late Pliocene—(*c*. 2·3 Ma) subsidence reaches present axis and continues to migrate towards the Australian continent. Shortening of cover rocks by thrusting involves southern Timor (Kolbano) and migrates towards Australia.

Quaternary—subsidence and shortening of the cover rocks by thrusting confined to the Timor–Tanimbar Trough and southern margin of the islands of the southern Outer Banda Arc (e.g. south coast region of Timor). The low degree of disturbance of the uppermost part of the Quaternary section in the Trough (Schluter & Fritsch, 1985) indicates that the amount of thrusting diminished significantly in the later half of the Quaternary. There appears to have been no late Quaternary thrusting onshore Timor.

COMPARISON WITH OLDER FOREDEEPS

The Timor–Tanimbar Trough differs from older foredeeps in that it is flooded by the sea. This limits direct access but it affords a view of how some of these older foredeeps (e.g. Himalayan) may have appeared at an early stage in their evolution. However, Gansser (1974) took the view that the foreland trough below the Indo–Gangetic Plain is filled with largely very young non-marine detritus and should not therefore be regarded as a foredeep. In his classification the Siwaliks of the folded and thrust Subhimalaya represent the deposits of the true foredeep (s.s.). Figure 4 compares foredeeps from several Mesozoic–Cenozoic orogenic belts by means of scale sections. The Timor–Tanimbar Trough, which developed during the last 3 Myr is not only much younger than the other foredeeps it has evolved over a much shorter period. For example, the Himalayan foredeep (s.l.) has reached its present form after more than 20 Myr,

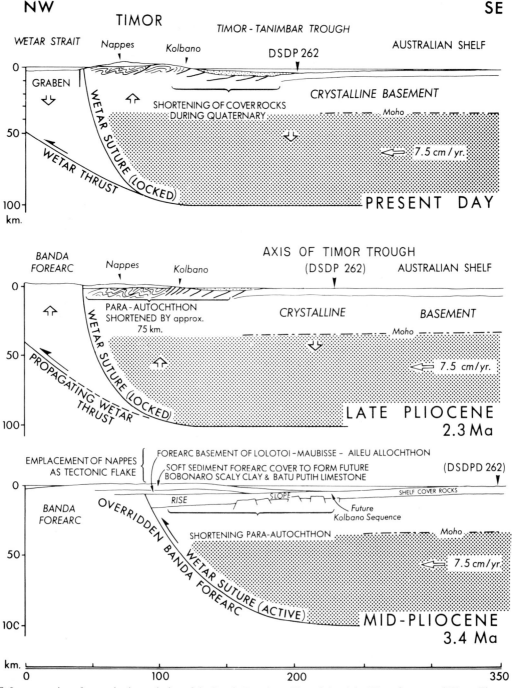

Fig. 8. Interpretation of stages in the evolution of the Banda Foredeep. The origin of the Wetar Suture and Wetar Thrust are discussed by Price & Audley-Charles (1983, 1986). Note that the axis of the Timor Trough migrated 75 km towards the Australian craton at 7·5 cm yr⁻¹ for 1 Myr after nappe emplacement on Timor. However, since then, throughout the Pleistocene the southward migration of the Trough has been much slower at about 2 cm yr⁻¹. Both reef growth and outward building of the shelf may have counteracted the southward migration of the Trough (Veevers, 1971; Edgerley, 1974). The Trough subsided at 1 mm yr⁻¹ for the last 2·3 Myr contemporaneously with uplift at 1 mm yr⁻¹ of Timor Island.

the foredeep to the Western Cordillera in North America developed over more than 120 Myr and the Alpine foredeep represents about 20 Myr of evolution.

In terms of scale, these older foredeeps are similar. It can be seen that if the Banda Orogen were involved in further crustal shortening associated with another collision against an Asian arc to the north then the development of a series of lower nappes below the existing nappes would involve slices of the Australian continental shelf moving south over the shelf and incorporating some of the foredeep deposits. This would tend to increase the similarity between Himalayan and Banda collisional orogens. According to Searle (1983) at the end of the Cretaceous, but before India collided with Tibet in the Eocene at about 40 Ma, the northern continental margin of India collided with a volcanic island arc. If we follow that interpretation we should compare the Banda orogen with the end-Cretaceous arc–continent collision of northern India and not the later continent–continent event which has obscured or greatly modified any Early Eocene foreland basin developed by the end-Cretaceous arc–continent collision.

ACKNOWLEDGMENTS

A preprint of the paper by Schulter & Fritsch, a helpful review by W. J. M. van der Linden, discussion with Neville Price and help with the artwork by Colin Stuart are gratefully acknowledged.

REFERENCES

ACHDAN, A. & TURKANDI, T. (1982) *Preliminary Geologic Map of the Kai (Tayandu and Tual) Quadrangles, Maluku.* Geological Research and Development Centre, Bandung, Indonesia.

AUDLEY-CHARLES, M.G. (1968) The geology of Portuguese Timor. *Mem. geol. Soc. London,* **4,** 1–76.

AUDLEY-CHARLES, M.G. (1983) Reconstruction of eastern Gondwanaland. *Nature,* **306,** 48–50.

AUDLEY-CHARLES, M.G. (1985) The Sumba Enigma: is Sumba a diapiric fore-arc nappe in process of formation? *Tectonophys.* **119,** 435–449.

AUDLEY-CHARLES, M.G. (1986) Rates of Neogene and Quaternary tectonic movements in the Southern Banda Arc based on micropalaeontology. *J. geol. Soc. London,* **143,** 161–175.

AUDLEY-CHARLES, M.G., CARTER, D.J., BARBER, A.J., NORVICK, M.S. & TJOKROSAPOETRO, S. (1979) Reinterpretation of the geology of Seram: implications for the Banda arcs and northern Australia. *J. geol. Soc. London,* **136,** 547–568.

BALLY, A.W. (1983) Seismic expression of structural styles. *Am. Ass. Petrol. Geol. Stud. Geol.* **15,** 3.

BALLY, A.W. & SNELSON, S. (1980) Facts and principles of World petroleum occurrence: realms of subsidence. *Mem. Can. Soc. Petrol. Geol.* **6,** 9–94.

BARBER, A.J., AUDLEY-CHARLES, M.G. & CARTER, D.J. (1977) Thrust tectonics in Timor. *J. geol. Soc. Aust.* **24,** 51–62.

BERNOULLI, D., LAUBSCHER, H.P., TRUMPY, R. & WENK, E. (1974) Central Alps and Jura Mountains. In *Mesozoic–Cenozoic Orogenic Belts* (Ed. by A. M. Spencer). *Spec. Publ. geol. Soc. London,* **4,** 85–108.

BOWIN, C., PURDY, G.M., JOHNSTON, C., SHOR, G., LAWVER, L., HARTONO, H.M.S. & JEZEK, P. (1980) Arc-continent collision in the Banda Sea region. *Bull. Am. Ass. Petrol. Geol.* **64,** 868–915.

BROWN, M. & EARLE, M.M. (1983) Cordierite-bearing schists and gneisses from Timor, eastern Indonesia: *P–T* conditions of metamorphism and tectonic implications. *J. Metamorph. Geol.* **1,** 183–203.

BRUNNSCHWEILER, R.O. (1978) Notes on the geology of eastern Timor. *BMR Bull.* **192,** 9–18.

CARTER, D.J., AUDLEY-CHARLES, M.G. & BARBER, A.J. (1976) Stratigraphical analysis of island arc-continental margin collision in eastern Indonesia. *J. geol. Soc. London,* **132,** 179–198.

CHAMALAUN, F.H., LOCKWOOD, K. & WHITE, A. (1976) The Bouguer gravity field and crustal structure of eastern Timor. *Tectonophys.* **30,** 241–259.

EARLE, M.M. (1983) Continental margin origin for Cretaceous radiolarian cherts in western Timor. *Nature,* **305,** 129–130.

EDGERLEY, D.W. (1974) Fossil reefs of the Sahul Shelf, Timor Sea. *Proc. 2nd int. Coral Reef Symp.,* Brisbane, **3,** 627–637.

GANSSER, A. (1974) Himalaya. In: *Mesozoic–Cenozoic Orogenic Belts* (Ed. by A. M. Spencer). *Spec. Publ. geol. Soc. London,* **4,** 267–278.

HAMILTON, W. (1979) Tectonics of the Indonesian region. *U.S. geol. Surv. Prof. Pap.* 1078.

JACOBSON, R.S., SHOR, G.G., KIECKHEFER, R. M. & PURDY, G.M. (1978) Seismic refraction and reflection studies in the Timor-Aru Trough system and Australian continental shelf. *Mem. Am. Ass. Petrol. Geol.* **29,** 209–222.

JOHNSTON, C.R. & BOWIN, C.O. (1981) Crustal reactions resulting from the mid-Pliocene to Recent continent-island arc collision in the Timor region. *BMR J. Aust. Geol. Geophys.* **6,** 223–243.

KENYON, C.S. (1974) *Stratigraphy and sedimentology of the Late Miocene to Quaternary deposits of Timor.* Unpublished Ph.D. Thesis. University of London.

McCAFFREY, R., MOLNAR, P. & ROECKER, S. (1985) Microearthquake seismicity and fault plane solutions related to arc-continent collision in the eastern Sunda Arc, Indonesia. *J. geophys. Res.* **90,** 4511–4528.

McCAFFREY, R. & NABELEK, J. (1984) The geometry of back-arc thrusting along the eastern Sunda arc, Indonesia: constraints from earthquake and gravity data. *J. geophys. Res.* **89,** 6171–6179.

MILSOM, J. & AUDLEY-CHARLES, M.G. (1985) Post-isostatic readjustment in the southern Banda Arc. In: *Collision Tectonics* (Ed. by M. P. Coward & Alison C. Ries). *Spec.*

Publ. geol. Soc. London, **19**, 353–364. Blackwell Scientific Publications, Oxford.

MINSTER, J.B. & JORDAN, T.H. (1978) Present-day plate motions. *J. geophys. Res.* **83**, 5331–5354.

POWELL, D.E. (1976) The geological evolution of the continental margin off Northwest Australia. *APEA J.* **16**, 13–23.

PRICE, N.J. & AUDLEY-CHARLES, M.G. (1983) Plate rupture by hydraulic fracture resulting in overthrusting. *Nature*, **306**, 572–575.

PRICE, N.J. & AUDLEY-CHARLES, M.G. (1986) Tectonic collision processes after plate rupture. *Tectonophys.* (submitted).

ROSIDI, H.M.D., SUWITODIRDJO, K. & TJOKROSAPOETRO, S. (1979) *Geologic Map of the Kupang-Atambua quadrangle, Timor*. Geological Research and Development Centre, Bandung, Indonesia.

SCHLUTER, H.U. & FRITSCH, J. (1985) Geology and tectonics of the Banda Arc between Tanimbar island and Aru island. *Geol. Jb.* **30**, 3–41.

SEARLE, M.P. (1983) Stratigraphy, structure and evolution of the Tibetan-Tethys zone in Zanskar and the Indus suture zone in the Ladakh Himalaya. *Trans. R. Soc. Edinb. Earth Sci.* **73**, 205–219.

SILVER, E.A., REED, D., McCAFFREY, R. & JOYODIRWIRYO, Y. (1983) Back arc thrusting in the eastern Sunda arc, Indonesia: a consequence of arc-continent collision. *J. geophys. Res.* **88**, 7429–7448.

VEEVERS, J.J. (1971) Shallow stratigraphy and structure of the Australian continental margin beneath the Timor Sea. *Mar. Geol.* **11**, 207–249.

VEEVERS, J.J. (1984) *Phanerozoic Earth History of Australia*. Clarendon Press, Oxford, 418 pp.

VEEVERS, J.J., FALVEY, D.A. & ROBINS, S. (1978) Timor trough and Australia: facies show topographic wave migrated 80 km during the past 3 m.y. *Tectonophys.* **45**, 217–227.

VON DER BORCH, C.C. (1979) Continent-island arc collision in the Banda Arc. *Tectonophys.* **54**, 169–193.

WHEELER, J.V., CHARLESWORTH, H.A.K., MONGER, J.W.H., MULLER, J.E., PRICE, R.A., REESOR, J.E., RODDICK, J.A. & SIMONY, P.S. (1974) Western Canada. In: *Mesozoic-Cenozoic Orogenic Belts* (Ed. A. M. Spencer). *Spec. Publ. geol. Soc. London*, **4**, 591–623.

Alpine basins of Europe and Asia

Spec. Publs int. Ass. Sediment. (1986), **8**, 105–139

The Oligocene to Recent foreland basins of the northern Apennines

FRANCO RICCI LUCCHI

Istituto di Geologia, University of Bologna, Via Zamboni 67-40127 Bologna, Italy

ABSTRACT

Elongated foreland basins were formed by compression and shortening of the African–Adriatic continental margin from the Oligocene to the Quaternary, and filled mostly by turbidite deposits. Unlike other foredeeps (e.g. Rocky Mountains), loci of deposition occurred on the fold belt itself. Thus, major basins (foredeeps) were coupled with minor basins carried piggy-back on thrust sheets. This system migrated, mostly stepwise, toward the craton, reflecting an alternation of thrust activity and quiescence. Subsidence and deposition cycles are well synchronized and coupled in major and satellite basins until the end of the Miocene, then decoupled. The backbone of the Apennines chain emerged, and the molasse stage replaced the flysch stage; the feeding changed from alpine to apenninic, but resedimentation in deep water was still dominant. The migration of subsident axes slowed down and vertical movements became more important. Older foredeep wedges were incorporated in the thrust belt and uplifted while the rates of subsidence and sedimentation increased in the Po Basin.

The stratigraphy of both foredeep and thrust-based deposits is revised in terms of depositional sequences; the Miocene ones reflect not only the regional tectonic control but also correlate with the global Vail curve. Examples of correlation between marginal and basinal sequences, of dispersal patterns, cannibalistic feeding, topographic control on gravity sediment flows, and peculiar clastic facies are discussed.

INTRODUCTION

The Apennines are made of nappes and thrusts accreted to a spur of the African continent, the Adriatic Foreland (Elter & Scandone, 1980). Though geometrically similar to subduction complexes (Fig. 1), the Apenninic thrust belt consists of continental rocks except for the uppermost tectonic unit (Ligurian Sheet) of oceanic affinity. The Ligurian forms the highest nappe in both the Apennines and the Alps. It is the remnant of a small ocean that was closed by the collision between Africa and Europe from the Late Cretaceous to the Eocene. The suture of this collision is indicated in Fig. 2.

According to majority opinion oceanic lithosphere was entirely consumed under the European craton (for an opposite view, see Reutter, 1981): the ensuing Apenninic orogeny (Oligocene to Recent) is thus called 'post-collisional' and explained by ensialic or A-type subduction (Kligfield, 1979; Bally & Snelson, 1980; Boccaletti *et al.*, 1980; Roeder, 1980). The

Ligurian terrane (LIG), already deformed by alpine thrusting, participated in the Apenninic thrust belt with opposite vergence (E–NE). Even after consumption of the Tethyan Ocean, relative movements of Europe and Africa continued and played a major role in Apenninic orogeny. Broadly speaking, this chain is also a product of continental collisions. During its migration towards the foreland, the Apenninic thrust system incorporated progressively younger foreland deposits, mostly clastic; the thrusting determined both their conditions of deposition and deformation. Thrust advance was accompanied by the opening of extensional basins in the 'wake'; the unconformable 'post-orogenic' sedimentation of the back-thrust area is thus coeval with the 'synorogenic' one in the *frontal part* of the belt. The category of hinterland and intermontane basins is not considered in this paper.

The basement of the Apennines is not exposed; its participation in thrusting is suggested by geophysical

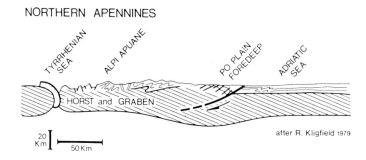

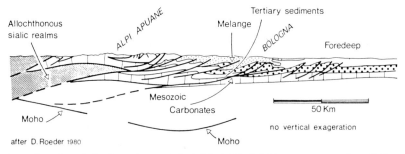

Fig. 1. Schematic sections through the Apennines and the Adriatic Block. Melange corresponds to LIG and ASC of this paper.

evidence (Fig. 2; La Vecchia, Minelli & Pialli, 1984). Surface structures reflect décollement and disharmonic folding of the sedimentary cover (Fig. 3).

A Triassic evaporitic level favoured the major detachment; others were provided by younger pelitic horizons.

Foreland deposits (the 'miogeosynclinal sequence' of Bortolotti *et al.*, in Sestini, 1970) are here referred as *autochthonous*, not to indicate undisplaced units but to differentiate foreland from LIG sediments. Thrusts made of this autochthon, unlike the LIG sheet, maintained their relative position from SW to NE: a Tuscan Sequence is thus juxtaposed, and partly superposed, to an external or Umbrian Sequence. An intermediate unit, the Modino–Cervarola (Figs 4 and 8), is assigned either to the outer Tuscan or an independent domain (Emilian).

Clastic bodies form the bulk of all these thrusts. Their degree of deformation decreases towards the foreland. The middle–upper Miocene Marnoso-arenacea is the one that provides the best opportunities for studying stratigraphy and facies.

Another group of scattered but relatively undeformed deposits is represented by 'rafts' floating on the LIG sheet (late geosynclinal sequence of Sestini, 1970; *epi-Ligurian*, this paper).

The LIG sheet consists of ophiolites, pelagic and turbiditic limestones and turbiditic sandstones embedded in shales; some see it as a sort of giant debris flow, in which the clayey units are the 'matrix'. The old name, 'Argille Scagliose' (ASC), and the newer one, Chaotic Complex, reflect this tendency, and also a certain perplexity about its classification as a stratigraphic or a tectonic unit. The concept of chaoticity, however, is not easy to define; it depends on scale, recent surficial movements and especially subjective judgement. An old school, revived today by German geologists (Günther, Reuter, etc.), regards LIG as a stratified pile of gravity flow deposits (olisthostrome concept); in other words not a single huge debris flow but as a stack of debris flows with an overall stratigraphic order. Others, although agreeing that LIG was emplaced by gravity, make a distinction between olisthostrome and nappe, or sheet, on a scale basis (see Fig. 5). The Dutch school, following the tradition of van Bemmelen, extends the landslide model (Fig. 6A) to the deeper thrusts ('everything by gravity'), whereas most Apenninic geologists interpret them as compressional features providing the conditions for a passive sliding of LIG (Fig. 6B). This second conception is reflected in the interpretation of seismic profiles of the Po Plain (Pieri & Groppi, 1981;

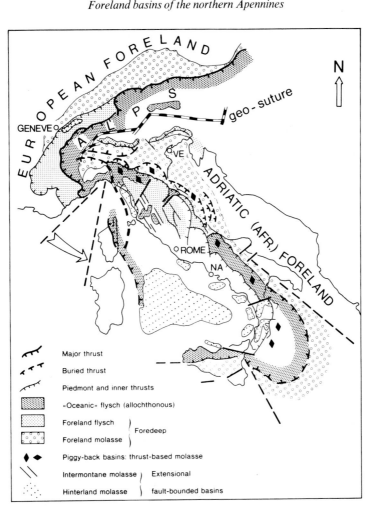

Fig. 2. Orogenic deposits and basins of the Apennines and the Alps; from various sources, compiled by Ricci Lucchi & Ori (1984).

Bally, 1984) and Adriatic (Abruzzi) foredeep (Crescenti *et al.*, 1980).

Whatever the cause, gravity is generally accepted as a mechanism of emplacement for LIG, as well as the fact that it operated underwater. For the latest displacements, however, (Pliocene) compression and uplift of LIG find increasing support in structural (Castellarin *et al.*, 1986) and stratigraphical data (this paper).

The main effects of the LIG overthrust in the foreland basins were: (1) to invade them, progressively or abruptly, and stop sedimentation, (2) to stimulate subsidence with its load, (3) to shield the overtopped clastics from erosion and resedimentation, (4) to shed blocks and debris flows that modified bottom topography.

The Ligurian invasion was stopped, in parts of the foreland, by rising 'autochthonous' structures and channelled along transversal lows. This is not fact, but interpretation. An alternative view is that the LIG sheet advanced over the whole front with success until the end of the Miocene, and was later removed by erosion and vertical faulting.

PALAEOGEOGRAPHIC FRAMEWORK

The surrection of the Apenninic chain has usually been attributed to a sudden, paroxysmal event, the 'Tortonian phase'. We now realize that it occurred in at least three stages spanning 8–9 Myr; firstly in the Tuscan area (Tortonian, 9–11 Ma), then in the Adriatic

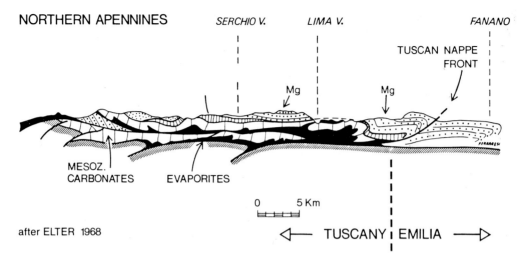

Fig. 3. A classical geologic section of the northern Apennines showing deformation style of the autochthonous sedimentary cover, and décollement planes lubricated by evaporites Triassic. Mg: macigno clastic wedge. Sand unit at right end (NE): Cervarola wedge.

side (Messinian and Plio–Pleistocene). Actually, thrusting is still going on in the outer Apennines. It can thus be said that no real post-orogenic sediments exist there. Post-orogenic can only refer to places where thrust activity definitely ceased, i.e. inner and hinterland areas (see Fig. 2).

Among the uplift phases, that of Messinian is here emphasized because it produced some radical and irreversible changes in the geologic setting. The chain became a permanent geomorphic feature; since the Messinian, a divide between western and eastern slopes, can be recognized in sedimentation. A tem-

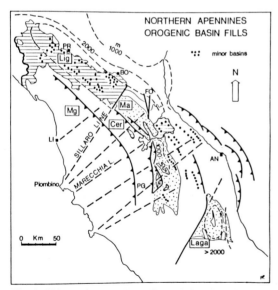

Fig. 4. Older and recent thrust lines of the Apennines and Adriatic Foreland, transversal structures and Ligurian (Lig) overthrust. Deformed wedges of the older foredeep (flysch stage): Mg, Macigno; Cer, Cervarola; Ma, Marnoso-arenacea (with AGIP isopachs in buried portion).

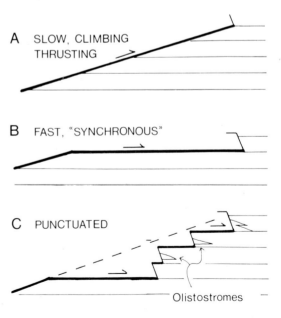

Fig. 5. Hypotheses on the mode of LIG thrusting over the foredeep sequence. Modified after de Jager (1979).

A

B

Fig. 6. Models for overthrusting of the Ligurian Sheet. (A) Gravity sliding (after de Jager, 1979). (B) Compressional shearing plus sliding (G. G. Ori, inspired from Gela Nappe, Sicily).

poral divide is consequently placed at this point between *pre-* and *post-emergence* depositional stages. The subsiding foredeep (Fig. 1) is flanked by a mostly submarine-thrust belt in the first stage, and its clastic supply must be provided by distant or extra-orogenic sources (Fig. 7). In the second stage the thrust belt becomes the major contributor. The two stages can thus be characterized on a petrological basis. They are here called *flysch* and *molasse*, respectively.

These old terms have been used in so many different ways that one more will do no harm. In the Apennines they work in connection with geomorphic stages; they do not work if tied to more speculative geotectonic concepts or purely sedimentological definitions (see Dzulinski & Smith, 1964). From Oligocene to Quaternary, most foredeep deposits are turbidites, for example; we should call them all flysch on a sedimentological basis. It is important to remember the differences in Apenninic and Alpine usage of these terms. The Apenninic flysch is a foreland deposit based on continental crust, the typical Alpine flysch is related to an oceanic setting of Mesozoic age (see original definition by Studer, revision by Hsü, 1970, recent interpretations summarized by Caron *et al.*, 1981; Homewood & Caron, 1982; Homewood, 1983 and Abbate & Sagri, 1984). The Alpine-type flysch is present in the Apennines as a *structural* unit, i.e. a part of the LIG complex. Its depositional setting is still quite speculative. As for the molasse, the Apenninic type is syn-orogenic and mostly resedimented in deep water, the classical Alpine deposit is of shallow water to continental environment.

Palaeogeographically, it is important to realize that the foreland of the Apennines coincides with the hinterland of the Alps. The embryonic Alpine chain was already emergent when Apenninic subsidence started. It could thus act as a source for Apenninic basins. In other words, Alpine molasse and Apenninic flysch were not only coeval but fed from the same chain. Alps-derived materials (low to high-grade metamorphic rocks and dolomite: see Cipriani & Malesani, 1963; Valloni & Zuffa, 1984) found more than one way to the Apenninic foredeep: through the

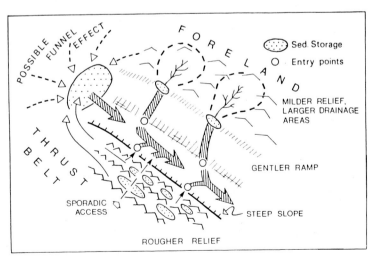

Fig. 7. Dispersal pattern in an idealized Apenninic foredeep. The slightly older alpine chain acted as a source on both sides and NW end. Longitudinal vector in the thrust belt can be inverted.

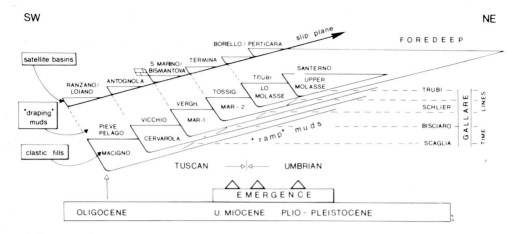

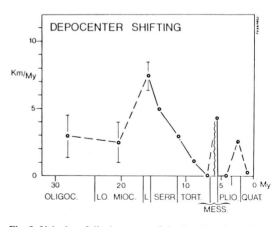

Fig. 8. Major and minor (satellite) clastic wedges of the migrating foredeep-thrust system in northern Apennines.

foreland (NE), longitudinally (NW ends of basins), and from the rear of the thrust belt (SW). The Alpine arc, in fact, continued in the Tyrrhenian area (see Fig. 2), and could occasionally feed the foredeep through breaches in the thrust belt (Fig. 7; see also Fig. 22). Also cratonic blocks adjacent to the Alpine suture (Corsica–Sardinia) could become accessible sources until the late Miocene when the Tyrrhenian Sea opened. As a consequence, the meaning of 'Alpine' and 'Apenninic' as provenance indicators (Gandolfi, Paganelli & Zuffa, 1981) should be reappraised. Alpine usually indicates northern provenance, Apenninic indicates SW ('inner orogenic').

The migration of tectonic and sedimentary events through the foreland is seen by different schools either as a slow, continuous process or a discontinuous one, with long phases of 'calm' interrupted by dramatic 'crises'. Stratigraphic analysis shows, on one hand, that syn-orogenic sediments do not form a regular sheet but show thickness maxima that can be interpreted as standing depocentres (Fig. 8). On the other hand, temporal overlaps and physical continuity do not allow a view of them as wholly discrete bodies. Closing one 'basin' was possibly concomitant with opening the next. The standpoint adopted here is that sedimentation and deformation went hand in hand, almost continuously but at a variable pace; the terms tectonic 'phase', or event, will indicate accelerations and decelerations of the process. The average velocity of displacement of subsident axes were calculated under this assumption. The points represent only rough estimates, especially for older stages (Fig. 9).

We can see, however, that the Apenninic foredeep 'climbed' the Adriatic margin at a relatively high speed (5–10 km per Myr) in the flysch stage, with a peak in the Burdigalian–Langhian that matches an event affecting a large area (CNR, structural model of Italy, in preparation; Sartori, personal communication). Significantly, subsidence shifted less in the post-emergence stage.

Sedimentation rates (Fig. 10), too, reflect the flysch-molasse transition with a radical increase; in the flysch stage, the highest values characterize the Langhian–Serravallian Marnoso-arenacea, but figures for older units are probably underestimated.

Fig. 9. Velocity of displacement of the foredeep (and thrust front). Only rough estimates available for older stages. Compare with Figs 4, 8, 12 and (for time-scale) 17.

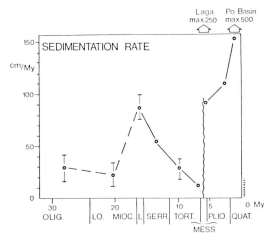

Fig. 10. Changes in sedimentation rate in the foredeep. Values drawn from recognized depocentres (only one is assumed for each of the two oldest wedges, for lack of data).

FORELAND BASINS AND THEIR FILLS

Syn-orogenic sediments are mostly detrital in the Apennines. Sedimentary bodies filling both present and 'fossil' basins are here called *clastic wedges*, under the assumption that ancient analogues accumulated in the same kind of asymmetrical basins. Supporting evidence is presented below.

Each clastic body had a 'panhandle' of hemipelagic sediments in the foreland ramp (Fig. 8).

Minor basins existed within and outside the foredeep, which designates the major depocentre. Their fills have volumes of 50–500 km³, as compared with the 3000–30,000 km³ of major wedges; subsidence rates were also lower.

Presently active or preserved foredeeps (Fig. 2) are the Po Plain, the north-central Adriatic, the Bradano Trough and its prolongation in the Ionian Sea. The geometry, as seen in cross-section, varies from basin to basin and in different parts of the same basin. An attempt to schematize it is made in Fig. 11 (see examples in Fig. 12).

Reference can be made to the Molasse Basin north of the Alps as the simpler case (A): elongated trough with asymmetrical section, the gentler side formed by the downflexed foreland, the steeper one based on the thrust belt and involved in deformation. Thrusting and thrust loading account for this basic shape. Sedimentary sequences are almost stacked without much migration of depocentre.

The Padan and Adriatic basins are examples of *complex* foredeeps with thrusts active in their substratum. Sub-basins are created either in a sequence (B) or at the same time (C), with subequal or different sizes.

The term *piggyback basin* was introduced by Ori & Friend (1984) for a thrust-based basin marginal to a foredeep (D). It is here included in a wider category of *minor* or *satellite basins* encompassing cases related to both allochthonous and autochthonous substratum. The most typical ones 'float' on the LIG Sheet.

A major problem, in palaeogeographic reconstruction, is to recognize whether foredeep splitting occurred during or after deposition (E). In modern basins, seismic records provide the evidence of synsedimentary tectonic mobility (convergence of reflectors, onlaps, unconformities: see Fig. 12). In pre-Pliocene analogues, such geometric features are

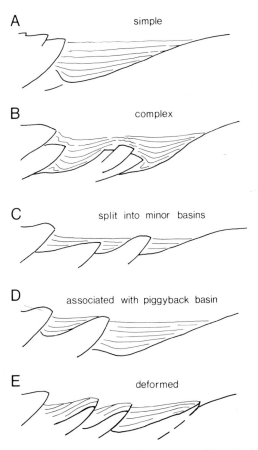

Fig. 11. Different foredeep profiles as suggested by seismic records.

MAIN MOLASSE CYCLES IN THE PADAN FOREDEEP

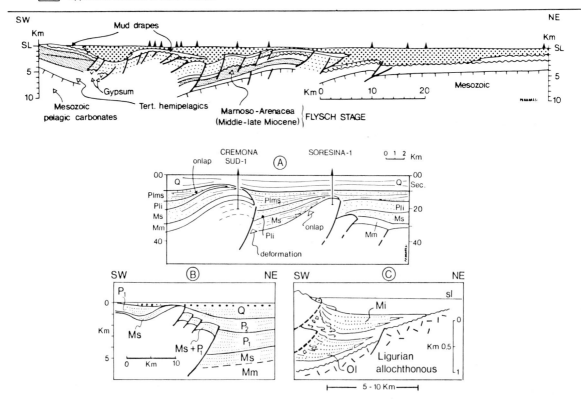

Fig. 12. A complete section of the Po Plain (SE portion, modified after Dondi *et al.*, in Cremonini & Ricci Lucchi, 1982), and details of minor, thrust-based basins (A and B). (C) is the inferred profile of an Oligo–Miocene minor basin (see epi-Ligurian sequence, Fig. 14). In (A) Plms = P2 of this paper, Pli = P1, Ms = upper Miocene, Mm = Middle Miocene.

masked or removed by later deformation, erosion and cover.

Another problem is to judge whether the base of clastic wedges marks a discontinuity in sedimentation. Seismically, it looks only slightly unconformable on foreland ramps. Onlaps of comparable scale are hardly visible in outcrops (see, e.g. the base of the Laga Formation in the Vomano Valley: Mutti, Nilsen & Ricci Lucchi, 1978). In many cases, all that can be seen is a gradual *or* abrupt *lithologic* transition; this is not a decisive aspect, considering the inadequacy of age control.

The amount of subsidence in Apennine foreland basins is not easily explained by thrust loading alone (Scandone, personal communication, 1982; Royden

& Karner, 1984). However, this problem of mass deficit exceeds the scope of this paper; the same can be said for thrust geometry and kinematics.

FROM PRESENT TO PAST FORELAND BASINS

The migratory nature of the foredeep and the persistence of turbiditic sedimentation in it invite the use of the present configuration as a model for previous stages. To check on this, we face various difficulties: reduced scale of observation in comparison with seismic sections, scarcity or imprecise definition of time lines, problems of matching lithofacies and seismic facies, limitation of stratigraphical analysis to unidimensional columns, and so on.

One major problem is to infer palaeotopography from ancient sediments. On an intermediate scale, this can be done to a certain extent: localized sand lenses, wedging of sands into pelites, intraformational slumps and other 'slope' indicators can reflect a mobile 'mesotopography' (Fig. 13) on the other hand, parallel layers of great lateral continuity, and particularly *megabeds* (Ricci Lucchi & Valmori, 1980; see Figs 14, 18, 19 and 23) are taken as evidence of a flat bottom. On a larger scale, the subdued gradients of depositional environments are hardly recognizable. Facies and thickness changes of turbidites are, in principle, utilizable, but their interpretation is not unequivocal. Turbidity currents react to topography in more than one way, and create for themselves a topography with deposition. Submarine slopes provide either impetus or obstacles according to situation of feeding routes and entry points (Fig. 7). Slope breaks and flat bottoms provide local base levels for gravity. They stimulate sand deposition by the effect of decreased gradient, expansion of previously channelled flow, or both; a combined 'spreading effect' builds submarine fans. In structurally controlled basins, however, channels can continue along the slope toe (up to the deepest available level), or the narrow basin itself acts like a channel. In this case, the spreading effect can be reduced by a 'confining effect' that carries sand along the basin axis. What about fans in this case? Are they distorted but still present? There is no effective way of determining if the upper surface of the sand body was sloping, flat, convex, etc. Sand can be plastered against the steeper flank of a foredeep by *longitudinal* flows if the asymmetrical section produces a channelling effect. Marginal equivalents (shalier beds) would be laid down on the ramp.

Moreover, variations of subsidence rate related to thrust activity could alternately constrict and expand the depositional section, possibly producing a vertical alternation of sand-rich, ribbon-like bodies and sheet-like deposits. Similar repeated motifs are common in both outcrop and subsurface record of Apenninic turbidite deposits, and have been originally interpreted as fan lobes interfingering with basin plain sediments (Mutti & Ricci Lucchi, 1972, 1975; Ricci Lucchi, 1975a, b). Of course, extrabasinal factors (changes of supply or sea-level) could as well account for these 'cycles' when they are correlated over wide distances.

COMPARING ANCIENT AND RECENT WEDGES OF THE FOREDEEP

Similarities

(1) Outward wedging of sedimentary bodies can be demonstrated at various scales (6–20 km distance, 10–300 m thickness) across basin axes. This can be taken as evidence of updip shalying on a foreland ramp. Such wedging is either bound by two adjacent thrust planes or can be traced across one or more thrusts (Figs 13 and 14). The former case is regarded as an evidence of a minor, syn-thrust depocentre, the latter one of a wider foredeep wedge deformed after deposition.

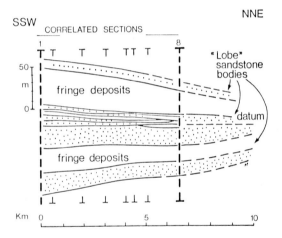

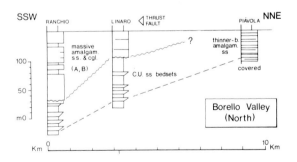

Fig. 13. Depositional asymmetry of foredeep sand bodies (Miocene, MA). Wedging is toward the foreland. Within-thrust correlation in (A) (on a bed-by-bed basis), across-thrust in (B). (B) shows a section of Tortonian palaeochannels (see text and Fig. 24). (A) from Ricci Lucchi & Pignone (1979).

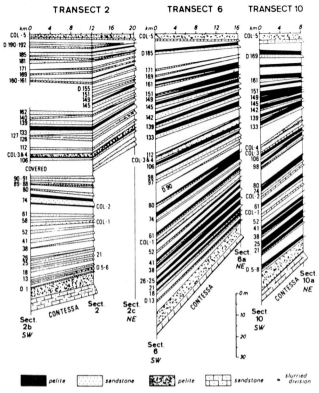

Fig. 14. Rampward wedging of a basin plain section (post-Contessa interval, Ricci Lucchi & Valmori, 1980); Miocene MA, inner and main stage of subsidence and deposition.

(2) Slide-slump bodies: their maximum thickness is observed in basinal areas close to thrust fronts but not directly *on* thrusts; their extension along basin strike can reach up to 20 km. These gravity-emplaced bodies, often referred to as *olistostromes* or chaotic levels, thin out rapidly towards the opposite basin side, as sand bodies do.

(3) Pelitic lenses and drapes: hemipelagic mudstones are associated in various manners (laterally or vertically, *in situ* or displaced, undisturbed or contorted) with turbidite sediments.

They are broadly interpreted as slope deposits and partly equalled to the mud drapes mantling top and flanks of structural highs in the Po Plain subsurface (Fig. 12). These pelitic bodies are important elements of correlation between the thrust belt and the foreland ramp, where sedimentation was also hemipelagic. Pelitic lenses of chaotic aspect are commonly associated with extrabasinal olistostromes at the top of clastic wedges. They form there a peculiar *closure*

facies, reflecting uplift or remoulding of submarine topography by tectonic movements.

(4) Sediment dispersal: longitudinal deflection after lateral input, and regardless of location of entry points, was an almost constant pattern in all stages of deposition including the present, alluvial one (Po Plain).

(5) Control of transversal structures: a segmentation of the orogenic front is documented, at least since the Miocene, by remarkable and abrupt differences of cumulative subsidence along strike (Figs 19 and 24; see also Cremonini & Ricci Lucchi, 1982).

What kind of structure is responsible for this (wrench faults, re-entrants of curved thrusts, or others), is still under discussion. Horizontal movements were apparently subordinate to vertical ones. Patterns of sedimentation show that some of these lineaments had a long life (basement faults?), and played an important sedimentary role in several respects:

allowing access of sediment from the orogenic side of the foredeep, where sources were usually locked by topographic barriers (Fig. 7);

providing a mechanical trigger (earthquakes) for failure of huge sediment masses stored in basin margins; as a result megaturbidites arrived in the foredeep (this is not proven for the molasse stage);

forming topographic constrictions in the basin with ensuing deviation, reflection or ponding of turbidity currents.

Differences

(1) Basin closure: it happened abruptly, for tectonic causes, in ancient basins (not necessarily at once, but anyway *interrupting* local sedimentation); recent basins are sediment filled.

(2) Vertical depositional trends: ancient sequences are uniform or 'stationary' in spite of fluctuations of sand/pelite ratio, rate of input and so on; sedimentary cycles and shoaling-up trends are more obvious in molasse deposits.

(3) Tectonic displacements: greater importance of horizontal movements in the flysch stage, of vertical movements and subsidence in the molasse stage (Figs 9 and 10).

(4) Magnitude of resedimentation events: there is no firm evidence at the moment, but some data suggest that size and frequency of gravity flows (which, as in other catastrophic phenomena, should be inversely related) changed in the molasse stage. A higher but more diffuse sand content, a poorer definition of individual beds ('streaky bedding'), a lesser lateral continuity, and the lack of discrete hemipelagic interbeds characterize late turbidites in contrast with earlier ones. Observations of recent cores from both the Po Plain (unpublished data) and minor prodelta basins of the Ionian Sea (Colella *et al.*, 1984) give the impression that thin-bedded turbidites are almost seasonal events. Considering, on the other hand, the data reported below for the Marnoso-arenacea, it seems that recharge times were longer and storage areas larger in flysch times (Mutti *et al.*, 1984). More data about repeat times of turbidites are needed, however, for both ancient and modern settings.

COMPARING MINOR AND MAJOR CLASTIC WEDGES

The deposits of satellite basins share many characteristics with their bigger equivalents: correlative trends of subsidence and deposition (Fig. 15), preva- lence of turbiditic facies, geometry of sand bodies and slumps-olistostromes (wedging, mobile topography, etc.), multisourced supply (in part from the same sources), predominance of longitudinal dispersal.

Differences are also evident: lateral changes were more drastic and rapid, subsidence and topography more differentiated, resedimented deposits more immature and disorganized, local sources and *cannibalistic input* (recycling of sand from previous wedges or basin margin) more important.

To continue the comparison, a line must be drawn between flysch and molasse stages also in minor basins (confirming that they were part of the same system). In the late Miocene, the LIG sheet passed from a condition of prevailing subsidence to one of dominating uplift. A new set of satellite basins (the piggy-back as originally defined) relayed the epi-Ligurian ones: they were seated on autochthonous sediments.

In the flysch stage, the foredeep and the epi-Ligurian basins were closely tied in terms of subsidence and deposition cycles ('positive' correlation: see Fig. 15 and Table 1). The underlying mechanism still awaits explanation. A hinge line existed instead between major and minor basins in the molasse stage ('negative' correlation; see also Fig. 16).

Distinct sequences, or cycles, are stacked in satellite basins (suggesting a single basin that was closed and reactivated several times), and juxtaposed in the foredeep (Fig. 8). In other words, subsidence was 'stationary' in the moving LIG substratum and migrating in the 'fixed' foreland. How was this accomplished? And why were the satellite basins decoupled in the late Miocene?

PRE-OROGENIC DEPOSITS

No major clastic bodies are present in the foreland sequence before the onset of Apenninic orogeny (Oligocene); one must go deep in the stratigraphy to find the continental paralic unit of 'Verrucano' (Triassic). The immediate precursors of clastic wedges are hemipelagic sediments. Beneath them, Mesozoic siliceous-carbonate pelagites and platform carbonates accompany the rifting and collisional events of the Alpine cycle. Carbonate turbidites and debrites are the main expressions of tectonic activity in the Mesozoic.

In late Eocene–early Oligocene, an increase of clay input turns deposition from pelagic to hemipelagic; this subtle change marks the start of Apenninic orogeny in the foreland. Since then, hemipelagic and

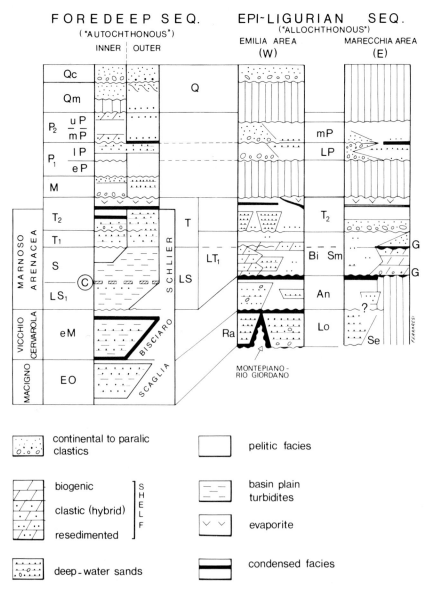

Fig. 15. Litho- and chronostratigraphic subdivisions used in this paper. Some hiatuses are underlined by vertical hatching. Q, Quaternary. Qc, continental Pleistocene. Qm, marine Pleistocene. P2, middle (mP) and upper (uP) Pliocene. P1, early (eP) and lower (lP) Pliocene. M, post-evaporitic ('upper') Messinian. T, Tortonian–'lower' Messinian. T2, late Tortonian–Messinian 'Sahelian' cycle. T1, early Tortonian. S, Serravallian. L, Langhian. S1, early Serravallian. eM, late Oligocene–early Miocene. EO, Eocene–Oligocene. Ra, Ranzano Formation. Lo, Loiano Formation. Se, Monte Senario Formation. An, Antognola Formation (with revisions). Bi, Bismantova Formation (with revisions). Sm, San Marino Formation. C, Contessa marker bed. G, glauconite.

terrigenous muds continue to blanket the foreland ramp and are progressively 'transgressed' by clastic facies (the term pre-orogenic can still apply on a local basis). Some slow-deposition facies are particularly useful for correlating foreland and thrust domains; they have the ideal requisites for stratigraphical markers (reduced thickness and wide areal distribution), and are here referred to as *condensed facies*. Included in them are dark and black shales, sapropels, silica-rich sediments, glauconite-rich sediments, vol-

Table 1. Comparison of depositional sequences in foredeep and satellite basins (epi-Ligurian)

Similarities	Differences
Common condensed facies: black shales in T2 (two levels), glauconite in LS, chert and ash layers in upper eM versicolour shales in EO	Shelf facies of Miocene age: present in satellite basins, absent in foredeep
Common depositional trends: passage from deep-water to shallow-water facies in latest stage	Earlier transition to shallow-water deposition in satellite turbidite basins (lower Pliocene) than in foredeep (Pleistocene)
upward decrease of sand content and size in turbidite sequences	Absence of sequence M in satellite basins
highest proportion of intrabasinal supply and CaCO₃ in sequence LS	Earlier exclusion from Alpine feeding (T2 versus M) in satellite basins
Erosional phase in upper Miocene ('flysch-molasse' transition)	Decoupled trends in sequences of the molasse stage (Fig. 16): erosion in marginal basin, resedimentation in foredeep
Similar vertical evolution in individual sequences of the flysch stage	

highest proportion of intrabasinal supply and $CaCO_3$ in sequence LS

canic horizons and volcanic-derived clays (bentonitic, smectitic, zeolitic). See references in Sestini (1970), Guerrera (1979), Mezzetti, Morandi & Pini (1980) and Lorenz (1984).

Pre-orogenic deposits are usually assumed as forming an almost continuous succession. A closer analysis will certainly reveal time gaps, whose location and meaning will be of paramount importance in assessing the vertical mobility of the foreland ramp.

GENERAL REMARKS ON STRATIGRAPHIC AND FACIES ANALYSIS

The Apenninic literature is overcrowded with local names; it is not simple, for a foreign geologist, to become familiar with them.

Lithostratigraphic subdivisions have been made with classical methods (groups, formations, members), more or less adapted to the regional setting (tectono-stratigraphic units, complexes). An attempt to identify sedimentary cycles or *depositional sequences*, as defined by Mitchum, Vail & Thompson (1977), was made for the molasse stage (Ricci Lucchi *et al.*, Dondi *et al.*, in Cremonini & Ricci Lucchi, 1982). The same approach is followed here. Using depositional sequences (i.e. recognizing unconformities that correlate with synchronous depositional surfaces, and bound *isochronous* sedimentary bodies and bedsets) reduces the old nomenclature to the minimum necessary. Unfortunately, also new terms are introduced (see acronyms in Fig. 15).

The scheme presented here is still tentative and subject to future revisions. The sequences are to be regarded as provisional, regional subdivisions. No hierarchical arrangement is attempted for the moment: only *basic cycles* (depositional motifs, cycloth-

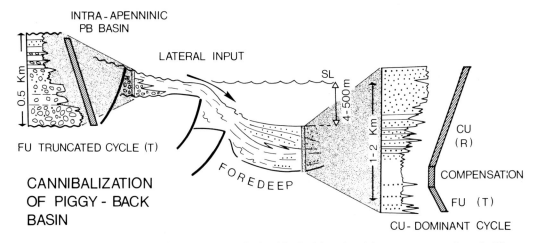

Fig. 16. Example of 'negative' correlation between marginal and basinal deposits of the same sequence, from the Pliocene of the Bologna area. Coarse clastics trapped in marginal satellite basin during transgressive phase (T), which develops a fining-up (FU) sequence; resedimentation in foredeep peaking in regressive phase (R) with coarsening-up (CU) trend.

ems) are distinguished at the lowest rank (possible autocycles controlled by sedimentary mechanisms only). A comparison with the Vail chart (itself under revision, see Fig. 17) shows some good correspondences in the Miocene but a significant mismatch in the Plio–Pleistocene.

Sequence boundaries were chosen where one or more of the following features occurs: time gaps in the biostratigraphic record, extensive erosional surfaces (subaerial or submarine), condensed horizons, emplacement of slides and olistostromes, onset of turbidite deposition, closure or shifting of the same. Table 2 summarizes the boundary events as resulting in part from a team project led by M. Boccaletti (in preparation for the Committee of Mediterranean Neogene).

Chronostratigraphic subdivisions are based on calibration of magnetic and biostratigraphical scales (based on foraminifera, Colalongo et al., 1978; and

nannofossils for the Pliocene and Pleistocene. The chart of Vail, Mitchum & Thompson (1977) has been followed for the underlying intervals with some modifications in the Miocene (see Borsetti & Cati, 1968 in Cremonini & Ricci Lucchi, 1982). The wide intervals of uncertainty in the Oligocene and lower part of the Miocene underscore some severe limitations of biostratigraphy. Owing to poor preservation of fauna in clastic facies, banality of many planktonic foraminifera, scarcity of macroforaminifera and amplitude of biozones, the resolution is low and the language problems acute. Different divisions are adopted, no calibrations are made, different names are given to the same interval and vice versa.

The column of the foredeep in Fig. 15 is split into two parts indicating, respectively, more marginal, less subsiding areas to the left, and depocentral areas (this part combines outcrop and subsurface information). Among the two 'allochthonous' sequences, only the

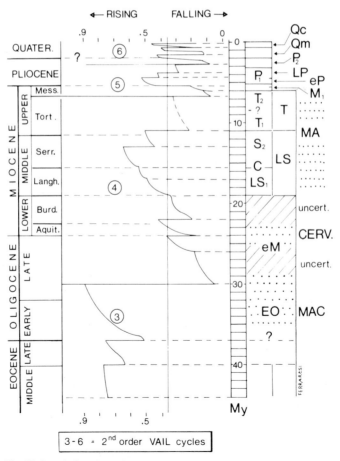

Fig. 17. Correlation chart of Apenninic sequences and Vail et al. (1977) curve.

Table 2. Approximate age of main sedimentary breaks in northern Apennines (Ma)

0·5–0·7	Middle Pleistocene	Qc/Qm	last coastal onlap
1·6–1·8	Pliocene–Pleistocene	Qm/P2	appearance of 'cold guests' in Mediterranean; late thrusting in Po–Adriatic area
3·0–3·4	Lower–middle Pliocene	P2/P1	cooling in northern hemisphere; thrusting, late displacement of LIG
4·3–5·0	Early–lower Pliocene	eP/1P	thrusting, local erosion, displacement of LIG
5·2–5·5	Miocene–Pliocene		refilling of Mediterranean by Atlantic waters; tectonic quiet
5·8–6·0	Lower–upper Messinian		thrusting, sea-level fall; widespread erosion (desiccation)
8–9(?)	'Lower'–'upper' Tortonian	T1/T2	subaerial erosion in marginal areas and LIG thrusts, submarine erosion in foredeep
11–12	Serravallian–Tortonian (mid–late Miocene)	S/T1	closure of inner MA
19–22	Burdigalian (lower Miocene)	eM/LS	start of MA; closure of Cervarola Basin; Bisciaro–Vicchio condensed episode
24	Oligo–Miocene		generalized regression-hiatus in ramp sediments
26–29	Middle–late Oligocene	EO/eM	closure of Macigno Basin, Pievepelago–Scaglia cinerea Scisti Varicolori condensed horizon
40	Middle–late Eocene		base of EO onset of Macigno Basin

Emilian one will be described. The Marecchia section is shown mainly for comparative purposes: it represents shelf sedimentation except for EO and eM.

The correspondence between sequences and classical stratigraphy for the foredeep is shown in Figs 8 and 15. No formal units are in use for the Plio–Pleistocene except on a local basis. The major clastic bodies are buried under the Po Plain and designated as formations (AGIP geologists recognize only two supersequences in the foreland: Oligo–Miocene and upper Messinian–Pleistocene). They are: Fusignano Formation for sequence M, Porto corsini for seq. P1, Porto Garibaldi for P2, Asti sands for Qm and Qc. Heteropic clays are all called Santerno.

Turbidites are described as *facies* with reference to the Mutti & Ricci Lucchi (1975) scheme. As for the environmental interpretation, the fan model will be quoted on occasion but not adopted as a guideline. After more than a decade of data acquisition in both modern and ancient sediments, the application of fan models meets with increasing difficulties (see Normark, Mutti & Bouma, 1984). Substantial revisions of the original Italian model have been recently made by Mutti (1985), who emphasizes the effects of sea-level changes on turbidite depositional systems. The emphasis is placed more on tectonics and basin topography in this paper.

REVIEW OF MAIN FOREDEEP DEPOSITS: EMILIA TO UMBRIA

The flysch stage is represented by the Oligocene Macigno Formation, the Oligo–Miocene Cervarola Formation, and the Miocene Marnoso-arenacea. Not much stratigraphical or sedimentological work has been done in recent years on the older, 'Tuscan' wedges; information can be found in Sestini (1970), Mutti *et al.* (1975) and Ghibaudo (1980). The Marnoso-arenacea (MA) is the best studied body, and will be described as the representative of northern Apennines *flysch wedges*.

Deposits of the post-emergence or molasse stage are described in a guidebook of the Italian Geological Society (Cremonini & Ricci Lucchi, 1982). Their main depocentres are still preserved in the Po Plain but good surface sections are also available in the Pede-Apenninic (piedmont) belt: a thick section of 'blue clays' with coarse-grained intercalations overlies Messinian (uppermost Miocene) evaporites and anoxic sediments. The base of the clay section is still Miocene and of fresh to brackish water environment. The basal unconformity marks not only the tectonic uplift of the chain but also an episode of a Mediterranean-scale evaporative drawdown and subaerial erosion (Hsü, Cita & Ryan, 1973).

Geologic setting

Syn-orogenic deposits of the emergent Apennines chain are folded and thrusted in a style that is comparable with that of the outer buried portion (Figs 11 and 28). This style was perceived and brightly described by 'pre-seismic' geologists (Migliorini, Signorini; in Sestini, 1970). Folds and steeply inclined longitudinal ('Apenninic') faults are the surface expression of thrusting. Shortening has been only guessed; balanced sections are not yet available.

Thrust lines separate the main wedges (Fig. 4) and also split individual wedges (Fig. 18); distensional block-faulting is superimposed on them in the SW part of the chain (Figs 1 and 2).

The sedimentary wedges are laterally bounded and internally split by transversal tectonic lines of controversial interpretation: deep faults driving the thrusts, post-thrusting faults (with vertical or transcurrent movement, or a combination of them), lateral ramps of thrusts.

It is here assumed that foreland thrusting started in the Oligocene (Dallan & Nardi, 1974; Kligfield, 1979;

Boccaletti et al., 1980), and formed the first foredeep, the Macigno Basin. It then progressed north-eastward along with the emplacement of the LIG sheet by gravity. Depocentres were activated and closed by the same tectonic mechanism, which is the old idea of the 'orogenic front' propagating as a compressional wave. This is no more than a conceptual frame, which has never been formulated and tested accurately. Neither is the connection with hinterland basins clear: their opening is not well dated, especially in the Tyrrhenian area, and a migratory pattern, coeval with frontal thrusting, is not precisely proven.

As for transversal lines, basin analysis favours an early development during Apenninic orogeny: vertical movements are particularly testified by differences of subsidence, facies and depth in adjoining sectors of the foredeep. Part of the evidence is documented in previous papers (Ricci Lucchi, 1975a, b, 1981; Cremonini & Ricci Lucchi, 1982); new data are presented in this paper. They concern two prominent lineaments crossing the MA area: the Sillaro and Forli lines.

The Sillaro Line (Figs 4 and 19) separates the MA in two parts: a western one, buried and overloaded by the LIG Sheet, and an eastern part that crops out from the Bologna area to Umbria region. Defined as a wrench fault by Bortolotti (1966, in Sestini, 1970), the Sillaro line played a mostly vertical role, determining a greater basement subsidence to the west. This favoured the encroachment of LIG masses, whereas 'autochthonous' thrusting and basement uplift created barriers in the other sector. Gravimetric data and the thickness profile of Marnoso-arenacea along the chain strike (Fig. 19) are in agreement with a deep-seated stepfaulting.

According to Boccaletti et al. (1982), the Sillaro would be but one element of the 'megashear' system active in the west Mediterranean during the Apenninic orogeny.

An alternative view (De Jager, 1979; ten Haaf & van Wamel, 1979) sees the Sillaro as the lateral ramp of an arcuated thrust that 'transgressed' Miocene and Pliocene deposits of the foredeep, with local enhancement of subsidence.

Whatever the right model is, an inversion of relief occurred along the line during the chain emergence: differences of subsidence on the two sides were roughly equalized during Sequence T2, up to the evaporitic phase, then an uplifting trend took place (Sequence M), which went on to the west in the Pliocene while the eastern side foundered at the impressive rate of about 1 mm yr^{-1} (see below, Fig. 27; and compare

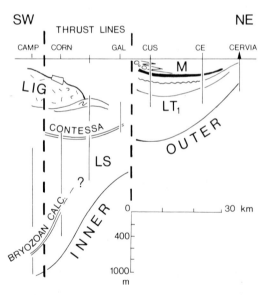

Fig. 18. A transversal cross-section of MA along Bidente valley (modified after Ricci Lucchi, 1975b). Notice shifting of subsidence from inner to outer depocentre, the two being separated by a thrust, and correlability of turbidite layers across thrusts in the inner portion. The point here is that thrust slices do not *necessarily* represent the fill of distinct basins or sub-basins as implied by some structural interpretations (see ten Haaf & van Wamel, 1979).
Localities: CAMP, Campigna; CORN, Corniolo; GAL, Galeata; CUS, Cusèrcoli; CE, Cesena.

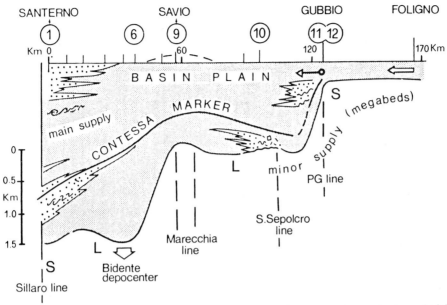

Fig. 19. Schematic longitudinal section of inner MA body. Datum taken at the uppermost recognized megabed. The dashed line in the Savio–Marecchia area points out a rising tendency (Verghereto High). Entry points and subsidence steps are taken as indicators of transversal tectonic lines active during sedimentation. Circled numbers refer to correlated sections of Ricci Lucchi & Valmori (1980).

Figs 24 and 29). The interesting point here is that the geomorphic and sedimentary transition between 'flysch' and 'molasse' stages of the orogen is reflected not only along its front but *also* transversally.

The Forli line is not parallel to the Sillaro (NE–SW); that is it does not follow the so-called anti-Apenninic trend but a N–S and NNW–SSE one (Figs 4, FO; 20). This a neotectonic lineament with active seismicity; however, an earlier record of its activity lies in Messinian sediments, when it separated different evaporitic basins (Rabbi & Ricci Lucchi, 1968; Ricci Lucchi, 1975a, b). It then formed a structural high with unstable slopes that stimulated gravity resedimentation of sabkha-type gypsum. But even before the Messinian 'salinity crisis', i.e. in the Tortonian (Figs 13B and 20), a lineament of this orientation can be recognized by the excavation and filling of linear submarine channels that headed into south Alpine structures beyond the Padan foreland (see base of Sequence T2, Fig. 15). Similar and parallel palaeochannels exist in the Sillaro–Santerno area. They all cut through previous axes of deposition and deformation; interesting enough, the same axes were restored after sequence T2. Furthermore, there is no evidence of thrusting at the base of T2 (it is only found after channel abandonment), nor advance of LIG

masses, nor shift of subsidence. On the other hand, vertical shear lines of NW–SE trend were active in the foreland and southern Alps (Massari, personal communication); a phase of strike-slip, interrupting the Apenninic thrusting, can thus be suggested. It could have been induced by a change in direction of convergence between Africa and Europe (inferred on structural ground but, unfortunately, not precisely dated by Letouzey & Tremolières, 1980) or other causes. In any case, the Forli line is the local expression of a wider structural rearrangement that is possibly unrelated to the thrust mechanism.

Flysch stage: the Marnoso-arenacea wedge

MA is a composite, migratory turbidite wedge with a volume of 28,000 km³ (including the buried part) and local thickness in excess of 3,000 m. It accumulated from Langhian (or late Burdigalian according to other schemes) to late Tortonian, in a time span of about 12 Ma. The outcrop belt is 180 km long, 40 km wide; the degree of shortening is not known. Three more persistent depocentres can be recognized: Langhian, Serravallian and Tortonian. They correspond, though not precisely, to sequences LS1, S and T of this paper. No usual unconformities separate the

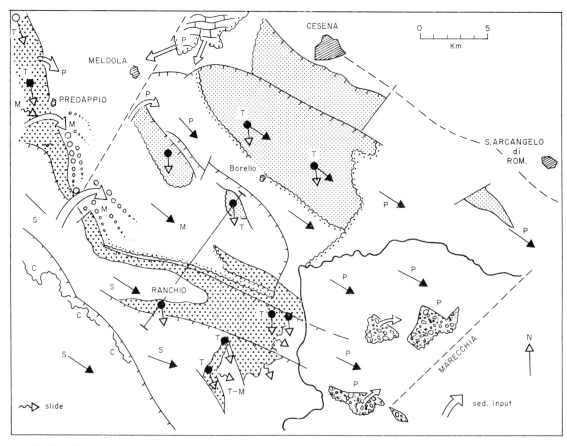

Fig. 20. Simplified geological map of the area between Forlì and Marecchia lines, compiled after Cremonini & Farabegoli (1981), Cremonini *et al.* (1982), Farabegoli (1983) and author's unpublished data. Near the end of the flysch stage, Tortonian turbiditic channels (T) cut through previous basinal trend (S), which is related to thrust lines and is restored in the molasse stage (P, Pliocene). In the SE corner, the margin of the Marecchia LIG sheet is contoured: it looks like a giant debris flow. After its emplacement in early Pliocene, it was covered by deep-water clastics of the foredeep, partly fed by longshore highs exhuming the Miocene MA. Blank arrows indicate the local sources of sediment (skeletal in thrust-top organic banks west of Cesena) of the molasse stage.

sequences (or subsequences) in the field: sedimentation apparently went on during switches of subsidence, except in the rear part of the basin where it was closed by sliding and thrusting. Sequence boundaries are therefore linked to advances of the LIG sheet; closure facies of the inner thrusted area are then correlated with the coeval, conformable turbidites to the NE.

The most important break in MA deposition is represented by the S/T boundary; an older and *inner stage* is thus distinguished from an *outer stage*. Sequence T, formerly considered continuous, is now split into T1 and T2 along the erosional surface that marks the onset of the previously mentioned submarine channels. Depocentre location is the same for T1

and T2. The two sequences are provisionally lumped into a sequence T (for Tortonian) where the discontinuity has not been recognized.

The base of MA is not exposed in Romagna, where the greatest accumulation occurred with reference to the basin axis (Fig. 19). It was reached by exploratory drilling, and consists of grey mudstones with chert similar to those cropping due SW (Vicchio Marls, in the thrust belt) and SE (Bisciaro in the Umbria–Marche foreland) (Merla, 1951). Going from Umbria to Marche (SW to NE), the base of MA youngs from Burdigalian–Langhian to Tortonian. This marked diachronicity (over 5 Myr) is well documented palaeontologically and provides the best example, in the northern Apennines, of syn-orogenic clastics 'climbing up' the foreland ramp.

The transition from inner to outer MA seems to match a regressive peak in the Vail curve (after a long interval of rising sea-level, which is matched, too, by the characters of inner MA).

The top of MA is a discontinuous, diachronous mud drape, as a whole; many discrete bodies were identified (see Bruni, 1974; De Jager, 1979, among others). Their names are omitted with the exception of the two more commonly used to indicate the inner closure (Verghereto, S-T) and the outer closure (Tossignano, T2), respectively.

The inner stage

The older MA is tectonically split into three parts that stratigraphically correspond to two sequences only (an inner depocentre with subsidence peaking in Langhian, an outer one in Serravallian). Correlation is, in fact, possible across thrusts (Fig. 18). The Contessa megabed, stratigraphically marking a decrease of sand input from the main source (the bed itself came from a different source), is taken as a sequence boundary in the basinal portion of the wedge. It is tentatively correlated with a transgressive phase in shelf areas all around Italy, within and outside the orogenic belts.

The post-Contessa interval S is characterized by distinctive calcareous turbidites ('colombine') of SE provenance (Fig. 24) interbedded in a vertically uniform and shale-rich section of siliciclastic turbidites fed by the main, Alpine source. The Sequence S is closed by an increase of sand in siliciclastic turbidites and the disappearance of limey turbidites (SE of Sillaro Line) or the opposite trend (upward shalying into hemipelgites in a bulging area across the Marecchia line, Fig. 19), and finally by the emplacement of the LIG, which caused a shift of the subsidence to the NE. LIG overrode the foredeep only west of the Sillaro and along the narrow breach of the Marecchia line (acting like a transversal channel amid autochthonous thrusts). The movements of the allochthonous sheet were preceded and/or accompanied by tectonic instability of the inner foredeep margin. Type and causes of such instability, which induced sliding and topographic irregularities, are still discussed (see some models in Fig. 21).

Depositional setting: The Miocene foredeep had its maximum development (extension and subsidence) in this period (Figs 19 and 22). Lithologically, MA stands out among older and younger clastic wedges for its

higher carbonate content (mostly in the form of cement derived from organic remains) and finer grain size (abundance of clay and silt). This makes it a bad reservoir, and is connected with the transgressive trend, of likely global nature, that accompanies its deposition. During high-stands, shelves are enlarged and organic production increased; on the other hand, erosion and terrigenous input decrease, which is apparently reflected in the curve of the sedimentation rate (Fig. 10).

The parallelism and lateral continuity of bedding are the most impressive characters of inner MA. Not only some outstanding markers (Contessa and similar

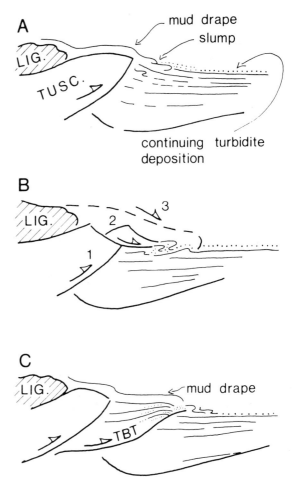

Fig. 21. The 'Verghereto closure' of the inner MA basin. Different explanations for the presence of the hemipelagic body on top of basinal turbidites. In C, TBT = thin-bedded turbidites, visible on the eastern 'shoulder' of Marecchia line (Badia Tedalda).

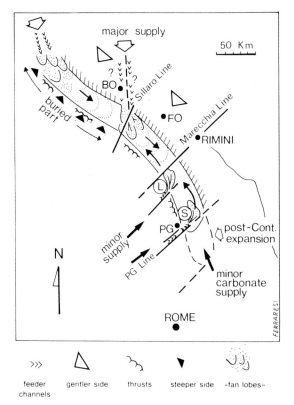

feeder   gentler side   thrusts   steeper side   «fan lobes»
channels

Fig. 22. Palaeogeographic sketch map of MA foredeep in Langhian–Serravallian (LS, inner stage). Contessa flow was introduced from an inferred Perugia (PG) line. From Ricci Lucchi (1984).

beds) but about 50% of visible layers can be traced for over 60 km (Fig. 23) and have a volume in excess of 1 km³. These are the megaturbidites and basin-wide turbidites of Ricci Lucchi & Valmori (1980). In spite of their size, megalayers are surprisingly fine-grained: a very loose correlation exists between grain size and bed thickness.

In terms of the Bouma model, most layers are base-missing and sand-missing. They regularly alternate with hemipelagites (10–20% of total volume).

The repeat time of mega-events varied from some millennia for the main, siliciclastic supply, to some 10^4 yr for carbonate and hybrid sources. Such low frequencies imply long recharge times in marginal storages, and fit in high stands of sea-level (Mutti, 1985). Collapses of shelf sediments could be produced by regressive pulsations (simple gravity failure) or seismic shocks and tsunami waves.

The multisourced deposits described above are interpreted as a *basin plain* association, no matter whether a deep-sea fan existed or not. No systematic CU or FU trends are observed except in local depocentres. Rather, 'basic cycles' (sequences of Ricci Lucchi, 1975a) occur repeatedly. They reflect changes of sand input, and consist of several metre thick bedsets of sandy turbidites with complete Bouma sequences, clay chips and local amalgamation surfaces (facies C and D1). These bodies stand out between the muddier turbidites of the proximal plain (NW) but lose their individuality distally over distances of 10–50 km (Fig. 23). Many, but not all, display a thickening-up of individual layers of aggradational kind (though some show a slightly offlapping pattern). Evidence of erosional channelling at the base is lacking. Sandstone bodies interbedded with basin plain deposits were originally explained as pulsating fan lobes (Mutti & Ricci Lucchi, 1972, 1975). Subsequent appreciation of their size and basin-wide extent suggested (Ricci Lucchi & Valmori, 1980) that some of them were accumulated by megaflows bypassing a fan system growing west of the Sillaro.

Instead of a fan, erosional unconformities and slide features should be found according to Mutti's most recent views. The sand 'lobes' would represent a low-stand turbidite system with a high efficiency of transport related to the clay content (Mutti, 1979). Topographic trapping of sand has already been suggested here as a possible alternative or complementary mechanism (Ricci Lucchi, 1985). Given flows of the same volume and energy, a restriction of depositional area would cause an increase of both local competence and local thickness, and *vice versa*.

Sediment supply and dispersal: The main detrital mode is siliciclastic; most indicative are the rock fragments, including serpentinite, glaucophane and dolomite, which testify to the Alpine provenance (Gandolfi *et al.*, 1981). Minor extrabasinal sources were active in L and S times along the SW side of the basin in Umbria (Figs 19 and 21); 'minor fans' and megaturbidites are the record of these entry points. They are composed of extrabasinal particles (from both LIG and 'autochthonous' thrusts) and carbonate particles from unstable shelves hosting organic remains more or less mixed with LIG-derived siliciclastics.

In summary, the basinal transport pattern of the inner MA is longitudinal-bipolar, with multiple entry points located along both the 'orogenic' and the ramp side (the latter providing the major input). Large marginal repositories were available for preparing big gravity flows, which largely bypassed the toe of

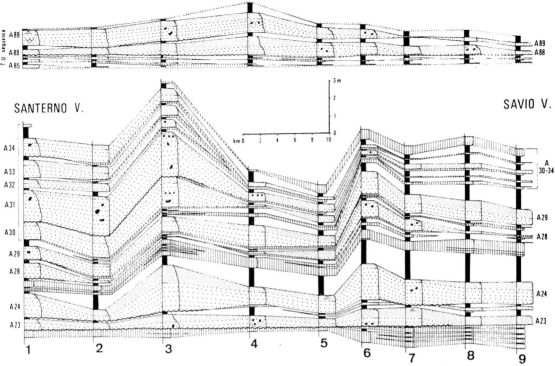

Fig. 23. Bed-by-bed axial correlation of megaturbidites and hemipelagites (hatched) in two depositional lobes of inner Marnoso-arenacea. From Ricci Lucchi & Valmori (1980).

feeding slopes and were 'channelled' by the basin flanks.

The outer Marnoso-arenacea (Figs 18 and 24)

The wide Serravallian plain was progressively narrowed and abandoned by thrusting, leaving some residual subsiding areas (satellite basins) in Umbria. In the Marecchia area, the Verghereto High, slowly rising in late Serravallian, collapsed or was breached by the LIG sheet.

The base of Sequence T1 is marked by an increase of sand content, which continues up to the summit in combination with a decrease of $CaCO_3$ and an increase of porosity (improvement of reservoir quality). The trend reflects a regression in marginal areas: not an episode of shelf progradation, but an erosional phase (see T1/T2 boundary in Fig. 17), i.e. a relative fall of sea-level. In the basin itself, T1 culminates with a phase of submarine erosion. Previously regarded by the writer as an 'autocyclic' event within the same 'progradational' sequence (encroachment of inner fan channels on outer fan deposits), this erosional surface is now reinterpreted as an unconformity of regional significance. It is present in the southern Apennines, Calabria, and Sicily, where it truncates overthrusted units. In essence, this surface seals different domains, from the foreland to the thrust belt, over a wide area, and marks the start of a generalized deepening trend (FU). This trend was eustatically controlled according to Farabegoli (1983). The coarse basal deposits grade rapidly into a pelitic drape that extended over the thrust belt and the epi-Ligurian basins. Intraformational and LIG-detached masses are embedded in the pelites, stressing a remobilization of the Apenninic front.

If the supra-regional character of the T1/T2 unconformity will be confirmed by further research, the implications are noteworthy in terms of eustatic lowering or structural re-arrangement. The common sedimentary trend of marginal and deep areas, previously decoupled (see 'negative' correlation as exemplified by Fig. 16), indicates in any case a considerable change of hinge lines between erosional and depositional domains.

Regressive facies of reduced thickness and peculiar character are present at the top of Sequence T2. They

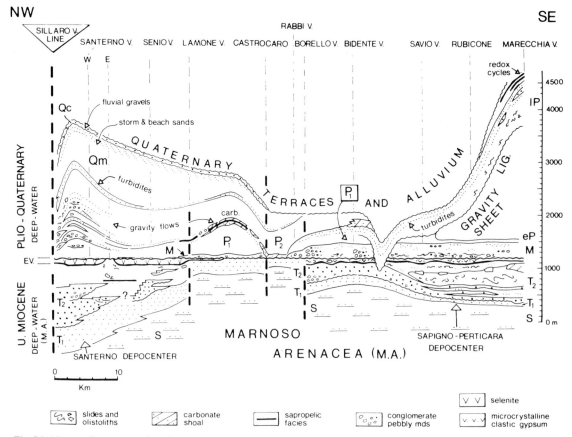

Fig. 24. Along-strike cross-section of upper Miocene to Pleistocene deposits of the Emilia Foredeep, encompassing the flysch-molasse unconformable transition (M-P). In this single section, the 'migratory' stratigraphy of the foothill belt, about 30 km wide, is actually projected.

reflect the insulation of the Mediterranean and the start of its 'salinity crisis'. A very large drop of sea-level (at least hundreds of metres) is recorded by a few metres of anoxic deposits (diatomites, black shales) capped by a stromatolitic limestone. It is amazing to find no evidence of coarse clastics, 'wild' erosion, unconformity, but only of local slumping. The erosional surfaces attributed to Messinian desiccation in the recent literature *do not* correlate with this level, which looks conformable in seismic sections, but with the top of gypsum (passage to clastic unit M). The Messinian evaporitic event, though palaeogeographically very important, is de-emphasized in this paper with respect to orogenic events. In another perspective, it might represent a separate sequence.

Depositional setting: Sequence T1: the thin-bedded turbidites are different from those of the previous stage. They have a higher sand/pelite ratio and are not interbedded with hemipelagites (facies D1). This change was interpreted as a transition from basin plain to a more proximal 'fan fringe' environment (Mutti & Ricci Lucchi, 1975; Mutti, 1977). Fringe-type turbidites should represent the margins of thicker beds, which effectively alternate vertically with them. Thick sand turbidites are grouped in 'lobe' bodies not dissimilar from those of the previous stage, but thicker and increasingly coarser up section. They mostly consist of facies C and B1 (high-density turbidity currents). Their cross-section is wedge-shaped (Fig. 13A), the longitudinal shape is unknown. Lateral thinning and thickening of individual beds tend to compensate mutually. Thicker beds have flat to slightly concave-up erosional base cutting up to 3 m of underlying deposits. No real, large-scale channelling is obvious. The products of intraformational, single-event erosion are visible as beds or intrastratal divisions of clay chips embedded in a chaotic, mud-rich sandstone.

Sequence T2: the lower portion is formed by multistorey channelled bodies of massive to crudely laminated sandstones with pockets of conglomerate. They form two distinct wedges (Fig. 24), one in the Santerno area, the other in the Borello–Savio area. Santerno channels are organized in two migrating sets. The direction of shifting is ENE. The depth of individual channels (some tens of metres) can be appreciated in some large exposures, already described (see ref. in Ricci Lucchi, 1981). Interchannel deposits are mudstones with thin, discontinuous sand interbeds; they grade upwards into sandstone-free, massive pelites with a dark condensed level (*a pre-Messinian anoxic event*) rich in fish remains, overlain by two sandy olistholiths.

The Borello-Savio sand wedge lies at the foot of the Forli line. It thins and fines out rapidly in both down-current and across-current directions (Fig. 13B). Massive, amalgamated sandstones of facies A are spectacularly developed in the narrow Ranchio depocentre. Palaeocurrents are parallel to those of Santerno (to SSE and south). The geometry of individual channels is not apparent. The composite body looks like a channel or valley itself.

Sediment supply and dispersal: Hybrid and carbonate turbidites are absent in T1 and T2 sequences. Siliciclastics are single-sourced; provenance is still from the Alps, but has shifted to the east (Alpine II, in Gandolfi *et al.*, 1981). Occasionally, turbidity currents swept the toe of LIG slides along the south-western slope and incorporated angular clasts of 'Apenninic' material (Bruni, 1974; De Jager, 1979). This contamination is visible also at the top of the 'lobe' body of (Fig. 13A) (granules and small pebbles). Entire beds made of 'cannibalistic' sand (derived from older MA or Cervarola uplifted by thrusting) are rare but present (Turrito layer in the Savio–Borello area: Gandolfi *et al.*, 1981).

Sand and gravel of unit T2 are the latest Alpine contribution to the Romagna foredeep. The gravel is polymict, with about 50% of crystalline rocks and 50% of sedimentary rocks representing typical units of the Southern Alps sequence (Ricci Lucchi, 1981). These particles were transported as far as the Marche region, beyond the Marecchia line, through the Savio channelway.

The molasse stage: post-evaporitic supersequence

Recently published seismic sections (Pieri & Groppi, 1981; Pieri, in Bally, 1984), well logs (Dondi, Rizzini & Mostardini, 1982) and outcrop sections (Ricci Lucchi *et al.*, 1981; Cremonini & Ricci Lucchi, 1982) provide a wealth of data for correlating deposits of the foothill belt and the subsurface. Correlations, still to be improved, show impressive differences of subsidence over short distances (gradients of up to 1 km km^{-1}) both along dip and strike (see Figs 12, 24, 25 and 28).

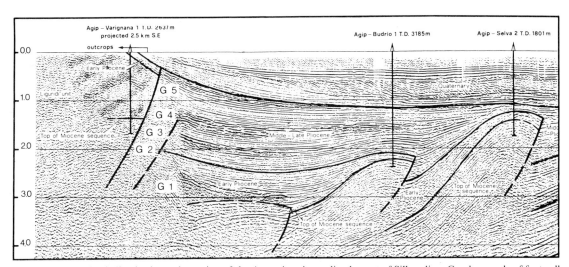

Fig. 25. AGIP seismic line in the prolongation of the Apennines immediately west of Sillaro line. Good example of footwall sunclinal basin. Irregular reflectors pass into smoother ones to the right (NE). They possibly correlate with olistostromes, pebbly mudstones and coarse-grained lenses of outcrop sections (see Figs 24, 26 and 27).

Sequence M (upper Miocene ; Di tetto and Colombacci Formations of Selli, 1954)

A phase of denudation (fluvial dissection, sliding, debris flows) marks the passage (once believed continuous) between evaporitic and subsequent clastic deposition. The unit is mostly clayey, with horizons of chemical limestones and organic clays of lacustrine character; the fauna is specialized and indicates brackish to freshwater conditions, with possible hypersaline episodes (Colalongo *et al.*, 1978). Unit M, therefore, represents a sort of 'brackish transgression' connected to a large water body of abnormal salinity, which occupied eastern Europe (Paratethys) and was isolated from the ocean. The sedimentation of clay was punctuated by coarse-grained episodes organized as cyclothems. They reflect transgressive-regressive fluctuations controlled very probably by factors operating at the scale of the western Mediterranean: their number (6–7) is the same in different environments (reefal, evaporitic, etc.) of peninsular Italy, Sicily and Spain. Sequence M corresponds to the upper cycle of the evaporitic sequence in the Sicily and Mediterranean basins (Montadert, Letouzey & Mauffret, 1978).

The sequence is thin and discontinuous in the western outcrop area; east of the Forlì line, it suddenly attains a thickness of 400 m. Main depocentres are, however, in the Po Plain (more than 1 km of clays and clayey-sandy turbidites). AGIP geologists do not consider it as a separate sequence because of the weak seismic expression of its upper boundary.

In the regional palaeographic context, this unit is important because it marks the 'feeding inversion' from Alpine sources (and the ram slope) to the Apenninic, local sources and thrust slopes. Lenses of alluvial fan and fan-delta conglomerates testify to this change: see Ricci Lucchi (1975a, b) and de Feyter & Molenaar (1984). Fan-deltas were mostly confined within synclines amid nearshore growing structures: in this early molassic stage, thrusts still had a damming effect on sediment dispersal. Both shallow-water and deeper, resedimentation-dominated systems existed (see Wescott & Ethridge, 1980).

Sequence P1 (lower Pliocene)

Sequence P1 is defined in the subsurface. It can be split into a lower (eP) and an upper (lP) sequence on basin margins, especially in outcrops of SE Romagna (Fig. 20). eP consists of banded clays with abundant deep-marine microfauna, similar to Sicilian 'trubi', which mark the return of Atlantic water in the

Mediterranean. This peculiar transgression is abrupt in terms of depth change (like the reverse one at the base of Messinian) but subtle in its lithological (clay on clay) and seismic expressions. Its nature was the object of lively debates in the 1970s: an abundant literature (not quoted here) is available.

In the Padan Foredeep, marine water flooded basins occupied by freshwater bodies. No desiccation or other unconformable surface can be proven in depocentral areas: as for marginal areas, nowhere have the basal Pliocene deposits been found. They probably onlapped steep slopes (the sea deeply indented the chain) and the next tectonic regression (eP/lP) could easily remove them. It must also be noted that coarse clastics did not reach basinal areas during early eP; this fact is not limited to the Apennines foredeep, but seems to be a Mediterranean-wide phenomenon, suggesting a tremendous reduction of sediment input to basins (Montadert *et al.*, 1978).

The eP/lP boundary is expressed by an unconformity and mud on highs and the inception of turbiditic sands in depocentres (Farabegoli, 1983). The sand input increases through lP, which shows the same CU trend and 'negative' correlation with shelf areas as the previous T1. Thickness of lP does not exceed 500 m in outcrop. In the subsurface, eP and lP are not distinguished (although the sand increase is apparent) and the aggregate thickness of P1 is 3,500–4,500 m.

Calcareous-skeletal and hybrid sands appear again in Pliocene turbidites after inner MA. While the terrigenous fraction derived chiefly from exhumation of MA and older flysch wedges, the intrabasinal fraction was provided by banks and shoals on top of thrusts and anticlines not yet attached to the mainland (see 'spungone' and 'amfistegina limestone' facies of local literature). Some shoals are preserved and display beach-shoreface sequences, coquina layers and storm layers.

East of the Sillaro line, channelized and unconfined lenses of disorganized gravel and massive sand, with remarkable examples of pebbly mudstones (Fig. 26), are intercalated in bathyal clays of lP. They are accompanied by isolated or aligned blocks of LIG limestones. These resedimented facies reflect the lateral palaeoslope of the Sillaro structure, with LIG units exposed in the western side (Fig. 27).

Sequence P2 (mid–upper Pliocene)

Marginal unconformities and variations of sand input in depocentres characterize also this transgressive-regressive cycle; the base is more gradual than in

Fig. 26. 'Dirty' conglomerates and pebbly mudstones embedded in bathyal clays. Inner part of the Pliocene foredeep, molasse stage. Pebbles from close sources (LIG terrane). Western side of Santerno Valley.

P1 in deep water, where a condensed horizon with repeated redox cycles is present. Facies and sedimentation patterns are the same as in the underlying sequence, with clays dominating in outcrop sections and turbidite sands in the subsurface (maximum thickness: 4,000 m).

The last phase of subsidence shifting occurred in the Po basin from P1 to P2. Thrusting continued through P2, but horizontal displacements were coming to an end.

Sequence Qm (Pleistocene)

The marine Quaternary is transgressive in marginal areas west of Bologna (beach facies, locally bouldery

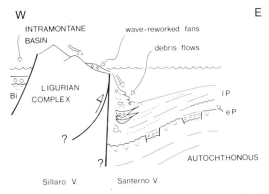

Fig. 27. Inferred situation across the Sillaro Line in the Pliocene. Modified after Cremonini & Ricci Lucchi (1982).

on LIG substratum) but conformable in depocentral areas where deposition was continuous since the end of the Miocene. Other marginal areas were uplifted by late Pliocene movements that continued, in the SE part of the Po Plain, in the Pleistocene (western side of Sillaro, Cesena). Resedimented sands and muds reach a thickness of 1,500 m in Po depocentres (Asti Sands) and 800 m in the Santerno section. The outcrop area is now restricted between the Sillaro and Forlì lines. The Santerno section was a candidate for the stratotype of the Plio–Pleistocene boundary, but the high sedimentation rate and the reworked state of fossils, among other things, discouraged this choice.

Sequence Qm ends with 'yellow sands' with a gradual transition; shallowing is at first revealed by the fauna in the upper clays, then sand storm layers appear. A passage from shoreface to foreshore is recorded in the overlying sheet of fine, well sorted sand locally enriched in stranded pebbles. This depositional regression is tentatively correlated with the offlapping set of Quaternary deposits of Fig. 25, which represent an extensive progradational episode in the Po and Adriatic basins (Dondi *et al.*, 1982; Ori, Roveri & Vannoni, this volume). The yellow sands are sharply truncated by fluvial gravels. This contact is well exposed in some quarries only.

Sequence Qc (Pleistocene)

This unit records the last Pleistocene high stand along the Apenninic margin, even though it consists mostly of continental deposits (alluvial, lacustrine, paludal, with pedogenic horizons: see Cremaschi, in Cremonini & Ricci Lucchi, 1982). The vertical trend is symmetrical (FU-CU) with minor oscillations. The sequence is only some metres to tens of metres thick but remarkably extensive and uniform in character from western Emilia to Abruzzi. No general consensus exists about the subdivision of upper Qm and Qc. Yellow sands have been also considered transgressive, and the overlying gravels as regressive. Some confusion arises because there are littoral and fluvial sands, even though not so clean, above the gravels. As for the gravels, they cannot be included in the same sequence as the yellow sands: they overstep the sands to the SW and lie on an unconformity atop of Qm blue clays. Qc deposits are correlated with the most recent coastal onlap detectable in Po Valley sections (Fig. 25).

THE DEPOSITS OF THE 'FLOATING' SATELLITE BASINS

These peculiar sediments lying on top of LIG overthrust (Fig. 28) are described here, after recent surveys, for the Bologna area, close to the Sillaro Line. Previous studies insisted on more western areas (Reggio and Parma Apennines: see, for references, Sestini, 1970 late geosynclinal sequence).

A cross-section parallel to Apenninic strike (Fig. 29) highlights that epi-Ligurian stratigraphy is not really of a layer-cake kind.

Effects of tectonics on sedimentation and basin topography were strong. More recent deformations do not make the analysis of depositional sequences easy, but original spatial relationships of sedimentary bodies are not destroyed. The traditional interpretation of this succession is that of a continuous

sedimentation with many facies changes (mosaic pattern). The present scheme emphasizes vertical discontinuities and de-emphasize lateral variability.

Formal rock units were established by AGIP geologists (Pieri, 1961; Lucchetti et al., 1962) in Emilia, by Ruggieri (1958, 1970) and Ricci Lucchi (1964, 1967, in Cremonini & Ricci Lucchi, 1982) in the Marecchia area. See Sestini (1970) for a review in English.

The formational names are here maintained but not always in their original meaning: some boundaries (e.g. Antognola–Bismantova) are ill-defined, and uncertainty exists in their position. Some revisions have already been made by Modena and Parma geologists (Bettelli & Bonazzi, 1979; Annovi, 1980; Bonazzi & Panini, 1981; Fornaciari, 1982) and are extended in this paper on the basis of lithologic markers, consistency with sedimentary trends of foredeep and Marecchia areas and sandstone petrology (Zuffa, 1969; Gazzi & Zuffa, 1970). The objective was to identify isochronous intervals (sechrons, see Vail et al., 1977) in conformable sections, and unconformities correlatable with sechron boundaries. A more substantial and thorough stratigraphic redefinition of these sediments is in the offing. This paper only starts it.

The succession basically consists of a vertical repetition of arenaceous and pelitic units. Changes of lithology in clastic sediments are often time-transgressive and cannot be relied upon as time lines; on the other hand, unconformities can be present *within* clastic bodies (erosional surfaces, angular contacts, wedging, etc.). Therefore we adopted the following method: use the condensed facies as time markers, and regard vertical lithofacies changes and every geometrical deviation from parallelism as a *suspect* or candidate unconformity.

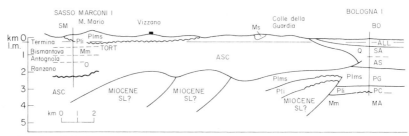

Fig. 28. Interpreted section (AGIP) of foredeep-allochthon relationships SW of Bologna, showing the stacked sequences of satellite basins crossed by Sasso Marconi 1 well. Some symbols as in Figs 11 and 14; others refer to informal lithostratigraphic units established by AGIP in the foredeep: ALL, alluvium; SA, Santerno Clays; AS, Asti Sands; PG, Porto Garibaldi Formation; PC, Porto Corsini Formation. After Lucchetti et al. (1962).

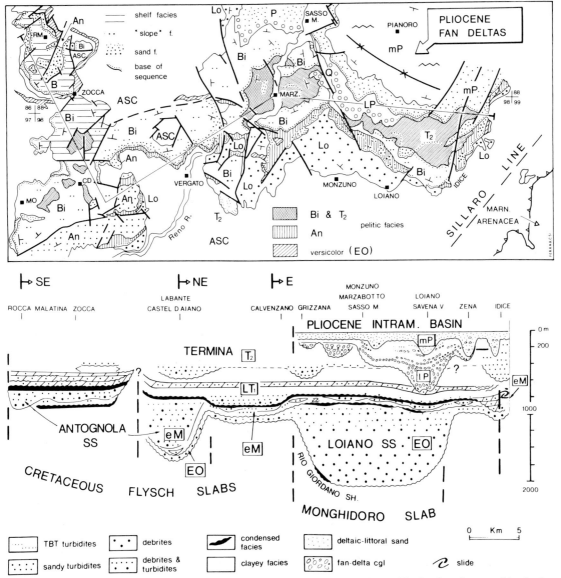

Fig. 29. Map and strike cross-section of satellite basin fills in the vicinities of Bologna. The flysch-molasse partition is also clearcut in these sequences. Molasse deposits, resembling in this case the classical ones (Alpine), comprise the Pliocene intramontane basin. After Lipparini (1966), Cremonini & Elmi (1971), Bettelli & Bonazzi (1979), Annovi (1980), Ricci Lucchi *et al.* (1981) and unpublished data.

An example will illustrate the case: the redefinition of Bismantova boundaries.

A condensed facies occurs at the top of the Antognola Marl: it is known as cherty marl, and referred to here as a *siliceous horizon*, for its abundant silica content (chert levels, radiolaria, volcanic glass). This horizon encompasses the Oligo–Miocene and is perfectly correlatable with the Bisciaro of the foreland

ramp and similar facies around the western Mediterranean (Guerrera, 1979; Lorenz, 1984). The siliceous horizon invariably separates pre-Miocene siliciclastic sandstones from overlying calcareous sandstones (more than 40% $CaCO_3$ in this area), a fact not adequately appreciated so far. This complementary criterion (appearance of hybrid clastics) can therefore be used where the siliceous horizon is lacking for

Table 3. Lithostratigraphy of epi-Ligurian sequences

	Emilia	Marecchia	Depositional sequences
Pliocene: no formal units: see 'parautochthous' or 'semi-allochthonous' Pliocene, 'intra-Apenninic' Pliocene			mP* 1P†
Erosional surface			
Messinian:		Gessoso Formation	
Tortonian:	Termina Marl	Ca Gessi Clay Acquaviva Conglomerate	T2
Erosional surface			
Langhian– Tortonian:	Bismantova	Fumaiolo and Montebello Clay San Marino	LT1
Erosional surface			
Upper Oligocene– Lower Miocene	Antognola Marl	Campaolo contact not clear	eM
Eocene–Oligocene	Ranzano (W) and Loiano (E) Montepiano Marl	Senario	EO
Basal unconformity			
Slabs of Cretaceous (Ligurian proper) or Eocene (Sicilide) flysch units			

* mP = UPP in Ricci Lucchi *et al.* (1981)
† lP = LP in Ricci Lucchi *et al.* (1981)

sedimentary or tectonic reasons. Higher up in the succession, the carbonate content decreases again everywhere (see T1). A simple rule is thus available for defining the Bismantova unit as a depositional sequence: it is a post-siliceous, carbonate-rich body. It has been applied by putting some sandstone bodies, previously mapped as Bismantova, in under- or overlying units. Notice that, after this little surgery, Bismantova is not only slimmer and less hummocky but also more obviously consanguineous with inner MA.

Other revisions will be mentioned below; they still await a biostratigraphic verification. Uncertainty exists, in particular, about the position of L/S, S/T, and T1/T2 boundaries in pelitic, apparently conformable sections.

Geologic setting

Structural and depositional axes are less parallel and regular, in Oligo–Miocene sequences of the flysch stage, than in coeval foredeep wedges. The palaeocurrent pattern is almost unknown. More obvious are the discordant attitude, the transversal transport direction, and the NW–SE synclinal axis of the following molassic sediments (Pliocene).

The depocentral locations were seemingly controlled by transversal structures, some of which very persistent or repeatedly reactivated. The almost coincident boundaries of Eo–Oligocene and Pliocene sequences in a WE direction (Fig. 29) is striking (the eastern limit is the Sillaro line). In spite of complexities, the subsidence pattern was more organized and systematic than one would expect on a plastic substratum subject to movements with a high erratic component like gravity sliding. Either the LIG complex had a relatively coherent behaviour or passively reflected a control exerted by deeper 'autochthonous' structures. A kinematic model that explains the history of epi-LIG basins is still lacking. This being the case, we do not know whether subsidence proceeded during displacement or during standstills, or what kind of movement, vertical or horizontal, produced the unconformities. As subsiding periods can be parallelized with those of the foredeep, we will assume that most sedimentation occurred during pauses of tectonic transport. As for unconformities, they are attributed for the moment to generic 'phases' or 'events'.

Concerning allochthony, Merla (1951, 1959) pointed out that the lowermost units (Eo–Oligocene) are quite exotic to the Padan foredeep domain, while younger ones are more similar and eventually attached to foreland sediments. He saw this as a proof of decreasing allochthony, in a framework of continuous deposition and continuous displacement, with a final molassic seal of Plio–Quaternary deposits. By realizing that there are many such seals in the succession, we are looking at a peculiar, 'multiple' molasse, where unconformities imply both vertical and horizontal movements. The problem is: what was the size of displacement in each 'leap', was it always the same or was it variable, and how can we assess it?

The problem is at present under study, under the stimulus and the assumption that epi-LIG deposits can provide a decisive key for correlating events whose record is more masked in other domains. The basic point is, that sequences preserved their stratigraphical relationships in the epi-LIG, whereas subsidence migrated and tectonic deformation was more severe in the foredeep. Provenance and dispersal data are badly needed, as well as refined facies and stratigraphical analysis.

Sediment supply

In their comprehensive study of Palaeogene epi-Ligurian sandstones, Gazzi & Zuffa (1970) concluded that the variability of composition does not reflect diagenetic changes but differences of provenance. They inferred a 'segmented' situation of source areas and basins, with lateral sills apt to differentiate the various systems. This is quite in agreement with the evidence discussed in this paper. Lateral transport from the SW provided a major contribution in this case, in contrast with the foredeep. This point reinforces the idea of the epi-Ligurian basins as traps dammed to the NE by thrusts. LIG rocks themselves fed the satellite basins, indicating that a hinge line existed between an eroded part and a subsiding part of the LIG sheet. The hinge persisted there until the end of the Miocene, then shifted at the front of LIG, probably as an effect of a change in the mechanism of LIG displacement.

In the Oligo–Miocene, a detrital mode of Alpine provenance is also present; this longitudinal input is common to the coeval foredeep.

Sequence EO (Loiano)

Sequence EO is represented by discontinuous basal shales (Rio Giordano; Abbate, 1969, in Bonazzi & Panini 1981), a major sand wedge (Loiano Sandstone), and upper versicolour shales ('redbeds' of Greig, 1937, in Lipparini, 1966). The shaly facies are interpreted

as condensed horizons of hemipelgic character and are restricted to topographic highs and to the western margin of the basin (where a sill separated Loiano from Ranzano, a sandstone of different composition). The upper shales are correlated with the 'versicolour shales' of the Tuscan sequence (base of Cervarola); they would have formed a single 'seal' on LIG and foreland thrusts during and after the late Oligocene tectonic phase.

The Loiano resedimented sand is of particular interest to sedimentologists and petroleum geologists. It has peculiar characteristics: compositionally mature (arkose), texturally immature and sedimentologically disorganized. Most beds look like ill-sorted mass flow deposits (facies A1 and A2). Where large clasts are abundant, one would call them typical debris flows. The problem is that the matrix content is quite low. It is therefore difficult to find a suitable genetic mechanism: grain flow, liquefied flow, inertial flow, sandy debris flows are possible names, but no one enjoys general acceptance. These terms, in fact, are not well defined in either a theoretical or experimental way (G. V. Middleton, pers. comm.). We use here *mass flow*, with specifications about grain size.

Internal organization and bedding are poorly defined. Pelitic interbeds are lacking, except for the upper part, where thinner sandstone beds, grading and Bouma sequences (mostly truncated: T_{ab}, T_{ac}) also occur. Lateral thickness changes are dramatic, from more than 1000 m to less than 100 m. No shallow-water features are found, contrary to some claims. In the western part of the area, the thickness is reduced but the facies are mostly conglomeratic and disorganized. Loiano fills irregularities in the LIG substratum, resembling a lag deposit. Considering the overlying thick lens of Antognola sandstone, the presence of a valley could be hypothesized in this area. The depression, however, could have resulted from subsequent subsidence.

Loiano palaeocurrents are virtually unknown. Bonazzi & Panini (1971) quote some NE-dipping laminae and contradictory sole markings. The provenance from a granite–gneissic source has already been mentioned. The pebble fraction also contains a local, substratal mode (angular clasts of LIG limestone and sandstone).

In summary, Loiano is a strongly lensoid body (shape and length in NE direction not defined yet) resembling a channel fill with a subtle fining-up trend. It could be called a marginal or proximal deposit, but not in the sense of a shelf or nearshore setting. Source area and basin were close to each other and probably separated by a narrow shelf and a steep slope (diad margin of Ricci Lucchi, 1985a). The depositional system could be the subaqueous portion of a gravity-dominated fan delta (truncated type of Wescott & Ethridge, 1980).

Sequence eM (Antognola)

With few exceptions the base of eM is considered transitional. The significance of condensed facies, which are normally diastemic, has been underestimated. Local channels truncate the boundary but were likely cut during eM ('Anconella sandstone'). What event precisely caused the interruption of Loiano sedimentation remains to be ascertained. The closure of Loiano is correlated with that of the Macigno Basin.

Antognola includes a pelitic facies and a sandy facies. The first is the classical one. It transgresses the boundaries of the Loiano Basin, and consists of hemipelagic mudstones with internal unconformities (slumps, slump scars), olistostromes, and thin to medium bedded sandy turbidites. It looks much like the slope and 'closure' facies associated with the foredeep (e.g. Verghereto). The sandstone facies is markedly lenticular, like Loiano, but more distributed (small discrete bodies with only one major depocentre to the west). Most lenses have an erosional base and could be called channels. It is not clear, however, if they constituted a distributive system of slope channels or only bottom irregularities enhanced by scouring and possibly related to differential compaction or tectonic mobility.

The carbonate content of the Antognola sandstone is very low; sand composition has not been studied in the Bologna area. By analogy with the outcrops of the type area (Enza valley and Carpineti; see Fornaciari, 1982), it is here suggested that part of the sand derived from cannibalization of Loiano (especially in disorganized beds) and partly from a fresh alpine supply (Zuffa, 1969).

Sequence eM is terminated by the siliceous horizon (which is possibly entitled to form a separate sequence). In spite of its supra-regional occurrence, this remarkable level is locally discontinuous and displays abrupt thickness changes and huge slumps (see the eastern end of the cross-section, Fig. 29), a clear indication of mobile topography and resedimentation.

The closure of sequence eM is correlated here with the remobilization of LIG that overrode the Cervarola Basin. The siliceous horizon should correspond to the lower Vicchio Marl, the closure facies of Cervarola.

Sequence LT1 (Bismantova)

This unit is a composite one transgressing the LIG substratum in both this area and in Marecchia. The LIG complex was uplifted by the tectonic phase that closed eM. The following transgression, however, seems unrelated to tectonism, not only because of the good correlation with the Vail chart, but also for intrinsic reasons. Consider the characterization of LT1 with respect to the other sequences: a finer texture, a high amount of intrabasinal material and carbonate, an abundance of glauconite, a more regular shape (almost tabular), a very low sedimentation rate ($2–5$ cm ka^{-1} !), the presence of shelf facies (western and eastern parts of the area in Fig. 29) and the occurrence of cyclothems. Except for the basal units, tectonic tilting and differential subsidence were at a minimum with respect to other sequences: only 130 m of turbidites were accumulated in the deepest local sand trap (Calvenzano).

The most widespread and uniform facies is a massive marl, fossiliferous and pervasively bioturbated, rich in silt and fine sand; the carbonate content is variable on a local basis, with local names indicating more (Schlier) and less (Pantano) calcareous varieties. Faint traces of wavy and hummocky bedding are preserved at places. A shallow basin, an outer shelf or a low-gradient slope could account for the depositional setting. The marly facies may have occupied an intermediate position between the basin margin and its centre. It is *intercalated* in its upper part by irregular surfaces (slump scars?) or grades up into a *calcarenite sheet* made of structureless, graded, and graded-to-laminated beds (up to 2 m thick) alternating with marls and siltstones. Trace fossils are commonly present. The body is organized in several thickening-up cycles. Biogenic debris are mixed in various proportions with siliciclastics of both local (ophiolitic) and distant (Alpine) provenance. Lenses of intraformational breccia embedded in the marl are also found.

Only reconnaissance work has been carried out on the calcarenite sheet. Graded beds could represent turbidites or storm layers; massive, brecciated and nodular units may be attributed to big storms, slope failure (often frozen at an incipient stage) or seismic-induced deformation. The shape and lateral continuity of the body, and the absence of major disruption and slumping, indicate resedimentation events of widespread occurrence, possibly unrelated to local tectonics. These deposits are thus interpreted as relatively shallow, regressive events, i.e. products of sea-level stand exposing shelf areas to 'cannibalism'. The correlative truncation surfaces, more pronounced near western and eastern highs, are the direct expression of the denudation process.

Correlation of the shelf-slope calcarenitic episode with foredeep sequences is still being tested: possible equivalents are the Contessa megaturbidite or the younger Verghereto mudstones (see MA).

The upper part of LT1 is clayey ('Mazzabotto' or 'Cigarello' facies), probably unconformable and associated with limey shelf olistoliths and thin-bedded turbidites. It seems to mark a sudden deepening.

Sequence LT1 is truncated by recent erosion or a palaeosurface, not yet well dated, which is tentatively correlated with the base of T2. Cycle T2 has its best expression in the Marecchia area.

Sequence T2 (Termina)

Sequence T2 consists of marine pelites and proximal turbidite sandstones. It is poorly studied, both stratigraphically and sedimentologically. The best characterization is a compositional one, showing a dominant, cannibalistic mode. Some beds are made mostly of Bismantova carbonate material, others of feldspar-rich Loiano.

Sandstone bodies are lensoid and patchily distributed (Fig. 29), which suggests a rejuvenated topography, exhumation of older deposits of the stack, and formation of localized sand traps.

As in the foredeep, the clastic influx displays a receding trend and is replaced by pelitic deposition. The parallelism of the sedimentary trends in all palaeogeographic domains during T2 is impressive, and needs an adequate explanation. The only differences consisted in depth of water (shelf versus basin) and local supply; the Termina basin, for instance, continued to trap, like previous ones, sand coming from the inner thrust belt, whereas the MA foredeep received an Alpine influx. The Termina sequence ends with the conformable Messinian gypsum near the piedmont line but is otherwise truncated by an erosional surface forming the base of Pliocene sequences.

Pliocene sequences: lP and mP

The so-called Apenninic 'gulf' represents the last subsident stage on the LIG sheet. More precisely, two stages exist, forming distinct wedges (Figs 29 and 30). The downdip thickening of both suggests the possibility that the basins were in a back-thrust position, and

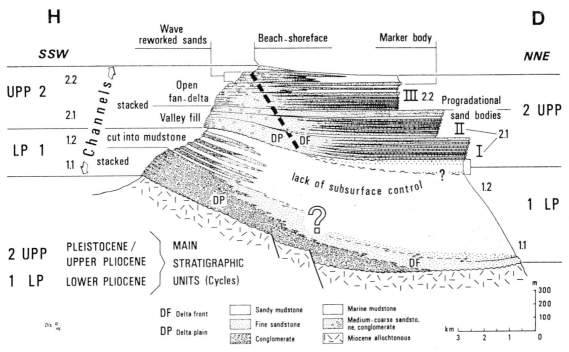

Fig. 30. Dip cross-section of piggyback Pliocene deposits along Sàvena Valley SW of Bologna. Fan delta deposits organized into two systems corresponding to the two sequences lP and mP (here marked UPP). From Ricci Lucchi *et al.* (1981).

were opened by extension acting on previous thrust planes (see La Vecchia, Minelli & Pialli, 1984). All minor bodies within the wedges display the same geometry. Open sea sediments are present only in the lower lP sequence; a clay body, devoid of turbidites, wedges out rapidly southwards and is replaced by nearshore and continental clastics. In the upper sequence (mP), distal deposits are shallow-water muds with abundant evidence of storm reworking (interbedded sands); proximal facies are finer-grained than in lP, and lateral transitions more gradual. These differences indicate that palaeoslopes were steeper in the first cycle.

Both sequences record a transgressive trend abruptly truncated by erosion. The erosional phases, well correlatable with those of the 'autochthonous' foredeep margin, left scars that Ricci Lucchi et al. (1981) interpreted as streamcut valleys. The palaeovalleys were filled by the transgressions, or, better, by rivers that tried to keep pace with a rising sea-level. Many cyclotherms record, in fact, fluctuations of relative sea-level superimposed on the main trend. Nowhere in Italy are these 'motifs' so well exposed and numerous as in this area, and particularly in Sequence mP. Although transgressive in general trend, most clastic sediments were deposited under the influence of torrential streams; they are preserved either as fluvial or wave-reworked deposits, comprehensively designated as fan-delta facies. Volumetrically subordinate are truly marine sediments, in the form of mud drapes that mark the transgressive peaks. During low stands, very likely related to local tectonic uplift, rivers became entrenched in the mud drapes, which favoured meandering. Coarse-grained point bars thus form the most typical marginal deposits of these basins. They are increasingly modified by organisms and wave action proceeding towards Bologna. Good examples of beach structures and longshore bars are exposed in the sandstone cliff that marks the approximate position of the southern shoreline. Littoral facies dominate on the less subsident shoulders to the west and east of the depocentral area (the abrupt thickness changes suggest lateral faults, parallel to the Sillaro line).

Alluvial facies are practically identical to those of modern alluvial fans of the same area; the width of palaeovalleys is also similar. It is concluded that the hydrology and sediment load of Pliocene streams were the same as today. The fan deltas, therefore, fall in the category of humid-region systems. Furthermore, they were built in a shallow basin there were no steep submarine slopes, and no foreset beds developed.

They are not comparable with systems of Gilbert type, which are dominated by gravity processes.

In summary, epi-Ligurian sequences show both deep-water, resedimented facies, including peculiar proximal sands, and shallow-water to fluvial clastics. In between, the unique Bismantova facies record a strong eustatic imprint on regional and local tectonic control.

Depocentres swung from east to west between Oligocene and Miocene, then back to east in late Miocene and Pliocene. Overall, the greatest subsidence occurred to the east, close to the Sillaro transversal line. This pattern, along with the presence of shelf facies only in the western part of the area, suggests a depositional asymmetry in an east–west direction ('Apenninic', or longitudinal).

ACKNOWLEDGMENTS

The author is grateful to G. G. Ori, L. Buldini, A. Argnani and M. Roveri for providing field data, to R. Barbieri and M. L. Colalongo for age determinations, to P. Ferraresi for drawings and S. Galli for typing. I express particular thanks to my colleagues involved in RCMNS team work: M. Boccaletti, R. Gelati, L. Dondi, M. Gnaccolini, E. Centamore, G. Deiana and F. Massari; their ideas and date were influential even though not always quoted.

A. W. Bally revised a previous draft and gave useful suggestions for its improvement.

REFERENCES

ABBATE, E. & SAGRI, M. (1984) Le unita torbiditiche cretacee dell'Appennino settentrionale ed i margini continentali della Tetide. Mem. Soc. geol. Ital. 24, 115–126.

ANNOVI, A. (1980) La Geologia del Territorio di Montese (Appennino Modenese). Mem. Sci. Geol. 34, 67–84, Padova.

BALLY, A.W. & OLDOW, J.S. (1984) Structural styles and the evolution of sedimentary basins. Short Course Notes, AAPG Fossil Fuels of Europe Conference, Genève, 238 pp.

BALLY, A.W. (ed.) (1984) Seismic expression of structural styles. A picture and work atlas. AAPG Stud. Geol. Ser. 15, 3.

BALLY, A.W. & SNELSON, S. (1980) Realms of subsidence. In: Facts and Principles of World Petroleum Occurrence (Ed. by A. D. Miall). Mem. Can. Soc. Petrol. Geol. 6, 9–94.

BETTELLI, G. & BONAZZI, U. (1979) La Geologia del teritorio tra Guiglia e Zocca (Appennino Modenese). Mem. Sci. Geol. 32, 24 pp., Padova.

BOCCALETTI, M., CONEDERA, C., DECANDIA, F.A., GIANNINI, E. & LAZZAROTTO, A. (1980) Evoluzione del l'Appennino settentrionale secondo un nuovo modello strutturale. Mem. Soc. geol. Ital. 21, 358–373.

BOCCALETTI, M., CONEDERA, C., DAINELLI, P. & GOCEV, P. (1982) The Recent (Miocene–Quaternary) regmatic system of the western Mediterranean region. *J. Petrol. Geol.* **5**, 31–49.

BONAZZI, U. & PANINI, F. (1981) Lineamenti geologici della zona a nord di Grizzana tra il F. Reno e il T. Setta (Appennino bolognese). *Atti Soc. Nat. Mat. Modena*, **112**, 1–19.

BRUNI, P. (1974) Considerazioni tettoniche e palaeogeografiche sulle serie del l'Appennino Bolognese tra le valli del l'Idice e del Santerno. *Mem. Soc. geol. Ital.* **12**, 157–185.

CARON, C., HESSE, R., KERCHOVE, C., HOMEWOOD, P., VAN STUJIVENBERG, J., TASSÉ, N. & WINKLER, W. (1981) Comparaison préliminaire des flyschs à Helmintoides sur trois transversales des Alpes. *Eclog. geol. Helv.* **74**, 369–378.

CASNEDI, R., MORUZZI, G. & MUTTI, E. (1978) Correlazioni elettriche di lobi deposizionali torbiditici nel sottosuolo abruzzese. *Mem. Soc. geol. Ital.* **18**, 23–30.

CASTELLARIN, A., EVA, C., GIGLIA, G. & VAI, G.B. (1986) Analisi strutturale del fronte Appenninico-Padano. *Giornale Geol.* Bologna.

CIPRIANI, C. & MALESANI, P.G. (1963) Ricerche sulle arenarie. IX: caratterizzazione e distribuzione geografica delle arenarie appenniniche oligoceniche e mioceniche. *Mem. Soc. geol. Ital.* **4**, 339–375.

COLALONGO, M.L., CREMONINI, G., FARABEGOLI, E., SARTORI, R., TAMPIERI, R. & TOMADIN, L. (1978) Evoluzione palaeoambientale della Formazione a Colombacci in Romagna. *Rend. Soc. geol. Ital.* **1**, 37–40.

COLELLA, A., DI GERONIMO, I., D'ONOFRIO, S., FORLANI, L., LOLLI, F. & CASALE, V. (1984) Sedimentologia, stratigrafia ed ecologia dei depositi superficiali della conoide lotto marina del Crati. *Giornale Geol.* **45**, 251–284.

CREMONINI, G. & ELMI, C. (1971) *Note illustrative della Carta Geologica d'Italia. Foglio 98 Vergato.* Servizio Geologico d'Italia, Roma.

CREMONINI, G. & FARABEGOLI, E. (1981) *Carta Geologica 1:25.000 della Regione Emilia-Romagna, tavv. Cusercoli-Borello (con note illustrative)*, 25 pp., Pitagora, Bologna.

CREMONINI, G., FARABEGOLI, E., MICCOLI, C. & PIERI, L. (1982) *Carta Geologica del l'Appennino emiliano-romagnolo, tavv.* Predappio e Bertinoro, Firenze.

CREMONINI, G. & RICCI LUCCHI, F. (eds) (1982) Guida alla geologia del margine appenninico-padano. *Guide Geol. Reg. Soc. geol. Ital.* Bologna, 248 pp.

CRESCENTI, U., D'AMATO, C., BALDUZZI, A. & TONNA, M. (1980) Il Plio-Pleistocene del sottosuolo abruzzese-marchigiano tra Ascoli Piceno e Pescara. *Geol. Rom.* **19**, 63–84.

DALLAN NARDI, L. & NARDI, R. (1974) Schema stratigrafico e strutturale dell'Appennino settentrionale. *Mem. Acc. Lunigianese Sci. 'G. Capellini'*, **42**, 1–212.

DONDI, L., RIZZINI, A. & ROSSI, P. (1982) Quaternary sediments of the Adriatic Sea from Po Delta to Gargano promontory. Manuscript, Advanced Research Institute, *Medit. Basin Conf.*, 14 pp.

DZULYNSKI, S. & SMITH, A.J. (1964) Flysch facies. *Ann. Soc. Geol. Pol.* **34**, 245–266.

ELTER, P. & SCANDONE, P. (1980) Les Apennins. In: *Geology of the Alpine Chains Born of the Tethys* (Ed. by J. Auboin, J. Debelmas & M. Latreille). *26th Int. Geol. Congr., Coll C5, Mem. B.R.G.M.* **115**, 99–102.

FARABEGOLI, E. (1983) *Note illustrative alla Carta Geologica della Regione Emilia-Romagna, tavv. Cesena e Sogliano sul Rubicone.* Patron Editore Bologna, 39 pp.

DE FEYTER, A.J. & MOLENAAR, N. (1984) Messinian Fanglomerates: the Colombacci Formation in the Pietrarubbia Basin, Italy. *J. sedim. Petrol.* **54**, 746–758.

FORNACIARI, M. (1982) Osservazioni litostratigrafiche sul margine sud-orientale della sinclinale Vetto-Carpineti (Reggio-Emilia). *Rend. Soc. geol. Ital.* **5**, 117–118.

GANDOLFI, G., PAGANELLI, L. & ZUFFA, G.G. (1981) Provenance and detrital-mode dispersal in Marnoso Arenacea Basin (Miocene, northern Apennines). *2nd Eur. Reg. Meeting, I.A.S., Bologna Abstr. vol.* pp. 65–68.

GAZZI, P. & ZUFFA, G.G. (1970) Le arenarie palaeogeniche del l'Appennino emiliano. *Miner. Petrol. Acta*, **16**, 97–137.

GHIBAUDO, G. (1980) Deep-sea fan deposits in the Macigno Formation (Middle-Upper Oligocene) of the Gordana Valley, northern Apennines, Italy. *J. sedim. Petrol.* **50**, 722–741.

GUERRERA, F. (1979) Stratigrafia e sedimentologia dei livelli tripolacei del Miocene inferiore-medio appenninico. *Boll. Serv. geol. Ital.* **94**, 233–262.

TEN HAAF, E. (1964) Flysch formations of the northern Apennines. In: *Turbidites* (Ed. by A. H. Bouma & A. Brouwers), pp. 127–136.

TEN HAAF, E. & VAN WAMEL, W.A. (1979) Nappes of the Alta Romagna. *Geologie Mijnb.* **58**, 145–152.

HOMEWOOD, P. (1983) Palaeogeography of Alpine flysch. *Palaeogeogr. Palaeoclim. Palaeoecol.* **44**, 169–184.

HOMEWOOD, P. & CARON, C. (1982) Flysch of the western Alps. In: *Mountain Building Processes* (Ed. by K. J. Hsü), pp. 157–168. Academic Press, London.

HSÜ, K.J. (1970) The meaning of the word Flysch—a short historical research. *Spec. Pap. geol. Ass. Can.* **7**, 1–11.

HSÜ, K.J., CITA, M.B. & RYAN, W.B.F. (1973) The origin of the Mediterranean Evaporite. In: *Init. Rep. D.S.D.P.* (Ed. by W. B. F. Ryan, K. J. Hsü *et al.*), **13**, 1203–1231.

DE JAGER J. (1979) The relation between tectonics and sedimentation along the 'Sillaro Line' (northern Apennines, Italy). *Geol. Ultraiectina.* **19**, 1–98.

KLIGFIELD, R. (1979) The northern Apennines as a collisional orogen. *Am. J. Sci.* **279**, 676–691.

LA VECCHIA, G., MINELLI, G. & PIALLI, G. (1984) L'Appennino Umbro-Marchigiano: tettonica distensiva e ipotesi di sismogenesi. *Boll. Soc. geol. Ital.* **103**, 467–483.

LETOUZEY, J. & TREMOLIERS, P. (1980) Palaeo-stress field around the Mediterranean since the Mesozoic derived from microtectonics: comparison with plate tectonic data. *Mem. B.R.G.M.* **15**, 261–273.

LIPPARINI, L. (1966) *Note illustrative della Carta Geologica d'Italia. Foglio 87 Bologna.* Servizio Geologico d'Italia, Roma.

LORENZ, C. (1984) Les silexites et les tuffites du Burdigalien, marqueurs volcano-sédim. entaires—corrélations dans le domaine de la Méditerranée occidentale. *Bull. Soc. géol. Fr.* **26**, 1203–1210.

LUCCHETTI, L., ALBERTELLI, L., MAZZEI, R., THIEME, R., BIONGIORNI, D. & DONDI, L. (1962) Contributo alle conoscenze geologiche del Pedeappennino padano. *Boll. Soc. geol. Ital.* **81**, 5–245.

MERLA, G. (1951) Geologia dell'Appennino settentrionale. *Boll. Soc. geol. Ital.* **70**, 95–382.

MERLA, G. (1959) Essay on the Geology of the northern Apennines. In: *Atti del Convegno sui Giacimenti Gassiferi dell'Europa Occidentale* (Ed. by Acc. Naz. Lincei), Milano.

MEZZETTI, R., MORANDI, N. & PINI, G.A. (1980) Studio mineralogico delle porzioni pelitiche nelle 'Marne di Antognola' della zona di Zocca (Modena). *Miner. Petrol. Acta*, **24**, 57–75.

MITCHUM, R. M., VAIL, P. R. & THOMPSON, S., III (1977) The depositional sequence as a basic unit for stratigraphic analysis. In: *Seismic Stratigraphy—Applications to Hydrocarbon Exploration* (Ed. by C. E. Payton). *Mem. Am. Ass. Petrol. Geol.* **26**, 63–82.

MONTADERT, L., LETOUZEY, J. & MAUFFRET, A. (1978) Messinian Event; seismic evidence. In: *Init. Rep. D.S.D.P.* **42**, 1037–1050.

MUTTI, E. (1977) Distinctive thin-bedded turbidite facies and related depositional environments in the Eocene Hecho Group (South-central Pyrenees, Spain). *Sedimentology*, **24**, 107–131.

MUTTI, E. (1979) Turbidites et cones sous-marins profonds. In: *Sédimentation Détritique* (Ed. by P. Homewood), **1**, 353–419. Fribourg.

MUTTI, E. (1985) Turbidite systems and their relations to depositional sequences. In: *Provenance of Arenites* (Ed. by G. G. Zuffa), pp. 65–94.

MUTTI, E., PAREA, G.C., RICCI LUCCHI, F., SAGRI, M., ZANZUCCHI, G., GHIBAUDO, G. & IACCARINO, S. (1975) Examples of turbidite facies and facies associations from selected formations of the northern Apennines. *9th Int. Congr. Sedim., Exc. Guidebook, Field Trip A-11, Modena*, 120 pp.

MUTTI, E. & RICCI LUCCHI, F. (1972) Le torbiditi del l'Appennino settentrionale: introduzione all'analisi di facies. *Mem. Soc. geol. Ital.* **11**, 161–199.

MUTTI, E. & RICCI LUCCHI, F. (1975) Turbidite facies and facies associations. *9th Int. Congr. Sedim., Exc. Guidebook, Field Trip A-11, Modena*, pp. 21–36.

MUTTI, E., NILSEN, T.H. & RICCI LUCCHI, F. (1978) Outer fan depositional lobes of the Laga Formation (Upper Miocene and Lower Pliocene), east-central Italy. In: *Sedimentation in Submarine Canyons, Fans and Trenches* (Ed. by D. J. Stanley & G. Kelling), pp. 210–223. Dowden, Hutchinson & Ross, Stroudsburg.

MUTTI, E., RICCI LUCCHI, F., SÉGURET, M. & ZANZUCCHI, G. (1984) Seismoturbidites: a new group of resedimented deposits. In: *Seismicity and Sedimentation* (Ed. by M. B. Cita & F. Ricci Lucchi). *Mar. Geol.* **55**, 103–116.

NORMARK, W.R., MUTTI, E. & BOUMA, A.H. (1984) Problems in turbidites research: a need for COMFAN. *Geo-Mar. Lett.* **3**, 53–56.

ORI, G.G. & FRIEND, P.F. (1984) Sedimentary basins formed and carried piggyback on active thrust sheets. *Geology*, **12**, 475–478.

ORI, G.G., ROVERI, M. & VANNONI, F. (1986) Plio-Pleistocene sedimentation in the Apenninic foredeep (central Adriatic Sea, Italy) (this volume).

PIERI, M. (1961) Nota introduttiva al rilevamento del versante appenninico padano eseguito nel 1955–59 dai geologi del l'AGIP Mineraria. *Boll. Soc. geol. Ital.* **80**, 3–34.

PIERI, M. & GROPPI, G. (1981) Subsurface geological structure of the Po Plain, Italy. *P.F. Geodinamica, C.N.R.* **414**, 23 pp.

RABBI, E. & RICCI LUCCHI, F. (1968) Stratigrafia e sedimentologia del Messiniano forlivese. *Giornale Geol.* **34**, 595–640.

REUTTER, K.J. (1981) A trench-forearc model for the northern Apennines. In: *Sedimentary Basins of Mediterranean Margins* (Ed. by F. C. Wezel). Tecnoprint, Bologna. pp. 433–444.

RICCI LUCCHI, F. (1975a) Depositional cycles in two turbidite formations of northern Apennines. *J. sedim Petrol.* **45**, 1–43.

RICCI LUCCHI, F. (1975b) Miocene palaeogeography and basin analysis in the Periadriatic Apennines. In: *Geology of Italy* (Ed. by C. Squyres), **2**, 129–236. PESL, Castelfranco-Tripoli.

RICCI LUCCHI, F. (1981) The Marnoso-arenacea turbidites, Romagna and Umbria Apennines. In: *Excursion Guidebook, with Contribution on Sedimentology of some Italian Basins* (Ed. by F. Ricci Lucchi), pp. 229–303. 2nd IAS European Meeting, Bologna.

RICCI LUCCHI, F. (1984) Deep-sea fan deposits on the Miocene Marnoso-arenacea Formation, northern Apennines. *Geo-Mar. Lett.* **3**, 203–210.

RICCI LUCCHI, F. (1985) Influences of transport processes and basin geometry on sand composition. In: *Provenance of Arenites* (Ed. by G. G. Zuffa), pp. 19–48.

RICCI LUCCHI, F., COLELLA, A., GABBIANELLI, G., ROSSI, S. & NORMARK, W.R. (1984) The Crati submarine fan, Ionian Sea. *Geo-Mar. Lett.* **3**, 71–77.

RICCI LUCCHI, F., COLELLA, A., ORI, G.G., OGLIANI, F. & COLALONGO, M.L. (1981) Pliocene fan deltas of the Intra-apenninic Basin, Bologna. In: *Excursion Guidebook, with Contribution on Sedimentology of some Italian Basins* (Ed. by F. Ricci Lucchi), pp. 79–162. 2nd IAS European Meeting, Bologna.

RICCI LUCCHI, F. & ORI, G.G. (1984) Orogenic clastic wedges of the Alps and the Apennines. *Abstract, AAPG Fossil Fuels of Europe Conference*, Geneva. *Bull. Am. Ass. Petrol. Geol.* **68**, 798.

RICCI LUCCHI, F. & PIGNONE, R. (1979) Ricostruzione geometrica parziale di un lobo di conoide sottomarina. *Mem. Soc. geol. Ital.* **18**, 125–133.

RICCI LUCCHI, F. & VALMORI, E. (1980) Basin-wide turbidites in a Miocene over-supplied deep-sea plain: a geometrical analysis. *Sedimentology*, **27**, 241–270.

RUGGIERI, G. (1958) Gli esotici neogenici della colata gravitativa della Val Marecchia. *Atti Acc. Sci. Lett. Arti Palermo*, **4**, 17, 169 pp.

ROYDEN, L. & KARNER, G.D. (1984) Flexure of lithosphere beneath Apennine and Carpathian Foredeep Basins: evidence for an insufficient topographic load. *Bull. Am. Ass. Petrol. Geol.* **68**, 704–712.

RYAN, W.B.F., CITA, M.B., DREYFUS, RAWSON M., BURCKLE, L.H. & SAITO, T. (1974) A paleomagnetic assignment of Neogene stage boundaries and the development of isochronous datum planes between the Mediterranean, the Pacific and Indian oceans in order to investigate the response of the world ocean to the Mediterranean 'salinity crisis'. *Riv. Ital. Paleont. Stratig.* **80**, 631–688.

SCANDONE, P. (1980) Origin of the Tyrrhenian Sea and Calabrian Arc. *Boll. Soc. geol. Ital.* **98**, 27–34.

SESTINI, G. (1970) Flysch facies and turbidite sedimentology. In: *Development of the Northern Apennines Geosyncline* (Ed. by G. Sestini). *Sedim. Geol.* **4**, 559–597.

VAIL, P. R., MITCHUM, R. M. & THOMPSON, S., III (1977) Global cycles of relative changes of sea level. In: *Seismic Stratigraphy—Application to Hydrocarbon Exploration* (Ed. by C. E. Payton). *Mem. Am. Ass. Petrol. Geol.* **26**, 83–98.

VALLONI, R. & ZUFFA, G.G. (1984) Provenance changes for arenaceous formations of the northern Apennines, Italy. *Bull. geol. Soc. Am.* **95**, 1035–1039.

WESCOTT, W.A. & ETHRIDGE, F.G. (1980) Fan-delta sedimentology and tectonic setting-Yallahs fan delta, southeast Jamaica. *Bull. Am. Ass. Petrol. Geol.* **64**, 374–399.

ZUFFA, G.G. (1969) Arenarie e calcari arenacei miocenici di Vetto-Carpineti (Formazione di Bismantova, Appennino settentrionale). *Miner. Petrol. Acta,* **15**, 191–219.

ZUFFA, G.G. (1980) Hybrid arenities—their composition and classification. *J. sedim. Petrol.* **50**, 21–29.

Spec. Publs int. Ass. Sediment. (1986) **8**, 141–168

A small polyhistory foreland basin evolving in a context of oblique convergence: the Venetian basin (Chattian to Recent, Southern Alps, Italy)

F. MASSARI*, P. GRANDESSO*, C. STEFANI* *and* P. G. JOBSTRAIBIZER†

*Istituto di Geologia dell'Universita', Via Giotto, 1 Padova, Italy; †Istituto di Mineralogia dell'Universita', Corso Garibaldi, 37 Padova, Italy

ABSTRACT

The Venetian basin evolved in two distinct stages: (a) from the Chattian to the Langhian it behaved as a foreland basin of the Dinaric range and evolved under rather weak tectonic control, due to low rates of thrust propagation. Positive features related to the growth or rejuvenation of structural highs acted as a focus for the localization of shelf sand ridges during transgressive stages, while major regressive stages are marked by progradation of significantly diachronous sandstone bodies (during the late Aquitanian and the late Burdigalian) and by evidence of partial basin isolation and faunal endemism. Such events can be correlated with major discrete pulses of thrust activity in the outer Dinarides. From Chattian to Langhian an important 'external' source of clastics was represented by the axial zone of the Alpine range (Austroalpine and Penninic units) subjected to rapid uplift and denudation; (b) from Serravallian onwards a drastic change in the tectonic framework of the Venetian basin occurred: the axis of subsidence shifted in position and trend, and the basin was thereafter incorporated in the South Alpine kinematic system. The latter evolved in a context of oblique convergence which resulted in a migration of thrusting and of foreland subsidence around the South Alpine compressional belt from the west (Lombard foredeep) toward the east (Venetian foredeep) during the time span from late Oligocene to Recent. This shifting pattern was accommodated by wrenching along the Insubric (Periadriatic) lineament and along sets of conjugate transverse faults which segmented the Southern Alps in a number of sub-areas with significantly diachronous tectono-sedimentary evolution. On a larger scale, the South Alpine and Apenninic domains show a close correspondence in timing of major events and in stepwise eastward shifting of deformation and foredeep subsidence, suggesting an evolution within the same megashear system.

From Serravallian onwards an imbricate stack of overthrusts rapidly 'prograded' SSE-wards in the eastern South Alpine area and shed a huge amount of clastics into the Venetian basin. Unconformity-bounded sequences developed in response to discrete pulses of thrust progradation, with localized angular unconformities and along-strike differential subsidence resulting from the interplay of thrust pulses with strike-slip motion along transverse faults.

Although tectonic control is pervasive, especially in the second stage of basin evolution, the effect of major eustatic events can still be recognized in the trend of sedimentation.

INTRODUCTION

Foreland basin analysis is commonly carried out within a cylindrical perspective, which may be the correct approach in the case of frontal convergence but may be misleading in the context of oblique convergence. In this case, diachronism of thrusting along the orogenic belt, widespread shear effects expressed by conjugate sets of strike-slip faults segmenting the foreland into compartments with diachronous evolution, and migration of foreland basin depocentres around the compressional belt would be expected.

Cases of oblique convergence are fairly common in the Mediterranean ranges during the post-collisional stage; in fact in this area the collisional event produced

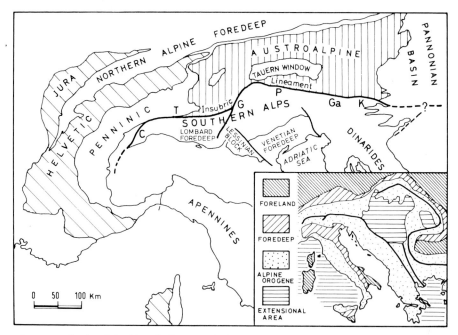

Fig. 1. The main structural units of the Alps. C: Canavese fault; T: Tonale fault; G: Giudicarie fault; P: Pusteria fault; Ga: Gail fault; K: Karavanken fault. Insert: simplified tectonic map of the Alpine, peri-Adriatic and Carpathian ranges (from Giese & Nicolich, 1982, slightly modified).

a number of small crustal blocks which later evolved more or less independently from one another during the post-collisional stage and were commonly involved in rotations and oblique convergence.

This paper aims to highlight the distinctive characters of foreland basin evolution in such a context, by describing the history of a small foredeep located in the South Alpine domain, the Venetian basin.

This foreland basin is located in the interference area of three orogenic belts, Dinarides, Alps and Apennines (Fig. 1). Due to the interplay of different orogenic 'polarities' the Venetian basin is a polyhistory basin with a complex evolution.

GEOLOGICAL SETTING

The South Alpine area is part of Adria, regarded by Hsü (1982) as an independent microplate during the Cenozoic. The central part of Adria, mostly submerged beneath the Adriatic sea, can be considered as the foreland of the orogenic belts made up of the Apennines, Southern Alps, Dinarides and Hellenides (Channell & Horváth, 1976).

Perhaps the most significant result of applying plate tectonic concepts is the discovery that most orogenic architecture is the product of post-collisional compression and most clastic sediments infilling the basins in orogenic segments is the product of denudation of the resulting relief. The Southern Alps are an example of a post-collisional range essentially built by Neogene deformations which are an expression of the late stages of convergence between Europe and Adria. They are generally considered the 'autochthonous hinterland' of the Austroalpine nappes from which they are separated by a major fault-system, the so-called Insubric (Periadriatic) lineament, including from the west to the east the Canavese, Tonale, Giudicarie, Pusteria, Gail and Karavanken faults. As pointed out by Bernoulli (in Trümpy, 1980) the geometry of these post-nappe, post-metamorphic, essentially Miocene faults implies large-scale allochthony in the Southern Alps as well (De Sitter & De Sitter Koomans, 1949). Deformation in the latter is mainly by south-verging thrusts (Fig. 2) involving the basement in the north. Décollement thrusts also occur in the south. Crustal shortening in the Southern Alps has been estimated to be at least 100% by Bernoulli (in Trümpy, 1980) but is not uniform, because considerably shortened areas exist in the west and in the east, lateral to a weakly deformed central block (the Lessinian block, Fig. 2). Such strong differences

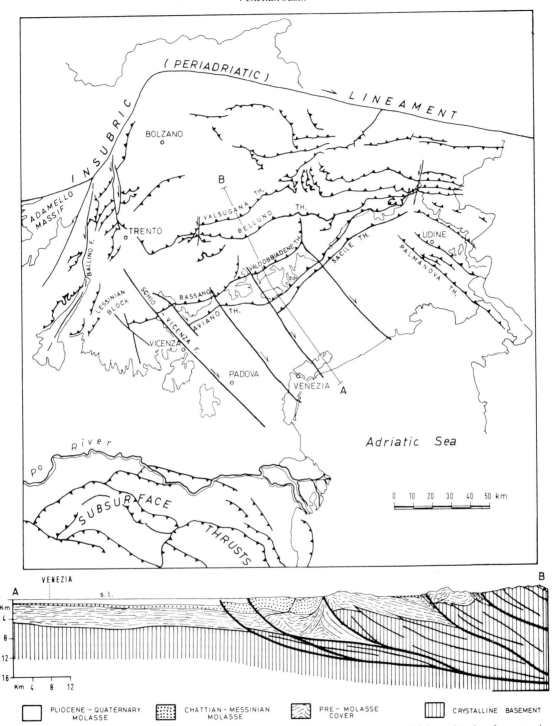

Fig. 2. Simplified structural map and cross-section of the eastern part of the Southern Alps and of the main subsurface tectonic features of the easternmost Po plain (partly from Castellarin, 1981, Pieri & Groppi, 1981, Ambrosetti *et al.*, 1986 and Massari *et al.*, 1986, slightly modified). The Sacile thrust, the western segment of Aviano thrust, the thrusts located in the Udine area and south of the Po river are blind thrusts.

are also reflected by sedimentary evolution since only the lateral areas (Lombard and Venetian areas) contain thick flysch and molasse sequences whereas the central block maintained the character of a structural high, with thin, condensed sedimentation throughout its history. Differences in behaviour may well be predetermined by an ancestral inhomogeneity of the lithosphere, most probably generated (or enhanced) by thinning of the crust during the Jurassic rifting stage. In fact, the South Alpine foredeep, like many foredeeps, is underlain by a sequence of the passive margin type (Bally & Snelson, 1980). An early rifting history during the Jurassic created a system of carbonate platforms and seamounts with intervening basins, more or less elongated on a N–S trend, probably inherited from older structural elements; it should be noted that Neogene South Alpine folds and thrusts are oriented at high angles to this trend. The isopic zones, resulting from the extensional phase can be at least partly traced outside the South Alpine domain in the Apennines and Dinarides (Aubouin, 1973). The birth of the Lessinian high in the central part of the South Alpine domain dates back to this stage; by contrast, lateral to this block, the Lombard and Venetian zones were affected by much stronger subsidence which gave rise to the Lombard and Belluno basins.

In a later stage, upper Cretaceous to lower Tertiary flysch deposits encroached on the Lombard and Venetian basins. Both of them were underlain by continental crust and were located in the rear of the northern margin of Adria which acted as the main source area for clastic sediments.

The dating of the continental collision is still under discussion: it is regarded as late Eo-Alpine by Dal Piaz, Hunziker & Martinotti (1972) and latest Eocene to early Oligocene by Swiss authors (the 'Meso-Alpine' orogeny of Trümpy, 1973). The 'Meso-Alpine' event had important effects even in the South Alpine domain. Faulting, thrusting and shearing are assumed by Gaetani & Jadoul (1979) to have taken place in the Bergamasc Alps (central South Alpine domain) before the deposition of the upper Oligocene–Miocene 'molasse' (Gonfolite) and a folding phase has been documented by Brack (1981) to have taken place before the Adamello intrusion. In the eastern South Alpine domain important pre-Aquitanian overthrusts were emplaced in the Dolomites (Doglioni, 1984) and might well be connected with the late Eocene climax of shortening which affected the Dinarides.

After the 'Meso-Alpine' event a marked stress relaxation effect can be recognized during the Oligo-

cene both in the northern foreland of the Alps (Rhine–Rhône graben system) and in the South Alpine domain ('Tongriano' conglomerates in the fault-bounded troughs of Ligure–Piemontese basin, middle Oligocene basalts of Thiene–Marostica graben, north of Vicenza) as well as along the Insubric belt as far as the Maribor area (Dal Piaz, 1976). The clustering of periadriatic late Alpine intrusions around 30 Ma should be remembered in this regard (Laubscher, 1986). As stressed by Laubscher, no compression was transmitted between the African and European plates at this time. In the eastern South Alpine and Dinaric domains evidence of this extensional phase is more scanty. It might have been responsible for the collapse of the Mesozoic platform in the area of the Venetian plain along a network of NE- and NW-trending faults (thick Oligocene sediments occur on the downthrown blocks south of this collapsed margin). Furthermore the Oligocene Dinaric molasses (Val Tremugna area north of Udine, Sava basin, etc.) seem to be confined within fault-bounded troughs crossing internal and external zones of the orogen irrespectively of the structures formed in the late Eocene main phase of deformation.

Further compressional shortening of the crustal stack from late Oligocene onwards could only be accommodated by consumption of the continental lithosphere. Continued continent–continent convergence led to the formation of a steeply northward dipping subduction zone (Laubscher, 1971, 1974) in addition to the southward-dipping one formed in the Cretaceous (Dal Piaz et al., 1972). According to Panza et al. (1980) and Miller, Müller & Perrier (1982) crustal sections across the Alps show that in a position displaced toward the inner side of the Alpine area, the asthenosphere is crossed by high velocity lithospheric material which reaches to depths of about 250 km and probably corresponds to two slabs of cold lower lithosphere subducted during the plate collision process. Severe shortening in the South Alpine domain can probably be regarded as a consequence of the northward A-type subduction of the South Alpine lithosphere (Castellarin, 1979; Castellarin, Frascari & Vai, 1980; Castellarin & Vai, 1982).

Laubscher (1971) and Laubscher & Bernoulli (1982) note that the flat Miocene thrusts of the Southern Alps and the steep, straight Insubric fault fit the division of roles often observed along obliquely convergent plate boundaries. In addition to a vertical throw along the Insubric line, which is considerable (up to 15 km at least in the west according to Trümpy, 1973) and significantly decreasing eastwards, they point out that oblique-dextral convergence led to significant dextral strike-slip along the line during the late Oligocene–Neogene.

Diachronism in the deformation of the Southern Alps was pointed out by Castellarin & Vai (1981) who showed that the main shortening event in the western area of the Southern Alps was of late Oligocene–early Miocene age, as marked by the coarse-grained clastics of the Lower Gonfolite Group, whereas the eastern area experienced the climax of deformation during the late Miocene and Pliocene. The diachronism, expressed by the eastward growth of regional tectonic elements, was later stressed by Castellarin (1984) who also recognized an intermediate, mostly Tortonian compressional arc, developed close to the Giudicarie line, extending along the southern border of the Lombard Alps.

From late Oligocene to middle Miocene a deep-water clastic sequence ('South Alpine molasse') was deposited in the Lombard foredeep (Gonfolite group, about 3600 m thick according to Günzenhauser, 1985). It unconformably overlies a substratum of variable age. Conglomerates are particularly abundant in the upper Oligocene–Aquitanian interval (with pebbles of South Alpine, Austroalpine and Penninic provenance, as pointed out by Bernoulli, in Trümpy, 1980) and in the Serravallian (Rizzini & Dondi, 1978). Occurrence of clasts of Bregaglia granitoids in the late Oligocene molasse of the Como–Chiasso area seem to imply that the Bregaglia body (dated about 30 Ma.) was emplaced, cooled and unroofed by the erosion of several (5?) km of overlying rock before the late Oligocene. A tremendous rate of uplift in the source area is involved (Trümpy, 1973; Rögl *et al.*, 1975).

Beyond the Lessinian block, unaffected by molasse sedimentation, another foreland basin can be recognized in the eastern part of the South Alpine domain, the Venetian basin. From Chattian to Langhian it may be considered a foreland basin of the Dinaric range (Massari *et al.*, 1986). Late orogenic thrusting occurring in the outer Dinarides is expressed by the progressive outward shifting of the kinematic front towards the Adriatic coast and into the offshore area (Aubouin, 1973; Horváth, 1984). Such deformations were accompanied by late tectonic molasses. According to Miljush (1973) the clastic molasse in the outer Dinarides was deposited mostly during the Oligocene and early–middle Miocene, with a total thickness in the Adriatic basins of up to 4000 m.

An important palaeogeographical and geodynamic change occurred in the middle Miocene. A major phase of extension took place in the Pannonian basin and internal Dinarides and was accompanied by major subcrustal thinning (Royden, Horváth & Burchfiel, 1982; Royden, Horváth & Rumpler, 1983). This extensional phase, coupled with a strong reactivation of dextral shear along the Periadriatic–Vardar transcurrent fault system, was interpreted by Channell & Horváth (1976) as the result of an eastward movement of the Pannonian unit, probably to accommodate the northward continuing convergence of Adria and Europe. The latter culminated in the major late Miocene compressional phase, whose effects are widespread all around the Alps (Laubscher, 1986) and Apennines. Major south-verging thrusting, uplift and southward tectonic progradation occurred in the eastern South Alpine area during this phase, concurrently with spasmodic subsidence and extreme sedimentation rates in the Venetian basin. Strike-slip faults outlined the basin and controlled the sedimentation in this stage (Massari *et al.*, 1986).

A further outward shifting of depocentres occurred during the Plio–Quaternary in the foredeep belt of the Periadriatic ranges (Po, Venetian and Adriatic basins), where subsidence coupled with compressional deformation continued, with the accumulation of a sequence several kilometres thick in the Po plain and in the Adriatic.

THE VENETIAN FORELAND BASIN

Biostratigraphic remarks

Detailed biostratigraphic analyses of some sections and outcrop areas of Venetian molasse were carried out in previous studies (Gelati, 1969; Massari, Iaccarino & Medizza, 1976; Grandesso, 1980; Cason *et al.*, 1981; Stefani, 1982) in which a number of biostratigraphic events were recognized, which fit well into the frame of zonal schemes proposed for the Italian Neogene by Borsetti *et al.* (1979), Iaccarino & Salvatorini (1982) and are calibrated against the standard zonation of Blow (1969) (Fig. 3). The biostratigraphic resolution of foraminifera in the upper Tortonian–Messinian sequence is rather low. In fact the planktonic assemblages are controlled by ecological factors and are very rare or absent, most of the species being long-ranging and not age-diagnostic. The Tortonian–Messinian boundary is placed somewhat below a brackish-water ostracod assemblage referred to the early Messinian by Bossio & Ciampo (personal communication).

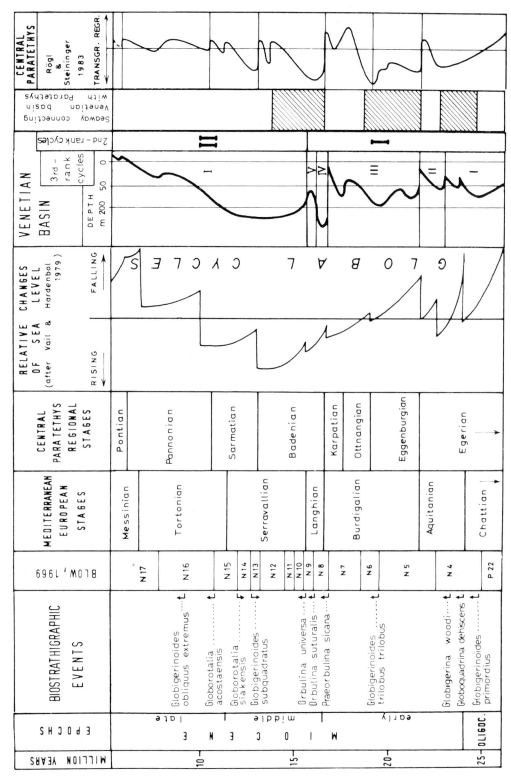

Fig. 3. Neogene biostratigraphy (with indication of main events in the Venetian basin), tentative correlation of Mediterranean and Central Paratethys stage systems (mostly after Rögl & Steininger, 1983) and comparison of the average trend of transgressions and regressions between the Venetian basin and Central Paratethys. Note the similarity of the trend in spite of the marked difference in tectonic framework of compared areas. The Plio–Quaternary evolution of the Venetian basin has not been taken into account due to scarcity of data.

In the time versus palaeobathimetry curve of Fig. 3 three main zones were distinguished following Wright's (1978) suggestion: inner neritic (0–50 m), outer neritic (50–200 m) and epibathial, based on percentage of planktonic foraminifera, Fischer's α diversity index and some significant benthic taxa. The use of benthic forms is mainly based on comparison with the depth distribution of similar recent assemblages.

The inner neritic assemblage is characterized by absence or very low percentages (under 10%) of planktonic foraminifera, α index under 5, presence of larger foraminifera, mainly of *Operculina*, *Heterostegina*, *Lepidocyclina* and *Miogypsinidae* and of *Ammonia*, *Elphidium* and *Florilus* among the smaller ones. The latter genus also occurs in outer neritic assemblages. *Spiroplectammina pectinata* and *S. deperdita* are regarded by Grünig (1984) as typical species of the inner-middle shelf.

The outer neritic assemblage is characterized by percentages of planktonic foraminifera ranging from about 10 to 50%, α index over 5 and common presence of the following benthic species: *Hanzawaia boueana*, *Fursenkoina schreibersiana* and *Spiroplectammina carinata*. *Heterolepa dutemplei* is generally very abundant. In the upper Eocene 'Marna di Possagno' this species was regarded by Grünig & Herb (1980) as typical of the outer neritic zone. Many other species such as *Cassidulina* spp., *Trifarina angulosa*, *Lenticulina* spp. have their upper depth limit within this zone and live today in the Mediterranean and the Gulf of Guascogna (Blanc-Vernet, Pujos & Rosset-Moulinier, 1984). The genera *Cyclammina*, *Karreriella* and *Martinottiella* are presently common below 100 m (Murray, 1973). Among the *Uvigerina* species the minute hispid forms like *U. farinosa* and *U. auberiana*, the ribbed forms like *U. peregrina* and *U. pigmaea* and the slightly striated to almost smooth forms like *U. semiornata* and *U. laviculata* are useful (Boersma, 1984).

High percentages (over 50%) of planktonic foraminifera and high species diversity are generally indicative of epibathyal depth. Among the benthic forms, *Bolivinidae* are well represented by species with sinuate sutures like *Bolivina dilatata*, *B. antiqua* and probably *B. ferasini*. *Bolivina aculeata* has its upper limit in this zone; *Gyroidinoides altiformis*, *Heterolepa pseudoungeriana* and *Hoeglundina elegans* are considered significant forms of this assemblage in the Gulf of California (Ingle, 1980) and in the Mediterranean (Wright, 1978) associated with *Gyroidinoides soldanii*. Among the *Uvigerina* species, *U. longistriata* and *U. barbatula* are indicative of upper bathial depths (Boersma, 1984).

The clastic fill of the basin

The clastic fill of the Venetian foreland basin ranges in age from Chattian to Recent; it is bounded at the base by a major unconformity and is most commonly underlain by Eocene flysch deposits, in places slightly deformed by the Dinaric shortening climax of the Priabonian. The basin is bounded at the western side by the Schio–Vicenza fault (Fig. 4) which also marks the eastern margin of the Lessinian block and is an old, probably deep-reaching, repeatedly reactivated fault.

The geologic section AB of Fig. 5, orthogonal to the South Alpine structural axes, shows a distinct clastic wedge which displays maximum thickness (up to 4000 m) against the leading edge of the South Alpine overthrusts and gradually wedges out southwards. However the isopach map (Fig. 4) and the geologic section CD (Fig. 5), orthogonal to the Dinaric overthrusts show that the above-outlined clastic wedge actually interfingers in the eastern area of the basin with a set of differently oriented clastic wedges, which thicken significantly north-eastwards, against the front of the Dinaric overthrusts. These were essentially active during the Oligo–Miocene, with some later reactivation (Amato *et al.*, 1976) and border at least two asymmetric molasse basins: an inner trough (in the Udine area) bounded at its inner side by the Tricesimo overthrust, and an external basin separated from the former by a positive structure bordered by the Udine–Buttrio and Palmanova thrusts (Fig. 5). These asymmetric troughs are closely comparable in position to other similar basins located on the outer margin of Dinarides, particularly to the north Dalmatian basin located in the offshore area of the northern Adriatic (Celet, 1977; Miljush, 1973).

The basin fill can be subdivided into cycles of different hierarchical level. These usually show a transgressive-regressive trend and are bounded by top and basal unconformities which can be traced on the basin scale except in the case of the lowermost-rank cycles. Especially in the lower part of the basin fill, which is dominated by siliciclastic shelf sediments, the basal unconformities are overlain by heavily bioturbated glauconitic beds, commonly bearing basal thin conglomerates with quartz and chert pebbles, shell fragments, shark teeth and occasional bone fragments. These beds typically form broad and thin blankets traceable on the basin scale and are regarded as 'condensed' units of reworked sediments (lags) deposited at a very slow rate and linked to transgressive stages. Although quantitatively insignificant, such

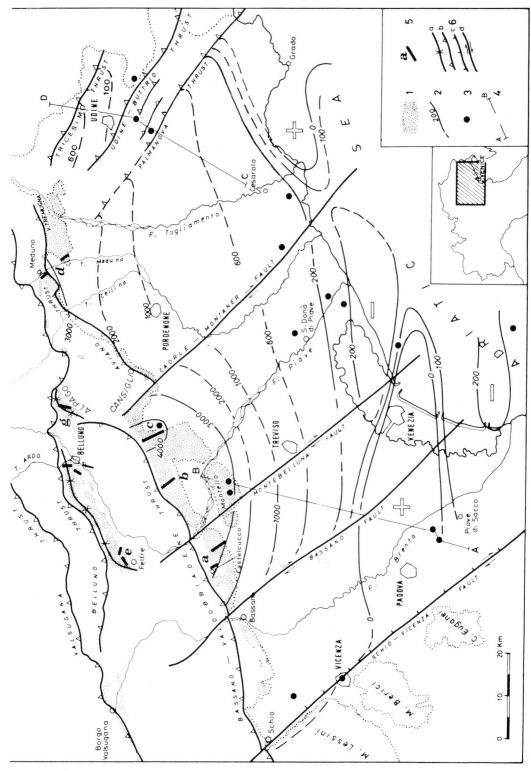

Fig. 4. Isopachs of the Chattian to Messinian sequence of the Venetian foreland basin and main tectonic elements controlling the sedimentation. (1) Molasse outcrops; (2) isopachs with inferred trends shown by dashed lines; (3) location of wells; (4) traces of geologic sections (see Fig. 5); (5) measured sections (a: Monfumo; b: Follina; c: Vittorio Veneto; d: T. Meduna; e: Feltre; f: Belluno; g: Alpago); (6) major structural elements bearing on the evolution of the foreland basin during the Chattian—Messinian time span: (a) syncline, (b) thrust, (c) normal fault, (d) strike-slip fault.

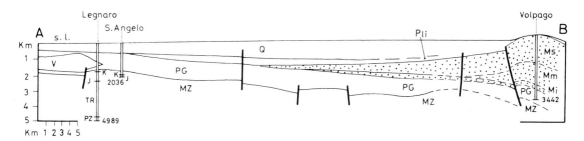

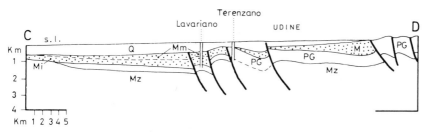

Fig. 5. Geological sections across the Venetian foreland basin (traces in Fig. 4). Section AB from Pieri & Groppi (1981), slightly modified. PZ: Palaeozoic, TR: Triassic, J: Jurassic, K: Cretaceous, MZ: undifferentiated Mesozoic, PG: Palaeogene, V: volcanics, Mi: lower Miocene, Mm: middle Miocene, Ms: upper Miocene, Pli: lower Pliocene, Q: Quaternary. Section CD from Pieri & Groppi (1981), Carobene & Carulli (1981), and Carobene, Carulli & Vaia (1981), slightly modified. Symbols as in section AB. M: Miocene.

beds are very important in regional correlation and vertical pattern analysis.

The factors controlling the cyclicity are often difficult to identify since cycles may commonly result from the interplay of several factors such as the synsedimentary tectonics (the role of lithospheric loading is probably of primary importance in the case of highest-rank cycles), the sea-level changes and the rate of sediment supply. Due to the complexity of factors involved, we prefer to adopt for the hierarchy of cycles an informal nomenclature unrelated to the Vail, Mitchum & Thompson (1977) terminology.

The whole basin fill can be regarded as a first-rank cycle since it is bounded at the base by a major unconformity of supraregional extent. Two second-rank cycles characterized by significantly different tectonic polarity, and respectively Chattian to Langhian and Serravallian to Recent in age (Fig. 6) are the sedimentary expression of two distinct stages in the history of the basin. The older cycle shows evidence of the Dinaric tectonic 'progradation', whereas the younger contains a full record of the foreland subsidence and deformation that accompanied the South Alpine orogeny. The foredeep axis changed in consequence from a Dinaric NW–SE trend to a South Alpine ENE–WSW trend. At a lower hierarchical level third- and fourth-rank cycles can be recognized;

the fact that most of them can be traced on the basin scale probably still reflects an allocyclic control. At a further lower rank, cycles a few metres thick and areally restricted exist in some stratigraphic intervals; at least in part they may be autocyclic in nature.

Chattian to Langhian cycle

This cycle is a multistorey complex up to 820 m thick which can be subdivided into five third-rank cycles or megasequences (Fig. 6) (the term 'megasequence' is used here as a synonym of third-rank cycle), and a number of lower-rank cycles. Unconformities bounding both major and minor cycles are often overlain by glauconitic key-beds and are often marked by a slightly erosional surface with burrows extending downwards into the underlying unit. In most cases the changes in thickness and grain size within the sandstone bodies and particularly the commonly recognizable trend of thickening and coarsening towards the NE suggest a link with thrusting activity in the outermost Dinaric domain. In addition the persistence of some structural highs, either passively inherited from the pre-molasse time or reactivated, can be inferred from some distinctive trends of sedimentation.

The first megasequence (Chattian–lower Aquitanian)

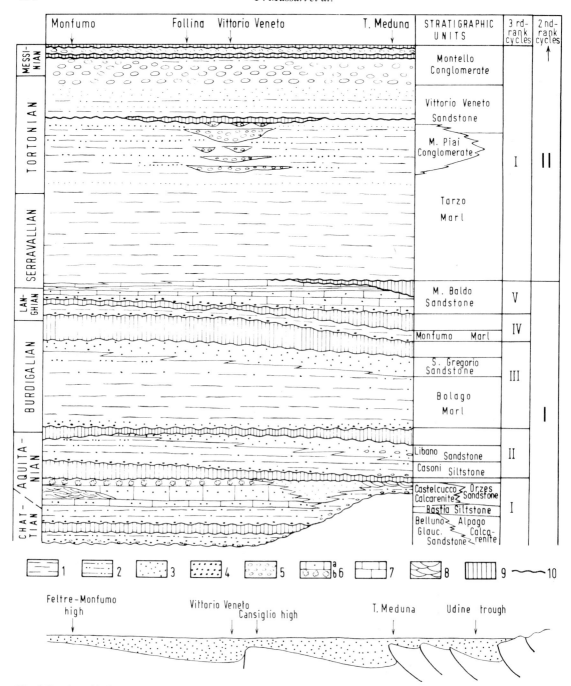

Fig. 6. Stratigraphic framework, informal units, second-rank, third-rank cycles of the Chattian to Messinian sequence of the Venetian basin. This scheme is based on surface data collected along an ENE-trending belt of outcrops; consequently it mostly depicts the diachronisms and lateral facies changes related to the Dinaric polarity. The location of measured sections with respect to the Dinaric structural elements is shown below, by a schematic ENE-trending section. The Messinian unconformity surfaces are only approximately shown; see Fig. 11 for details. (1) Marl, mudstone, (2) siltstone, (3) sandstone, (4) glauconitic sandstone, (5) conglomerate, (6) hybrid arenite (a), with rhodolites (b), (7) limestone, (8) trough cross-bedding, (9) hiatus, (10) unconformity surface.

shows a distinct trend shoaling upwards and includes two facies associations, partly coeval: the first (best developed in the Belluno and Alpago areas, Fig. 7) records the southward progradation of a tide- and wave-dominated delta system, whereas the second is a mixed siliciclastic-carbonate facies association characterized by a foramol assemblage and local abundance of rhodolites, and is inferred to represent a shelf complex of offshore banks and bars with relatively slow sedimentation rate.

The evidence of tidal activity during the early Aquitanian probably reflects some constriction of the Venetian basin which acted to amplify the tidal wave. Active distributary channels opened into the northern and eastern margins of the basin. The components of the *first facies association* are: (i) distributary channel fill (sequence a in Fig. 8), (ii) tidal bars (sequence b in Fig. 8) and (iii) nearshore wave-built bars (sequence d in Fig. 8).

(i) A channel about 8 m deep with trough cross-bedded sandy-conglomeratic fill exists in the Alpago area, where it erosively overlies subtidal bar deposits. It shows well-developed epsilon cross-stratification suggesting a certain channel sinuosity, and average palaeocurrent direction towards 215°. Tidal influences in the fill are suggested by local evidence of bipolar palaeocurrents in some of the in-channel bedforms, presence of mud drapes on sandstone layers, sparse marine fauna (fragments of *Ostrea* in the channel lag) and *Ophiomorpha* burrows in the uppermost part of the fill. Mudstone clasts and plant debris are abundant in places.

(ii) Sequences interpreted as built up by migrating subtidal bars (sequence b, Fig. 8) typically show a trend of thickening and coarsening upwards and are up to 40 m thick. Outer shelf mudstones with diverse foraminiferal fauna are overlain by massive heavily bioturbated siltstones in which the diversity of the foraminiferal assemblage progressively decreases: as the grain size tends to coarsen from siltstone to sandy-siltstone, the genera *Elphidium*, *Ammonia* and *Florilus* together with *Haplophragmoidinae* predominate, suggesting significant freshwater influx. This facies in turn grades upwards into a plane-bedded sequence with more and more distinct layering due to a decrease in the intensity of bioturbation. The depositional theme of this division is represented by a vertically aggrading rhythmic alternation of thicker bioturbated sandstone layers 10–20 cm thick (sometimes with small-scale trough cross-bed sets and remains of mud drapes almost completely obliterated by the bioturbation) and bundles of thinner sandstone units (usually laminae) sometimes rippled, with mud drapes occasionally rich in small plant debris. Cyclic variations in the energy of the transporting medium are suggested by occasionally observed rhythmic sequences of thickening-thinning bundles which can be regarded as 'vertically accreted' bundle sequences. Within this bedded division simple tubes and meniscus-filled burrows decrease in abundance upwards and are accompanied in the upper part by *Ophiomorpha*. The last division consists of sets of predominantly unidirectional trough cross-beds of upward increasing scale and grain size (from fine sandstone to coarse and granule-bearing sandstone). In the lower part of this division the sets (20–100 cm thick) are characterized by irregularly developed thickening-thinning bundle sequences, with thinner bundles represented by long tangential foresets grading downdip into well-developed wave- and current-rippled bottomsets with mud drapes. The larger-scale sets in the upper part of this division are up to 3 m thick, and are characterized by poorly developed or absent bottomsets, upslope fining trend in the foreset laminae (effect of size sorting during avalanching), common normal grading in the foreset laminae and the presence of small-scale isolated ripple trains climbing up the lee side of the bedforms, sometimes associated with mud drapes; the latter are however rare in this facies and are usually replaced by mudstone clasts due to break-up. Reactivation surfaces can be identified mostly in the thicker cross-bed sets and may show a thin mud drape. Bedforms involved are probably large, ebb-dominated, sinuous-crested or barchanoid dunes. Trough axes dip towards 215°, consistently with the direction of the channel axis; this suggests longitudinal migration of bars in the direction of predominant tidal flow. The effect of high-energy waves can occasionally be recognized only as a later reworking of the bar top, to form swash-platforms (sequence c in Fig. 8).

In conclusion, the lack of traces of storm influence in the middle and upper divisions of the sequence, the inferred relatively high subtidal flow velocities and the fact that palaeocurrents show little variance and are diametrically opposed suggest that the bars were confined within the still channelized distributary mouth rather than migrating on unconfined shelves. The lower muddy facies of the sequence, in turn, was probably deposited seawards in front of the mouth. The large size of the bedforms may depend on highly asymmetrical tides and seems to confirm the view of De Mowbray & Visser (1984) that the size of bedforms is essentially a function of relative magnitude of the dominant and subordinate components of flow.

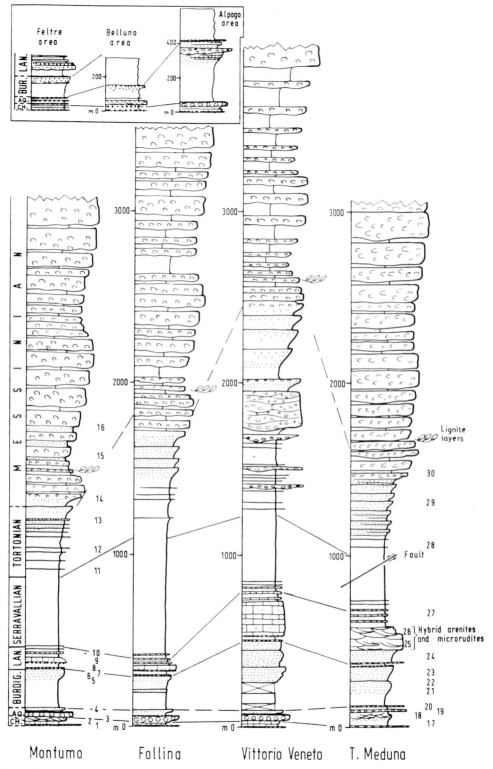

Fig. 7. Logs of measured sections (location in Fig. 4). Symbols as in Fig. 6 except for pelitic intervals which are indicated in blank. Numbers in Monfumo and T. Meduna logs refer to samples for petrographic analyses.

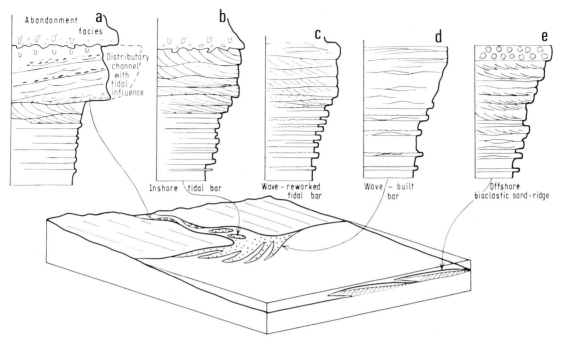

Fig. 8. Suggested sedimentary model for the interpretation of the lateral facies changes in the lower Aquitanian depositional system of the eastern Venetian basin (not to scale).

(iii) Sequences interpreted as representing wave-built near-shore bars coarsen upwards from outer to inner shelf mudstones interbedded with hummocky-stratified storm-layers, and eventually to amalgamated hummocky-stratified beds and low angle cross-beds (sequence d in Fig. 8).

The *second facies association* is typically represented by a shoaling upwards sequence mostly consisting of hybrid and sometimes glauconitic arenites, with a foramol assemblage comprising larger foraminifera (*Miogypsinidae, Operculina, Heterostegina, Amphystegina, Nephrolepidina, Acervulinidae*), Coralline algae (*Lithothamnium, Lithophyllum, Archaeolithothamnium, Mesophyllum*), bryozoans, echinoderms, bivalves (mostly pectinids) and serpulids. The sequence is usually plane-bedded, with interbeds of silty mudstone which tend to decrease in thickness upwards and finally disappear in the uppermost part of the sequence. The shoaling trend is suggested by the trend thickening and coarsening upwards, by the occasional presence of hummocky stratifications towards the top and by the common presence of rhodolites in the upper part of the sequence, first with sparse and irregularly shaped branching forms, and then with tightly packed, usually rounded nodules. In the Castelcucco area the sequence shows evidence of a

higher-energy setting: an upward transition can be observed from wave-rippled layers locally with hummocky stratification, to almost purely biocalcarenitic sets of unidirectional trough cross-beds (average dip towards 340°), up to 70 cm thick, rarely bearing ripple trains climbing up the foreset laminae, and finally to a bank full of rounded rhodolites (sequence e, Fig. 8). This type of sequence may represent storm- and/or tide-influenced shelf sand ridges formed on a submarine topographic high. According to Swift & Rice (1984) an initial topographic relief may act as a focus for sand deposition, resulting in positive feedback; as the bedform becomes larger it extracts more sand from the transported load and the resulting sequence would reflect progressive shoaling and increasing perturbation of flow as the amplitude of sand ridge increases. It is worthwhile stressing that the Castelcucco sequence, as well as the condensed, glauconite-rich sequences of the Feltre area, north of Castelcucco, are in fact located on a roughly N–S trending basement high which has a history of tectonic activity that extends back into the Jurassic. This palaeohigh was clearly a positive area also during Miocene times as indicated by the thinning of major units across its top.

In conclusion, a transition can be postulated from a tidally influenced distributary channel to subtidal inshore bars of the distributary mouth, to nearshore storm-dominated bars, and finally to shelf sand-ridges consisting of mixed siliciclastic-carbonate sediments and localized on palaeohighs. A similar depositional system was described by Homewood & Allen (1981) in the Burdigalian molasse of western Switzerland.

The reactivation of palaeohighs and their influence on sedimentation has been recognized in the outer (western) area of the foreland basin, presumably affected by an extensional regime. On the other hand synsedimentary compressional tectonics at the inner side is documented by the geometry of sedimentary bodies in the eastern, thrust-controlled troughs.

The *second megasequence* (upper Aquitanian) is simple, while the *third megasequence* (Burdigalian) is composite since it consists of a number of lower-rank cycles. They both show an overall transgressive-regressive trend with transition from outer shelf mudstones with diverse foraminiferal fauna to pro-delta mudstones and finally to highly bioturbated and locally wave-rippled delta-front sandstone bodies which become younger and fine-grained SW-wards marking a progradation of delta systems in this direction and suggesting a link with discrete pulses of movement of the Dinaric thrust front at the north-eastern side of the basin. The Burdigalian transgression is particularly marked and is a basin-wide event correlatable with the Eggenburgian transgression of the Paratethys and probably linked to the well-documented world-wide Burdigalian sea-level rise (Vail *et al.*, 1977).

In both megasequences environmental deterioration accompanying the regressive trend is documented by: (a) occurrence of oligotypic euryaline foraminiferal assemblages in the prodelta mudstones and of an endemic fauna of Cetacea in the same unit; (b) presence of *Uvigerina* assemblages in the prodelta mudstones of the third megasequence pointing to oxygen-depleted bottom waters. Such evidence, coupled with lack of preserved high-energy physical structures in the sandstone bodies suggests that in both cases the basin evolved from an open-shelf setting towards partial isolation. It should be noted in this regard that the late Burdigalian is marked by strong regressions in the Central Paratethys too (Fig. 3) (Rögl & Steininger, 1983).

Fourth megasequence. After the deposition of the third megasequence, a rapid deepening of the basin occurred, which brought about the diachronous deposition of a transgressive glauconitic layer rapidly followed by outer shelf to epibathyal marls with up to 70% of planktonic foraminifera (Fig. 3); the base of

this unit becomes distinctly younger SW-wards (from uppermost Burdigalian in the NE to lower Langhian in the SW) marking an onlapping relationship with respect to the Dinaric foreland ramp (Figs 6 and 7). Moreover, the megasequence is truncated at the top, the regressive stage being represented by a very sharp diachronous unconformity surface, which becomes an erosional surface on submarine highs. The rapid bathymetric changes coupled with the diachronous character of the sedimentation suggests a thrust-controlled subsidence.

The *fifth megasequence* is represented by a diachronous body of litharenites sometimes containing an abundant carbonate intrabasinal fraction. Even in this case a SW-wards progradation is indicated by the fact that the base of the body becomes younger in this direction (lower Langhian in the NE, middle Langhian in the SW) (Figs 6 and 7); this trend, coupled with a distinct SW-wards decrease in grain size and in overall thickness, indicates the persistence of tectonic control by 'prograding' Dinaric structures. The facies association shows significant changes in the different parts of the basin and is clearly influenced by positive and negative features of the basin topography.

At the north-eastern side of the basin, in the Meduno area (which was probably part of a Dinaric thrust-controlled positive structure, as suggested by the presence of a distinct erosional unconformity at the top of the megasequence) the sandstone body consists of two shoaling upwards cycles followed by a sequence fining and thinning upwards; the latter marks the beginning of a stage of rapid deepening recognizable throughout the basin. The first, more complete, shoaling sequence is about 56 m thick. A lower division consists of hummocky-stratified bioturbated fine-grained sandstone layers with siltstone interbeds, suggesting an alternation of high-energy storm-influenced deposition and post-storm emplacement of fines from suspension. This facies is overlain by trough cross-bedded fine to coarse granule-bearing sandstones, sometimes glauconitic, displaying a net upward increase of grain size and scale of the cross-bed sets. The latter typically consists of couplets of carbonate-poor and carbonate-rich layers containing debris of echinoderms, bryozoans and reworked Nummulites. The sets are occasionally paved by abundant mudstone clasts and range in thickness from some tens of centimetres to over 2 m. Trough cross-bedding is dominant, indicating palaeoflow approximately towards 300° and is locally accompanied by isolated planar cross-bed sets. Foresets of cross-beds are sometimes overlain by isolated sets of small-scale ripple-drift cross-lamination or by trains of wave-ripples, locally bearing mud drapes. Bioturbation is nearly absent in this division. The sequence is inferred to represent a sand ridge deposited by geostrophic storm flow, possibly with superimposed tidal activity, on a positive feature of the shelf bottom.

This interpretation may account for facies distribution within the basin; in fact cross-bedded sandstone bodies are localized either on a culmination above a blind or emerging thrust (Meduna area), or on a rejuvenated palaeo-high (Feltre area) (profile in Fig. 6) and grade laterally into sequences of fine-grained, sheet-like sandstone layers generated by storms, with increasing amounts of mudstone interbeds. Only in some intermediately positioned areas can occasional layers be found showing high-angle cross-bedding in isolated sets about 30 cm thick recording the extension of the dune field into the inter-ridge areas during infrequent severe storms.

Due to SW-wards migration of the sand body, the south-western-most area of the basin was reached by sand deposition only in the middle Langhian; as a result the lower shoaling part of the sandstone body is lacking here, and only a reduced sandstone sequence occurs, showing a distinct trend of deepening upwards.

The fact that shelf sands directly overlie a sharp basal unconformity substantiates the idea of a distinct transgressive event accompanied by the onset of an open shelf circulation. This probably implies the flooding of seaways which had previously been closed. The widespread middle Miocene extensional event which created a number of fault-bounded basins in the inner Dinarides and in Central Paratethys may well be at least partly responsible for this re-opening of a previously interrupted seaway. But it should also be remembered that a world-wide sea-level rise is known to have occurred during the Langhian and early Serravallian (Badenian transgression in the Paratethys). This transgression spread all over the Mediterranean and Paratethys, re-opening the seaways to the Indo–Pacific which had been closed as a result of the collision of the Arabian and Turkish plates in middle Burdigalian time (Rögl & Steininger, 1983).

Serravallian to Recent cycle

The upper second-rank cycle (Serravallian to Recent) is much thicker than the lower one and contains a full record of the foreland subsidence that accompanied the South Alpine orogeny. A prominent unconformity corresponding to the Miocene–Pliocene boundary separates a third-rank cycle (Serravallian–Messinian megasequence), which is well known in outcrops, from the Plio–Quaternary succession, for which there are little data, and will be only briefly mentioned. Stratigraphic evidence implies that the thrust emplacement in the eastern Southern Alps was rapid and may have required no more than 10 Ma from the Serravallian to the late Pliocene.

The *Serravallian to Messinian megasequence* basically consists of three divisions: (1) a lower sequence of basin floor hemipelagic mudstones and marls deposited as active subsidence related to the structural load of thrust sheets lowered the basin floor from outer-shelf to epibathyal depths, (2) an intermediate basin-fill sequence which records the rapid slope progradation and (3) an upper sequence of stacked fan-delta and alluvial fan deposits coming from the erosional denudation of the advancing thrust front.

The foraminiferal assemblages of the thick hemipelagic marls of the lower part of this succession suggest that the basin reached epibathyal depths in the early Serravallian. The percentages of planktonic foraminifera remain very high (60–70%) until the late Serravallian and then drastically drop to 10–20% in the early Tortonian. This sudden change is accompanied by the occurrence of repeated organic-rich and pyrite-bearing, sometimes varved horizons with a specialized microfauna very rich in the genera *Bolivina*, *Bulimina* and *Uvigerina*, suggestive, according to Phleger & Soutar (1973), of oxygen-depletion in bottom waters. This might reflect basin confinement and/or a density stratification due to low salinity water masses near the surface. With the onset of slope progradation a lot of reworked foraminifera comprising both upper Cretaceous and lower Tertiary forms occurs as well as repeated evidence of gravity-displaced coeval forms, resedimented from shallow-water settings. Very fine-grained turbidites sparsely interbedded with the mudstones show the thin 'streaky' bedding which is typical of molassic turbidites and compares well with recent examples of prodelta turbidites (Ricci Lucchi, 1986).

During the early and middle Tortonian most areas of the foredeep chiefly received prodelta-slope mudstones with only sparse thin-bedded turbidites, except the Vittorio Veneto area, where the progradational muddy sequence contains a number of channelized bodies consisting of resedimented conglomerates and sandstones mostly emplaced by mass flow (M. Piai Conglomerate in Figs 6 and 9). In the progradational sequence of this area the transition from basinal mudstones to channelized bodies is accompanied by

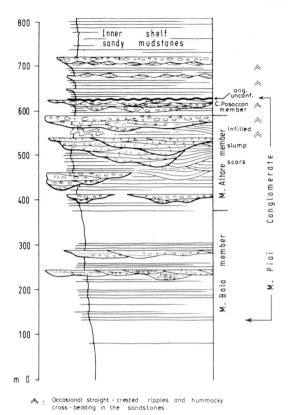

Fig. 9. Scheme of the Tortonian progradational sequence of
the Vittorio Veneto area (M. Piai Conglomerate). The
channelized bodies of M. Bala, M. Altare and C. Posoccon
members consist of gravity resedimented conglomerates
(mass flows) accumulated in front of an actively growing fan-
delta. The large bodies of M. Altare member are interpreted
as the filling of large, composite, shelf-edge slump-scars.

sparse thin-bedded turbidites which are not organized
in distinct sequences. In fact, sandstone beds are not
volumetrically very significant and seem mostly
confined to the overbank areas lateral to the channels.
The width/thickness ratio of channelized bodies tends
to decrease upwards significantly: in the lower part of
the progradational sequence, which is dominated by
prodelta mudstones, the bodies are relatively thin (8–
22 m) and laterally persistent, and are actually
multilateral channel systems (M. Bala member in Fig.
9); in the upper part (M. Altare member) they become
significantly thicker (up to 100 m) and volumetrically
much more important with respect to the surrounding
deposits, which show the first evidence of shallow-
water environments and filled slump scars, especially

in the lower part of the member (Fig. 9). The larger
bodies of resedimented conglomerates are interpreted
as the fill of wide depressions produced on the highest
part of the slope by the lateral coalescence of several
slump scars. The largest one shows an infilling pattern
with oblique progradational units showing a tangential
lower contact.

These coarse-grained lithosomes may be compared
to the Mississippian ravine-filling conglomerate and
sandstone bodies described by Harbaugh & Dickinson
(1981) in a similar setting of slope progradation related
to actively growing fan-deltas. As suggested by
Harbaugh & Dickinson (1981), submarine ravines
that scar the upper slope are sites for local deposition
of gravel and sand transported downward by sediment
gravity flows. The resulting bodies are more common
in the upper slope, so that the prograding fan-delta
slope can form a thick coarsening upward sequence.
In the delta setting a variety of subaqueous instability
processes may take place near the shelf edge, especially
in the case of rapid sediment introduction. Several
adjoining slump scars in front of the prograding
system may produce intersecting arcuate patterns that
outline a much larger feature than the single slump
scars. The large evacuated composite scar may be
later infilled with oblique progradational units (Coleman, Prior & Lindsay, 1983). In a thrust-controlled
setting slope-instability processes are probably enhanced by the virtual lack of a shelf and storage of
large amounts of coarse clastics directly from the
nearshore area into the slope by sediment gravity
flows.

Resedimented conglomerates associated with the
Tortonian slope progradation are restricted to the
Vittorio Veneto area and show a sharp termination
against the Cansiglio high; this suggests that the M.
Piai Conglomerate was accumulated in a structural
low related to transverse tectonic lines. In fact there is
some evidence of an extensional structure trending
north–south, probably generated by a transtensile
motion along a conjugate couple of transverse faults
(NNE sinistral and NW dextral). This structural
depression may have acted as a preferential route
along which coarse-grained sediment was funnelled
(Fig. 10). In foreland basins transverse tectonic lines
commonly provide access of sediment to the foredeep
from the orogenic side of the basin, i.e. from sources
usually 'locked' in the thrust belt itself (Ricci Lucchi,
1986), such as minor basins bounded at the outer side
by rising thrust fronts. The fact that the M. Piai
Conglomerate is bounded at the top by a localized
angular unconformity of about 10° suggests tectonic

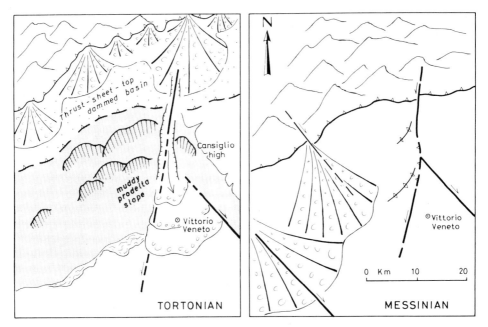

Fig. 10. Sedimentary model for the Vittorio Veneto area in Tortonian and Messinian time. Fan deltas derived from the leading edge of the South Alpine thrusts pass southwards into a muddy prodelta slope (Tortonian). Local connection between a thrust-sheet-top basin (Nichols & Ori, 1983) (dammed by a rising thrust) and the foredeep, and downslope funnelling of mass flows, are made possible in this stage by the development of a structural depression as a result of transtensile motion along a conjugate couple of transverse faults (NNE sinistral and NW dextral). Subsequent Messinian folding and thrusting associated with the onset of transpression resulted in the gradual occlusion of this preferential pathway.

control through the interplay of thrusting and strike-slip motion along transverse faults.

The rapid slope outbuilding during the Tortonian brought about a substantial shoaling of the basin and the subsequent late Tortonian–early Messinian fan-delta sedimentation took place on a shallow shelf created by the former progradational episode. The Tortonian slope complex (M. Piai Conglomerate) characterized by important mass movements and by rapid progradation due to the storing of most of the clastic sediment on the slope, is thus later replaced by shelf-type fan-deltas characterized by the storing of the majority of gravels near the shoreline. These different types seem to fit well into the two stratigraphic models developed by Wescott & Ethridge (1980) for coarse-grained, humid-region, dominantly progradational fan-deltas of island arc or continent-collision coasts. The Yallahs-type fan-delta is characteristic of truncated fans which build directly on to relatively steep submarine slopes where gravity-resedimentation processes are most important. The second model, based on fans along the SE Alaska

coastline, is characteristic of more completely developed subaerial fans that prograde on to continental or island shelves. The Alaska-type fan-delta stage is expressed in the Venetian basin by stacked and interfingering nearshore sequences including mouth-bar and gravelly-beach cycles, and distributary-channel conglomerates encased in pelites with brackish-water fauna deposited in shallow interdistributary bays. Such sequences were probably built up in a shallow, partially enclosed, microtidal sea.

Later, under a spasmodic subsidence, a sequence about 1400 m thick of alluvial-fan conglomerates interfingering with lacustrine and flood-plain mudstones was laid down during the Messinian. At this stage a more or less continuous belt of alluvial fans with interfingering radial palaeocurrent patterns encroached on the basin along the whole length of its edge. An overall trend of coarsening upwards and the inverted stratigraphy seen in clasts implies that the conglomerates are a progradational sedimentary response to the southward shifting of thrusting in the South Alpine domain. A South Alpine source is clearly indicated by the palaeocurrent pattern and by the clast composition (Massari, Rosso & Radicchio, 1974). Recycling of molasse deposits in the younger molasse indicates that older proximal molasse at the inner, orogenic side was folded and faulted beneath the thrusts and cannibalized.

In the thick Messinian conglomerate sequence a number of fourth-rank cycles have been identified, which can be followed along strike all over the basin margin. They are bounded by distinct unconformities (best recognizable from aerial photos) which gradually change in character along-strike from angular unconformities near transverse faults trending NW–SE or north–south, to simple disconformities away from them (Fig. 11). For the most part the sequences coarsen and thicken upwards, with a final trend of fining-thinning upwards in the uppermost part. Moreover the first of them is also capped by a bundle of lignite layers which can be followed along strike with very high lateral persistence (Fig. 7). The basal unconformity and the coarsening trend is thought to reflect a definite thrusting pulse and uplift in the source area leading to a major progradation of the fan-system, whereas the fining trend and lignite complex may record the retrogradation of the system due to the lowering of relief in the source area and progressive decrease in subsidence rate. Relatively low along-strike persistence of angular unconformities and differential along-strike subsidence rates are thought to reflect interaction of thrusting with strike-slip motion along transverse faults. Apparently the latter also controlled the location of fan apices. On the other hand, folding and thrusting associated with the onset of transpression along a transverse line which had acted as a preferential route of sediment supply during a previous transtensile stage may result in the gradual occlusion of the pathway and cutting off of coarse clastic supply, as suggested by the evolution of the Vittorio Veneto area during the Tortonian and Messinian (Fig. 10). Strike-slip faulting along a set of conjugate shears trending NW–SE and north–south or NNE–SSW probably played an important role during the younger stage of the Venetian basin history. It should be particularly stressed that the Dinaric overthrusts bounding the basin at the eastern side were probably reactivated in this stage with a significant strike-slip component of movement.

A striking characteristic of the alluvial facies associations is the fact that both isolated channel-fill conglomerates of the interfan areas interpreted as relatively steep-slope meandering stream sequences, and stacked conglomerate units interpreted as braided-river outer-fan deposits (Massari, 1984) are associated with an anomalously high proportion of overbank fines. Similar situations were documented by Hayward (1983) and by McLean & Jerzykiewicz (1978) in similar tectonic settings. Preservation of thick overbank fines requires rapid subsidence preventing reworking of floodplains and such a situation may be common in foreland basins affected by thrust-controlled loading subsidence. In conclusion there is abundant evidence of intra-Messinian tectonics in the Venetian basin.

The complete lack of Messinian deposits in a belt located in the southern area of the Venetian foreland (Fig. 5) and elongated parallel to the strike of the fold-and thrust-belt suggests a definite upwarping of this area during the Messinian. The formation of a peripheral bulge as a response to thrust loading (Quinlan & Beaumont, 1984) can be suggested. This positive feature may have acted as a threshold bounding and isolating the foredeep. Due to the excess of freshwater influx linked to drainage down the South Alpine slopes, the foredeep was kept free from evaporite precipitation during the Messinian and evolved into a freshwater pensile basin like the padan basins (Dondi, Mostardini & Rizzini, 1982).

The *Plio–Quaternary* succession is known only from subsurface data and its detailed analysis is beyond the scope of this paper. It undoubtedly represents the continuation of the trend initiated in the Serravallian. After the far-reaching Pliocene transgression, a strong thrusting phase took place in the late Pliocene leading to the shifting of the basin axis south of the Aviano thrust (Fig. 2). Still younger deformation can be detected in the subsurface as blind thrusts (Sacile thrust) also involving the Quaternary (Zanferrari *et al.*, 1982). It should be remembered in this regard that the dislocation of the fracture plane of the main 1976 Friuli earthquake corresponds to a shallow overthrust directed from south to north (348°) and dipping 13° (Müller, 1977); the radiation pattern reveals a tectonic stress direction from SSE to NNW, consistent with Serravallian to Pliocene compression and perpendicular to the strike of the Eastern Alps, indicating that convergent plate movements are still active.

Petrography of arenites

A quantitative analysis of arenites, including both the gross composition and heavy mineral associations, was made in order to identify the source areas of clastics.

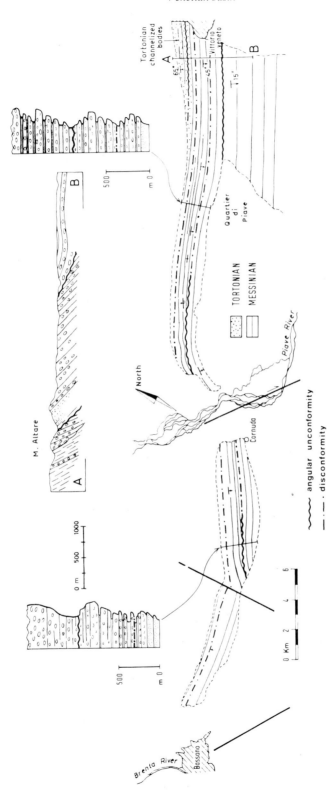

Fig. 11. Schematic photogeological map of the upper Tortonian–Messinian sequence outcropping between Vittorio Veneto and Bassano. A number of thrust-controlled depositional sequences are separated by unconformities. As a result of interaction between wrenching along transverse lines and thrusting, the sequence boundaries are seen to change along strike from angular unconformities (close to the transverse faults) to disconformities (away from them). Angles and the truncation pattern of unconformities are shown in the geological section; schematic logs indicate the trend of sequences.

160   F. Massari et al.

Results

Analytical data* are plotted in Figs 12 and 13. Quantitative data were integrated by qualitative observations on mineralogical features.

The gross composition diagram (Fig. 12) clearly shows that two main petrofacies can be distinguished (previously recognized by Stefani, 1984 and by Massari *et al.*, 1986). The first one (Chattian to Langhian samples) is characterized by sublitharenites and litharenites (*sensu* Pettijohn, Potter & Siever, 1972) with more than 50% quartz. The second group (Tortonian to Messinian) consists of litharenites with a very high amount of lithic fragments (carbonate terrigenous fragments). In both petrofacies the feldspars are represented by albite-oligoclase and rare K-feldspar.

The relative percentages of the main heavy minerals are represented in Fig. 13, where 'others' include: zoisite, clinozoisite, rare pistacite, sphene and chloritoid. Garnet and zircon are characterized by high variability in roundness and colour. In the heavy mineral suites the persistence of some minerals typical of medium-grade (e.g. kyanite and staurolite) and low-grade (e.g. chloritoid) metamorphic rocks along the whole succession should be noted. Among them, kyanite and staurolite could be related to an Austroalpine and/or Penninic source considering that they have not been recorded in the South Alpine basement. One of the most important features is the variability of the amphibole population: common hornblende and actinolite-tremolite are the most abundant, while glaucophane and crossite (both identified by microprobe analysis) make up a less abundant but highly significant fraction. The occurrence of these Na-amphiboles in association with relatively abundant chromite and Mg-chromite suggests a provenance from ophiolitic units. In the Eastern Alps such minerals have been recorded in the ophiolitic units of the Tauern window [Miller (1974, 1977); a blue sodic amphibole (gastaldite) was described by Bianchi (1934) and Dal Piaz (1934) in the Picco dei Tre Signori

*Thirty samples (mostly medium- to fine-grained arenites) from Monfumo (1–16) and T. Meduna (17–30) sections, ranging in age from the Chattian to the Messinian, were studied. Gross composition modal analysis was carried out according to the criteria proposed by Gazzi (1966), Dickinson (1970) and Zuffa (1980). An average of 250 terrigenous grains were counted on R-alizarine stained thin sections. For heavy mineral analysis, the 44–177 μm fraction, obtained by gentle crushing of the whole rock in water, was used. Different grains, transparent and opaque, were counted in order to identify about 300 transparent grains (except for phyllosilicates) for each sample.

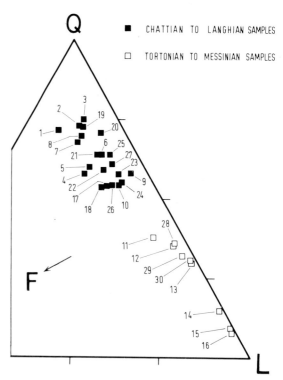

Fig. 12. Q–F–L diagram showing the composition of arenites in Monfumo (1–16) and T. Meduna (17–30) sections (positions of samples in Fig. 7). Q: quartz grains; F: feldspars; L: fine-grained lithic fragments, including carbonate terrigenous grains. Data on T. Meduna samples are partly deduced from Stefani (1984).

area], which seems to be their most likely source area. Detrital glaucophane has been recorded in some sequences of the so-called Rand Cenoman in the Lower Austroalpine by Woletz (1967). However a recycling from this source can be regarded as unlikely taking into account that (1) the Austroalpine nappes were already emplaced long before the late Oligocene, (2) the watershed between the northern and southern sides of the Alps was probably located along the axial zone during the late Oligocene (Frisch, 1976).

The second petrofacies, including samples of Tortonian–Messinian age, is well-defined by a high amount of carbonate terrigenous debris, mostly represented by dolomite grains. These data, coupled with pebble counts of conglomerate framework (Massari *et al.*, 1974), clearly indicate that the main source was the sedimentary cover and, to a small extent, the basement of the eastern South Alpine domain.

In conclusion, integrated analysis of both heavy and light mineral data points out that, in agreement with proposed bipartition of the molasse succession,

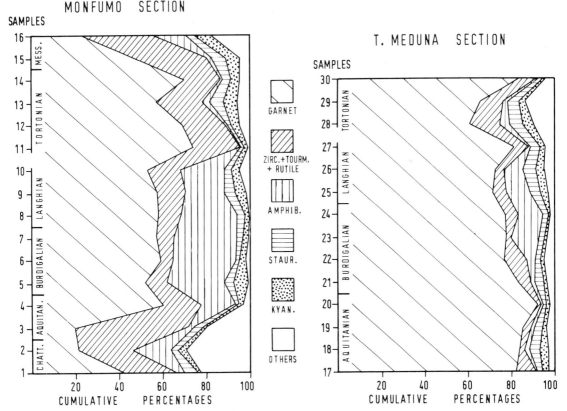

Fig. 13. The heavy mineral suite of Monfumo and T. Meduna arenites (position of samples in Fig. 7). The Serravallian sequence has not been sampled due to its marly lithology.

essentially two petrofacies may be differentiated, respectively characterized by Alpine and South Alpine sources. A contribution from a Dinaric source cannot be definitely demonstrated. The lack of minerals or clasts diagnostic of this source may reflect the fact that the Dinaric basement was not exposed to erosion during the Neogene, or that relief of the Dinaric chain was rather low.

The relatively high compositional maturity of Chattian–lower Aquitanian arenites suggests recycling of sediments related to the middle Oligocene pause separating the 'Meso-Alpine' event from the 'Neo-Alpine' orogeny. The increase in percentage of stable minerals (zircon + tourmaline + rutile) in the Messinian samples (Monfumo section) suggests moreover a recycling and cannibalizing of the molasse. In addition the identification of high-pressure index minerals in sediments as old as Chattian testifies to the time when rocks formed by early phases of Alpine high-pressure metamorphism (probably the Tauern area) were uplifted and subjected to erosion.

EVOLUTION OF THE VENETIAN BASIN

A marked differentiation of the South Alpine area during the Mesozoic rifting stage controlled both later deformation during the collisional and post-collisional stages and the pattern of foreland basins. The onset of molassic sedimentation in the post-collisional stage was preceded by a middle Oligocene widespread extensional phase which is held responsible for the major unconformity at the base of the molassic clastic fill. The widespread character of this event suggests the connection with a supraregional process, in agreement with Bally's (1980) statement that major unconformities bounding at the base the clastic wedges of foreland basins may be related to a supraregional geodynamic process. It should also be remembered that the late Oligocene is characterized by a series of important palaeogeographic changes in the Mediterranean area, such as the beginning of the late orogenic stage in the Dinaric–Hellenic and Alpine belts and the beginning of compressional deformation in the Apennines (Fig. 14). Furthermore it corresponds to a very pronounced world-wide sea-level drop (Vail et al., 1977), marking the boundary of a Vail et al. supercycle. The above facts seem to agree with Bally's (1980) contention that flat 'shoulders' of the Vail et al. supercycles mark short periods of plate reorganization resulting in a new plate tectonic regime.

As pointed out by Laubscher (1971), plate kinematics in the Alps and Apennines call for the superimposition of at least two kinematic tendencies i.e. north–south and east–west. This statement is in good agreement with the Neogene evolution of the Apenninic–South Alpine system. In fact the eastward time-shifting of flysch-basin depositional axes and of

thrusting in the Apennines (Ricci Lucchi, 1984a) seems to match the history of deformation and of molasse sedimentation in the South Alpine domain. Eastwards shifting of the main shortening events in the Southern Alps (Castellarin & Vai, 1981) is documented by surface and subsurface (Pieri & Groppi, 1981) structural relationships and by the eastward migration of molasse depocentres during the time span from late Oligocene to Recent. There is a surprising correlation between this shifting pattern and the eastward switching of the apices of the main flysch fans in the Apenninic domain (Ricci Lucchi, 1984a). In fact there is a close feeding relationship between the South Alpine molasses and the main Apenninic flyschs, the migration of flysch depocentres

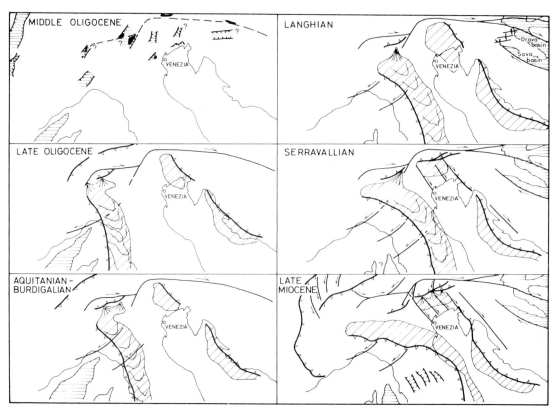

Fig. 14. Speculative and simplified palaeogeographic maps showing the evolution of the Periadriatic ranges and major foreland basins from the middle Oligocene to the late Miocene. Note the eastward shifting of deformation and of foreland-basin depocentres within the South Alpine and Apenninic domains. The Venetian basin behaved as a Dinaric foreland basin up to the Langhian. From Serravallian onwards it was involved in the South Alpine orogenesis developing in a context of oblique convergence. Horizontal ruling: areas of active extension; oblique ruling: foredeeps; indented lines: major thrust belts; arrows: direction of motion along major transcurrent faults; barbed lines: normal faults; black: main late Alpine intrusions. These maps are partly deduced from the results attained by the working group RCMNS coordinated by Boccaletti. The contributions of Gelati and Zanferrari are particularly acknowledged.

being strictly correlated to the eastward shifting of source areas located in the South Alpine and Alpine domains (Ricci Lucchi, 1984b) (Fig. 14).

It should be pointed out that, although continuous over a long time span, the deformation is punctuated by a number of major shortening pulses which took place during distinct, relatively brief periods as the locus of thrust faulting encroached in a stepwise way on the foredeep. The presence of syntectonic angular unconformities within thick and coarse-grained clastic wedges suggests that major thrusting phases were accompanied or quickly followed by rapid foredeep subsidence and uplift of source area without a significant time delay.

The above statements imply a number of consequences:

(1) The shifting of deformation around the orogenic arc could not be achieved without the concurrent mutual release of adjoining compartments along transverse tectonic lines acting as strike-slip faults. The strike-slip component of the late Oligocene–Miocene compression is clearly shown by the style of deformation in the South Alpine domain, and by locally developed transpressive effects, as suggested by the thrusting of cover slices with an 'en échelon' pattern. Boccaletti & Dainelli (1984) stressed the importance of conjugate shear systems during the post-collisional history of the Mediterranean area and pointed out the role of transcurrent lines in segmenting the thrust belts and foreland areas into compartments of different crustal thickness which show a diachronous and partly independent tectono-sedimentary evolution.

The shortening pattern is obviously influenced by the previous history of crustal thinning during the Mesozoic rifting stage and the major lines allowing mutual release of adjoining blocks mostly correspond to reactivated ancestral elements, and in some cases to deep-reaching faults (e.g. the Giudicarie and Schio–Vicenza lines). The role of transverse lines in addition is important in molasse sedimentation since it allows an intermittent supply of coarse clastic sediments to the foredeep from minor thrust-sheet-top basins located in the inner area via submarine channels (Fig. 10) (see also Ricci Lucchi, 1986);

(2) The time-shifting of major thrusting and of molasse depocentres around the South Alpine orogenic arc might reflect the eastward migration of A-type subduction, as suggested by Burchfiel & Royden (1982) for the Carpathians. In a context of dextral wrenching along the Insubric line (Laubscher, 1971; Laubscher & Bernoulli, 1982) this might have implied that progressively more eastern segments of this lineament were involved in a transpressive regime concurrently with eastward migration of A-subduction. The gradual eastward decrease in vertical displacement observable along the Periadriatic lineament may reflect this evolution.

Both (1) and (2) might in turn have resulted from the oblique convergence which caused a superimposition of normal compression and wrenching (Laubscher & Bernoulli, 1982). Wrenching is most prominently expressed in the Insubric fault zone but is distributed throughout the Alps during the Neogene. Oblique dextral convergence, accommodated by wrenching along a set of conjugate strike-slip faults may explain the fact that there is not a single molasse basin in the South Alpine range developed along its whole length, but there are in fact two separate basins (Fig. 1), bounded by faults which are strongly suspected to have acted as strike-slip faults during the basin evolution. In a certain sense the South Alpine foreland basins can be regarded as shear basins in the sense of Kingston, Dishroon & Williams (1983), linked to the obliqueness of convergence. The main stages of evolution of the South Alpine foredeeps should be considered within this structural framework. Due to the eastward shifting of deformation, only from Serravallian onwards is the Venetian basin actively involved in the South Alpine kinematic system. In the first phase of its history, from the Chattian to the Langhian, the basin has to be regarded as part of the Dinaric system, i.e. a weakly deforming foreland basin linked to the outermost and latest thrusting activity of the northern Dinarides.

The Venetian basin is thus a polyhistory basin in which a South Alpine 'polarity' is superimposed on an older Dinaric 'polarity', as indicated by the thickness pattern of sedimentary wedges, the clear petrofacies differentiation within the arenites and the pattern of clastic dispersal. The two-stage evolution of the Venetian basin is also indicated by the curve of cumulative thickness versus time (Fig. 15).

The first stage of basin evolution (Chattian to Langhian)

During the first stage of basin development tectonic control was rather weak due to low rates of tectonic progradation. It should be noted that a comparison between the curve of transgressions and regressions in the Venetian basin and that of Central Paratethys elaborated by Rögl & Steininger (1983) shows a striking similarity, in spite of the fact that the

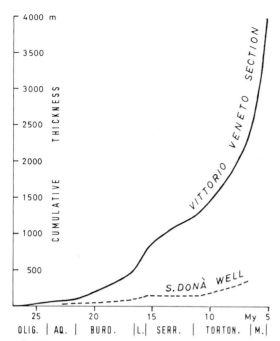

Fig. 15. Time versus cumulative thickness of the sedimentary column; the Vittorio Veneto section is located near the leading edge of the thrust stack while the S. Donà well is located in the outer area of the foreland basin. The pattern of the curves highlights the differentiation of basin evolution into two distinct stages.

compared areas (Fig. 3) have different types of tectonic framework. Strengthening of shelf circulation during the early Aquitanian and Langhian in the Venetian basin (as witnessed by storm- and/or tide-influenced deposits) coincide with major transgressions in the Central Paratethys. Such events imply the flooding of seaways connecting the Venetian basin and Central Paratethys (Rögl & Steininger, 1983). Positive features related to growth or rejuvenation of structural highs within the Venetian basin acted as the focus for sand deposition and localization of shelf sand ridges during these transgressive stages. On the other hand some regressive stages, accompanied by evidence of partial isolation and of faunal endemism, as well as by progradation of significantly diachronous sandstone bodies (during the late Aquitanian and late Burdigalian) can be correlated with discrete pulses of thrust activity in the outer Dinarides.

During the Chattian to Langhian stage of basin evolution the rapidly rising Alpine axial zone represented an important 'external' source of clastic sediments, as indicated by the conspicuous contribution from Penninic and Austroalpine units. The clastic

influx from the Dinarides cannot be estimated due to the fact that this source was essentially made up of poorly distinctive sedimentary rocks (mostly Tertiary flysch). Examples of basins primarily fed by an 'external' source represented by an adjoining rising range are quite common in the flysch basins of Mediterranean ranges (Ricci Lucchi, 1984b), and reflect the intricate interplay of adjacent ranges with diachronous evolution, a typical situation in the Mediterranean area. The Miocene flyschs of the Apennines (primarily fed by Alpine sources) may be mentioned as an example. The case is however more anomalous in the molasse foredeeps, which are usually flanked by a high-relief range.

The second stage of basin evolution (Serravallian to Recent)

During the middle Miocene a drastic change occurred in the tectonic framework of the basin: the axis of the foredeep shifted in position and trend and the basin was thereafter incorporated into the South Alpine kinematic system. It should be stressed in this regard that in the Mediterranean orogenic belt the middle Miocene was a time of major geodynamic and palaeogeographic changes, including the major phase of extension and subcrustal thinning in the Pannonian area and inner Dinarides, and the related strong reactivation of the dextral shear along the Periadriatic/ Vardar transcurrent fault system (Horváth, 1984). These facts may represent the response to a large-scale geodynamic event which also had some important effects in the South Alpine–Apenninic system, such as the above-outlined behavioural change of the Venetian basin and the spectacular acceleration of depocentre shifting in the Apennines during the Langhian (Ricci Lucchi, 1986).

From the Serravallian to the Recent the tectonic control on sedimentation and on the subsidence pattern in the Venetian basin was much stronger than during the first stage, due to the rapid SSW-ward 'progradation' of an imbricate stack of overthrusts (Fig. 16). Localized angular unconformities merging laterally into disconformities are thought to reflect the interplay of thrusting pulses with strike-slip motion along transverse faults.

The Serravallian–Tortonian megasequence clearly reflects the tectonic control on subsidence and sedimentation. In an initial stage accelerated subsidence leads to deposition of epibathyal marls (Serravallian) followed by organic-rich lower Tortonian mudstones.

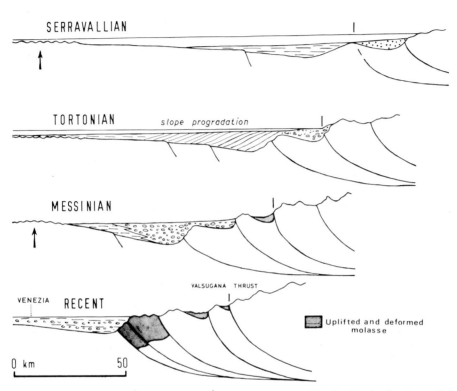

Fig. 16. Idealized transect across the eastern South Alpine orogenic belt and foreland basin showing evolution from the Langhian to the Messinian. An emergent imbricate fan gradually moved across the molasse basin and the inner molasse was cannibalized and recycled. No vertical scale implied.

A second stage is characterized by rapid and spectacular slope progradation (middle Tortonian) due to prominent clastic influx, which brought about the up- and out-building of a thick clastic wedge mostly consisting of prodelta mudstones. Localized conglomeratic bodies within this sequence (Vittorio Veneto area) represent mass-flow deposits probably funnelled along a structural depression connecting an inner thrust-sheet-top basin with the foredeep and genetically related to transverse strike-slip faults (Fig. 10). Eventually in the third stage (late Tortonian) the subsidence is overtaken by sedimentation and a stack of vertically aggrading fan-delta sequences is laid down on the platform created by the previous progradation. The sequence is interpreted as resulting from (1) accelerated subsidence under thrust-loading, (2) a strong clastic influx related to a thrusting surge

and finally (3) a reduced sedimentation rate under decreased, dominantly sediment-load subsidence.

Sediment dispersal pathways during the Tortonian–Messinian were mostly related to strike-slip transverse lines segmenting the thrust belt. The overall architecture of the Messinian alluvial system was very sensitive to the influence of diastrophism, as suggested by the internal organization of megasequences, bounded by distinct unconformities.

Palaeocurrent indicators and the composition of Tortonian–Messinian sandstones and conglomerates reflect a derivation of clastic sediments from a source which lies entirely (or almost entirely) within the eastern South Alpine domain. The abundant evidence of reworking at the expense of older molasse points to cannibalistic erosion of folded and faulted inner molasse involved in the thrusting.

ACKNOWLEDGMENTS

The ideas in this paper have been discussed with Professors Zanferrari, Sassi, G. V. Dal Piaz (Padova), Gelati (Milano), Brack (Zurich), Doglioni (Ferrara), and we have benefitted from their comments. We are grateful to Dr G. G. Ori and Professor G. V. Dal Piaz who carefully reviewed the manuscript and contributed to its improvement by helpful critical comments.

We gratefully acknowledge Professor A. Dal Negro for his valuable assistance in the mineralogical analysis. Dr L. Dondi kindly provided some information about the subsurface stratigraphy in the Padan plain. We are also indebted to Dr Brogiato and Mr Todesco for the careful execution of drawings and to Mrs Garbo for typing the manuscript. The work was financially supported by the Consiglio Nazionale delle Ricerche of Italy, grant C.T. 83.02203.05 and by 'Centro di Studio per i Problemi dell'Orogeno delle Alpi Orientali' of C.N.R.—Padova.

REFERENCES

AMATO, A., BARNABA, P.F., FINETTI, I., GROPPI, G., MARTINIS, B. & MUZZIN, A. (1976) Geodynamic outline and seismicity of Friuli-Venetia Julia Region. *Boll. Geofis. teor. appl.* **72**, 217–256.

AMBROSETTI, P., BOSI, C., CARRARO, F., CIARANFI, N., PANIZZA, M., PAPANI, G., VEZZANI, L. & ZANFERRARI, A. (1986) Neotectonic map of Italy 1:500,000 sheets 1 and 2. *Quad. Ricerca Scient.* **114/4** (in press).

AUBOUIN, J. (1973) Des tectoniques superposées et de leur signification par rapport aux modèles géophysiques: l'example des Dinarides; paléotectonique, tectonique, tarditectonique, neotectonique. *Bull. Soc. géol. Fr.* **15**, 426–460.

BALLY, A.W. (1980) Basins and subsidence—a summary. In: *Dynamics of Plate Interiors* (Ed. by A. W. Bally, P. L. Bender, T. R. McGetchin & R. I. Walcott). *Geodyn. Ser.* **1**, 5–20.

BALLY, A.W. & SNELSON, S. (1980) Realms of subsidence. In: *Facts and Principles of World Petroleum Occurrence* (Ed. by A. D. Miall). *Can. Soc. Petrol. Geol.* **6**, 9–94.

BIANCHI, A. (1934) Studi petrografici sull'Alto Adige orientale e regioni limitrofe. *Mem. Ist. Geol. R. Univ. Padova,* **10**, 243 pp.

BLANC-VERNET, L., PUJOS, M. & ROSSET-MOULINIER, M. (1984) Les cénoses de Foraminifères benthiques des plateaux continentaux francais (Manche, Sud-Gascogne, Ouest-Provence). In: *Benthos '83, 2nd int. Symp. Benthic Foraminifera,* Pau, April 1983 (Ed. by H. J. Oertli), pp. 71–79. Elf Aquitaine, Esso REP and Total CFP, Pau.

BLOW, W.H. (1969) Late middle Eocene to Recent planktonic foraminiferal biostratigraphy. In: *Proc. first int. Conf. Planktonic Microfossils,* Geneva 1967 (Ed. by P. Brönniman & H. H. Renz), pp. 199–422. Brill, Leiden.

BOCCALETTI, M. & DAINELLI, P. (1984) Il sistema regmatico neogenico-quaternario nell'area mediterranea: esempio di deformazione plastico-rigida post-collisionale. *Mem. Soc. geol. Ital.* **24** (1982), 465—482.

BOERSMA, A (1984) *Handbook of Common Tertiary Uvigerina.* Microclimates Press, Stony Point, New York, 207 pp.

BORSETTI, A.M., CATI, F., COLALONGO, M.L. & SARTONI, S. (1979) Biostratigraphy and absolute ages of the Italian Neogene. *Ann. Géol. Pays Hellén.* **1**, 183–197.

BRACK, P. (1981) Structures in the southwestern border of the Adamello intrusion (Alpi Bresciane, Italy). *Schweiz. miner. petrogr. Mitt.* **61**, 37–50.

BURCHFIEL, B.C. & ROYDEN, L. (1982) Carpathian foreland fold and thrust belt and its relation to Pannonian and other basins. *Bull. Am. Ass. Petrol. Geol.* **66**, 1179–1195.

CAROBENE, L. & CARULLI, G.B. (1981) Foglio 40 Palmanova. In: *Carta Tettonica delle Alpi Meridionali (alla scala 1:200.000)* (Ed. by A. Castellarin), *C.N.R., P.F. Geodin.* **441**, 51–54.

CAROBENE, L., CARULLI, G.B. & VAIA, F. (1981) Foglio 25 Udine. In: *Carta Tettonica delle Alpi Meridionali (alla scala 1:200.000)* (Ed. by A. Castellarin). *C.N.R., P.F. Geodin.* **441**, 39–45.

CASON, C., GRANDESSO, P., MASSARI, F. & STEFANI, C. (1981) Depositi deltizi nella molassa Cattiano-Burdigaliana del Bellunese (Alpi Meridionali). *Mem. Sci. Geol.* **34**, 325–354.

CASTELLARIN, A. (1979) Il problema dei raccorciamenti crostali nel Sudalpino. *Rend. Soc. geol. Ital.* **1** (1978), 21–23.

CASTELLARIN, A. (1981) *Carta Tettonica delle Alpi Meridionali (alla scala 1:200.000). C.N.R., P.F. Geodin.* **441**, 220 pp.

CASTELLARIN, A. (1984) Schema delle deformazioni tettoniche sudalpine. *Boll. Oceanol. teor. appl.* **2**, 105–114.

CASTELLARIN, A., FRASCARI, F. & VAI, G.B. (1980) Problemi di interpretazione geologica profonda del Sudalpino Orientale. *Rend. Soc. geol. Ital.* **2** (1979), 55–60.

CASTELLARIN, A. & VAI, G.B. (1981) Importance of Hercynian tectonics within the framework of the Southern Alps. *J. struct. Geol.* **3**, 477–486.

CASTELLARIN, A. & VAI, G.B. (1982) Introduzione alla geologia strutturale del Sudalpino. In: *Guida alla Geologia del Sudalpino centro-orientale* (Ed. by A. Castellarin & G. B.Vai). *Guide geol. region. S.G.I.,* Bologna, pp. 1–22.

CELET, P. (1977) The Dinaric and Aegean Arcs: the geology of the Adriatic. In: *The Ocean Basins and Margins. 4A: the Eastern Mediterranean* (Ed. by A. E. M. Nairn, W. H. Kanes & F. G. Stehli), pp. 215–261. Plenum Press, New York.

CHANNELL, J.E.T. & HORVÁTH, F. (1976) The African/ Adriatic promontory as a palaeogeographical premise for Alpine orogeny and plate movements in the Carpatho-Balkan region. *Tectonophys.* **35**, 71–101.

COLEMAN, J.M., PRIOR, D.B. & LINDSAY, J.F. (1983) Deltaic influences on shelfedge instability processes. In: *The Shelfbreak: Critical Interfaces on Continental Margins* (Ed. by D. J. Stanley & G. T. Moore). *Spec. Publ. Soc. econ. Paleont. Miner.,* Tulsa, **33**, 121–137.

DAL PIAZ, G.B. (1934) Studi geologici sull'Alto Adige orientale e regioni limitrofe. *Mem. Ist. Geol. R. Univ. Padova,* **10**, 245 pp.

DAL PIAZ, G.V. (1976) Alcune riflessioni sull'evoluzione geodinamica alpina delle Alpi. *Rend. Soc. Ital. Miner. Petrol.* **32**, 380–385.

DAL PIAZ, G.V., HUNZIKER, J.C. & MARTINOTTI, G. (1972) La zona Sesia-Lanzo e l'evoluzione tettonico-metamorfica delle Alpi nordoccidentali interne. *Mem. Soc. geol. Ital.* **11**, 433–466.

DE MOWBRAY, T. & VISSER, M.J. (1984) Reactivation surfaces in subtidal channel deposits, Oosterschelde, southwest Netherlands. *J. sedim. Petrol.* **54**, 811–824.

DE SITTER, L.U. & DE SITTER KOOMANS, C.M. (1949) The geology of Bergamasc Alps. Lombardia, Italy. *Leidse Geol. Med.* **14B**, 1–257.

DICKINSON, W.R. (1970) Interpreting detrital modes of graywacke and arkose. *J. sedim. Petrol.* **40**, 695–707.

DOGLIONI, C. (1984) *I sovrascorrimenti nelle Dolomiti: sistemi di ramp-flat.* Tecnoprint, Bologna, 22 pp.

DONDI, L., MOSTARDINI, F. & RIZZINI, A. (1982) Evoluzione sedimentaria e paleogeografica nella Pianura Padana. In: *Guida alla Geologia del Margine Appenninico-padano* (Ed. by G. Cremonini & F. Ricci Lucchi). *Guide geol. region.* S.G.I., Bologna, pp. 47–58.

FRISCH, W. (1976) Ein Modell zur alpidischen Evolution und Orogenese des Tauernfensters. *Geol. Rdsch.* **65**, 375–393.

GAETANI, M. & JADOUL, F. (1979) The structure of the Bergamasc Alps. *Rend. Acc. Naz. Lincei*, **65**, 411–416.

GAZZI, P. (1966) Le arenarie del flysch sopracretaceo dell'Appennino modenese; correlazioni con il flysch di Monghidoro. *Mineral. Petrogr. Acta*, **12**, 69–97.

GELATI, R. (1969) Nuove osservazioni sulla successione stratigrafica di età miocenica affiorante sul torrente Meduna in provincia di Pordenone. *Riv. Ital. Paleont. Stratigr.* **75**, 165–182.

GIESE, P. & NICOLICH, R. (1982) Explosion seismic crustal studies in the Alpine Mediterranean region and their implications to tectonic processes. In: *Alpine-Mediterranean Geodynamics* (Ed. by H. Berckhemer & K. Hsü). *Geodyn. Ser.* **7**, 39–73.

GRANDESSO, P. (1980) Dati preliminari sulla stratigrafia della serie molassica del Vallone Bellunese. *Paleont. Stratigraf. Evol.* **1**, 131–134.

GRÜNIG, A. (1984) Phenotypic variation in Spiroplectammina, Uvigerina and Bolivina. In: *Benthos '83, 2nd int. Symp. Benthic Foraminifera.* Pau, April 1983 (Ed. by H. J. Oertli), pp. 249–255. Elf Aquitaine, Esso REP and Total CFP, Pau.

GRÜNIG, A. & HERB, R. (1980) Paleoecology of Late Eocene benthonic Foraminifera from Possagno (Treviso—northern Italy). In: *Studies in Marine Micropaleontology and Paleoecology. A Memorial Volume to O. L. Bandy* (Ed. by W. V. Sliter). *Cush. Found. Spec. Publ.* **19**, 68–88.

GÜNZENHAUSER, B.A. (1985) Zur Sedimentologie und Palaogeographie der oligo-miocaenen Gonfolite Lombarda zwischen Lago Maggiore und der Brianza (Südtessin, Lombardei). *Beitr. geol. Karte Schweiz*, Neue Folge, **159**, Lieferung, 114 pp.

HARBAUGH, D.W. & DICKINSON, W.R. (1981) Depositional facies of Mississippian clastics, Antler foreland basin, central Diamond Mountains, Nevada. *J. sedim. Petrol.* **51**, 1223–1234.

HAYWARD, A.B. (1983) Coastal alluvial fans and associated marine facies in the Miocene of SW Turkey. In: *Modern and Ancient Fluvial Systems* (Ed. by J. D. Collinson & J. Lewin). *Spec. Publs int. Ass. Sediment.* **6**, 323–336. Blackwell Scientific Publications, Oxford.

HOMEWOOD, P. & ALLEN, P. (1981) Wave-, tide-, and current-controlled sandbodies of Miocene Molasse, western Switzerland. *Bull. Am. Ass. Petrol. Geol.* **65**, 2534–2545.

HORVÁTH, F. (1984) Neotectonics of the Pannonian basin and the surrounding mountain belts: Alps, Carpathians and Dinarides. *Annls Géophys.* **2**, 147–154.

HSÜ, K.J. (1982) Editor's introduction. Alpine-Mediterranean geodynamics: past, present and future. In: *Alpine-Mediterranean Geodynamics* (Ed. by H. Berckhemer and K. J. Hsü). *Geodyn. Ser.* **7**, 7–14.

IACCARINO, S. & SALVATORINI, G. (1982) A framework of planktonic foraminiferal biostratigraphy for Early Miocene to Late Pliocene Mediterranean area. *Paleont. stratigr. Evoluz.* **2**, 115–125.

INGLE, J.C. (1980) Cenozoic paleobathymetry and depositional history of selected sequences within the southern California continental borderland. In: *Studies in Marine Micropaleontology and Paleoecology. A Memorial Volume to O. L. Bandy* (Ed. by W. V. Sliter). *Cush. Found. Spec. Publ.* **19**, 163–195.

KINGSTON, D.R., DISHROON, C.P. & WILLIAMS P.A. (1983) Global basin classification system. *Bull. Am. Ass. Petrol. Geol.* **67**, 2175–2193.

LAUBSCHER, H.P. (1971) The large-scale kinematics of the western Alps and the northern Apennines and its palinspastic implications. *Am. J. Sci.* **271**, 193–226.

LAUBSCHER, H.P. (1974) Evoluzione e struttura delle Alpi. *Scienze*, **72**, 264–275.

LAUBSCHER, H.P. (1986) The late Alpine (Periadriatic) intrusions and the Insubric line. *Mem. Soc. geol. Ital.* **26**, 21–30.

LAUBSCHER, H. & BERNOULLI, D. (1982) History and deformation of the Alps. In: *Mountain Building Processes* (Ed. by K. J. Hsü), pp. 169–180. Academic Press, London.

MASSARI, F. (1984) Resedimented conglomerates of a Miocene fan delta complex. Southern Alps, Italy. In: *Sedimentology of Gravels and Conglomerates* (Ed. by E. H. Koster & R. J. Steel). *Can. Soc. Petrol. Geol.* **10**, 259–278.

MASSARI, F., GRANDESSO, P., STEFANI, C. & ZANFERRARI, A. (1986) The Oligo–Miocene molasse of Veneto-Friuli region, Southern Alps. *Giornale Geol.* (in press).

MASSARI, F., IACCARINO, S. & MEDIZZA, F. (1976) Depositional cycles in the Tortonian–Messinian of the Southern Alps (Italy): transition from fan-delta to alluvial fan sedimentation. *Messinian Seminar 2, Field Trip Guidebook*, pp. 17–37.

MASSARI, F., ROSSO, A. & RADICCHIO, E. (1974) Paleocorrenti e composizione dei conglomerati tortoniano-messiniani compresi tra Bassano e Vittorio-Veneto. *Mem. Ist. Geol. Min. Univ. Padova*, **31**, 1–20.

McLEAN, R.J. & JERZYKIEWICZ, T. (1978) Cyclicity, tectonics and coal: some aspects of fluvial sedimentology in the Brazeau–Paskapoo Formations, Coal Valley area, Alberta, Canada. In: *Fluvial Sedimentology Canadian* (Ed. by A. D. Miall). *Soc. Petrol. geol. Calgary*, **5**, 441–468.

MILJUSH, P. (1973) Geologic-tectonic structure and evolution of outer Dinarides and Adriatic area. *Bull. Am. Ass. Petrol. Geol.* **57**, 913–929.

MILLER, C. (1974) On the metamorphism of the eclogites and high-grade blueschists from the Penninic Terraine of the Tauern Window, Austria. *Schweiz. miner. petrogr. Mitt.* **54**, 371–384.

MILLER, C. (1977) Chemismus und phasenpetrologische Untersuchungen der Gesteine aus der Eklogitzone des

Tauernfensters, Österreich. *Tschermaks miner. petrogr. Mitt.* **24**, 221–277.

MILLER, H., MÜLLER, S. & PERRIER, G. (1982) Structure and dynamics of the Alps. A geophisical inventory. In: *Alpine–Mediterranean Geodynamics* (Ed. by H. Berckhemer & K. Hsü). *Geodyn. Ser.* **7**, 175–203.

MÜLLER, G. (1977) Fault-plane solution of the earthquake in northern Italy, 6 May 1976, and implications for the tectonics of the Eastern Alps. *J. Geophys.* **42**, 343–349.

MURRAY, J.W. (1973) *Distribution and Ecology of Living Benthic Foraminiferids.* Heinemann, London, 274 pp.

NICHOLS, G. & ORI, G.G. (1983) Sedimentation along the compressive margin of the Ebro basin (NE Spain). *4° IAS Regional Meeting* (abstract), pp. 83–86.

PANZA, G., CALCAGNILE, G., SCANDONE, P. & MÜLLER, S. (1980) La struttura profonda dell'area mediterranea. *Scienze*, **141**, 276–285.

PETTIJOHN, F.J., POTTER, P.E. & SIEVER, R. (1972) *Sand and Sandstone.* Springer-Verlag, New York.

PHLEGER, F.B. & SOUTAR, A. (1973) Production of benthic foraminifera in three east Pacific oxigen minima. *Micropaleont.* **19**, 110–115.

PIERI, M. & GROPPI, G. (1981) *Subsurface Geological Structure of the Po Plain, Italy.* C.N.R., P.F. Geodinam. **414**, 13 pp.

QUINLAN, G.M. & BEAUMONT, C. (1984) Appalachian thrusting, lithospheric flexure, and the Paleozoic stratigraphy of the Eastern Interior of North America. *Can. J. Earth Sci.* **21**, 973–996.

RICCI LUCCHI, F. (1984a) The deep-sea deposits of the Miocene Marnoso Arenacea Formation, Northern Apennines. *Geo-Mar. Lett.* **3**, 203–210.

RICCI LUCCHI, F. (1984b) Flysch, molassa, cunei clastici: tradizione e nuovi approcci nell'analisi dei bacini orogenici dell'Appennino Settentrionale. In: *Cento Anni di Geologia Italiana. Vol. giub. I Centenario S.G.I.*, Bologna, pp. 279–295.

RICCI LUCCHI, F. (1986) The foreland basins system of Northern Apennines and related clastic wedges: a preliminary outline. *Giornale Geol.* (in press).

RIZZINI, A. & DONDI, L. (1978) Erosional surface of Messinian age in the subsurface of the Lombardian Plain (Italy). *Mar. Geol.* **27**, 303–325.

RÖGL, F., CITA, M.B., MÜLLER, C. & HOCHULI, P. (1975) Biochronology of conglomerate bearing Molasse sediments near Como (Italy). *Riv. Ital. Paleont. Stratigr.* **81**, 57–88.

RÖGL, F. & STEININGER, F.F. (1983) Vom Zerfall der Tethys zu Mediterranean und Paratethys. Die neogene Palaeogeographie und Palinspastik des zirkum-mediterranean Raumes. *Ann. Nat. Museum Wien*, **85**, 135–163.

ROYDEN, H.L., HORVÁTH, F. & BURCHFIEL, B.C. (1982) Transform faulting, extension, and subduction in the Carpathian Pannonian region. *Bull. geol. Soc. Am.* **93**, 717–725.

ROYDEN, L., HORVÁTH, F. & RUMPLER, J. (1983) Evolution of the Pannonian basin system, 1. Tectonics. *Tectonics*, **2**, 63–90.

STEFANI, C. (1982) Geologia dei dintorni di Fanna e Cavasso Nuovo (Prealpi Carniche). *Mem. Sci. Geol.* **35**, 203–212.

STEFANI, C. (1984) Sedimentologia della molassa delle Prealpi Carniche occidentali. *Mem. Sci. Geol.* **36**, 427–442.

SWIFT, D.J.P. & RICE, D.D. (1984) Sand bodies on muddy shelves: a model for sedimentation in the Western Interior Seaway, North America. In: *Siliciclastic Shelf Sediments* (Ed. by R. W. Tillman & C. T. Siemers). *Spec. Publs Soc. econ. Paleont. Miner., Tulsa*, **34**, 43–62.

TRÜMPY, R. (1973) The timing of orogenic events in the Central Alps. In: *Gravity and Tectonics* (Ed. by K. A. De Jong & R. Scholten), pp. 229–251. Wiley, New York.

TRÜMPY, R. (1980) *Geology of Switzerland. A Guide-book.* (Ed. by Schweizerische Geol. Kommission), pp. 24–93. Wepf & Co., Basel.

VAIL, P.R. & HARDENBOL, J. (1979) Sea-level changes during the Tertiary. *Oceanus*, **22**, 71–79.

VAIL, P.R., MITCHUM, R.M. & THOMPSON, S. (1977) Seismic stratigraphy and global changes of sea level. 4. Global cycles of relative changes of sea level. In: *Seismic Stratigraphy. Applications to Hydrocarbon Exploration* (Ed. by C. E. Payton). *Mem. Am. Ass. Petrol. Geol.* **26**, 83–97.

WESCOTT, W.A. & ETHRIDGE, F.G. (1980) Fan-delta sedimentology and tectonic setting, Yallahs fan-delta, Southeast Jamaica. *Bull. Am. Ass. Petrol. Geol.* **64**, 374–399.

WOLETZ, G. (1967) Schwermineralvergesellschaftungen aus ostalpinen Sedimentationsbecken der Kreidezeit. *Geol. Rdsch.* **56**, 308–320.

WRIGHT, R. (1978) Neogene paleobathymetry of the Mediterranean based on benthic foraminifera from DSDP Leg 42A. In: *Init. Rep. Deep Sea drill. Proj.* **42** (Ed. by K. Hsü et al.), pp. 837–846. U.S. Government Printing Office, Washington.

ZANFERRARI, A., BOLLETTINARI, G., CAROBENE, L., CARTON, A., CARULLI, G.B., CASTALDINI, D., CAVALLIN, A., PANIZZA, M., PELLEGRINI, G.B., PIANETTI, F. & SAURO, U. (1982) Evoluzione neotettonica dell'Italia Nord-Orientale. *Mem. Sci. Geol.* **35**, 355–376.

ZUFFA, G.G. (1980) Hybrid arenites: their composition and classification. *J. sedim. Petrol.* **50**, 21–29.

Spec. Publs int. Ass. Sediment. (1986) **8**, 169–182

A comparison between a present-day (Taranto Gulf) and a Miocene (Irpinian Basin) foredeep of the Southern Apennines (Italy)

TULLIO PESCATORE *and* MARIA ROSARIA SENATORE

Dipartimento di Scienze della Terra, Università di Napoli, Largo S. Marcellino, 10 80138 Napoli, Italy

ABSTRACT

The Gulf of Taranto, located between an orogenic belt (Apennine Chain and Calabrian Arc) and the Apulia foreland, is the present-day foredeep area of the Southern Apennines.

Several Plio–Pleistocene depocentres exist overlying different structural units. From west to east, these are: basins on the allochthonous thrust sheets (piggyback basins); the foredeep *sensu stricto*, located at the front of the thrust sheets, and minor basins on the Apulia Unit.

An analogous distribution of the areas of sedimentation has been inferred from analysis of the Irpinian Units in the Apennine Chain. These units were deposited from Burdigalian to Tortonian in a basin called the Irpinian Basin. The diachronism of the facies and the regressive trend of the Irpinian Units allow us to consider an evolution of the basin with an eastwards shifting of the foredeep.

The Gulf of Taranto would seem to be a foredeep in a mature stage in which the front of the thrust sheets is close to the Apulia Platform; the Irpinian Basin, in its initial stage (Burdigalian), represents a wide foredeep tending to shrink; it closed when the margin of the thrust sheets reached the Abruzzi–Campania Platform (Tortonian), provoking a major phase of deformation.

Structural trends of the Irpinian Basin were controlled by features acquired during the tensile regime of the Adria continental margin.

INTRODUCTION

The Southern Apennine Chain is composed of units deriving from the deformation of the continental margin of the Adria plate (D'Argenio, Horvarth & Channel, 1980). During the Mesozoic, this margin was characterized by a prevalently tensile tectonic regime and the sedimentary domains were represented by a succession of carbonate neritic platforms and deep basins (D'Argenio, Pescatore & Scandone, 1975); some authors hypothesize transcurrent movements as a major structural control (Horvarth & D'Argenio, 1985).

During the Palaeogene, compressive deformation affected both the Adria and European continental margins of the oceanic Tethys. In the Neogene, as a consequence of the opening of the Tyrrhenian Basin (Scandone, 1979), the most external domains of the Adria plate were involved in the orogeny.

In these external zones, terrigenous sedimentation was at first mainly quartzose (Numidian Flysch) in the late Oligocene and Aquitanian, but subsequently arkosic-lithic from the Burdigalian to the Tortonian. During the Burdigalian–Tortonian interval the arkosic-lithic sedimentation occurred both upon the thrust sheets and in front of them. At the same time, calcareous clastic sedimentation occurred in more external basins adjacent to the Apulia Platform. The sedimentary domain including all these depocentres has been called the Irpinian Basin (Cocco *et al.*, 1972; Pescatore, 1978).

Tectogenesis shifted eastwards and southwards in the Neogene and Quaternary; as a result the Southern Apennines has assumed its present configuration (Fig. 1) with the Gulf of Taranto as the foredeep, located between the allochthonous thrust sheets of the chain to the west and the Apulia foreland to the east (Selli & Rossi, 1975; Belfiore *et al.*, 1981).

The aim of this paper is to point out the analogies and the differences between the Irpinian Basin, which

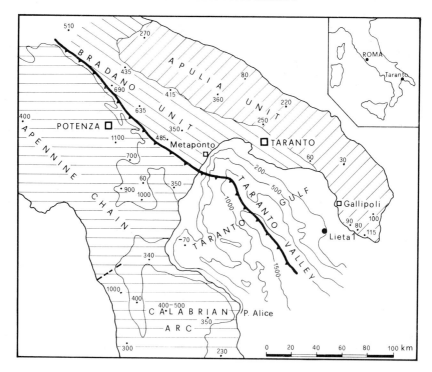

Fig. 1. Location map with values of uplift and subsidence (−) in metres for the last million years (from Ciaranfi *et al.*, 1983). Bathymetry is in metres. The toothed line corresponds to the front of the allochthonous thrust sheets of the Apennines. Lieta 1 is a borehole (AGIP, 1977).

represents the Miocene foredeep of the Southern Apennine Chain, and the Gulf of Taranto.

The Irpinian Basin and the Taranto Gulf can be considered *sensu lato* as foredeep areas. Different depocentres can be distinguished: piggyback basins (on the thrust sheets), foredeep *sensu stricto* (at the toe of the thrust sheets) and minor foreland basins.

THE GULF OF TARANTO

Regional geology

The structural framework of the land surrounding the Gulf of Taranto is portrayed in Fig. 1. The allochthonous thrust sheets of the Apennine Chain and the Calabrian Arc are derived from the deformation of palaeogeographic elements of the continental margin of the Southern Tethys (Haccard, Lorenz & Grandjaquet, 1972; Amodio Morelli *et al.*, 1976; Catalano *et al.*, 1977; Pescatore, 1978; etc.). The deformation in these sectors lasted until the Miocene

or the Pliocene, followed by intense uplift in the Quaternary with maximum estimated values of 1000 to 1200 m (Bousquet, 1972; Ciaranfi *et al.*, 1983). The Crati Valley (Figs 1 and 2), on the other hand, underwent subsidence and sedimentation during the Quaternary; the Crati River today represents the major stream in the zone, collecting and carrying a large quantity of terrigenous material to the Corigliano Basin (Crati Group, 1981; Ricci Lucchi *et al.*, 1983–84).

The Bradano Unit (Fig. 1) is a terrigenous succession (D'Argenio, Pescatore & Scandone, 1973), several thousand metres thick, accumulated in a Plio-Pleistocene foredeep (Bradano foredeep, Crescenti, 1975). This unit was affected, in early Pleistocene, by differential uplift, decreasing towards the Apulia foreland (Cotecchia & Magri, 1967; Ricchetti, 1967; Vezzani, 1967).

Some rivers of this zone cross the Apennine thrust sheets (Sinni River and Agri River; Fig. 2) while others cross the Bradano Unit (Basento and Bradano Rivers); the terrigenous materials transported by these rivers are deposited on the continental shelf or reach

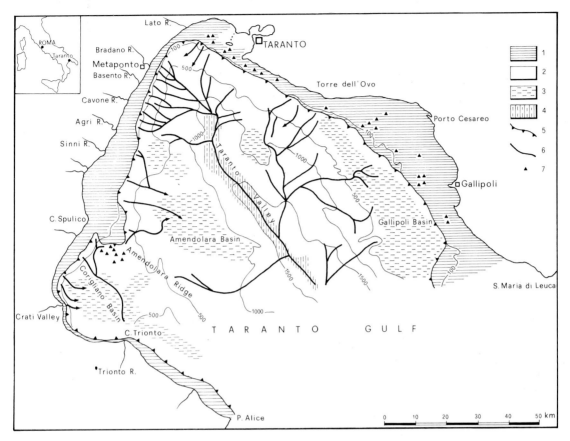

Fig. 2. Bathymetric features of the Gulf of Taranto. (1) Continental shelf. (2) Continental slope and ridges. (3) Flat areas (basins). (4) Taranto Valley (foredeep basin). (5) Shelf break. (6) Main channels. (7) Coralline algal banks.

bathyal depths through channels on the slope (Belfiore, 1984).

The Apulia foreland (Fig. 1) is a stable area affected essentially by vertical movements. It has a continental basement (Morelli *et al.*, 1975) covered by a Ceno–Mesozoic carbonate succession exceeding 6000 m (Apulia Unit; D'Argenio *et al.*, 1973). The Apulia Unit extends below the Plio–Pleistocene deposits of the Bradano foredeep as far as the thrust belt, where it was found below the thrust sheets (Carissimo *et al.*, 1963); Servizio Geologico d'Italia, 1969).

Apulia consists of karst land with low relief and without a stream system. Terrigenous materials which reach the shoreline are scarce.

Morphology and sedimentation

Three sectors can be distinguished in the Gulf of Taranto, each with specific morphological characters and different areas of sedimentation (Fig. 2):

Western sector

This sector lies between the margin of the Taranto Valley and the coast. It covers the southern continuation of the Apennines.

Two types of shelves are recognizable in this sector:

(1) A prograding shelf with frontal accumulation (Mougenot, Boillot & Rehault, 1983) between Capo Spulico and the mouth of the Agri River. Its width is between 9 and 14 km and the average inclination is 1°. The shelf break is located at a depth of *c.* 125–150 m.

(2) A regrading shelf (Mougenot *et al.*, 1983) which is found between the mouths of the Agri and Lato Rivers and between C. Spulico and Punta Alice. Its width varies from some kilometres down to 1 km. The shelf break is located at a depth varying between 100 and 30 m. Various channels from the shelf break transfer the neritic material to an epibathyal depth.

In the western sector, submarine reliefs separated by transverse valleys make up the Amendolara Ridge, which has a depth of *c.* 25 m in its northern sector (Amendolara Bank), continuing due SW down to bathyal depths. Two basins, Corigliano and Amendolara, are separated by this ridge; the former is chiefly supplied by Calabrian rivers and by the Crati River in particular (Crati Group, 1981); the latter basin is supplied by the Sinni and Agri Rivers (Figs 2 and 3).

Central sector

The central sector is the Taranto Valley, a NW–SE trough located between the allochthonous thrust sheets to the west and the Apulia foreland to the east. It constitutes the southern continuation of the Bradano foredeep.

The valley heads on the shelf between Metaponto and the mouth of the Agri River.

Shelf terrigenous materials accumulate at the base of a steep slope where the tributary channels merge in a valley about 6 km wide. To the SE, seawards of Gallipoli, the width of the valley diminishes to 2 km. Erosion and sedimentation alternate in this area; moreover, some channels coming from the Apulia sector bring in bioclastic calcareous deposits. Continuing SE, the valley widens again, reaching a width of 8 km seawards of S. Maria di Leuca where another depocentre is located. The inclination of the valley is between 0·4° and 3° (Fig. 4).

The recent sediments are composed of sand and silt intercalated with pelitic deposits. The coarser materials are transported in the valley by turbidity currents (Senatore *et al.*, 1982).

Eastern sector

The eastern sector is situated between the Valley of Taranto and Apulia; it constitutes a stepfaulted foreland, characterized by a shelf with terraces linked to the post-Wurmian transgression (depths of 20–30, 50–60, 100–110 m). The shelf width varies from 4 to 22 km; the average inclination is between 1° and 2°. The slope, with an inclination between 4° and 6°, is cut by several channels and more than 30% of its total surface is covered by slump deposits (Senatore *et al.*, 1982).

This sector comprises two Plio–Quaternary depocentres. The most extended one, the Gallipoli Basin, is located seawards of Torre dell'Ovo and S. Maria di Leuca; the Plio–Quaternary deposits are more than 500 m thick and lie unconformably on the Cretaceous carbonate rocks of the Apulia foreland (Lieta 1 Well; AGIP, 1977).

Structural framework of the Gulf of Taranto

Our knowledge of the tectonic structures of the Gulf of Taranto (Fig. 5) is based principally on the interpretation given by different authors of the seismic profiles (Finetti & Morelli, 1971; Finetti, 1976; Grandjaquet & Mascle, 1978; Biju-Duval, Letouzey & Montadert, 1979; Barone *et al.*, 1982; Rossi, Auroux & Mascle, 1983; Tramutoli *et al.*, 1984) and on deeper crustal profiles interpreted by Cello *et al.* (1981).

The first interpretation of seismic profile MS 25 given by Finetti & Morelli (1971) clearly distinguishes the continuation of three main geological units of Southern Italy: chain, foredeep and foreland in the Gulf of Taranto.

A more recent interpretation of the MS 25 profile (Finetti, 1976) indicates imbricated thrusts characterizing the allochthonous deposits. The deformation affects the Messinian and Pliocene deposits whereas the Quaternary sedimentary cover remains undeformed (Fig. 5-2a).

Grandjaquet & Mascle (1978), in interpreting the same profile, added Quaternary normal faults which define a horst and graben structure involving the recent cover (Fig. 5-2b).

A deep crustal profile through the Gulf, presented by Cello *et al.* (1981), identifies the structural framework of the various allochthonous units of the Apennine Chain. The Pleistocene successions are undeformed in the western sector of the Gulf but are thrusted in the eastern zones (Fig. 5-3).

Rossi *et al.* (1983) have recently presented a structural scheme of the Gulf of Taranto in which the shallow structure of the eastern Calabrian margin is interpreted as a large gravitational slide produced by the progressive uptilt, towards the east, of the margin itself. This structure is considered to be a consequence of the convergence between Apulia and Calabria. This mechanism does not exclude the existence, at great depth, of imbricate structures, even though the subduction is believed to have been active only until the early Pliocene (Fig. 5-1).

Tramutoli *et al.* (1984) in their structural scheme of the Gulf of Taranto, point out the influence of Quaternary compressive tectonics. The same phases are recognized in onshore exposures (Metaponto Thrust Sheet; Ogniben, 1969).

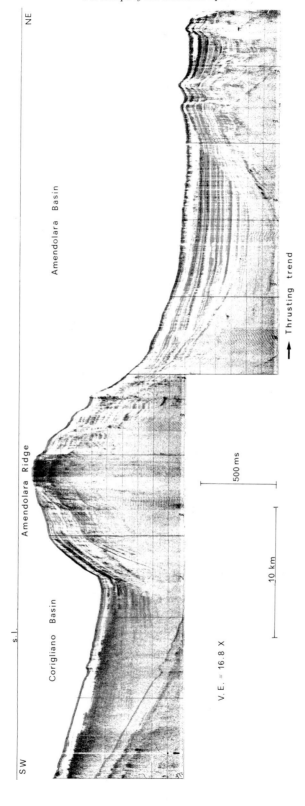

Fig. 3. Sparker profile across the Gulf of Taranto showing piggyback basins of the western sector. Location is on line 3 of Fig. 5.

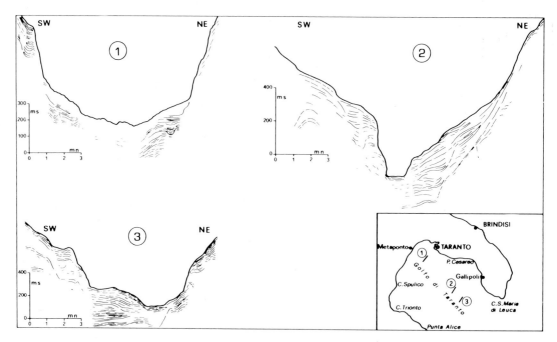

Fig. 4. Transverse profiles across the Taranto Valley from sparker profiles. (1) Northern sector; depocentre at the base of the slope corresponds with the confluence of the tributary channels. (2) Middle sector; seaward to Gallipoli, showing the marked reduction of the section of the valley. (3) Southern sector showing progressive widening.

Using Finetti's (1976) interpretation and the observations by Tramutoli *et al.* (1984), it is possible to recognize, on a structural basis, the same zones already described in morphological terms.

(1) Western sector: characterized by an imbricate structure with two main fronts of thrusting, the first one being localized at the eastern margin of the Amendolara Ridge and the second at the western margin of the Taranto Valley. The latter also constitutes the eastern boundary of the Apennine Chain.

The emergence of these thrusts forms structural and morphological highs. Towards the west they limit asymmetrical basins where the sedimentation occurred during deformation (Fig. 3).

(2) Central sector (Taranto Valley): the foredeep. This is a trough at the toe of the allochthonous thrust sheets.

(3) Eastern sector (Apulia margin): the foreland, with Plio–Quaternary basins.

IRPINIAN BASIN

The Irpinian Units, or Irpinids, include several tectonic units of the Southern Apennine Chain and, at present, outcrop in Campania and Lucania, along two major belts, both trending NW–SE (Pescatore, 1978, 1986). The south-western belt includes terrigenous deposits resting unconformably on the older allochthonous thrust sheets (Castelvetere Formation—upper part—and Gorgoglione Flysch) while the north-eastern belt is formed by calcareous and terrigenous turbidites conformably following the underlying deposits (Serra Palazzo Formation and Faeto Formation).

The stratigraphic columns referring to the Irpinian Units and their sedimentary substrata are given in Fig. 6.

Several studies detail the geological and sedimentological characteristics of the Irpinian Units (Boenzi & Ciaranfi, 1970; Cocco *et al.*, 1972; Crostella & Vezzani, 1964; Lojacono, 1975; Palmentola, 1970; Pescatore, 1971, 1978; Pescatore & Tramutoli, 1980).

The palaeogeographic domains existing during the Mesozoic and Early Cenozoic in this sector of the Adria continental margin (Fig. 7A), from west to east, have been named as follows: Campania–Lucania Platform, Lagonegro Basin, Abruzzi–Campania Platform, Molise Basin and Apulia Platform (Ippolito *et al.*, 1975; D'Argenio *et al.*, 1980).

An upper Oligocene–lower Miocene terrigenous

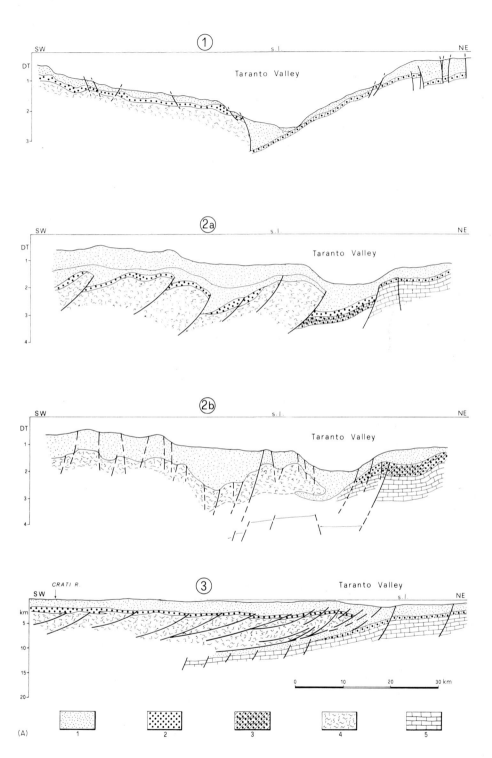

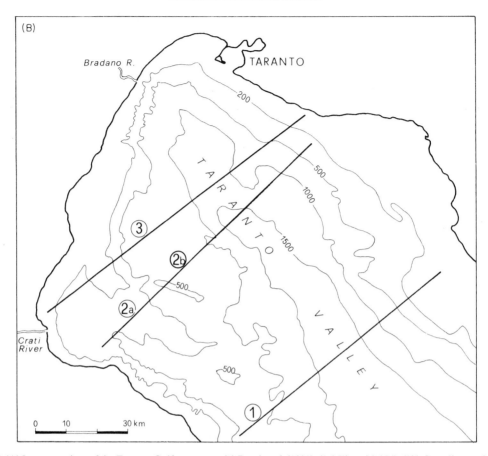

Fig. 5. (A) Interpretations of the Taranto Gulf structures. (1) Rossi *et al.* (1983). (2a) Finetti (1976). (2b) Grandjaquet & Mascle (1978). (3) Cello *et al.* (1981). 1—Terrigenous deposits (Quaternary); 2—Terrigenous deposits (Miocene and Pliocene); 3—Carbonate and marl deposits (Cenozoic), Apulia foreland; 4—Apennine thrust sheets (Mesozoic and Cenozoic); 5—Neritic carbonate deposits (Mesozoic), Apulia foreland. (B) Location of the seismic profiles.

succession consisting of an alternation of quartzose sandstone and pelites (Numidian Flysch; Palmentola, 1967; Ogniben, 1969; Wezel, 1970), represents the beginning of the terrigenous synorogenic sedimentation in the external domains of the Southern Apennines. These deposits, in the western sector, lie unconformably on the neritic carbonate deposits of the Campania–Lucania Platform, while, eastwards, they overlie the deposits of the Lagonegro and Molise Basins. The quartzose material was deposited mainly in basins between carbonate platforms (particularly in the Lagonegro Basin). The mature quartzose clastics probably derived from the African continental margin (Ogniben, 1969; Wezel, 1974).

In the early Miocene, the Campania–Lucania Platform progressively collapsed towards the foredeep as the front of the thrust sheets was approaching and received the quartzose sediments of the Numidian Flysch followed by the immature arkosic lithic deposits coming from the thrust sheets (Perrone & Sgrosso, 1981; Sgrosso, 1981, 1983).

At the same time, breccias and calcareous olistoliths from the higher parts of the Campania–Lucania Platform reached the foredeep (lower part of Castelvetere Formation), while in the Lagonegro and Molise Basins the sedimentation continued without immature terrigenous deposits.

In the Burdigalian, an important tectonic phase significantly modified the pre-existing domain. In particular, it deformed the terrains of the Campania–

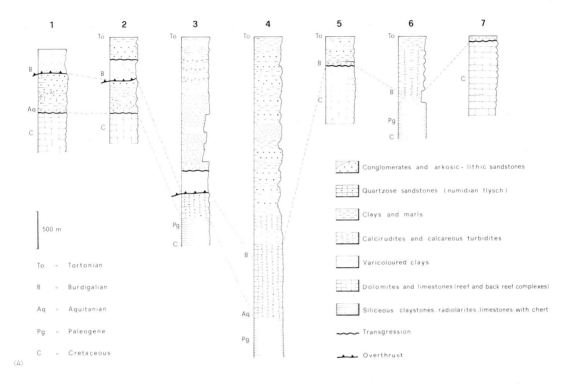

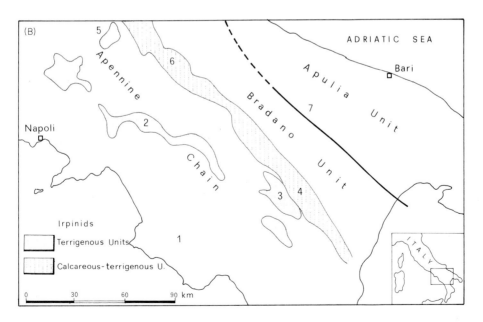

Fig. 6. (A) Stratigraphic columns of the Irpinian Units and their sedimentary substrata. (B) Location of the stratigraphic columns.

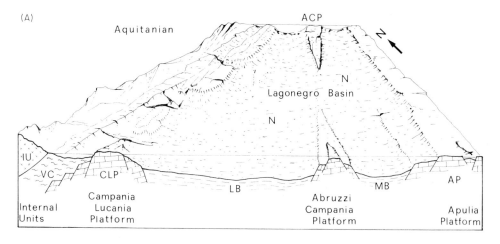

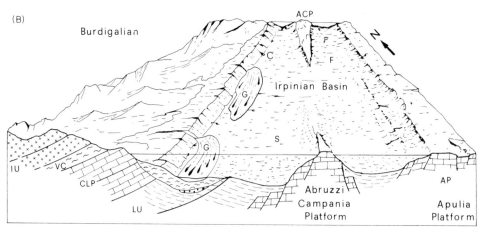

Fig. 7. Palaeogeographic scheme before (A) and after (B) the Burdigalian tectonic phase. IU—Alpine and Apennine Units. VC—Varicoloured clays. CLP—Campania-Lucania Platform. LB—Lagonegro Basin. ACP—Abruzzi-Campania Platform. MB—Molise Basin. AP—Apulia Platform. N—Numidian Flysch (quartzose sandstones). LU—Lagonegro Unit. G—Gorgoglione Flysch. C—Castelvetere Formation (upper part). S—Serra Palazzo Formation. F—Faeto Formation. Not to scale.

Lucania Platform and in part those of the Lagonegro Basin, while structuring the Irpinian Basin (Fig. 7B). Depocentres may be separated into three groups.

(1) Upon the allochthonous thrust sheets. Terrigenous successions were deposited at the base of the slope (upper part of Castelvetere Formation) or in elongate deep sea fans parallel to the structural belts (Gorgoglione Flysch).

The terrigenous deposits (arkosic-lithic) were supplied by erosion of the thrust sheets and probably from an emergent basement farther to the west (in the area of the present Tyrrhenian Sea, a northwards prolongation of the Calabria

thrust complex at that time). The supply was lateral with longitudinal transport.

These basins are piggyback basins (Ori & Friend, 1984) or satellite basins (Ricci Lucchi, this volume).

(2) At the front of the thrust sheets. A calcareous and terrigenous succession (Serra Palazzo Formation) was deposited at the toe of the thrust front. The calcareous material came from the Abruzzi–Campania and Apulia Platforms, while terrigenous clastics derived from the thrust sheets. Thus the supply was from both margins of the foredeep, while the transport was longitudinal.

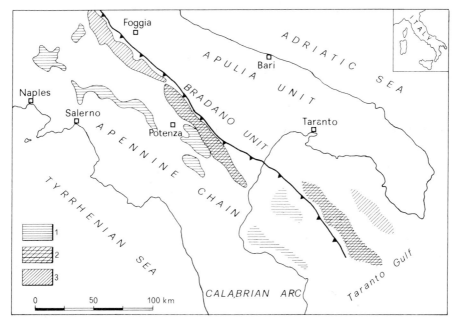

Fig. 8. Facies distribution in the Irpinian Units and the homologous sediments in the Gulf of Taranto. (1) Terrigenous sediments deposited in piggyback basins. (2) Calcareous and terrigenous sediments of the foredeep. (3) Calcareous and terrigenous sediments deposited on the Apulian foreland, unrelated to thrusting.

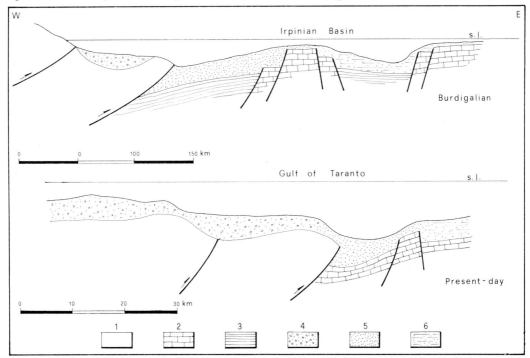

Fig. 9. Comparative sections across the Gulf of Taranto (interpreted) and the Irpinian Basin after the Burdigalian tectonic phase (reconstructed). (1) Allochthonous thrust sheets. (2) Neritic platforms. (3) Pelagic basins. (4) Piggyback basins. (5) Foredeep. (6) Basins located on foreland unrelated to thrusting. The scale of length is greatly approximated for the Irpinian Basin; not to scale vertically.

(3) In the areas not yet deformed, i.e. the Molise Basin, located between the Abruzzi–Campania and the Apulia Platforms, turbiditic and hemipelagic calcareous successions were deposited (Faeto Formation). The clastic materials were fed from both platforms.

An important point is that the sedimentation occurred while the axis of the foredeep migrated eastwards. Evidence of this migration is in the diachronism of the terrigenous facies (eastwards becoming younger) and the regressive trend of the successions (Pescatore, 1978).

Sedimentation in the Irpinian Basin terminated as a result of a new and important tectonic phase in the early Tortonian, which again displaced the sedimentary substratum and the overlying basin deposits eastwards. As a result, Irpinian Units were thrust over the sediments of the Abruzzi–Campania Platform.

SUMMARY AND CONCLUSIONS

The distribution of the Irpinian Units and the Plio–Quaternary depocentres in the Gulf of Taranto shows, from west to east, the same succession of basins (Figs 8, 9 and Table 1):

Table 1. Irpinian Basin units as compared with the Gulf of Taranto sedimentary domains

	Irpinian Basin	Gulf of Taranto
Piggyback basins	Castelvetere Formation—upper part	Amendolara Basin
	Gorgoglione Flysch	Corigliano Basin
Foredeep basin	Serra Palazzo Formation	Taranto Valley
Foreland basin	Faeto Formation	Gallipoli Basin

Factors which weaken the comparison between the Irpinian Basin and the Gulf of Taranto are the different sizes of these areas and their respective life spans. The width of the Irpinian Basin can be evaluated as two or three times greater than the Gulf of Taranto; there are no references for a definition of the length. There is a lack of stratigraphic data on the successions of the Gulf of Taranto, except for the Plio–Pleistocene succession of the Gallipoli Basin. The other basin successions have, therefore, been attributed to the Plio–Pleistocene period.

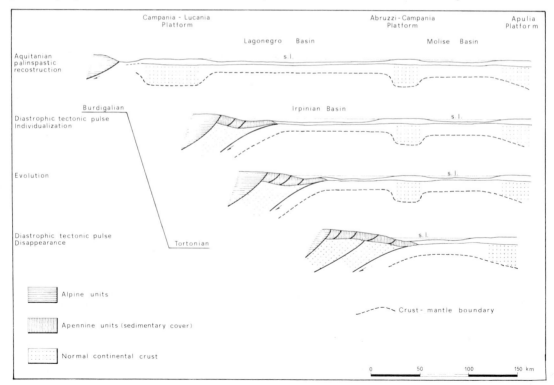

Fig. 10. The evolution of the Irpinian Basin.

Another difference is the significance of the Faeto Formation and Gallipoli Basin succession. In the former, Tertiary subsidence continues the pattern established in the Mesozoic of the Molise Basin (Fig. 6); contrarily, the Gallipoli Basin is a narrow Plio–Pleistocene depocentre resting on the Mesozoic Apulia Platform.

Lastly, the evolution of the Irpinian Basin occurred during a period of *c*. 10 Myr from the Burdigalian to Tortonian. Contrarily, for the Gulf of Taranto, only 2 Myr are taken into consideration.

During the compressive regime, features acquired in the previous crustal stretching should have influenced the development and evolution of the Irpinian Basin. This is marked by three stages (Fig. 10):

(i) Individualization: Burdigalian tectonic phase, collision of two crustal segments of different thicknesses.
(ii) Development: the western margin (orogenic) moved toward the opposite stable margin, which sank. This stage is characterized by a diachronism of the facies and a regressive trend in the successions.
(iii) Closure: Tortonian tectonic phase, which determined the collision with the normal crust of the Abruzzi–Campania Platform.

It would seem possible to conclude that:

(1) The discontinuities of the crustal thickness along the margin of the Adria, linked to the tensile regime of the Mesozoic, controlled the development of the flysch-like basins during the compressive regime. The flysch-like basins seem to be localized in areas with thinned crust.
(2) During a continuous compressive regime, the collision evolved in a discontinuous manner according to the crustal thickness of colliding blocks. Normal crust/normal crust collision gives rise to diastrophic phases, whereas normal crust/thinned crust collision provides flysch-like basins.

The Gulf of Taranto is in its final stage, since the front of the thrust sheets is very close to the margin of the Apulia Platform with its thick normal crust. A preceding stage is documented towards the west by A-type subduction (Bally & Snelson, 1980) of thinned crust below the Molise Basin.

ACKNOWLEDGMENTS

This investigation has been supported by Consiglio Nazionale delle Ricerche of Italy (P. F. Oceanografia e Fondi Marini) and Ministero della Pubblica Istruzione of Italy (M.P.I.60%).

REFERENCES

AGIP (1977) *Temperature Sotterranee* (Ed. by Brugora). Segrate, Milano, 1390 pp.

AMODIO MORELLI, L. *et al.* (1976) L'arco calabro-peloritano nell'orogene appenninico-maghrebide. *Mem. Soc. geol. Ital.* **17**, 1–60.

BALLY, A.W. & SNELSON, S. (1980) Realms of subsidence. In: *Mem. Can. Soc. Petrol. Geol.* **6**, 9–94.

BARONE, A., FABBRI, A., ROSSI, S. & SARTORI, R. (1982) Geological structure and evolution of the marine areas adjacent to the Calabrian Arc. *Earth Evolut. Sci.* **3**, 207–221.

BELFIORE, A. (1984) La dispersione dei sedimenti nel settore occidentale del Golfo di Taranto. *Boll. Soc. geol. Ital.* **103**, 415–424.

BELFIORE, A. *et al.* (1981) La sedimentazione recente del Golfo di Taranto (Alto Ionio, Italia). *I.U.N. Napoli, Ann. Fac. Sc. Nautiche, app. n.* 3, **49–50**, 1–96.

BIJU-DUVAL, B., LETOUZEY, J. & MONTADERT, L. (1979) Variety of margins and deep basins in the Mediterranean. In: *Geological and Geophysical Investigations of Continental Margins* (Ed. by J. S. Watkins, L. Montadert & P. W. Dickerson). *Mem. Am. Ass. Petrol. Geol.* **29**, 293–317.

BOENZI, F. & CIARANFI, N. (1970) Stratigraphia di dettaglio del Flysch di Gorgoglione (Lucania). *Mem. Soc. geol. Ital.* **9**, 68–80.

BOUSQUET, J.C. (1972) *La tectonique recente de l'Apennin calabrolucanien dans son cadre geologique et geophysique.* Thesis, Acc. Motpellier, 172 pp.

CARISSIMO, L., D'AGOSTINO, O., LODDO, C. & PIERI, M. (1963) Petroleum exploration by Agip Mineraria and new geological information in Central and Southern Italy from the Abruzzi to the Taranto Gulf. *Sixth World Petroleum Congr.*, Frankfurt Main, 12—26.6.1963, sect.1, **27**, 267–292.

CATALANO, R., CHANNEL, J.E.T., D'ARGENIO, B. & NAPOLEONE, G. (1977) Mesozoic paleogeography of Southern Apennines and Sicily. Problems of paleotectonic and paleomagnetism. *Mem. Soc. geol. Ital.* **15**, 95–118.

CELLO, G., TORTORICI, L., TURCO, E. & GUERRA, I. (1981) Profili profondi in Calabria Settentrionale. *Boll. Soc. geol. Ital.* **100**, 423–431.

CIARANFI, N. *et al.* (1983) Elementi sismotettonici dell'Appennino Meridionale. *Boll. Soc. geol. Ital.* **102**, 201–222.

COCCO, E., CRAVERO, E., ORTOLANI, F., PESCATORE, T., RUSSO, M., SGROSSO, I. & TORRE, M. (1972) Les facies sedimentaires du Basin Irpinien (Italie Meridionale). *Atti Acc. Pontaniana*, Napoli, **21**, 1–13.

COTECCHIA, V. & MAGRI, G. (1967) Gli spostamenti delle linee di costa quaternarie del Mar Ionio fra Capo Spulico e Taranto. *Geol. Appl. Idrogeol.* **3**, 3–27.

CRATI GROUP (1981) The Crati submarine fan, Ionian Sea. A preliminary report. *Int. Ass. Sediment. 2nd European meeting*, Bologna (abstract), pp. 34–39.

CRESCENTI, V. (1975) Sul substrato pre-pliocenico dell'avanfossa appenninica dalle Marche allo Ionio. *Boll. Soc. geol. Ital.* **94**, 583–634.

CROSTELLA, A. & VEZZANI, L. (1964) La geologia dell'Appennino Foggiano. *Boll. Soc. geol. Ital.* **83**, 121–142.

D'ARGENIO, B., PESCATORE, T. & SCANDONE, P. (1973) Schema geologico dell'Appennino Meridionale (Campania-Lucania). *Atti del Conv. 'Moderne vedute sulla geologia dell'Appennino',* Acc. Naz. Lincei, Quaderno, **183**, 49–72.

D'ARGENIO, B., PESCATORE, T. & SCANDONE, P. (1975) Structural pattern of the Campania-Lucania Apennines. In: *Structural Model of Italy* (Ed. by L. Ogniben, M. Parotto & A. Praturlon). *Quaderni de 'La Ricerca Scientifica',* C.N.R., **90**, 313–327.

D'ARGENIO, B., HORVARTH, F. & CHANNEL, J.E.T. (1980) Paleotectonic evolution of Adria, the African promontory. *XXVI Int. Geol. Congr.,* Mem. BRGM No. 115, 331–351.

FINETTI, I. (1976) Mediterranean ridge: a young submerged chain associated with the Hellenic Arc. *Boll. Geof. teor. appl.* **13**, pp. 1–31.

FINETTI, I. & MORELLI, C. (1971) Wide scale digital seismic exploration of the Mediterranean Sea. *Boll. Geof. teor. appl.* **14**, 291–342.

GRANDJAQUET, C. & MASCLE, G. (1978) The structure of the Ionian Sea, Sicily and Calabria-Lucania. In: *The Ocean Basins and Margins* (Ed. by A. E. M. Nairn, W. H. Kanes & F. G. Stheli), **4B**, 257–286. The Western Mediterranean.

HACCARD, D., LORENZ, C. & GRANDJAQUET, C. (1972) Essai sur l'evolution tectogenetique de la liaison Alpes-Apennines (da la Liguire à la Calabrie), *Mem. Soc. geol. Ital.* **11**, 309–431.

HORVARTH, F. & D'ARGENIO, B. (1985) Subsidence history and tectonics of the Adria margin. *Acta Geol. Hungarica,* **28**, (1–2), 107–115.

IPPOLITO, F., D'ARGENIO, B., PESCATORE, T. & SCANDONE, P. (1975) Structural-stratigraphic units and tectonic framework of Southern Apennines. In: *Geology of Italy* (Ed. by C. Squyres), pp. 317–328. Earth Science Society, Lybian Arabic Republic.

LOJACONO, F. (1975) Osservazioni sulle direzioni delle paleocorrenti nel Flysch di Gorgoglione (Lucania). *Boll. Soc. geol. Ital.* **93**, 1127–1155.

MORELLI, C., GIESE, P., CASSINIS, R., COLOMBI, B., GUERRA, I., LUONGO, G., SCARASCIA, S. & SCHUTTE, K.G. (1975) Crustal structure of Southern Italy. A seismic refraction profile between Puglia–Calabria–Sicily. *Boll. Geof. teor. appl.* **17**, (67), 183–207.

MOUGENOT, D., BOILLOT, G. & REHAULT, J.P. (1983) Prograding shelf-break types on passive continental margins: some European examples. *Spec. Publ. Sur. econ. Paleont. Miner.,* Tulsa, **33**, 61–77.

OGNIBEN, L. (1969) Schema introduttivo alla geologia del confine calabro-lucano. *Mem. Soc. geol. Ital.* **8**, 453–763.

ORI, G.G. & FRIEND, P.F. (1984) Sedimentary basins formed and carried piggyback on active thrust sheets. *Geology,* **12**, 475–478.

PALMENTOLA, G. (1967) Sui rapporti tra la Formazione di Stigliano e la Formazione di Serra Palazzo nei dintorni di Tolve (Potenza). *Boll. Soc. Natur. Napoli,* **76**, 291–297.

PALMENTOLA, G. (1970) Nuovi dati e considerazioni sulla Formazione di Serra Palazzo in Lucania. *Mem. Soc. geol. Ital.* **9**, 81–90.

PERRONE, V. & SGROSSO, I. (1981) Il Bacino Pre-Irpino: un nuovo dominio paleogeografico miocenico dell'Appennino Meridionale. *Rend. Soc. geol. Ital.* **4**, 365–368.

PESCATORE, T. (1971) Considerazioni sulla sedimentazione miocenica nell'Appenino Campano-Lucano. *Atti Acc. Pontaniana,* Napoli, **20**, 1–17.

PESCATORE, T. (1978) Evoluzione tettonica del Bacino Irpino (Italia Meridionale) durante il Miocene. *Boll. Soc. geol. Ital.* **97**, 783–805.

PESCATORE, T. (1986) Evolution of the flysch basins and continental collision: Irpinian Basin (Southern Italy). *Vesprem 1983* (in press).

PESCATORE, T. & TRAMUTOLI, M. (1980) I rapporti tra i depositi del Bacino Irpino nella media valle del Basento (Lucania). *Rend. Acc. Sci. Fis. Matem. della Soc. Naz. Sci., Lettere e Arti,* Napoli, serie IV, **47**, 19–41.

RICCHETTI, G. (1967) Lineamenti geologici e morfologici della media valle del Fiume Bradano. *Boll. Soc. geol. Ital.* **86**, 607–622.

RICCI LUCCHI, F., COLELLA, A., GABBIANELLI, G., ROSSI, S. & NORMARK, W.R. (1983–84) The Crati Submarine Fan, Ionian Sea. *Geo-Mar. Lett.* **3**, 71–77.

ROSSI, S., AUROUX, C. & MASCLE, J. (1983) The Gulf of Taranto (Southern Italy): seismic stratigraphy and shallow structure. *Mar. Geol.* **51**, 327–346.

SCANDONE, P. (1979) Origin of the Tyrrhenian Sea and Calabrian Arc. *Boll. Soc. geol. Ital.* **98**, 27–34.

SELLI, R. & ROSSI, S. (1975) The main geologic features of Ionian Sea. *Rapp. Comm. Int. Mer. Medit.* **23**, (4a), 115–116.

SENATORE, M.R., DIPLOMATICO, G., MIRABILE, L., PESCATORE, T. & TRAMUTOLI, M. (1982) Franamenti sulla scarpata continentale pugliese del Golfo di Taranto (Alto Ionio-Italia). *Geol. Romana,* **21**, 497–510.

SERVIZIO GEOLOGICO D'ITALIA (1969) *Foglio 201—Matera—della Carta geologica D'Italia* 1:100.000. Libreria dello Stato, Rome.

SGROSSO, I. (1981) Il significato delle calciruditi di Piaggine nell'ambito degli eventi del Miocene Inferiore nell'Appennino Campano-Lucano. *Boll. Soc. geol. Ital.* **100**, 129–137.

SGROSSO, I. (1983) Alcuni dati sulla possibile presenza di una quarta piattaforma carbonatica nell'Appennino Centro-Meridionale. *Rend. Soc. geol. Ital.* **6**, 31–34.

TRAMUTOLI, M., PESCATORE, T., SENATORE, M.R. & MIRABILE, L. (1984) Interpretation of reflection high resolution seismic profiles through the Gulf of Taranto (Ionian Sea, Eastern Mediterranean). The structure of Apennine and Apulia deposits. *Boll. Oceanol. teor. appl.* **2**, 33–52.

VEZZANI, L. (1967) I depositi plio-pleistocenici del litorale ionico della Lucania. *Atti Acc. Gioenia, Sci. Natur.,* Catania, **18**, 159–180.

WEZEL, F.C. (1970) Geologia del Flysch Numidico della Sicilia nord-orientale. *Mem. Soc. geol. Ital.* **9**, 225–280.

WEZEL, F.C. (1974) Flysch successions and the tectonic evolution of Sicily during the Oligocene and Early Miocene. In: *Geology of Italy* (Ed. by C. Squyres), pp. 1–23. Earth Science Society, Libyan Arabian Republic.

Spec. Publs int. Ass. Sediment. (1986) **8**, 183–198

Plio–Pleistocene sedimentation in the Apenninic–Adriatic foredeep (Central Adriatic Sea, Italy)

GIAN GABRIELE ORI, MARCO ROVERI *and* FABIO VANNONI

Istituto di Geologia e Paleontologia, Universita' di Bologna, Via Zamboni 67, 40127 Bologna, Italy

ABSTRACT

The Plio–Pleistocene foredeep in the Central Adriatic Sea exhibits a number of syndepositional tectonic structures. The structures are elongated NW–SE, and match the depositional axis of the Adriatic basin. Three depositional seismic units have been recognized: Unit 1, Pliocene in age, is divided into Sub-units a and b, Unit 2 of Pleistocene age, and Unit 3 (Pleistocene) lateral to Unit 2. Unit 1 consists of an onlap turbiditic and hemi-pelagic sequence mainly deposited parallel to the tectonic strike. Unit 2 is a deltaic and slope system that prograded from the Apennines. Unit 3 is deltaic too but prograded parallel to the basin axis and perpendicular to the trend of Unit 2. Structural highs, produced by thrusting, were subsequently eroded and detritus was shed (as channel deposits or slumps) into the basin forming denudation complexes. The faulting in the external area consists of blind faults and associated fold drapes that had a negligible influence on sedimentation. The foredeep migration was coupled with the deformation movements in the basin itself, and these, in turn, matched with the lateral and vertical building of Unit 1. From the structural point of view the following arise: (a) the foredeep is segmented by structural discontinuities (tear faults, lateral ramps, etc.), (b) the foredeep is strongly affected by thrust deformation, (c) the foredeep is detached from the basement by the most external fault.

INTRODUCTION

The Adriatic Sea is the southward extension of the Po Basin. Both are segments of the youngest foredeep of the Apenninic mountain chain (Fig. 1). Intensive hydrocarbon exploration by the AGIP Oil Company has revealed a number of structures below the surface of the Po Basin, mainly thrust faults, thrust-produced folds and extensional lag faults. The subsurface geology of the Po Basin has been extensively described and discussed in various papers (Rocco & Jaboli, 1958; Pieri & Groppi, 1981; Pieri, 1983).

The Adriatic Basin is the central segment of the Apenninic foredeep. To the south the foredeep turns into the Italian peninsula whereas in the southern Adriatic Sea only the external stable platform occurs. Despite the large amount of data collected by the oil companies in the extensively explored Adriatic Sea, little has been published (Dondi, Rizzini & Rossi, 1982).

This paper deals with the Plio–Pleistocene sedimentary sequences (Figs 2 and 3) that occur in the central segment of the Adriatic Sea. These sediments, like those of the adjacent segments, appear to have been strongly affected by synsedimentary tectonism (Fig. 4). For a comparison with the other areas the reader can refer to Dondi, Mostardini & Rizzini (1982), Pieri & Groppi (1981) and Pieri (1983) for the Po Plain; and to Casnedi, Crescenti & Tonna (1982) and Morlotti *et al.* (1982) for the Southern Apennines.

The basis of this study consists of about 1,500 km of unmigrated seismic lines and 26 well logs (Fig. 5A). The seismic lines were shot by G.S.I. for AGIP and the Ministry of Industry of Italy. Most wells were drilled by AGIP as the operator of an AGIP-Shell joint venture, and a few were drilled by Total and Elf. The well data consist of (a) descriptions of cuttings and cores, (b) spontaneous potential, resistivity and some gamma ray logs, and (c) distribution of foraminifera and biostratigraphic subdivisions, according to AGIP stratigraphy (see Dondi *et al.*, 1982). Depending on the amount and quality of the available data some approximations and simplifications had to be made.

To avoid terminology confusion we emphasize that, as far as this paper is concerned, the foredeep is the asymmetric basin adjacent to the main fold belt on

Fig. 1. Index map of the area of study.

Mesozoic carbonates (the passive margin sediments representing the stable platform of the Adriatic Sea, Fig. 2) are strongly faulted and probably involved in duplex fault systems (Roeder, 1984 and Fig. 4). Anticlines and synclines in the Mesozoic sedimentary cover could be interpreted as the product of the underlying ramp-flat morphology of the thrusts. The Mesozoic sedimentary pile is faulted and stacked vertically along several detachment surfaces, the most important of which probably consists of Triassic evaporites (Anidriti di Burano) that acts as the lowermost décollement surface for the thrusts. Additional décollements occur at various levels in the sequence. Other possible detachment surfaces are the Marne a Fucoidi (a thin marly unit embedded in the Cretaceous carbonates), the Schlier Formation (a Miocene marly deposit), and the Messinian evaporites. In this internal area, foredeep deposits are represented by Oligo–Miocene turbiditic units (e.g. Marnoso Arenacea; Fig. 4, see Sestini, 1970 and Ricci Lucchi, 1975).

To the east, in and near the study area, the thrust faults assume an imbricate fan aspect (Roeder, 1984; ref. Boyer & Elliot, 1983; Price, 1981). From the sole fault (Triassic evaporites), the thrusts spread up without evident flattenings and reach the Plio–Pleistocene deposits (Fig. 4). Most of these faults remain blind below the upper Pliocene and Pleistocene cover. Here, the Plio–Pleistocene sequence represents foredeep deposition whereas the underlying Oligo–Miocene sequence consists of hemi-pelagites (Schlier and Bisciaro Formations) laid down on the foreland slope (Fig. 4). This situation clearly illustrates the migration of the foredeep. As regards the sedimentary units as a whole, the migration appears to have occurred in discrete steps.

one side, and to the foreland slope on the other. The foreland slope is the inclined (towards the foredeep) margin of the stable platform (cratonic area). The foredeep deposits consist of clastic sediments, whereas the foreland-stable platform is composed mainly of carbonates and fine grained sediments.

GEOLOGICAL SETTING

The Apennines are a complex thrust belt, made up of tectonic units emplaced since the Oligocene and still moving. Mesozoic–Early Tertiary carbonate deposits represent passive margin sequences, whereas Tertiary clastic sediments are related with the foredeep formation. Tectonic activity and foredeep depocentres moved toward the foreland (north-eastwards and eastwards). Ophiolites and oceanic-related rocks are the remains of the Europe–Africa collision (Alpine) which occurred during the Cretaceous (Sestini, 1970; Roeder, 1984).

Each segment of the belt displays its own structure-basin setting (Ricci Lucchi & Ori, 1984; Ori, 1985; Ricci Lucchi, 1984). In this paper only the major structures present near and in the studied area are described. In the inner (onshore) part of the fold belt,

STRUCTURES

The structures present in the study area are the most external expression of the thrusts of the Apenninic fold belt. They die out against the stable platform that lies in the middle of the Adriatic Sea (Figs 4 and 5B). The internal (western) faults show the highest displacements (Fig. 6). These structures cannot be investigated in detail because they are situated at the edge of the seismic survey. For example, the Conero fault zone in Fig. 5(B) is simplified with a straight line, as its actual shape is not known.

The external faults are blind, and are capped by fold drapes (Fig. 6). These faults are defined blind

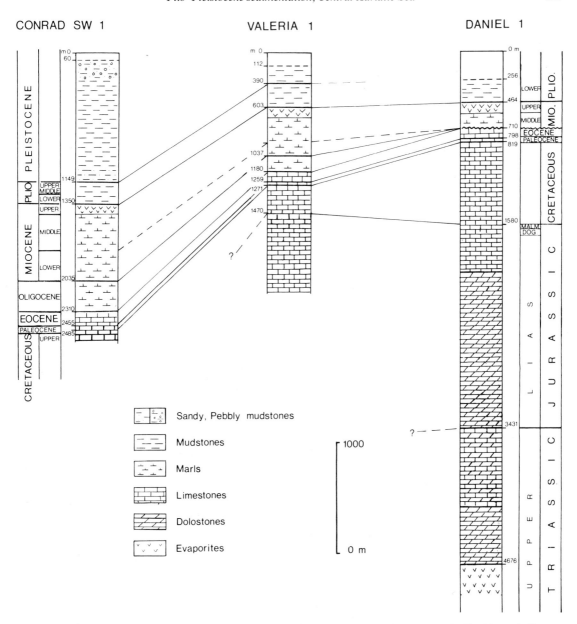

Fig. 2. Correlation of wells (AGIP) across the Adriatic Foredeep (see well locations in Fig. 5A). The Mesozoic–Eocene limestones represent the foreland platform. Oligo–Miocene fine-grained sediments are in turn the eastern (external) extension of a foredeep exposed in the Apenninic chain; they pinch out against the foreland platform: the boundary, as seen in seismic lines, consists of onlap reflectors. The Plio–Pleistocene deposits represent the last foredeep.

because they show displacements (even small) only in limestone units; upwards, into pelitic units, the displacements die out, and folds are produced. The external faults (Carlo, Edmond and Edgar faults: Fig. 5B) form the tip line of the deformed belt except for a central area where the front curves inwards, and the most external deformation is transferred to the Contessa fault (Fig. 5B). The shifting of the tip line appears to be related to the eastern pinchout of the Oligo–Miocene clastic deposits (Figs 2, 4 and 5B).

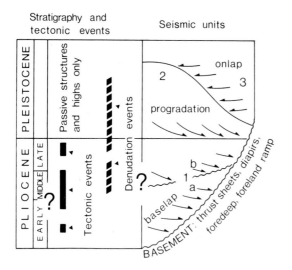

Fig. 3. Biostratigraphic correlation (data from AGIP), tectonic and seismic units. Unit 3 is shown to onlap against Unit 2, but it also shows a downlap pattern, perpendicular to this section.

Where the Contessa fault represents the tip line, the stable platform protrudes into the basin (Fig. 5B): the Plio–Pleistocene sequences are thin and no basin was formed (Figs 5C and 7). An additional complication arises from the presence of a structural high (Fig. 7): this structure is interpreted as a diapir of Triassic evaporites (Figs 5B and 7) which was emplaced just between the Contessa fault and the pinchout zone of the Oligo–Miocene Basin. The diapir is elongated parallel to the tectonic strike, and probably formed by

compressional forces, even if it may have been present early in the basin history. However, we have no direct evidence (well data) of the presence of evaporites in the structural high.

The basin-structure picture is summarized from Fig. 5(B, C). The foredeep is rimmed to the west by a set of faults (Ancona, Conero and Ester), but it is not a laterally continuous basin: it is interrupted by the discontinuity formed on one side by the westward extension of the stable platform, and on the other side by a set of thrusts (Contessa and Colosseo faults, Figs 5C, 7 and 8). The foredeep is also affected by a number of blind faults (Figs 5B and 6). They appear to produce gentle fold drapes on the overlying sediments. The tops of the folds represent the last deformed sediments, thus showing, in the seismic sections, the relative age of the deformation. Reflectors of the undeformed sediments onlap against the folds (Carlo fault, Fig. 6). Some of the blind faults are present in the foreland slope (Figs 5B, 13 and 15). They seem to be the youngest structures of the area and could be related to former extensional faults present in the foreland margin (see below). The presence of thrust faults to the east of the foredeep depocentres suggests that the basin is detached by a sole fault emplaced in Triassic evaporites.

Post-Messinian tectonic activity occurred in various pulses during the Pliocene, the most important of which is thought to have been during the middle Pliocene. The Pleistocene deposits are affected only by the already formed structures, acting as passive highs (Fig. 3).

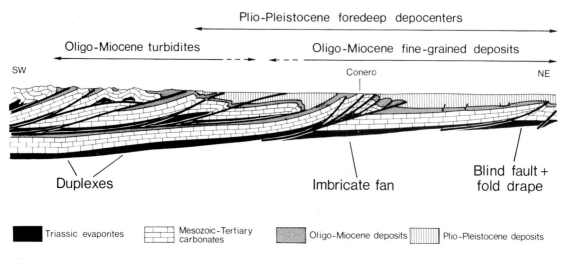

Fig. 4. Geological section across the eastern slope of the Apennines. Location in Fig. 1. Modified from Roeder (1984).

SEISMIC STRATIGRAPHY

The Plio–Pleistocene sediments can be subdivided into three units (Fig. 3 and Table 1). Unit 1, of Pliocene age, consists of parallel continuous reflectors; in places two sub-units have been identified. Unit 2 is a Pleistocene progradational system which advanced from the Apenninic chain and migrated eastward. Unit 3, also Pleistocene in age, is a progradational system that advanced southward from the Po Plain parallel to the Adriatic axis. These units form two cycles. The first is Pliocene in age (Unit 1), and is characterized by onlap boundaries and turbiditic sedimentation. The second cycle (Units 2 and 3) has the geometry of an offlap.

The base of the Plio–Pleistocene sequence is the Messinian evaporites. The evaporites are rather well defined in seismic sections by high-amplitude reflectors. In the north-eastern part of the area the evaporites are absent, but the boundary is still strongly marked on the seismic profiles. The Messinian evaporites are also easily detected in well logs where the spontaneous potential, resistivity and gamma ray curves show many sharp and frequent peaks. The Messinian horizon varies in thickness up to 100 m. The contact between the Messinian evaporites and the overlying Plio–Pleistocene sequence is a major unconformity, that marks a stage in the foredeep migration (Fig. 4). In the eastern area, the Messinian evaporites can be absent. In this case Pliocene reflectors directly overlie Early Tertiary carbonate deposits.

Unit 1

This unit is composed of parallel, laterally continuous reflectors, exhibiting variable amplitudes, but ranging within medium values (Figs 5D and 6–11, Table 1). Well data indicate a variable thickness, from 2,500 (in the depocentre) to 200 m. The unit is thinner on the structural highs. The reflectors are gently inclined eastward at 1–3° in the lower part of the unit and become horizontal upward. As the foreland platform merges westward at an angle of 3–10°, the basal boundary results in an onlap of inclined reflectors (Figs 6 and 14). In some cases, the dip of the reflectors increases due to uplifting of compressional features (see the Carlo fault in Fig. 6). However, the effects of tilting are not pronounced and disappear downdip over a short distance (commonly 3–4 km from the structure). The upper boundary is marked by the downlap of units 2 or 3 (Figs 6 and 7). Due to the

tangential base of the overlying units and the apparent absence of any break in sedimentation, the boundary is difficult to map in places.

Electric logs exhibit rather monotonous sequences of small peaks, indicating the presence of fine-grained sediments with thin bedded sandstones. Micropalaeontological data (AGIP) show a dominantly bathyal environment passing upwards to a neritic one.

The seismic record and electric logs, coupled with palaeoecological and lithologic (cores and cuttings) data, indicate that Unit 1 accumulated in a deep sea environment characterized by relatively low-energy processes. The fine-grained sediments consist of hemipelagites (pelites) intercalated with thin turbiditic sandstones. These deposits could be very similar to the blue-grey clays of the same age occurring in the Apenninic foothills. It is probable that the turbidity currents that deposited the sand layers flowed longitudinally in the basin (that is, parallel to the Adriatic axis). Pliocene fan-deltas, now exposed along the Apenninic margin, are likely to have been the point sources for the deep sea sediments of the foredeep. These entry points of sediments seem to occur at major discontinuities of the fold belt (Ricci Lucchi, 1984), such as could be the lateral ramps of large thrusts, strike-slip faults perpendicular to the tectonic strike, and zones of transfer faulting (Dahlstrom, 1969). It is noteworthy that the sedimentation of Unit 1 occurred on a gently inclined surface vertically and laterally built by turbidity currents likely flowing along strike. This is a particular aspect of the foredeep that may be common to other similar basins.

Unit 1 includes all the Pliocene, and its base is marked by the presence of *Sphaeroidinellopsis* spp. The upper boundary is given by the appearance of *Hyalinea balthica* (Dondi *et al.*, 1982). In places the unit can be subdivided into two sub-units, separated by a discontinuity (unconformity, break in sedimentation, or erosional surface) produced by the major tectonic pulse of the Middle Pliocene (Fig. 3). This discontinuity between the sub-units is stronger near the structures that generated it. In the more distant area, the boundary is conformable (see Fig. 6), and can be recognized by mapping and correlations.

In front of some thrust structures the seismic reflectors assume a disorganized pattern (Fig. 5B) characterized by discontinuity and variable amplitude, and are chaotically disposed. They form small lenses 0·3 s thick and 1 km wide, extending from the thrust structures into the basin. The overlying reflectors gently drape them. By analogy with similar features in Unit 2, these lenses are interpreted as the

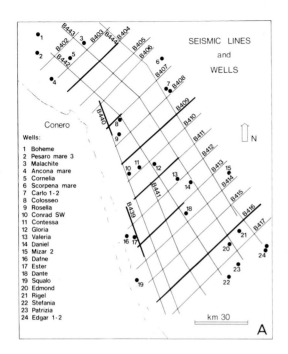

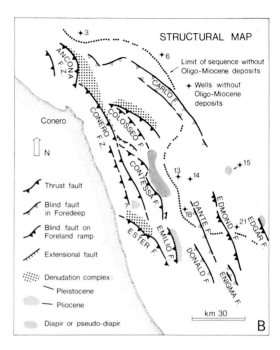

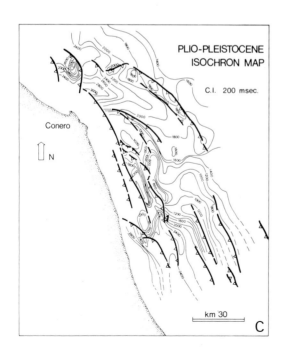

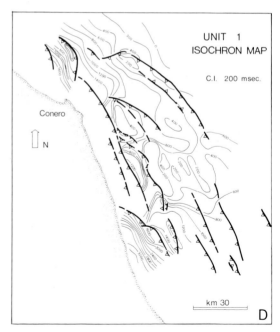

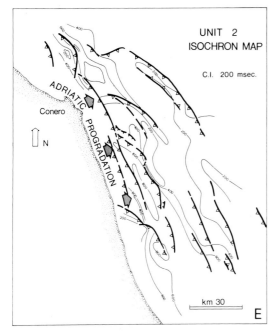

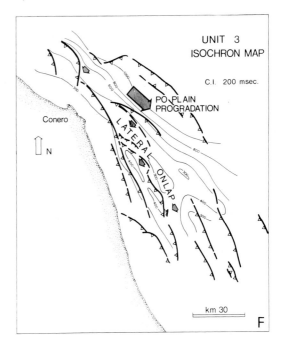

Fig. 5. Maps of the study area. (A) Seismic lines and well locations, the bold lines indicate segments shown in other figures. (B) Main tectonic structures. (C) Isochron of the Plio–Pleistocene sequence. (D) Isochron map of Unit 1, the thickness to the east of Ancona and of the Ester faults is due mainly to Sub-unit 1a. (E) Isochron map of Unit 2. (F) Isochron map of Unit 3. F: fault, FZ: fault zone.

product of erosion of the uplifted structures. They are called 'denudation complexes' (see below).

The isochron map of Fig. 5(D) shows that the maximum thickness of Unit 1 occurs in the proximal (western) part of the basin, which is mostly west of the Ancona and Ester faults. The high thickness values are due to Sub-unit 1a, which is very thick in this area (Fig. 6).

Unit 2

Unit 2 is a wedge of clinoform-type reflectors that are elongated along the basin and thin eastward (Figs 5E, 6 and 13, Table 1). The thickness of this unit is not as variable as that of Unit 1; it ranges between 800 and 1,000 m. As the previous unit, Unit 2 thins dramatically on the structural highs. The base of the unit is generally conformable with the underlying Unit 1, whereas the top can be either represented by topset reflectors, thereby reaching the surface, or represented by the foresets reflectors lapped by Unit 3 reflectors (Figs 6 and 8). The bottomset reflectors are laterally continuous, high in amplitude and form the tangential base of the foresets that display a more

variable pattern. The foresets are characterized by high-amplitude, continuous reflectors. They are in continuity to the topset reflectors, forming a sigmoidal pattern. Intercalated with them, discontinuous, low-amplitude reflector packages occur. This type of foresets is truncated by the topset forming an oblique pattern. This alternation is thought to reflect energy variation (respectively low and high) in depositional processes. Dondi *et al.* (1982) interpret this alternance as due to Quaternary sea-level changes. Lateral continuity, high amplitudes and high frequencies are the characteristics of the topset reflectors that converge westwards to produce a thinning of the package as a whole.

The electric logs display all the features present in deltaic sequences and are in perfect agreement with the geometries and facies recorded in the seismic sections. The bottomset and the lower part of the foresets, in the well logs (spontaneous potential and resistivity), appear as a monotonous, non-sequential near-baseline pattern; pelites with rare sands have been recorded in cuttings and cores, and are interpreted as prodelta sediments. The upper part of the foresets corresponds in the electric logs to funnel-type

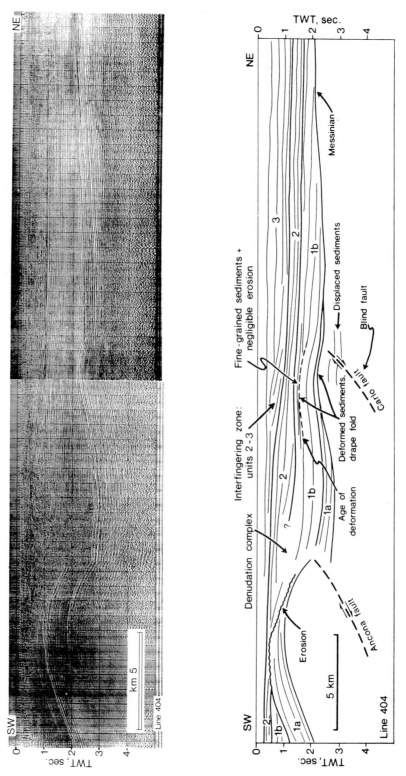

Fig. 6. Seismic line and line drawing of line 404 (location in Fig. 5A). The foredeep is well-developed, but faulting is present in the middle. Note the relations between the denudation complex and the erosional surface on the Ancona fault.

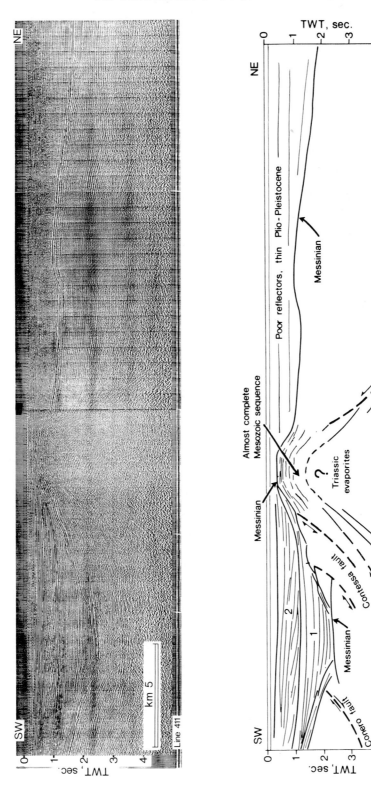

Fig. 7. Seismic line and line drawing of line 411 (location in Fig. 5A). The Contessa fault is schematic. Note the absence of a true foredeep.

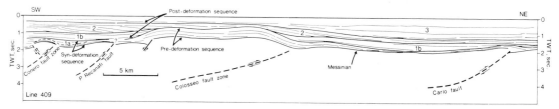

Fig. 8. Line drawing of line 409 (location in Fig. 5A). Note the relationship between the seismic units and the structure formed by the Colosseo fault and the ill-defined foredeep.

Table 1. Characteristics of the seismic units

Depos. cycles	Seismic units	Boundary geometry	Seismic facies	E-log facies	Environments
Pliocene	1a, b	Onlap (of inclined refl.)	Continuous medium amplitude and frequency	A-sequential monotonous	Deep-sea
Pleistoc	2	Offlap:			
		topset	Continuous, high amplitude	Blocky pattern	Fluvial, transitional
		foreset	Discontinuous variable amplitude or continuous high amplitude	Funnel-type	Mouth bar strandline slope
		bottomset	Continuous, medium to high amplitude	A-sequential	Basin floor
Pleistoc	3	Offlap ill-defined	Discontinuous variable amplitude	As for Unit 2	Deltaic as for Unit 2

sequences 20–30 m thick and there is an upwards increase of sand content detected in the lithologs. These deposits are interpreted as coarsening upward sequences due to the progradation of deltaic mouth bars or strand lines. The topsets exhibit 20–50 m thick units without a sequential arrangement. Sandstone is the lithology present. This facies is interpreted as the vertical stacking of channel deposits. Marine intercalations are common, with marine, shallow water forams present throughout the section.

Unit 2 forms a coarse deltaic-coastal apron over all of the Apenninic rim, and is believed to be the product of the coalescence of many fan deltas. Pleistocene fan deltas outcropping in the Apenninic foothills could be regarded as surface analogues (Ricci Lucchi *et al.*, 1981). The progradation is disturbed by structural

highs produced by former tectonic movements (Figs 7, 8 and 11): in this case the reflectors directly onlap against the structures.

Unit 2 displays a remarkable amount of chaotic or lenticular reflectors adjacent to the major thrust structures (Fig. 5B). They are interpreted as denudation complexes well developed (up to 1 s thick and several kilometres long) in this unit. The seismic facies are very complex, but two end-members can be recognized: (1) Low to medium amplitude reflectors, discontinuous and concave-up in shape; in north–south profiles (Figs 11 and 12) the reflectors assume an undulating pattern with many concave-up surfaces stacked on each other (Figs 6 and 11). (2) Next, the Ester fault discontinuous reflectors produce channel-like features (Fig. 10); they form a relatively thin

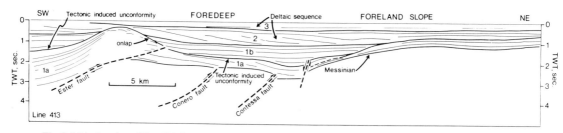

Fig. 9. Line drawing of line 413 (location in Fig. 5A). The Ester fault seems to flat into the Messinian evaporites.

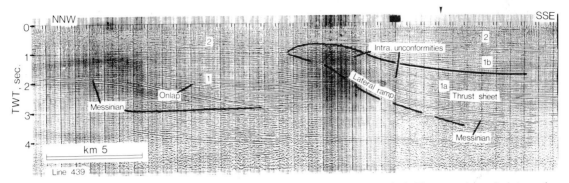

Fig. 10. Portion of line 439 (location in Fig. 5A) showing the lateral ramp of the Ester fault. The arrow shows the intersection with line 413 (Fig. 9). Notice the pull-up beneath the thrust ramp.

horizon 0·2 s thick. No wells have been drilled through the first facies; the second one has been drilled, it displays conglomeratic deposits forming units 10–20 m thick with sharp bases.

Unit 3

This unit is Pleistocene in age. Its base and top correspond to those of Unit 2 and it is part of the same offlap cycle. Low-angle southwards-dipping reflectors characterize this unit (Table 1), but they are not easily recognizable due to their poor continuity and high

variability in amplitude and frequency. In east–west profiles (that is, perpendicular to the foreset dips) reflectors are horizontal or dip at very low angles westward, pinching out against the slope of the offlap sequence of Unit 2 (Figs 5F, 6 and 13). This geometry is, however, even more complicated, as internal wedges of thin reflectors sets are present. Units 2 and 3 interfinger, mostly near the base of the sequence where packages of horizontal reflectors of Unit 3 are in between the bottomset reflectors of Unit 2. Thus, the downlap of Unit 2 is disturbed by onlap termination of the reflectors of Unit 3 (Fig. 6).

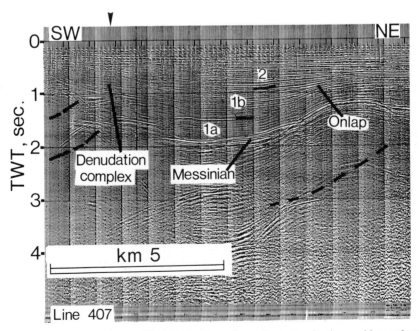

Fig. 11. Seismic line (line 407, for location see Fig. 5A) showing onlap, reflector terminations and internal unconformities in the ill-defined foredeep. Arrow shows intersection with line 440 (Fig. 12).

Electric logs and lithologies of Unit 3 display sequences very similar to the deltaic pattern of Unit 2, except for a general finer grain size of the sediments. The electro-facies and the offlap geometry are good evidence for the deltaic origin of this unit, which is thought to be the extension of the Po Plain system. A minimum estimation of the rate of progradation is 0.16 m yr^{-1}, whereas the maximum progradation rate of Unit 2 is 0.019 m yr^{-1}. The high rate of progradation could be responsible for the difference in seismic facies between the two progradational units.

DISCUSSION

The studied portion of the Apenninic foredeep is a crucial area as it corresponds to a sharp bend in the basin axis. The northern Po Basin axis is oriented NW–SE, whereas the portion south of Conero is roughly oriented north–south. This bend corresponds to a major transversal structure of the Apennines, the Ancona–Anzio line, which acted in a variety of ways during the Apenninic tectonic evolution and even earlier (Cantelli et al., 1982; Castellarin et al., 1982). During Neogene (Upper Miocene–Lower Pliocene) this structure acted as a dextral strike-slip fault, as well as a compressive reverse fault. After the Lower Pliocene the distribution of tectonic structures seems to be no more influenced by the Ancona–Anzio discontinuity. In fact the thrust belt is continuous across the Ancona–Anzio northern termination. The

main cycles (Unit 1: onlap, and Units 2 and 3: offlap) correlate well with the stratigraphy recorded in the surface geology of the Apenninic foothills (Cremonini & Ricci Lucchi, 1982). The depositional cycles do not seem to be influenced by the local tectonic movements. The emplacements of thrust sheets, with vertical displacements up to 1,000 m do not modify the nature of the depositional geometries (onlap, offlap, etc.). At the time of fault movements, only local unconformities are produced; then thrust highs only act as sources of detritus (Fig. 6). Thus it seems that the tectonic overprint does not modify the boundary types generated by major sea-level changes (Cremonini & Ricci Lucchi, 1982). The foredeep in this area is formed by two depocentres separated by structural features transversal to the basin axis (as it is observable in Figs 5C and 8).

Deformation and foredeep shifting

The deformation of the Apennines moved outwards through time, and resulted in the basins and thrusts becoming younger toward the foreland. This is well documented in the profile of Fig. 6: the Ancona fault produces the unconformity that is the boundary between sub-units 1a and 1b. The age of the Ancona fault hiatus is near the Middle Pliocene, and it corresponds to the major movement pulse. The same boundary is virtually conformable downdip. A more external structure (Carlo fault) deformed the sediment column up to the upper part of Sub-unit 1b. The age

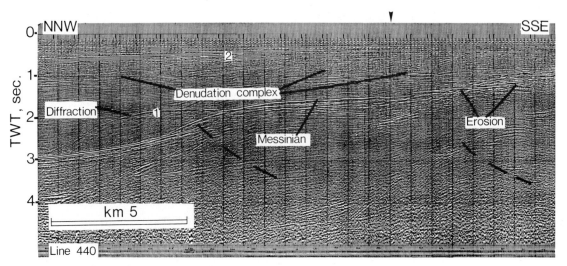

Fig. 12. Portion of line 440 (location in Fig. 5A). A denudation complex is shown in a north–south section. Note the variability of facies as compared with those of Figs 6 and 11. Arrow shows intersection with line 407 (Fig. 11).

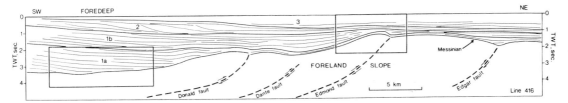

Fig. 13. Line 416 (location in Fig. 5A). The foredeep is well-developed, but thrust faults are present at its eastern portion and on the foreland, suggesting a detachment of foredeep basin from its basement.

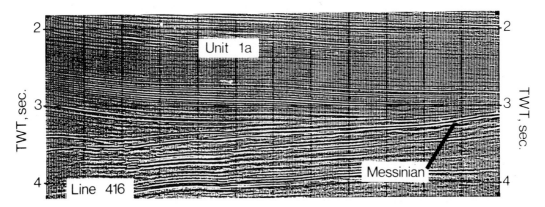

Fig. 14. Onlap of the inclined reflectors (Unit 1b) against the Messinian surface (close up of Fig. 13).

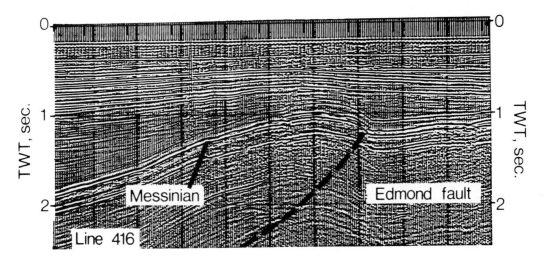

Fig. 15. Detail of the external faults (close up of Fig. 13).

of the Carlo fault relates to a minor movement during the Late Pliocene (Fig. 3). The Middle Pliocene age is supported by the correlation of the unconformable boundary of Sub-units 1a and 1b with the frequent absence of all or part of the Middle Pliocene in the well logs.

The migration of the edge of the fold belt is matched by the migration of the foredeep. The depocentre of Sub-unit 1a is to the west, in part off the survey area (Figs 6 and 9). The depocentre of Units 1b and 2 and 3 is in a more central position (Fig. 5C, E and F). In detail, the foredeep migration consists of the mixture of, at least, three processes: (1) the tectonic one that changes the structural settings and moves the basin, (2) the formation of thrust faults affecting the foredeep, or, even bypassing it and producing a detachment of the basement, (3) the sedimentary one that fills the basin controlled by the sea-level position (onlap, offlap, turbiditic or deltaic sedimentation). These processes seem to be independent of one another.

Denudation complexes

Special attention has been given to mapping the complex seismic facies believed to be a denudation complex. We believe they represent the erosional products of the thrust highs: extensive erosional surfaces deeply cutting the sedimentary sequences are seen on the thrust highs adjacent to denudation complexes (e.g. the Ancona fault in Fig. 6). The complexes could consist of coarse-grained channels or slumped units, and seismic record cannot resolve the interpretation. Wells drilled in the complex next to the Ester fault show conglomeratic channels, and similar deposits are seen adjacent to the thrust sheets exposed in the central Apennines (Cantalamessa *et al.*, 1981). However, the denudation complexes show a wide variability in seismic facies and many more wells would be required to interpret the lithology (and consequently the formation) of these deposits. The formation of conglomerate or coarse sand deposits may be related to subaerial erosion, slumping could in submarine environments. The relationship between thrust units, denudation complexes, and the surrounding sediments, are summarized in Fig. 16(A, B). The term 'denudation complex' is used rather than 'synorogenic conglomerates' or similar terms, because all the foredeep sediments accumulated along with active tectonism. The complexes are indeed closely related with the denudation of single thrust highs (Willemin, 1984).

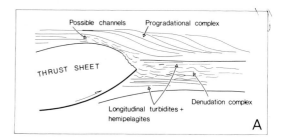

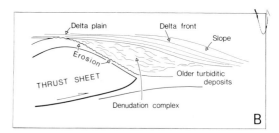

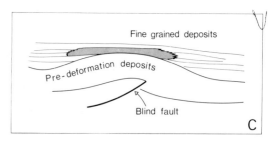

Fig. 16. Influence of structures on sedimentation. (A) Denudation complex of Unit 1 embedded in deep-sea facies. (B) Denudation complex in Unit 2 at the base of a progradational sequence, showing poorly defined reflectors concordant with the offlap pattern (cf. Figs 6 and 11). (C) Relations between blind fault-fold drape and the overlying undeformed sediments (cf. Fig. 6).

CONCLUSIONS

Some of the features that can be seen in the Apenninic foredeep along the central Adriatic coast are relevant to the knowledge of this type of basin.

(1) The foredeep may be affected by strong synsedimentary deformation, with the formation of thrust faults, folds and diapirs.
(2) The structures do influence sedimentation, but do not change the geometry of the boundaries of the sedimentary sequences as recognized in seismic records.
(3) Thrust highs can be eroded and supply sediment, and form denudation complexes.

(4) Sediments display a variety of facies and depositional trends, either perpendicular or parallel to the basin axis.

(5) The foredeep can be detached from its basement by thrust faults affecting the foreland slope.

(6) The foredeep can be segmented in a number of depocentres, the location of which is controlled by features of the fold belt and foreland, or diapirism.

These features are not present in all the foredeeps, but are thought to occur, at least in part, in some segments of the Mediterranean fold belt, such as the southern margin of the Pyrenees (Nichols & Ori, 1983; Ori & Friend, 1984), the Moroccan Rif (Michard, 1976), and the Northern Hellenides.

ACKNOWLEDGMENTS

F. Ricci Lucchi (University of Bologna) and N. Zitellini (CNR, Bologna) provided continuous help during the seismic interpretation. A. Castellarin (University of Bologna) gave useful suggestions. Various drafts of the paper have been reviewed by C. Schenck (U.S.G.S., Denver), A. Castellarin, G. J. Nichols (University of Liverpool), L. Dondi and A. Rizzini (AGIP, Milano), F. Ricci Lucchi, G. Sestini (Firenze) and N. Zitellini. The final part of the work has been carried out by the senior author during a stay at the U.S. Geological Survey (Denver) of which the use of the computing and other facilities is acknowledged. D. Roeder (Anschutz Corp., Denver) allowed us to show the section of Fig. 4. P. Ferraresi and P. Ferrieri provided technical support.

REFERENCES

BOYER, S.E. & ELLIOT, D. (1983) Thrust systems. *Bull. Am. Ass. Petrol. Geol.* **66**, 1196–1230.

CANTALAMESSA, G., CENTAMORE, E., CHIOCCHINI, U., DI LORITO, L., GIARDINI, G., MARCHETTI, P., MICARELLI, A., PENTONI, F. & POTETTI, M. (1981) Analisi dell'evoluzione tettonico-sedimentaria dei "bacini minori" torbiditici del Miocene medio-superiore nell'Appennino Umbro-Marchigiano e Laziale-Abruzzese: 9) il bacino della Laga tra il F. Potenza ed il F. Fiastrone-T. Fiastrella. *Studi geol. Camerti*, **7**, 17–79.

CANTELLI, C., CASTELLARIN, A., COLACICCHI, R. & PRATURLON, A. (1982) La scarpata tettonica mesozoica lungo il settore nord della "Linea Ancona-Anzio". *Mem. Soc. geol. Ital.* **24**, 14–154.

CASNEDI, R., CRESCENTI, U. & TONNA, M. (1982) Evoluzione della avanfossa adriatica meridionale del Plio-Pleistocene, sulla base di dati del sottosuolo. *Mem. Soc. geol. Ital.* **24**, 243–260.

CASTELLARIN, A., COLACICCHI, R., PRATURLON, A. & CANTELLI, C. (1982) The Jurassic-lower Pliocene history of the Ancona–Anzio line (Central Italy). *Mem. Soc. geol. Ital.* **24**, 325–336.

CREMONINI, G. & RICCI LUCCHI, F. (eds) (1982) Guida alla Geologia del margine appenninico-padano. *Soc. geol. Ital.*, *Guida geol. Regionale*, 247 pp.

DAHLSTROM, C.D.A. (1969) Balanced cross sections. *Can. J. Earth Sci.* **6**, 743–757.

DONDI, L., MOSTARDINI, F. & RIZZINI, A. (1982) Evoluzione sedimentaria e paleogeografica della Pianura Padana. In: *Geologia del Margine Appenninico-Padano* (Ed. by G. Cremonini & F. Ricci Lucchi), Soc. geol. Ital., Guida geol. Regionale, pp. 47–58.

DONDI, L., RIZZINI, A. & ROSSI, P. (1982) Quaternary sediments of the Adriatic Sea from Po delta to Gargano promontory. *Adv. Res. Inst., Mediterranean Basin Conf.* 23 pp.

MICHARD, A. (1976) Elements de Geologie Marocain. *Notes Mem. serv. geol.* **252**, 398 pp.

MORLOTTI, E., SARTORI, R., TORELLI, L., BARBIERI, F. & RAFFI, I. (1982) Chaotic deposits from the External Calabrian Arc (Ionian Sea, Eastern Mediterranean). *Mem. Soc. geol. Ital.* **24**, 261–275.

NICHOLS, G.J. & ORI, G.G. (1983) Sedimentation along the compressive margin of the Ebro basin (NE Spain). *IAS 4th Eur. Mtg, Abstr. Vol.* pp. 117–119.

ORI, G.G. (1985) Discussion on accretion and tectono-sedimentary progradation: the Taranto Gulf (Southern Italy), by G. Mascle & J. Mascle. *Geo-Mar. Lett.* **5**, 199–201.

ORI, G.G. & FRIEND, P.F. (1984) Sedimentary basins, formed and carried piggyback on active thrust sheets. *Geology*, **12**, 475–478.

PIERI, M. (1983) Three seismic profiles through the Po Plain. In: *Seismic Expression of Structural Styles* (Ed. by A.W. Bally). *Am. Ass. Petrol. Geol. Stud. Geol.* **15**, 3.4.1/8–3.4.1/26.

PIERI, M. & GROPPI, G. (1981) Subsurface geological structure of the Po Plain, Italy. *Progetto Finalizzato Geodin. CNR*, **414**, 23 pp.

PRICE, R.A. (1981) The Cordilleran foreland thrust and fold belt in the Southern Rocky Mountains. In: *Thrust and Nappe Tectonics* (Ed. by K. McClay & N.J. Price), pp. 427–448. *Spec. Publ. geol. Soc. London*, **9**, Blackwell Scientific Publications, Oxford.

RICCI LUCCHI, F. (1975) Miocene paleogeography and basin analysis in the Periadriatic Apennines. In: *Geology of Italy* (Ed. by C. Squyres), pp. 129–162. Earth Science Society of Libya.

RICCI LUCCHI, F. (1984) Flysch, molassa, cunei clastici: tradizione e nuovi approcci nell'analisi dei bacini orogenici dell'Appennino settentrionale. *Cento Anni geol. Ital. Volume Giubilare*, pp. 279–295.

RICCI LUCCHI, F., COLELLA, A., ORI, G.G., OGLIANI, F. & COLALONGO, M.L. (1981) Pliocene fan deltas of the Intrapenninic Basin, Bologna. *IAS 2nd Eur, Mtg, Excursion Guidebook*, **4**, 79–162.

RICCI LUCCHI, F. & ORI, G.G. (1984) Orogenic clastic

wedges of the Alps and Apennines (abstract). *Bull. Am. Ass. Petrol. Geol.* **64**, 798.

ROCCO, T. & JABOLI, D. (1958) Geology and hydrocarbons of the Po basin. In: *Habitat of Oil* (Ed. by L.G. Weeks), pp. 1153–1167. American Association of Petroleum Geology.

ROEDER, D. (1984) Tectonic evolution of the Apennines (abstract). *Bull. Am. Ass. Petrol. Geol.* **68**, 798.

SESTINI, G. (ed.) (1970) Development of the Apennines Geosyncline. *Sediment. Geol.* **4**, 201–642.

WILLEMIN, J.H. (1984) Erosion and the mechanics of shallow foreland thrusts. *J. struct. Geol.* **6**, 425–432.

Spec. Publs int. Ass. Sediment. (1986) **8**, 199–217

Dynamics of the Molasse Basin of western Switzerland

P. HOMEWOOD*, P. A. ALLEN† *and* G. D. WILLIAMS‡

*Institut de Géologie, Université de Fribourg, Pérolles, CH-1700 Fribourg, Switzerland; †Department of Earth Sciences, University of Oxford, Oxford OX1 3PR, U.K.; ‡Department of Geology, University College Cardiff, PO Box 78, Cardiff CF1 1XL, U.K.

ABSTRACT

The north Alpine foreland basin extends from French Savoy in the west to the Linz area of Austria in the east and parallels the Alpine front. In eastern Switzerland, Bavaria and Austria the basin is relatively intact apart from imbrication in the Subalpine zone. In western Switzerland, however, Alpine deformation has progressed north-westwards into the Jura along a décollement in Triassic salt, detaching the basin from the bulk of its crystalline basement.

Foreland basin deposits in western Switzerland are found in a number of structural positions: (1) within the external zone of the Alps where they are interleaved with north-westward-translated thrust sheets of the Helvetics; (2) in a zone of closely spaced thrust faults known as the Subalpine zone; (3) in a broad and little deformed region, the 'Plateau' or 'Mittelland' Molasse, and (4) in synclines in the folded and thrusted Jura province.

Sedimentation in the foreland basin continued from earliest Oligocene to late Miocene, a period of approximately 30 Ma. In western Switzerland post-Miocene uplift has resulted in erosion of the upper units and no sediments younger than *c.* 11 Ma are found. The earliest deposits are mostly turbiditic and were deposited in sub-basins (probably extensional but possibly piggy-back) ahead of the encroaching Alpine thrust sheets. The overlying sediments represent a regressive sequence from a wave dominated sea to shoreline sands to alluvial fans, floodplains and lakes. A transgression at *c.* 20 Ma flooded the continental basin and a tide-swept seaway was established along the alpine front.

The pre-Oligocene history of the European (Helvetic) margin of Tethys is dominated by extension and thermal subsidence. Early Tertiary continent–continent collision inverted the extended Helvetic margin and flexed the European lithosphere at the same time as the development of the Rhine–Rhône graben system. Subsidence rates increased markedly at this time. Decompacted subsidence curves suggest that a wave of subsidence migrated northwards across the Molasse Basin of western Switzerland in the same direction as thrust propagation. Locations close to the orogen subsided more rapidly and for a longer period than those near the feather-edge of the basin. Isostatically corrected tectonic subsidence curves are essentially linear for the relatively small time period analysed (15 Myr) and average 0.1 mm yr^{-1} close to the Subalpine zone. Rates of convergence obtained from the progressive pinch-out of stratigraphic units, rates of thrust propagation derived from the restored structural cross-section and rates of migration of depocentres estimated from the subsidence curves are all closely comparable.

INTRODUCTION

The Molasse Basin is a classical peripheral foreland basin (Dewey & Bird, 1970; Dickinson, 1974) which extends from Haute Savoie, France in the west through Switzerland and Bavaria to the Linz–Vienna area of Austria in the east, a distance of approximately 700 km (Fig. 1). The basin widens substantially to the east, attaining a maximum present-day width of about 150 km in southern Germany (Lemcke, 1973).

The present southern limit of the basin is generally taken to be marked by the frontal thrusts of the Alpine belt. However, in Austria, the earliest (basal Oligocene) Molasse formations overlie units of the Northern

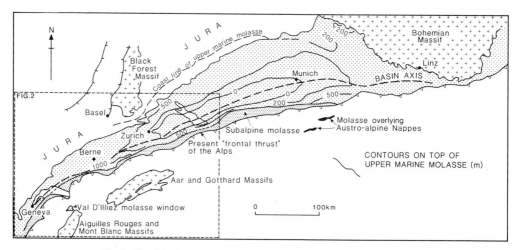

Fig. 1. Simplified map of the entire Molasse basin, after Lemcke (1974).

Calcareous Alps (Fuchs, 1976), and seismics and borehole data also reveal that the Molasse extends southwards, below the basal Helvetic thrust (e.g. Bally, 1983). In Switzerland, the older deposits of the basin are seen to be involved within the external fold-thrust nappes of the Helvetics. To the west, the Molasse underlies the far-travelled Prealpine and Ultrahelvetic nappes and locally appears in tectonic windows through overthrust units, as in the Val d'Illiez (Fig. 2). The northern, feather-edge margin of the basin is found in the Jura province or is delimited by blocks of crystalline basement such as the Bohemian Massif. In central and western Switzerland where the foreland cover is folded, thrusted and backthrusted

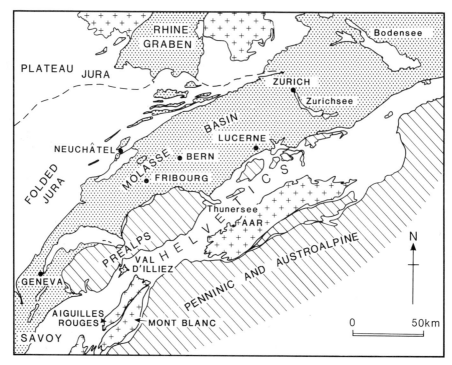

Fig. 2. Major structural divisions of western Switzerland and adjacent regions. Stippled area represents Molasse of the Subalpine zone, Mittelland, Jura and Rhine graben, but does not include north Helvetic 'flysch' found in the Helvetic province.

(Jura mountains), the distal deposits of the foreland basin occur as isolated remnants, generally preserved in synclines, and the exact position of the northern limit of the basin has been to some extent erased. The geometry of the basin fill is therefore wedge-shaped with the floor of the basin now being found at depths of greater than 4 km below sea-level in the south and with an attenuated feather-edge in the north.

The Molasse Basin is believed to represent a telescoped major zone of subsidence dynamically linked to the adjacent orogen (Laubscher, 1978; Karner & Watts, 1983; Mugnier, 1984; Mugnier & Vialon, 1986; Allen, Homewood & Williams, 1986). This coupling of the evolution of the foreland basin and its adjacent mountain belt through loading and lithospheric flexure (Price, 1973; Beaumont, 1981; Schedl & Wiltschko, 1984) allows numerical modelling of loads necessary to explain subsidence (Watts, Karner & Steckler, 1982). So far, it has been tacitly assumed that there is a direct relation between tectonic loading and basin history. In the case of the Alps, Karner & Watts (1983) found it necessary to involve the existence of a 'hidden' subsurface load to explain the observed gravity anomalies, although Stockmal, Beaumont & Boutilier, 1986) invoked the effect of continental margin topography to explain the apparent discrepancy. An important feature of the Molasse Basin west of Bern, shown by the basement structure (Rybach *et al.*, 1980), is the relationship to the contemporaneous Rhine Graben rift. Simple flexural subsidence of the foreland, in this part of the basin, is obviously precluded.

The history of the Molasse Basin can only be understood by considering the evolution of the Alpine chain during its collisional phase, as Molasse deposits are involved within the Alpine front, and overlapped the Jura prior to folding. It is necessary therefore to study not only the structure and stratigraphy of Molasse deposits *per se*, but also that of the Jura and the Alpine fold-thrust belt to some extent.

Our aim here is to summarize the stratigraphical and structural background which provides both the basis for studies of basin history and a critical template for geophysical models.

STRUCTURE OF THE MOLASSE BASIN, WESTERN SWITZERLAND

A transect across the Molasse Basin in western Switzerland, perpendicular to the thrust front, runs from the *Helvetic nappes* through the *Subalpine Molasse*

with the overlying *Prealps*, the '*Plateau*' Molasse (flat-lying), into the *Jura Mountains* (Figs 2, 3 and 4).

The *Helvetic Alps* form the northern or external zone of the Alpine belt in Switzerland. The Helvetic nappe belt consists of blocks of Palaeozoic basement such as the Aiguilles Rouges, Mont Blanc, Aar Massif (locally with infolded Permo–Carboniferous sediments), with their essentially un-detached cover of Triassic and younger sediments, interleaved with large scale fold-nappe structures. It is clearly beyond our scope to enter into a discussion of the complexities of the nappe geometries and strain histories (Ramsay, 1981; Masson, Herb & Steck, 1980), as what is relevant here is the relationship of Helvetic thrust sheets to foreland basin deposits and the timing of emplacement of the Helvetic sheets.

The earliest deposits of the embryonic foreland basin are found within and in front of the Helvetic nappe structures (Figs 3 and 4). A palinspastic restoration of the Morcles, Diablerets and Wildhorn nappes suggests the possibility that within the Helvetic realm separate compartments existed in this early (basal Oligocene) foreland basin, for the Taveyannaz and Val d'Illiez sandstones (North Helvetic 'flysch'), with their distinctive petrographies (Vuagnat, 1952; Sawatski, 1975; Homewood & Caron, 1982), appear to occupy different depocentres (Fig. 5). The proximal edge of the basin in early Oligocene times must have lain somewhere between the Diablerets and Wildhorn provinces, since only a thin unit of Eocene Flysch is present on the latter.

The *Prealps* represent a series of klippen, remnants of Penninic and Austro-alpine units, which have been translated far northwards from the central alpine zone. Décollement nappes of Mesozoic and Palaeo-gene rocks are thus now found near or at the outer margin of the Alps in western Switzerland, tectonically overlying both Helvetic nappes and deformed Molasse. No foreland basin deposits are found on the Prealpine nappes today, although Lemcke (1974) has argued, from subsidence curves, that there may have been a Molasse 'cover' before Plio–Quaternary uplift.

The *Subalpine Molasse* consists of a series of closely imbricated thrust slices adjacent to the Helvetic Zone and tectonically overlain, in part, by the Prealps (Fig. 4). Less competent horizons may show locally intense isoclinal folding. The stack of thrusts typically shows increasingly old stratigraphic sequences with proximity to the front of the fold-belt. Reflection seismics (unreleased) suggest that the thrusts and reverse faults may have a common sole at the base of the Molasse, above the Mesozoic carbonates. Neoalpine deforma-

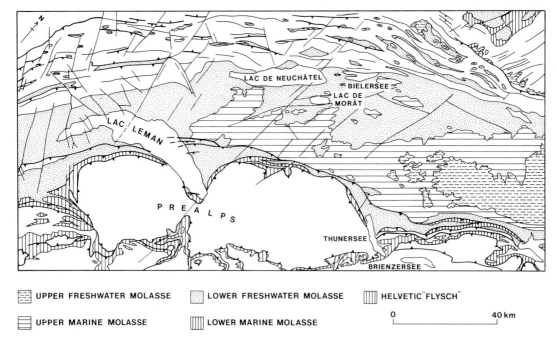

UPPER FRESHWATER MOLASSE	LOWER FRESHWATER MOLASSE	HELVETIC "FLYSCH"
UPPER MARINE MOLASSE	LOWER MARINE MOLASSE	0 40 km

Fig. 3. Distribution of Molasse Groups in western Switzerland. Note that Lower Marine Molasse (UMM) and north Helvetic 'flysch' are shown in identical ornament. Little Upper Freshwater Molasse is preserved west of a line joining Lake Thun and the Rhine graben.

tion was progressive and advanced apparently north-wards with time. This interpretation is supported by the composition of material incorporated within 'wildflysch' and 'mélange' units. The olisthostromes interstratified with the turbidites of the Lower Marine Molasse are composed of olistholiths from the Ultra-helvetic and Gurnigel (Penninic) décollement nappes (Weidmann, Homewood & Fasel, 1982), indicating that thrusts in the proximal part of the basin were active in early Oligocene times. The outermost thrust of the Subalpine Molasse in the Lucerne area of central Switzerland is lined by the Hornbuel Mélange (Matter *et al.*, 1980). This tectonic mélange comprises blocks of Aquitanian but not younger strata. The outermost 'subalpine' thrust was therefore active during early to mid-Miocene times.

The *Plateau or Mittelland Molasse* comprises relatively undeformed sequences between the reverse fault or frontal thrust of the Subalpine zone and the foot of the Jura mountains. Gentle anticlines and synclines as well as strike-slip faults characterize this sector (e.g. Schuppli in Althaus, 1947, Rigassi in Matter *et al.*, 1980).

The *Jura* province comprises the Plateau Jura and

the Folded Jura and corresponds to a tectonic high which existed in the early Oligocene between the Tertiary basins of the Molasse, the Rhine Graben to the north and the Bresse Graben to the west. However, during the later Oligocene, alpine detritus was shed into the Rhine Graben via a north-trending depression. In the folded Jura, the Mesozoic cover has been detached from its substratum, presumably along incompetent Triassic formations, since the extent of the deformation almost exactly matches the distribution of Triassic evaporite in the subsurface (Rigassi, 1977a). This Mesozoic cover has moved 2–25 km north or north-westwards with respect to the basement during the latest Miocene and early Pliocene. Molasse sediments which were deposited on top of the as yet undeformed Jura area of the foreland thus became eroded and isolated remnants of this feather-edge margin are now preserved in synclines (Fig. 4).

The picture of progressive compressional deformation during the Oligocene to Pliocene is complicated by oblique and strike-slip movement on oblique/lateral ramps that have been recorded in the Prealps as well as the Subalpine Molasse (Plancherel, 1979). The structures in the Jura are also believed to be

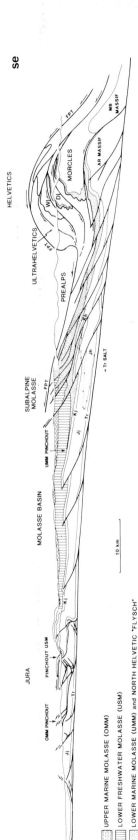

Fig. 4. Simplified structural cross-section from the Jura (NW) to the rear of the Helvetic nappes (SE). Inner edge of foreland basin is found in the Diablerets nappe whilst the distal feather edge margin progressively onlaps the Jura basement. Based essentially on Laubscher (1978), Rigassi, in Matter *et al.* (1980) and Boyer & Elliott (1982).

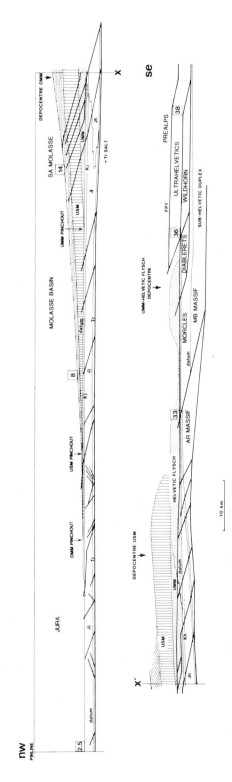

Fig. 5. Restored version of Fig. 4 with speculative timing of thrust movements in Ma (square boxes) and palinspastic positions of depocentres and pinch-outs in foreland basin deposits. Forms basis for Figs 16 and 17.

intimately related to strike-slip faulting on lateral structures (Rigassi, 1977a). Small-scale N–S-trending faults (with dominantly left-lateral movement) cross the entire Molasse Basin (Rigassi, 1977a) causing anomalous thicknesses of Oligocene foreland basin deposits locally to be developed and demonstrating their syntectonic nature (Diem, 1985).

Eocene extensional faulting in the Helvetic province (Trümpy, 1980, p. 50) influenced the deposition of shallow marine algal limestones, sandstones and shales. The influence of extension on the deposition of the slightly younger north Helvetic 'flysch' and lowermost Molasse is less well understood. Künzi *et al.* (1979) report basement granitic material derived from a steep south-facing scarp face, possibly representing a normal fault, in the north Helvetic 'flysch' of the Bernese Oberland, but the information is sparse. Tensional features in the distal part of the basin (Ziegler, 1982; Rigassi, in Debrand-Passard, 1984) may have been related to flexural extension in a forebulge region. These features were active growth faults (in eastern Switzerland) during Oligocene and early Miocene times (Pfiffner, this volume). The relative importance of the Mio–Pliocene inversion of these structures, compared with thin-skinned décollement in the Subalpine Molasse, has yet to be evaluated.

STRATIGRAPHY OF THE MOLASSE OF WESTERN SWITZERLAND

Although the major features of the stratigraphy of the Molasse were established over a century ago, the documentation of the stratigraphical details within a basin-fill of such rapid and ubiquitous lateral facies changes has proved difficult. The scarcity of macro-fossils in the dominantly continental succession has limited the application of classical biostratigraphical methods and oxidation has too often destroyed or altered the palynological assemblages. More recently, much progress has been made using ostracods, micromammals and charophytes for biostratigraphy and correlation (e.g. Engesser, Mayo & Weidmann, 1984).

Four groups are generally distinguished (Matter *et al.*, 1980), in ascending stratigraphic order (Figs 3, 4 and 5)

(1) Lower Marine Molasse (UMM) and North-Helvetic 'Flysch'.

(2) Lower Freshwater Molasse (USM).

(3) Upper Marine Molasse (OMM).

(4) Upper Freshwater Molasse (OSM).

Because the basin plunges eastwards, only the older Molasse groups are preserved in western Switzerland and no Upper Freshwater Molasse is found west of the city of Fribourg. This is apparently due to uplift and erosion, rather than to non-deposition.

Lower Marine Molasse

Outcrops of the Lower Marine Molasse are only found in the Subalpine zone and within the Helvetic nappes.

Analysis of the Eocene–Oligocene stratigraphy of the Diablerets and Morcles nappes, together with that of the underlying cover sequence reveals an abrupt onset of rapid subsidence at the Eocene–Oligocene boundary. The relatively thin Eocene transgressive sequence (littoral, shelf, open marine), arguably the result of global sea-level rise, is overlain by 500 m or more of turbidites deposited in moderately deep water over a short time span. Two distinct elongate troughs ran along the front of the Oligocene 'Swiss' Alps. The inner trough was filled with early Oligocene 'Tavey-annaz' sandstones (andesitic turbidites eroded from almost contemporaneous volcanoes) which can be seen capping the upper limb of the Morcles fold-nappe and the Diablerets nappe above (Masson *et al.*, 1980). In spite of claims of preservation of volcanic vents, andesitic debris invariably has proved to be resedimented turbiditic deposits (Cas & Lateltin, pers. comm., 1985). An outer trough, possibly of slightly younger age, was filled by clastic detritus derived predominantly from encroaching Austroalpine and Penninic thrust sheets. Lithoclasts such as radiolarian chert and spilitic fragments were derived from the Simme and Gêts units of the Prealps. These sediments are known as the Val d'Illiez formation (Fig. 6) and comprise coarse-grained, channellized turbidites incised into a thick background of muds and fine sands. Palaeocurrent indicators show that the turbidites fanned out from three main source areas in the Swiss Alps, one at the Rhône valley transect, one at the east of the Prealps (Lake Thun) and one in central Switzerland (Altdorf area).

Higher in the succession in the middle Oligocene occurs a regressive sequence passing from offshore mudstones at the base to a beach and terrestrial unit at the top. The presence of hummocky cross-stratification in the lower part of the sequence and the

Fig. 6. Val d'Illiez Formation: (A) View of the Tourche area above Morcles in the Rhône valley transect. 1, Basement; 2, Mesozoic thrust slices; 3, Parautochthonous Val d'Illiez Formation turbidites; 4, Overturned flank of the Morcles nappe; 5, Peak of Dents de Morcles. (B) Tight alpine folds in thin bedded turbidites. (C) Thin-bedded, fine-grained turbidites. (D) Slump features.

ubiquity of wave-ripple marks (Diem, 1985) suggest a wave-dominated seaway at this stage, although the palaeoecology of ostracods indicate highly fluctuating salinities in this residual water body (Carbonnel, in Weidmann *et al.*, 1982). Diem (1985) postulated a relatively small sea with a roughly WNW–ESE oriented shoreline experiencing storm waves of 4–5 s period. In the Rhône valley transect through the external Alps the regressive sequence is documented by the successive occurrence of littoral, beach, lagoonal and fluviatile deposits, and similar sequences are found throughout the region between Savoie, France and Austria.

This regressive sequence coincides in time with a postulated marked global sea-level fall (Vail, Mitchum & Thompson, 1977; Lemcke, 1983). The eustatic drop apparently drained the Alpine 'foredeep' very rapidly, with regression taking place from SW to NE. This process also explains the apparently anomalous orientation of the shoreline, trending perpendicular to the Alpine front in western Switzerland.

Whereas the normal thickness of the shoreface to shoreline sequence between Savoy and Austria is about 20 m, the transects at Monthey (Rhône valley), near Lucerne and near the Rhine show a marked increase in thickness (40 m). More rapid subsidence at these localities may be explained by compartmentalization of the thrust front by lateral or oblique faults.

A chaotic facies, one of the 'wildflysch' formations, occurs interstratified with Val d'Illiez Formation turbidites, immediately beneath the Prealpine thrust overriding the Lower Marine Molasse, and between the larger packets of Ultrahelvetics. The regional relationships, which are admittedly based on rather poor outcrop conditions, suggest similar conclusions to those which are well documented in the Ubaye Alps, France (e.g. Col de la Cayolle, Kerckhove, 1969). In the latter case, pebbly mudstones containing blocks and clasts of Subbriançonnais facies are interstratified with the foreland turbidites (Annot sandstones). Following this argument, the intercalations of wildflysch in the Val d'Illiez Formation should be interpreted as olisthostromes derived laterally from the southern, 'active' flank of the foreland toe trough. The composition of the wildflysch, contrasting with that of the Val d'Illiez turbidites, shows that the submarine flank of the trough was composed of slices of Ultrahelvetics and Gurnigel Flysch. The higher portion of the thrust stack, unroofed by subaerial erosion to feed the turbidites, was built up by the Gêts and Simme nappes (sub-units of the Prealps).

The Lower Freshwater Molasse

In late Oligocene times the seas were displaced from the Molasse depression and the landscape became dominated by at least seven gravel fans situated at the northern margin of the rising Alps (Büchi & Schlanke, 1977). These major sediment feeder systems such as the Pélerin fan in western Switzerland (Fig. 7), were characterized by steep proximal areas traversed by braided streams carrying boulders and pebbles, mostly cherts, radiolarites, siliceous limestones and 'Mocausa' sandstones and conglomerates derived from the Simme unit of the Prealps. They passed downstream into sinuous rivers which meandered freely over low gradient floodplains (Fasel, 1984). During late Chattian times, lakes and swamps were sufficiently important in western Switzerland to accumulate the 'Molasse à Charbon'. Otherwise fluvial systems drained into terminal fans.

In the Pélerin fluvial system exposed along the Léman lakeside, the downstream (SE to NW) facies changes from pebble conglomerates of the alluvial fan to multistorey sand bodies of sinuous rivers to lignites, marls and sands of lakes and floodplains is accomplished in a distance of less than 10 km. Even allowing for a shortening of 50% in this part of the Subalpine zone it seems unlikely that the Pélerin system (as we see it today) occupied the entire flexural depression north of the Alps in late Oligocene times. Such inconsistencies between depositional system size and anticipated widths of the flexural basin point to a far more complex inner margin to the basin than has hitherto been assumed. Discrepancies can be explained by either a combination of growth faults and inversion, or by as yet unrecognized thrusts causing additional shortening.

The relatively simple pattern of radial drainage evolved in the upper part of the Lower Freshwater Molasse (Aquitanian, Bersier, 1958) into a longitudinal system. In general, during wetter climates large river systems flowed north-eastwards, whereas during drier periods rivers ended in endorheic playas and saline lakes. The faunas of ostracods and foraminifera are brackish suggesting proximity to fully marine environments, but no marine strata of this age (late Oligocene to early Miocene) are known in the vicinity at present.

Progressive unroofing of the Alpine thrust-stack had by early Miocene times led to erosion of Austroalpine basement and Penninic units. The evolution of source terrains is marked by epidote dominating the heavy mineral spectrum at the expense of apatite

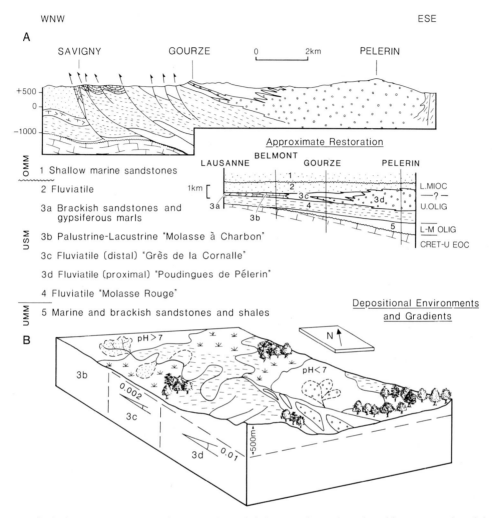

Fig. 7. The Pélerin fan system, Lower Freshwater Molasse: (A) Cross-section and stratigraphic reconstruction of the USM exposed along the lakeside between Vevey and Lausanne, after Rigassi in Matter *et al.* (1981). (B) General reconstruction of the Oligocene Pélerin alluvial fan with estimates of floodplain gradient, after Fasel (1984).

(Maurer, 1983). No Helvetic clasts are yet found in the Molasse, suggesting that either the Helvetic units remained undeformed in the early Miocene, or, more likely, the Helvetic antiformal stack had not yet reached the orogenic surface.

The Lower Freshwater Molasse attains its greatest thickness (*c.* 4 km) in the Subalpine Molasse where its most conspicuous formations consist of enormous masses of conglomerates known as Nägelfluh, mostly composed of Austroalpine clasts (e.g. the Rigi mountain near Lucerne). The group thins to the north and is just 300 m thick or less close to the Jura province (Figs 4 and 5).

Upper Marine Molasse

At the Aquitanian-Burdigalian boundary the Molasse basin was flooded by a shallow tide- and wave-dominated seaway, linking the Rhône Basin through the peri-Alpine foredeep with the Vienna Basin and thus with the easterly Paratethys.

In western Switzerland the transgressive base of the Upper Marine Molasse is a conformable contact with the underlying Lower Freshwater Molasse Group. Berger (1984) has constructed detailed models of the environments during the transgression based on an essentially palaeoecological analysis. Other workers (e.g. Homewood, 1981) have described the proximal to distal variations in sedimentary facies and environments and Homewood & Allen (1981) provided a semi-quantitative analysis of prevailing sea conditions.

The four major facies belts which overlie the basal transgressive sequence are as follows:

(1) A proximal fan delta facies (600–1000 m thick) is located near the main fluvial distributaries in the south.

(2) A coastal facies (300–700 m thick), more extensively developed between the major fan deltas, consisting of channellized tidal sandwaves and intertidal sand flats (Fig. 8).

(3) A nearshore facies (200–500 m thick) comprising thick (5–15 m) and extensive composite sandbodies (Fig. 8) composed mainly of draped megaripples locally incised by tidal channels or reworked into swash bars or flood ramps.

(4) An offshore facies (relative to the Alpine coast) where the Upper Marine Molasse is much thinner (<200 m) composed of thick pebbly and sandy coquinas with very large scale (up to 20 m) tabular cross-stratification (Allen *et al.*, 1985).

Homewood & Allen (1981) postulated a micro- to meso-tidal range, a fetch limited sea of perhaps 100 km width generating average wave periods of about 4 s. A reanalysis of the same data by Homewood, Allen & Yang (1985) using time series analysis (Yang & Nio, 1985) suggested meso- to macro-tidal ranges. Allen (1984), Allen & Homewood (1984) and Benkert (1984) have provided more detailed or local analyses.

The offshore coquina facies still presents some fascinating enigmas. Various palaeontological arguments suggest fairly deep water, some 100 m within the basin (Berger, 1984), whereas vadose cements suggest emersion of the shell bank crests and wave ripple mark geometries and dimensions likewise suggest very shallow water depths (Allen *et al.*, 1985).

Yet unrecognized sea-level fluctuations may thus have affected the lower Miocene Upper Marine Molasse. The appearance of coarse clastics at the base of the Group may well indicate orogenic activity in the fold-thrust belt at this time, corresponding to the emplacement phase of Helvetic nappes into the Austrian Molasse Basin established from seismic and borehole data (cf. Ziegler, 1982; Bally, 1983). Subsidence due to flexure as a response to structural activity in the Alps together with global sea level rise (Vail *et al.*, 1977) could be responsible for the flooding of the western Molasse Basin.

The younger part of the Upper Marine Molasse is characterized by regressive cycles of fan-delta deposits fed laterally to the basin from the Alps, and also contains an abundant macrofossil fauna. This is in stark contrast to the earlier part of the OMM and has led to the creation of a 'Helvetian' stage (Rütsch, 1958). Some clasts in the Helvetian OMM come from the Helvetic nappes, indicating uplift and unroofing of the Helvetic antiformal stack, perhaps related to the incorporation of a slice of basement into the Helvetic pile.

The Upper Freshwater Molasse

In middle Miocene times the Molasse Basin once again reverted to continental sedimentation with proximal alluvial fans such as the Napf discharging vast quantities of sediments and water into the basin. These alluvial fans fed a longitudinal river system but this time with a NE to SW palaeodrainage direction. The Upper Freshwater Molasse measures over 1500 m in thickness in the central parts of the gravel fans but only a few hundreds of metres further north. Very little of the OSM is found in western Switzerland, but the continuing presence of Helvetic pebbles suggests further unroofing of the Helvetic antiformal stack.

SUBSIDENCE HISTORY

Subsidence diagrams along the entire Molasse basin were provided over 10 years ago (Lemcke, 1974) (Fig. 9). Although these subsidence diagrams fail to take

Fig. 8. Facies of the Upper Marine Molasse in the vicinity of Fribourg: (A) Transgressive facies near Heitenreid. Beach (1), tidal flat (2) and distributary channel (3) deposits. Note slight angular discordance between beach stratification and overlying deposits. (B) Channel-shoal sequence from the coastal facies at Illens. 0, ponded intertidal channel fill; 1, erosive subtidal channel base; 2, subtidal channel deposits, trough cross-stratification; 3, shallow trough to tabular cross-bedded sandstones with neap-spring bundle sequences, intertidal shoal; 4, intertidal shoal top with neap-spring bundle sequences. (C) Intertidal sandwave form-set (1), Illens showing shallow stage run-off ripples (2). Dominant current (ebb tide) from right to left. (D) Subtidal sandwave bundles from nearshore facies, Bois du Devin, near Marly. Dominant current (ebb) to left. (E) Bundle sequence, Bois du Devin. Dominant current (ebb) to left.

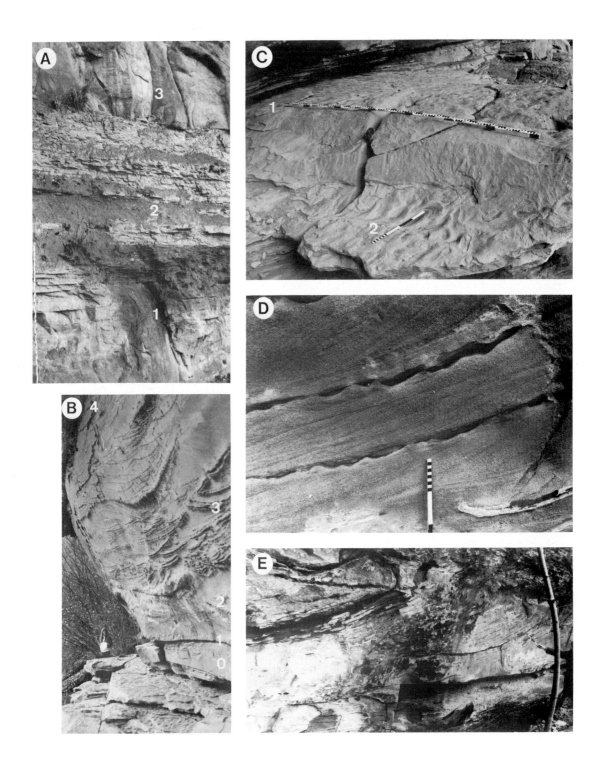

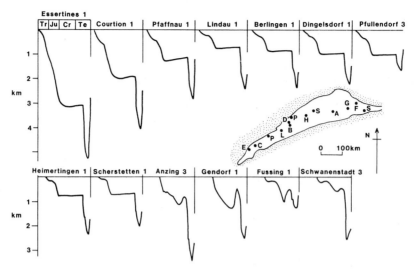

Fig. 9. Simplified subsidence curves from 13 boreholes in the Molasse basin, after Lemcke (1974). Note the very high rates of 'rebound' in western Switzerland compared with southern Germany.

into account the compaction history of the sedimentary units and palaeotopographic, palaeobathymetric and eustatic sea-level variations through time, they illustrate very clearly (1) the sudden onset of subsidence at the end of the Eocene or early in the Oligocene and (2) the high rates of post-Miocene 'rebound' in western Switzerland compared with southern Germany. This post-Eocene subsidence history is in marked contrast with the earlier style of subsidence in the north Alpine (Helvetic) region (Fig. 10).

The pre-Molasse subsidence history of the outer Alpine terrains has classically been linked with the development of the Tethyan ocean (ophiolites in the western Alps are dated from Middle Jurassic to Early

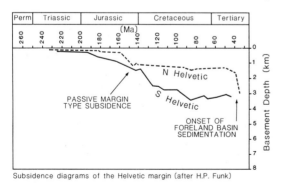

Subsidence diagrams of the Helvetic margin (after H.P. Funk)

Fig. 10. Subsidence diagrams of the Helvetic margin in the Central Alps. The onset of foreland basin sedimentation in the north Helvetic region is marked by an acceleration in subsidence, presumably representing lithospheric loading.

Cretaceous times, indicating the duration of oceanic spreading) (Lemoine, in Boillot *et al.*, 1984). Sediment accumulated in half graben sub-basins defined by listric normal faults. Throw was generally to the south (Swiss Alps) or to the east (Dauphiné Alps). Subsidence curves for the sequences below the Molasse (Lemcke, 1974; Funk, 1985) show a typical passive margin style with an initial high subsidence rate decreasing roughly exponentially with time (Fig. 10). Whereas the pre-Molasse subsidence history in the north Alpine region can be confidently related to lithospheric stretching and thermal contraction (McKenzie, 1978; Sclater & Christie, 1980), the abrupt onset of subsidence at the end of the Eocene or early in the Oligocene has been interpreted as being due to a major flexural event reflecting lithospheric loading of the European plate (Karner & Watts, 1983).

Considerable use has been made of subsidence or geohistory curves (van Hinte, 1978; Guidish *et al.*, 1985) in predicting oil or gas windows in hydrocarbon provinces and in the investigation of tectonic driving mechanisms for subsidence. The majority of studies concern basins which have originated through lithospheric stretching and subsequent thermal contraction (e.g. Keen, 1978; Steckler & Watts, 1978). Borehole data in western Switzerland (Fig. 11) have been used to construct subsidence history diagrams. In order to carry out an accurate decompaction, an additional overburden has been applied to the boreholes similar to the amount of rebound suggested by Lemcke (1974).

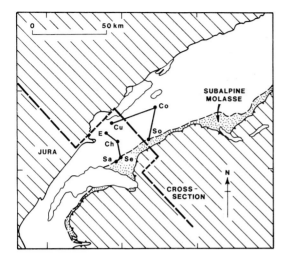

Fig. 11. Location of boreholes. E, Essertines 1; Cu, Cuarny 1; Co, Courtion 1; Ch, Chapelle 1; So, Sorens 1; Sa, Savigny 1; Se, Servion 1. Line of cross-section of Figs 4 and 5 also shown.

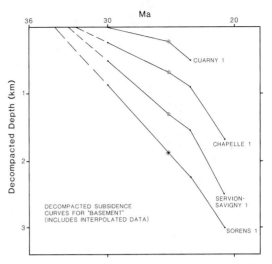

Fig. 12. Decompacted subsidence curves for the base of the Molasse Basin extrapolated to obtain onset of subsidence. Based on stratigraphy and heavy mineral picks (Maurer, 1983) from borehole-data. No eustatic or palaeobathymetric/palaeotopographic corrections have been applied. Boreholes are located on Fig. 11.

Clay mineral assemblages in Molasse deposits (Monnier, 1982) broadly support Lemcke's analysis.

Our decompaction technique follows that used by Sclater & Christie (1980) for North Sea wells. Porosity variations with depth are assumed to be exponential and given by

$$f = f_0 \, e^{-cZ} \qquad (1)$$

where f is porosity at depth Z, f_0 is the surface porosity and c is a coefficient related to lithology. Sediments are then decompacted by moving them vertically up the exponential curve. The equation

$$Z_2' - Z_1' = Z_2 - Z_1 - f_0/c(e^{-cZ_1} - e^{-cZ_2})$$
$$+ f_0/c(e^{-cZ_1'} - e^{-cZ_2'}) \qquad (2)$$

(where Z_1 and Z_2 are the present depths of the sedimentary section and Z_1' and Z_2' the equivalent depths following decompaction) can be solved numerically by computer iteration. The effects of the sediment load can also be removed to obtain the true tectonic or flexural subsidence (Steckler & Watts, 1978; Sclater & Christie, 1980; Bond & Kominz, 1984).

Decompacted subsidence curves for the 'basement' and for a base Chattian datum are shown in Figs 12 and 13. Current absolute age scales (Hardenbol & Berggren, 1977; Harland *et al.*, 1982) vary considerably for the Cenozoic. For consistency, the scale proposed by Berggren (in Debrand-Passard, 1984) has been adhered to throughout. The decompacted subsid-

ence curves show the proximal-distal variations across the basin. Localities close to the Subalpine zone (Sorens, Savigny, Servion) experienced strong subsidence at an earlier date and subsided more rapidly than localities close to the feather edge in the Jura. The result is a strongly asymmetrical basin cross-section

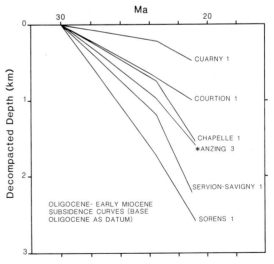

Fig. 13. Decompacted subsidence curves using the base Chattian (30 Ma) as a datum. No eustatic or palaeobathymetric/palaeotopographic corrections. Anzing 3 borehole, near Münich, included for comparison.

with a progressive pinch-out of younger stratigraphic units on to the European foreland (Figs 4 and 5). This pattern of successive pinch-outs has already been illustrated by Büchi & Schlanke (1977), Rigassi (1977b), and Matter *et al.* (1980) (Fig. 14A, B).

The offset of distal curves from proximal curves suggests that subsidence occurred in a northwards migrating 'wave' across the foreland. Since the boreholes are separated spatially and since the amount of shortening in the autochthonous Molasse is mini-

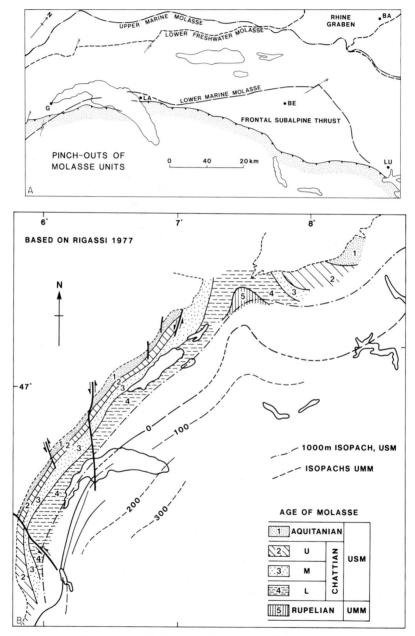

Fig. 14. Pinch-outs of stratigraphic units on to the European foreland. (A) General map of pinch-outs of UMM, USM and OMM in western and central Switzerland, after Rigassi in Matter *et al.* (1980). (B) Detail within Lower Freshwater Molasse (USM), after Rigassi (1977b).

mal, it is possible to calculate the rate of migration of this 'wave' of subsidence. Based on the extrapolated onset of subsidence (Fig. 12) the rate is approximately 6 mm yr^{-1}.

The average decompacted subsidence rate of the basement for wells close to the Subalpine zone is approx. 0·2 mm yr^{-1} whereas wells close to the feather edge margin are characterized by decompacted subsidence rates of only 0·05 mm yr^{-1}.

True tectonic subsidence is obtained after removal of the subsidence due to sediment load and after corrections for variations in water depth and eustatic sea-level fluctuations (Fig. 15). The porosity of the new sediment layers is

$$f = f_0/c \left(\frac{e^{cZ'_1} - e^{cZ'_2}}{Z'_2 - Z'_1} \right). \quad (3)$$

The bulk density of the sedimentary column is then

$$\bar{\rho}_s = \sum_i \left\{ \frac{\bar{f}_i \rho_w + (1 - \bar{f}_i) \rho_{sgi}}{s} \right\} Z'_i \quad (4)$$

where f_i is the mean porosity of the ith layer, ρ_{sgi} is the sediment grain density of the same layer, Z'_i is the individual sediment layer thickness and S is the total thickness of the column corrected for compaction.

The loading effect of the sediment is then given by

$$Y = S \left(\frac{\rho_m - \bar{\rho}_s}{\rho_m - \rho_w} \right) \quad (5)$$

where Y is the depth of the basement due to tectonic forces only, ρ_m, ρ_w and ρ_s are the mantle density, water density and mean sediment density respectively. Incorporating eustatic and palaeobathymetric/palaeotopographic effects, according to the general equation

presented by Steckler & Watts (1978) and modified by Bond & Kominz (1984)

$$Y = \Phi \left\{ S \left(\frac{\rho_m - \bar{\rho}_s}{\rho_m - \rho_w} \right) - \Delta_{SL} \left(\frac{\rho_w}{\rho_m - \rho_w} \right) \right\} \\ + (W_d - \Delta_{SL}) \quad (6)$$

where Δ_{SL} is the palaeosea-level relative to present, W_d is the palaeowater depth and Φ is a basement response function equal to unity for local (Airy) type isostasy. The flexural response of a thin elastic plate (Hetenyi, 1974) is determined principally by its flexural rigidity, which for the Molasse basin has been estimated by Watts *et al.* (1982) to be 10^{30}–10^{31} dyne cm^{-2}. Bearing in mind the uncertainties regarding changes in flexural rigidity through time and the anomalous value (Kusznir & Karner, 1985, fig. 3) given by Watts *et al.* (1982) for the Molasse Basin, we set $\Phi = 1$ in this study.

The tectonic subsidence curves with no eustatic or palaeobathymetric corrections are almost linear (Fig. 15). With corrections they become convex-upward, suggesting an increasing rate of tectonic subsidence through the Oligo–Miocene. There is no suggestion of the exponentially decreasing tectonic subsidence rate which characterizes the period of thermal contraction following lithospheric stretching. With a local isostatic model, the time-averaged rate of tectonic subsidence is approximately 0·1 mm yr^{-1} for wells close to the Subalpine zone. The rate of 0·06 mm yr^{-1} for Anzing 3, near Münich, serves as a comparison.

RELATION OF SUBSIDENCE TO TECTONIC DEVELOPMENT

The palinspastic restoration of the basin (Fig. 5) and knowledge of the approximate timing of deformation allows rates of deformation to be calculated. The approximate rate of thrust tip propagation is about 7 mm yr^{-1} when averaged across the Helvetic, Molasse and Jura belts (Fig. 16). The rate of horizontal shortening would be somewhat less, dependent on the overall percentage shortening in these structural zones.

The successive positions of pinch-outs and postulated depocentres for the Lower Marine Molasse (UMM)–north Helvetic 'flysch', Lower Freshwater Molasse (USM) and Upper Marine Molasse (OMM) on the restored section suggest a decreasing convergence rate through the Oligo–Miocene (Fig. 17). Oligocene rates from these stratigraphic methods are in the range 7–9 mm yr^{-1} decreasing to just 2 mm yr^{-1}

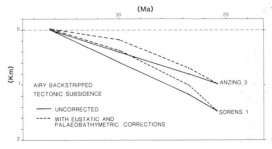

Fig. 15. Backstripped tectonic subsidence using an Airy model of isostatic compensation. Subsidence curves with eustatic and palaeobathymetric corrections are noticeably convex-up.

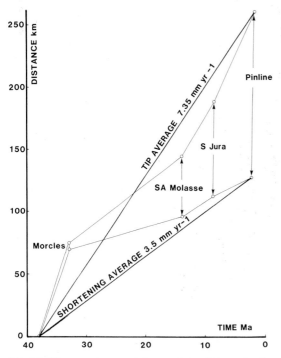

Fig. 16. Distance–time graph for neoalpine deformation in western Switzerland showing the tip propagation rate and shortening rate assuming a 50% shortening across the belt. Based on simplifications and assumptions shown in Figs 4 and 5.

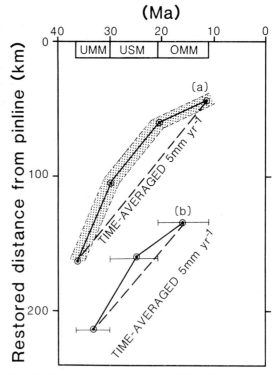

Fig. 17. Convergence rate during the Oligo–Miocene based on (A) the successive pinch-out of stratigraphic units on the restored section, and (B) the migration of postulated depocentres on the restored section.

in the Miocene. The average rate for migration of pinch-outs during the period 36·5–11 Ma is 5 mm yr^{-1}. The time-averaged rate for the migration of depocentres over a smaller time span is similarly 5 mm yr^{-1}. It must be emphasized that the importance of these figures is in their comparative rather than absolute values. It can readily be appreciated, for example, that the rates of horizontal migration of the basin derived from stratigraphy closely match the rates of shortening derived from structural-strati-graphic analysis. Both techniques necessarily rely on a reasonable restored section.

DISCUSSION

Lemcke (1974, fig. 9) published the most characteristic signature of the Molasse Basin in its regional setting, the Mesozoic to present subsidence curve. Typical features are the abrupt Oligocene onset, the strong and fairly regular rate, and the sudden 'rebound'. However, the subsidence curve cannot be

taken in itself as a general signature of foreland basins. The southern Rhine Graben, for instance, shows a very similar history to the Molasse Basin (Robert, 1985; Allen, Homewood & Williams, this volume).

In the case of the Molasse, it is reasonable to ascribe initial subsidence to the emplacement of an allochthonous load—the Austroalpine nappe complex. This is justified by the occurrence of Lower Marine Molasse lying unconformably on the Austro-alpine nappes, some 40 km behind the present thrust front (Fuchs, 1976). Subsidence must have been very rapid compared with the rate of infilling at this stage.

The classical Molasse stratigraphy (four groups, alternately marine and freshwater) may be followed some distance from Switzerland into Austria and Bavaria. However, east of the Münich traverse, facies remain marine and turbiditic in the subsurface throughout the section until the middle Miocene (Fuchs, 1976; Betz & Wendt, 1983). An underfilled equilibrium was thus established, with Alpine detritus draining north-eastwards.

The middle Miocene saw the reversal of palaeocurrent patterns, major rivers now flowing longitudinally but towards the south-west. There is little readily available evidence to suggest that this major switch in depocentres is due to a corresponding migration of thrust loads from east to west. Most data (summarized in Platt, 1986) suggest a principally radial thrust transport direction around the Alpine arc with no discernable discontinuity between eastern and western Switzerland. Laubscher (1982) demonstrated the markedly different structural style of the Helvetic nappes between western and central-eastern Switzerland, signifying different kinematics and therefore possibly timing of deformation between the two regions. Gourlay's (1986) study in the Helvetic nappes of western Switzerland suggests first a northward direction of movement, followed by a translation to the west, but these data originate from local steep zones around basement massifs. What, then, controlled the large along-strike variations in subsidence and sedimentary facies and the switch in palaeodrainage pattern in the Molasse Basin? The precise roles of inherited structural and bathymetric features and of lateral variations in orogenic wedge accretion and erosion have yet to be fully evaluated.

We end by commenting on the tedious discussion of the terms Flysch versus Molasse. The Bavarian Molasse Basin is mostly composed of turbidites and it would seem appropriate, in this context at least to call the deposits of the Molasse Basin 'molasse', as opposed to 'Flysch'. The fact that the early deposits of the Molasse Basin in western Switzerland, as defined by geohistory plots, are the turbiditic Taveyannaz and Val d'Illiez formations, and that these are called North Helvetic *Flysch* in Switzerland, is the result of a historical development somewhat divorced from advances in the field of geodynamics and basin analysis. If the term 'Molasse' is merely to be used to describe continental and shallow marine siliciclastics and the term 'flysch' to describe deep water turbidites, then they surely have no *explanatory* function in basin analysis and geodynamics and should be avoided.

ACKNOWLEDGMENTS

Much of the background study for this paper was carried out with the support of the Swiss National Science Foundation (Project 2.415–0.82) to whom we are grateful. We acknowledge colleague John Platt for his comments on the manuscript and Mary Marsland and Claire Pope for typing and draughting.

REFERENCES

ALLEN, P.A. (1984) Reconstruction of ancient sea conditions with an example from the Swiss Molasse. *Mar. Geol.* **60**, 455–473.

ALLEN, P.A. & HOMEWOOD, P. (1984) Evolution and mechanics of a Miocene tidal sandwave. *Sedimentology*, **31**, 63–81.

ALLEN, P.A., HOMEWOOD, P. & WILLIAMS, G.D. (1985) Structural and sedimentological evolution of the molasse basin of western Switzerland. *Abstr. 6th Int. Ass. Sediment. Reg. Mtg.* Lleida, pp. 15–17.

ALLEN, P.A., MANGE-RAJETZKY, M.A., MATTER, A. & HOMEWOOD, P. (1985) Dynamic palaeogeography of the open Burdigalian seaway, Swiss Molasse basin. *Eclog. geol. Helv.* **78**, 351–381.

ALTHAUS, H.E. (1947) Erdölgeologische Untersuchungen in der Schweiz. *Beitr. zur Geol. der Schweiz. Geotechnische serie 26/1*, 79 pp.

BALLY, A. (1983) Seismic expression of structural styles: a picture and work atlas. *Am. Ass. Petrol. Geol., Studies in Geology Ser.* **15**.

BEAUMONT, C. (1981) Foreland basins. *Geophys. J. R. astr. Soc.* **65**, 291–329.

BENKERT, J.P. (1984) Die litorale Faziesentwicklung des Luzerner Sandsteins im Entlen-Querschnitt. *Eclog. geol. Helv.* **77**, 363–381.

BERGER, J.P. (1984) *Transgression de la Molasse Marine Supérieure (OMM) en Suisse occidentale.* Ph.D. thesis, University of Fribourg, Switzerland, 311 pp.

BERSIER, A. (1958) Sequences détritique et divagations fluviales. *Eclog. geol. Helv.* **51**, 854–893.

BETZ, D. & WENDT, A. (1983) Neuere Ergebnisse der Aufschlussund Gewinnugstatigkeit auf Erdöl und Erdgas in Süddeutschland. *Bull. ver. Schweiz. Petroleum Geol. u. Ing.* **49**, 9–36.

BOILLOT, G., MONTADERT, L., LEMOINE, M. & BIJU-DUVAL, B. (1984) *Les Marges Continentales Actuelles et Fossiles Autour de la France.* Masson, Paris.

BOND, G.C. & KOMINZ, M.A. (1984) Construction of tectonic subsidence curves for the Early Paleozoic miogeocline, southern Canadian Rocky Mountains: implications for subsidence mechanisms, age of break-up, and crustal thinning. *Bull. geol. Soc. Am.* **95**, 155–173.

BOYER, S.E. & ELLIOTT, D. (1982) The geometry of thrust systems. *Bull. Am. Ass. Petrol. Geol.* **66**, 1196–1230.

BÜCHI, U.P. & SCHLANKE, S. (1977) Zur Paleogeographie der schweizerischen Molasse. *Erdöl-Erdgas-Z.*, **93**, 57–69.

DEBRAND-PASSARD, S. (1984) Synthèse Géologique du Sud-Est de la France. Stratigraphie et Paléogeographie. *Mém. Bur. Réch. géol. Min.*, **125**, 615 pp.

DEWEY, J.F. & BIRD, J.M. (1970) Mountain belts and the new global tectonics. *J. geophys. Res.* **75**, 2625–2647.

DICKINSON, W.R. (1974) Plate tectonics and sedimentation. In: *Tectonics and Sedimentation* (Ed. by W. R. Dickinson). *Spec. Publ. Soc. econ. Paleont. Mineral.*, Tulsa, **22**, 1–27.

DIEM, B. (1985) Analytical method for estimating palaeowave climate and water depth from wave ripple marks. *Sedimentology*, **32**, 705–720.

ENGESSER, B., MAYO, N. & WEIDMANN, M. (1984) Nouveaux gisements de mamifères et biostratigraphie de la Molasse subalpine vaudoise et fribourgeoise. *Schw. Pal. Abh.* **107**, 1–39.

FASEL, J.M. (1984) Paléohydraulique dans l'épandage fluviatile du Mt. Pélerin, Oligocene suisse. *Abstr. 5th Reg. Mtg. int. Assoc. Sedimentl.*, Marseille, p. 163.

FUCHS, W. (1976) Gedanken zur Tectogenese der nördlichen Molasse zwischen Rhône und March. *Jahr. Geol. B.–A. Wein* **119**/2, 207–249.

FUNK, H.P. (1985) Mesozöische Subsidenzegeschichte im Helvetischen Schelf der Oöstschweiz. *Eclog. geol. Helv.* **78**, 249–272.

GOURLAY, P. (1986) La déformation du socle et des couvertures delphino-helvétiques dans la région du Mont Blanc (alpes occidentales). *Bull. Soc. géol. Fr.* **8**, 159–169.

GUIDISH, T.M., KENDALL, C.G. St C., LERCHE, I., TOTH, D.J. & YARZAB, R.F. (1985) Basin evaluation using burial history calculations: an overview. *Bull. Am. Ass. Petrol. Geol.* **69**, 92–105.

HARDENBOL, J. & BERGGREN, W.A. (1977) A new Paleogene/ numerical time scale. *Am. Ass. Petrol. Geol. Stud. Geol.* **6**, 213–234.

HARLAND, W.B., COX, A.V., LLEWELLYN, P.G., PICTON, C.A.G., SMITH, A.G. & WALTERS, R. (1982) *A Geologic Time Scale.* Cambridge University Press.

HETENYI, M. (1974) *Beams on Elastic Foundation.* University of Michigan Press, Ann Arbor, 255 pp.

HINTE, J.E. van (1978) Geohistory analysis—application of micropaleontology in exploration geology. *Bull. Am. Ass. Petrol. Geol.* **62**, 201–222.

HOMEWOOD, P. (1981) Faciès et environnements de dépôt de la Molasse de Fribourg. *Eclog. geol. Helv.* **74**, 29–36.

HOMEWOOD, P. & ALLEN, P.A. (1981) Wave-, tide- and current-controlled sandbodies of Miocene Molasse, western Switzerland. *Bull. Am. Ass. Petrol. Geol.* **65**, 2534–2545.

HOMEWOOD, P. & CARON, C. (1982) Flysch of the western Alps. In: *Mountain Building Processes* (Ed. by K. J. Hsü), pp. 157–168. Academic Press, London, 263 pp.

HOMEWOOD, P., ALLEN, P.A. & YANG, C.S. (1985) Palaeotidal range estimates from the Miocene Molasse. *Abstr. 6th int. Ass. Sediment. Reg. Mtg.* Lleida, pp. 200–201.

KARNER, G.D. & WATTS, A.B. (1983) Gravity anomalies and flexure of the lithosphere at mountain ranges. *J. geophys. Res.* **88**, 10,449–10,477.

KEEN, C.E. (1978) Thermal history and subsidence of rifted continental margins—evidence from wells on the Nova Scotian and Labrador shelves. *Can. J. Earth Sci.* **16**, 505–522.

KERCKHOVE, C. (1969) La "Zone du Flysch" dans les nappes du l'Embrunais-Ubaye. *Geol. alp.* (Grenoble) **45**, 5–204.

KÜNZI, B., HERB, R., EGGER, A. & HÜGI, T. (1979) Kristallin-Einschlüsse im nordhelvetischen Wildflysch des Zentralen Berner Oberlands. *Eclog. geol. Helv.* **72**, 425–437.

KUSZNIR, N. & KARNER, G.D. (1985) Dependence of the flexural rigidity of the continental lithosphere on rheology and temperature. *Nature,* **316**, 138–142.

LAUBSCHER, H.P. (1978) Foreland folding. *Tectonophys.* **47**, 325–337.

LAUBSCHER, H.P. (1982) A northern hinge zone of the arc of the western Alps. *Eclog. geol. Helv.* **75**, 233–246.

LEMCKE, K. (1973) Die Lagerung der jüngsten Molasse im nördlichen Alpenvorland. *Bull. ver. Schweiz. Petrol. Geol. u. -Ing.,* **39**, 29–41.

LEMCKE, K. (1974) Vertikalbewegungen des vormeso-

zöischen Sockels im nördlichen Alpenvorland vom Perm bis zur Gegenwart. *Eclog. geol. Helv.* **67**, 121–133.

LEMCKE, K. (1983) Indications of a large sea level fall at the Rupelian/Chattian boundary in the German Molasse Basin. *Bull. ver. Schweiz. Petrol. Geol. u. -Ing.* **49**, 57–60.

MASSON, H., HERB, R. & STECK, A. (1980) Helvetic Alps of western Switzerland. In: *Geology of Switzerland, a Guide Book* (Ed. by R. Trümpy), pp. 109–153, Schweiz. Geol. Komm.

MATTER, A., HOMEWOOD, P., CARON, C., VAN STUIJVENBERG, J., WEIDMANN, M. & WINKLER, W. (1980) Flysch and Molasse of western and central Switzerland. In: *Geology of Switzerland, a Guide Book* (Ed. by R. Trümpy), pp. 261–293. Schweiz. Geol. Komm.

MAURER, H. (1983) Sedimentpetrographische analysen an Molasse-abfolgen der Westschweiz. *Jb. Geol. B-A.,* **126**, H.1, Wein.

McKENZIE, D. (1978) Some remarks on the development of sedimentary basins. *Earth planet. Sci. Lett.* **40**, 25–32.

MONNIER, F. (1982) Thermal diagenesis in the Swiss Molasse basin: implications for oil generation. *Can. J. Earth Sci.* **19**, 328–342.

MUGNIER, J.L. (1984) *Déplacements et Déformations dans l'Avant Pays d'une Chaîne de Collision.* Ph.D. thesis, University of Grenoble, 163 pp.

MUGNIER, J.L. & VIALON, P. (1986) Deformation and displacement of the Jura cover on its basement. *J. struct. Geol.* **8**, 373–388.

PLANCHEREL, R. (1979) Aspects de la déformation en grand dans les Préalps medianes plastiques entre Rhône et Aar. Implications cinématiques et dynamiques. *Eclog. geol. Helv.* **72**, 145–241.

PLATT, J.P. (1986) Dynamics of orogenic wedges and the uplift of high pressure metamorphic rocks. *Bull. geol. Soc. Am.* (in press).

PRICE, R.A. (1973) Large scale gravitational flow of supracrustal rocks, southern Canadian Rockies. In: *Gravity and Tectonics* (Ed. by K. A. de Jong & R. Scholten), pp. 491–502. Wiley, New York.

RAMSAY, J.G. (1981) Tectonics of the Helvetic nappes. In: *Thrust and Nappe Tectonics* (Ed. by K. R. McClay & N. J. Price), pp. 293–309, *Spec. Publ. geol. Soc. London,* **9**, Blackwell Scientific Publications, Oxford.

RIGASSI, D. (1977a) Genèse tectonique du Jura: une nouvelle hypothèse. *Paleolab. News* **2**, Terreaux du Temple, Geneva.

RIGASSI, D. (1977b) Subdivision et datation de la molasse "d'eau douce inferieure" du plateau Suisse. *Paleolab. News* **1**, Terreaux du Temple, Geneva.

ROBERT, P. (1985) Histoire Géothermique et diagenèse organique. *Bull. Cent. Réch. Expl.-Prod. Elf-Aquitaine Mém.* **8**, 345 pp.

RÜTSCH, R.F. (1958) Das Typusprofil des Helvetian. *Eclog. geol. Helv.* **51**, 107–118.

RYBACH, L., BÜCHI, U., BODINER, P. & KRUSI, H.R. (1980) Die Tiefengrundwasser des schweizerischen mittelandes aus geothermischer Sicht. *Eclog. geol. Helv.* **73**, 293–310.

SAWATSKI, G.G. (1975) *Etude géologique et minéralogique des Flyschs à grauwackes volcanique du synclinal de Thônes (Haute Savoie, France). Grès de Taveyanne et grès du Val d'Illiez.* Unpublished Ph.D. thesis, University of Genève.

SCHEDL, L. & WILTSCHKO, D.V. (1984) Sedimentological effects of a moving terrain. *J. geol. Chicago* **92**, 273–288.

SCLATER, J.G. & CHRISTIE, P.A.F. (1980) Continental stretching: an explanation of the post-Mid Cretaceous subsidence of the central North Sea Basin. *J. geophys. Res.* **85**, 3711–3739.

STECKLER, M.S. & WATTS, A.B. (1978) Subsidence of the Atlantic type continental margin off New York. *Earth planet. Sci. Lett.* **41**, 1–13.

STOCKMAL, G.S., BEAUMONT, C. & BOUTILIER, R. (1986) Geodynamic models of convergent margin tectonics: transition from rifted margin to overthrust belt and consequences for foreland basin development. *Bull. Am. Ass. Petrol. Geol.* **70**, 181–190.

TRÜMPY, R. (1980) *Geology of Switzerland: a Guide Book. Part A; an Outline of the Geology of Switzerland.* Schweiz. Geol. Komm., Wepf & Co., Basel, 104 pp.

VAIL, P.R., MITCHUM, R.M., JR & THOMPSON, S. (1977) Global cycles of relative changes of sea level. In: *Seismic Stratigraphy—Applications to Hydrocarbon Exploration* (Ed. by C. E. Payton). *Mem. Am. Ass. Petrol. Geol.* **26**, 83–91. Tulsa, Oklahoma.

VUAGNAT, M. (1952) Pétrographie, répartition et origine des microbrèches du Flysch nordhelvétique. *Matér. Carte géol. Suisse*, N.S. **97**, 1–103.

WATTS, A.B., KARNER, G.D. & STECKLER, M.S. (1982) Lithospheric flexure and the evolution of sedimentary basins. *Phil. Trans. R. Soc. A*, **305**, 249–281.

WEIDMANN, M., HOMEWOOD, P. & FASEL, J.M. (1982) Sur les térrains subalpins et le Wildflysch entre Bulle et Montreux. *Bull. Soc. Vaud Sci. nat.* **362**, 151–183.

YANG, C.S. & NIO, S.-D. (1985) The estimation of palaeo-hydrodynamic processes from subtidal deposits using time series analysis methods. *Sedimentology*, **32**, 41–58.

ZIEGLER, P.A. (1982) *Geological Atlas of Western and Central Europe.* Shell Int. Petroleum Maatsch., B.V., 130 pp.

Spec. Publs int. Ass. Sediment. (1986) **8**, 219–228

Evolution of the north Alpine foreland basin in the Central Alps

O. ADRIAN PFIFFNER

Institut de Géologie, Rue Emile-Argand 11, CH-2000 Neuchâtel, Switzerland

ABSTRACT

Based on a geometric thrust tectonic model the evolution of the Eo–Oligocene flysch and Oligo–Miocene molasse basin of eastern Switzerland can be correlated to the deformational history of the Helvetic zone of the adjoining Alps. The foredeep migrated in the direction of the foreland at the front of the advancing nappe pile. Downwarp of the foredeep was accompanied by normal faulting. Uplift of the hinterland was mainly due to movement on hinterland dipping thrust faults. During the infill of the North Helvetic flysch basin, deformation in the hinterland included the emplacement of exotic strip sheets (Pizol phase) associated with thrusting of the Penninic–Austroalpine nappes on to the internal Helvetic nappes, as well as nappe internal (Calanda phase) deformation of the latter. Sedimentation of the Lower Marine and Freshwater molasse is contemporaneous with basement shortening, coupled with the development of the Glarus thrust, in whose footwall flysch suffered already penetrative (Calanda phase) deformation. The Upper Marine and Freshwater molasse can be correlated to (Ruchi phase) post-metamorphic movement on the Glarus thrust and deformation of the internal (Sub-alpine) molasse. The erosion stage seems to correspond to an isostatic adjustment to the thickened and depressed crust.

INTRODUCTION

The foreland basin of eastern Switzerland comprises a sequence of flysch and molasse sediments and is located on the north side of the Alpine orogen (Fig. 1). It represents a peripheral basin overlying thickened lithosphere and its internal part is now overlain by various thrust sheets from the mio- and eugeosynclinal parts of the Alps.

The stratigraphy of the sedimentary sequence is summarized in Fig. 2. The older, internal part of the basin infill consists of the Late Eocene to Early Oligocene North Helvetic flysch (cf. Siegenthaler 1974). North Helvetic flysch is defined as a stratigraphic unit and the term flysch will be used in that sense throughout this paper. The Eocene Taveyannaz and Elm formations are characterized by mainly trench-parallel (from WSW) sediment transport and the presence of volcanic detritus. Around the turn of the Eocene–Oligocene epochs volcanic debris disappears and sediment transport changed to radial with a source in the south, coupled with a rapid decrease in water depth. The younger, external part of the basin infill represents a sequence of alternating shallow

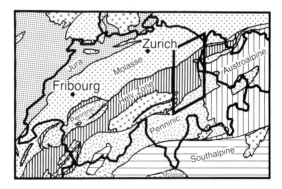

Fig. 1. Study area.

marine sediments (offshore, delta front and wave- and storm-dominated facies; cf. Frei, 1979) and freshwater gravel fan deposits (e.g. Frei, 1979; Habicht, 1945) and is referred to as molasse in this paper.

The flysch sequence now forms part of the Helvetic zone. This zone underlies the Penninic nappes and is subdivided into the Infrahelvetic complex underlying, and the Helvetic nappes overlying the Glarus thrust, a major thrust fault with up to 50 km of displacement. The Helvetic nappes are separated into the Lower and Upper Glarus nappe complex by the Säntis thrust, a detachment horizon at the level of lowermost Cretaceous marls. The Infrahelvetic complex consists essentially of the deformed Aar massif basement wedge with its cover sediments and some exotic strip sheets. The Lower Glarus nappe complex in the cross-section considered here consists, from the bottom to the top, of the Glarus thrust sheet s.s., the Mürtschen thrust sheet, and a number of imbricates of mainly Upper Jurassic limestones. The latter represent the substratum of the detached Cretaceous limestones forming the overlying Upper Glarus nappe complex (essentially the Säntis thrust sheet in this cross-section). The molasse is separated into the Plateau molasse, which is only slightly tilted, and the Sub-alpine molasse in which deformation lead to the existence of imbricate thrust sheets and folds. For a more detailed discussion of the tectonics of the Helvetic zone in this cross-section the reader is referred to, e.g. Trümpy (1969), Pfiffner (1978, 1981, 1985) and Funk *et al.* (1983), which also contain the principal references. For the Sub-alpine molasse see Habicht (1945).

Deformation of the basin and its hinterland includes an intricate succession of folding and thrusting events. The local sequence of deformational events can in part be followed over large domains and allowed for the recognition of individual phases of deformation [Pizol, Cavistrau, Calanda and Ruchi phases of Pfiffner (1977, 1978) and Milnes & Pfiffner, 1977, 1980)]. These phases are defined by local structural style and are not meant to be 'chronostratigraphic' units. In fact, it has been pointed out (Groshong, Pfiffner & Pringle, 1984) that they are likely to be diachronous. The dating of these phases as proposed by Milnes & Pfiffner (1980) is only valid for the Infrahelvetic complex, as will be discussed below. The degree of metamorphism that the internal part of the foreland basement underwent reached up to epizonal conditions and post-dates the main nappe-internal deformation.

The aim of this paper is to set up a geometric-kinematic model of the evolution of this foreland basin by means of a series of cross-sections at various geological times. The construction of these cross-sections involved arguments from different disciplines, such as structural geology, stratigraphy and sedimentology, petrology and geochronology. Emphasis was put on a 'to scale model' to emerge, as far as this is possible under the circumstances.

EVOLUTION

The tectonic evolution of the foreland basin will be discussed on the basis of a series of cross-sections (Fig. 4) representing a sequence in time. The specific geological restorations include an initial state, followed by various stages marking the infill of the flysch and molasse basins and conclude with erosion stages.

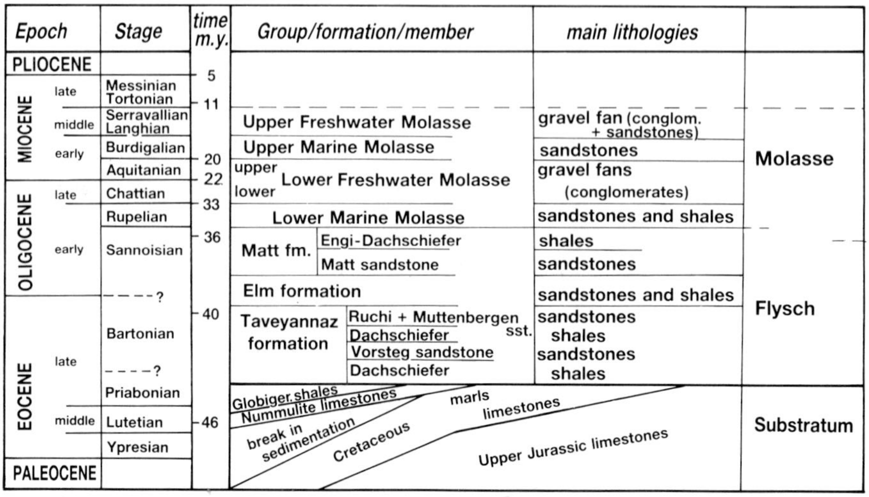

Epoch		Stage	time m.y.	Group/formation/member			main lithologies	
PLIOCENE			5					
MIOCENE	late	Messinian Tortonian	11					Molasse
	middle	Serravallian Langhian		Upper Freshwater Molasse			gravel fan (conglom. + sandstones)	
	early	Burdigalian	20	Upper Marine Molasse			sandstones	
		Aquitanian	22	upper Lower Freshwater Molasse lower			gravel fans (conglomerates)	
OLIGOCENE	late	Chattian	33	Lower Marine Molasse			sandstones and shales	
		Rupelian	36	Matt fm.	Engi-Dachschiefer		shales	
	early	Sannoisian			Matt sandstone		sandstones	
		– – – –?	40	Elm formation			sandstones and shales	Flysch
		Bartonian		Taveyannaz formation	Ruchi + Muttenbergen sst. Dachschiefer		sandstones shales	
EOCENE	late	– – – –?			Vorsteg sandstone Dachschiefer		sandstones shales	
		Priabonian		Globiger. shales Nummulite limestones break in sedimentation Cretaceous marls limestones				Substratum
	middle	Lutetian	46				Upper Jurassic limestones	
		Ypresian						
PALEOCENE								

Fig. 2. Stratigraphic column.

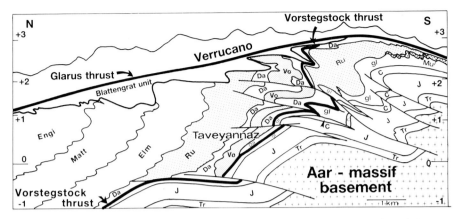

Fig. 3. Detailed cross-section through the internal part of the flysch foreland basin (in part after Siegenthaler, 1974 and Styger, 1961). Note the truncation of the flysch formations and the Vorstegstock thrust by the Blattengrat basal thrust. Subsequent out-of-sequence thrusting includes the Calanda phase slicing up of the Aar massif basement and cover, which deformed these two thrust faults, and the development of the Glarus thrust, which put originally more external Permian Verrucano on originally more internal Cretaceous–Tertiary of the Blattengrat strip sheet. Tr: Triassic, J: Jurassic, C: Cretaceous, Gl: Globigerina shales, Da: Dachschiefer, Vo, Ru, Mu: Vorsteg, Ruchi, Muttenberge sst.

The actual reconstructions were made so as to balance the sections by conserving bed length and volume and by keeping displacements on thrust faults somewhat consistent (see also Pfiffner, 1985). True balancing implies that the line of section be parallel to the transport direction of the nappes, which is impossible to achieve for this cross-section. Another useful constraint is that after sedimentation the overburden of any point in the profile can only increase if this overburden deforms, either by nappe emplacement or by internal shortening. The maximum overburden for the Aar massif was of the order of 12 km, judging from studies of metamorphism (e.g. Frey, 1978) and using a $30°C\ km^{-1}$ geothermal gradient, and occurred when all the Helvetic, Penninic and Austroalpine nappes were overlying it. Finally, pebbles and heavy mineral spectra in the foreland basin sediments indicate which thrust sheets were being eroded at various times and thus constrain the uppermost parts of the cross-sections to some degree.

Initial state (46 Myr)

Regarding the entire Central Alps it would be arbitrary and misleading to set the beginning of the evolution at this *Middle Eocene (46 Myr)* time mark, since flysch sedimentation in the more internal, Penninic domain started already in Cretaceous times (e.g. Trümpy, 1973). Concentrating on the Helvetic zone and the Molasse, however, this marks the starting

point for the 'Neogene orogeny' (Trümpy, 1973). No attempt is made here to elucidate the relation between deformation of the Penninic and Helvetic zones.

In Middle Eocene times a shallow water (5–200 m) Nummulite limestone was deposited on the cover of the future Aar massif, whereas to the south, in the realm of the future Helvetic nappes, sedimentation of pelagic Globigerina shales took place. In the cross-section a fault separates these two domains. This growth fault, as well as the one north of reference point T, represent the Eocene faulting documented by conglomerates containing limestone pebbles of the underlying formations (Brückner, 1946; Styger, 1961). The horst-like structure south of the realm of the future Säntis thrust sheet represents the 'South Helvetic ridge' of Leupold (1942), Rüfli (1959) and Wegmann (1961). This ridge separates the future Blattengrat thrust sheet (in the north) from the Sardona thrust sheet (in the south), both of which were about to travel far north on to the future Aar massif as 'exotic strip sheets' (Milnes & Pfiffner, 1977). The phase of emplacement of exotic strip sheets, involving the top of the cover only (Upper Cretaceous and Tertiary), and lacking penetrative structures such as, e.g. cleavage associated with thrust faults, is the earliest, Pizol phase of deformation recognizable in the Helvetic zone. In this reconstruction the future Helvetic nappes are taken as cover of the future Tavetsch and Gotthard massifs, a point discussed in more detail in Pfiffner (1985).

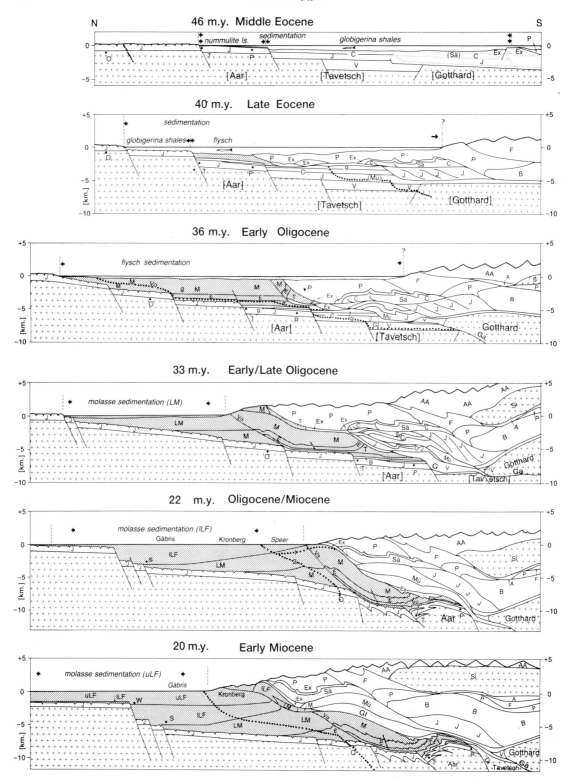

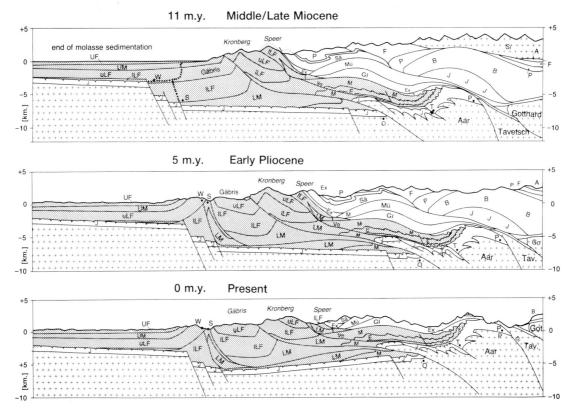

Fig. 4. Series of cross-sections showing the tectonic evolution of the foreland basin. Dotted lines indicate traces of future faults. Crosses: crystalline basement, shaded: foreland basin sediments. *Reference points:* O (point in subsurface where Sub-alpine molasse thrusts may run into basement), P (Panix), S (Scheftenau), T (Tierfehd), W (Wattwil).

Nappes, formations and thrusts: A: Arosa ophiolitic melange, AA: Austroalpine cover sediments, B: Bündnerschiefer, C: Cretaceous, Di: Disentis thrust, E: Engi fm., Ex: Exotic strip sheets, F: Falknis nappe, G: Globigerina shales, Ga: Garvera thrust, Gl: Glarus thrust, J: Mesozoic (mainly Jurassic), LF and LM: Lower Freshwater and Marine molasse (u, l: upper, lower), M: Matt fm., Mü: Mürtschen thrust, P: Penninic flysch, Sä: Säntis thrust, Si: Silvretta basement, T: Taveyannaz fm., UF and UM: Upper Freshwater and Marin molasse, V: Verrucano, Vo: Vorstegstock thrust.

Flysch basin (40 and 36 Myr)

In *Late Eocene (40 Myr)* times flysch sedimentation (Taveyannaz formation) occurred in the flysch trough above the future Aar massif. How far south this zone of sedimentation extended is unknown, but it cannot be ruled out that the Penninic flysch received sediments which were subsequently eroded. Sedimentation north of the fault near reference point T (Tierfehd) may have been of pelagic Globigerina shale type. The existence of this fault is indicated by the rather abrupt northerly termination of the Taveyannaz formation. This is illustrated in Fig. 3 by the northward ending of the Ruchi sandstone in the footwall of the Vorstegstock thrust. This Ruchi sandstone is the more

or less autochthonous cover of the Aar massif, whereas the Ruchi sandstone in the hangingwall is allochthonous and was originally situated just south of the Ruchi sandstone in the footwall (Siegenthaler, 1974).

The realm of the Helvetic nappes, the Glarus, Mürtschen, Säntis and Blattengrat thrust sheets, is lacking sediments of the Late Eocene. This is taken to indicate that this domain was already covered by exotic strip sheets and Penninic flysch. Furthermore this domain was already suffering internal deformation, namely a detachment at the levels of the Triassic and/or Lower Jurassic and the lowermost Cretaceous. The reason for this is that the trailing edge of the Helvetic nappes, marked by the imbricates made up of Upper Jurassic rocks, shows a comparatively very low degree of metamorphism (Metagenesis or Anchizone; Groshong *et al.*, 1984) and must therefore have escaped substantial overburden. The reconstruction showed that this implies most probably an early northward transport of these thrust sheets. The internal deformation of the Helvetic nappes is in part associated with a thrust-parallel and axial planar cleavage and a stretching lineation. This phase of deformation is called the Calanda phase (the Cavistrau phase between Pizol and Calanda phase is of local importance and not included in the present discussion). It will be noted that the Calanda phase is taken to have been initiated in the Säntis thrust sheet already at a time when the cover of the Aar massif, including the type locality of the phase, was still receiving sediments.

Early Oligocene (36 Myr) times mark the close of sedimentation in the cover of the future Aar massif. The sediments of the Elm formation show the transition from deep to shallow water type sedimentation. The rise in the south of the basin is documented by slumping in the Matt formation. Furthermore volcanic detritus ceases and sediment transport directions were radial with a source in the south (Siegenthaler, 1974). In the areas adjoining this study the first conglomeratic facies appear in the form of the Eggberge fan (shedding of pebbles from South Helvetic and Penninic units) to the west, in the cover of the Aar massif, and the Deutenhausen beds to the east, underlying the leading edge of the Helvetic nappes (Resch, 1976).

In Fig. 4 the Calanda phase internal deformation is shown to have progressed: the Säntis thrust sheet travelled further north and in its footwall the Mürtschen thrust sheet is individualized. The internal geometry of the units in the hanging wall of the Säntis thrust sheet is drawn to have changed compared to the preceding cross-section, implying that these Penninic units underwent deformation. Finally, in the uppermost part, the arrival of the Austroalpine nappes is to be noted. Their emplacement on the Penninic zone mainly took place in Cenomanian–Albian times (Trümpy, 1980; Laubscher & Bernoulli, 1982) and during subsequent events they seemed to have acted passively as 'orogenic lid' (Laubscher & Bernoulli, 1982; Laubscher, 1983).

Molasse basin (33, 22, 20 and 11 Myr)

The turn *Early–Late Oligocene (33 Myr)* marks the end of the Lower Marine molasse. The northern limit of the foreland basin is shown to coincide with a set of normal faults. Although the decrease in thickness and eventual disappearance of the Lower Marine molasse could equally well be described as onlap, for the next following beds, the lower Lower Freshwater molasse, thickness changes seem too rapid to be explained by onlap. The slope of the southern end of the basin is interpreted as being controlled by active thrusting: it is the emplacement of the Vorstegstock thrust sheet which can be observed in the Helvetic zone (see Fig. 3 and Siegenthaler, 1974). This thrust sheet comprises the whole internal foreland basin infill consisting of flysch, except for the remnants directly overlying the Aar massif cover in the footwall of the Vorstegstock thrust.

In the Helvetic zone internal, Calanda phase deformation continued and the Glarus thrust sheet s.s. is individualized. The future Aar massif and its cover are being buried in the process but still undeformed, such as to allow for the Vorstegstock thrust sheet detachment. This interpretation is compatible with geochronological data (Hunziker *et al.*, 1986): K/Ar and Rb/Sr ages from the Helvetic nappes cluster in the interval 35–30 Myr, whereas the underlying Aar massif cover yields younger ages (25–15 Myr for K/Ar). The internal deformation of the Helvetic nappes included the thrusting of the Gotthard on the Tavetsch massif, and this in turn provoked the high relief of the Alps at this time, elevations exceeding maybe 5 km.

During the following period up to about the *Oligocene–Miocene boundary (22 Myr)* the foredeep migrated further outwards with the thickest sediment pile near the marker point S (Scheftenau) consisting of the gravel fans Gäbris, Kronberg and Speer. The thickness of this lower Lower Freshwater molasse increases from Speer to Gäbris (cf. Habicht, 1945) and then rapidly decreases to the north (borehole Hünenberg 1; Lemcke, Büchi & Wiener, 1968). The exotic strip sheets at the southern margin allow the explanation of the occurrence of pebbles derived from the Tertiary of the South Helvetic zone (Leupold, Tanner & Speck, 1942).

The Infrahelvetic complex below the Glarus thrust is deformed, the Gotthard massif has overthrust the Tavetsch massif, and the latter is about to overthrust the Aar massif. The internal deformation in the Helvetic nappes was largely achieved, i.e. the internal structures were about to be transported passively. These interpretations are based on the fact that metamorphism within the Infrahelvetic complex post-dates its internal, Calanda phase deformation (Pfiffner, 1982; Groshong *et al.*, 1984) and occurred at the

Oligocene–Miocene turn (Frey *et al.*, 1974; Hunziker *et al.*, 1986), on considerations of displacements on the Glarus thrust (Pfiffner, 1985), the degree of metamorphism within the various thrust sheets and intuition.

By *Early Miocene (20 Myr)* times sedimentation of the Lower Freshwater molasse reached farther north on to the foreland. Drill hole data show that the Mesozoic substratum is eroded down to Upper Jurassic limestones overlain by continental Eocene Bohnerz formation (Büchi, Wiener & Hofmann, 1960). This erosion is also evident in the cover of the Aar massif, where Middle Eocene (Middle Lutetian) Nummulite limestones rest with an angular unconformity on their Mesozoic substratum (Upper Cretaceous in the south, Upper Jurassic in the north; see Fig. 3).

The southernmost, Speer gravel fan in the foreland basin does not contain Miocene sediments and most probably never received any sediments younger than Chattian (lower Lower Freshwater molasse). The neighbouring Kronberg delta in the north received younger sediments of at least up to Aquitanian age (upper Lower Freshwater molasse). In the next following, Gäbris delta, the younger sediments reach up to at least the Burdigalian Upper Marine molasse, and in the adjoining Plateau molasse sedimentation persisted up to the Langhian–Seravallian (Bürgisser, 1980) Upper Freshwater molasse. Each of the gravel fans mentioned forms today a thrust sheet of its own (Fig. 4). This situation is interpreted as a stack of imbricate thrust sheets whose formation progressed towards the foreland, the end of sedimentation in each individual imbricate marking the time it was overthrust by the overlying thrust sheet. All this deformation included very shallow conditions as indicated by local erosional features in the footwall of thrust faults. These erosional features lead, e.g. Heim (1932) to the conclusion that thrust sheets, including the toe of the Helvetic nappes, were moving on a palaeo-relief of considerable amplitude and that the arrival of the Helvetic nappes therefore post-dated thrusting within the Sub-alpine molasse. Habicht (1945) corrected this view in that he could show that the abrupt lateral termination of some thick conglomerate sequences represent in fact rapid facies changes at the limit of delta fans and not erosional features. Some local erosion (e.g. of the Speer gravel fan; see Heim, 1910–1917) however do exist, and also arguments from displacement variations on the basal thrust of the Helvetic nappes indicate that the thrust must have broken surface early in its activity (Pfiffner, 1985).

These constraints were used to reconstruct the very shallow features in Fig. 4.

For the basal thrust of the Speer thrust sheet two solutions can be put forward (see dotted lines 1 and 2 in the cross-section at 22 Myr in Fig. 4). The first solution is to connect the Speer thrust backwards to the Glarus thrust, a solution which could be used for all the Sub-alpine molasse thrusts. This option was used in Milnes & Pfiffner (1980) and Groshong *et al.* (1984). The second solution is to run the thrust fault down into the basement, a solution that again is possible for all the Sub-alpine molasse thrusts. Solution 2 was used and its consequences discussed in Pfiffner (1985). In this paper, the Speer thrust follows solution 1, while the Kronberg and Gäbris thrusts follow solution 2, the reason being that the Glarus thrust (solution 1) offsets the metamorphic zonation during its last activity in the Ruchi phase. The displacement post-dating metamorphism is probably less than 10 km, possibly around 5 km, but the situation is difficult to assess since cut-off points of metamorphic isolines are ill-defined and age determinations of illites yield mixing, deformation and cooling ages.

The high relief of the Alps at this time resulted from movements on the Disentis (lifting the Tavetsch relative to the Aar massif) and the Glarus thrusts, both dipping towards the hinterland, thus bringing hangingwall rocks nearer to the surface. Moreover the pebble spectra in the Lower Freshwater molasse indicate extensive erosion of Austroalpine cover sediments (namely Triassic dolomites) at this time (Habicht, 1945; Frei, 1979). This argument must be used with caution however, as the Miocene distributive province was subsequently completely eroded, rendering its reconstruction (as shown in Fig. 4) rather speculative, and the relatively shallow dips of the Penninic and Austroalpine thrust faults give more freedom than constraints on the elaboration of the geometry of this distributive province.

At the turn *Middle/Late Miocene (11 Myr)* molasse sedimentation, which had essentially shifted north on to the domain of the present-day Plateau molasse, the Hörnli fan, came to a halt (Bürgisser, 1980). The youngest sediments, the Upper Freshwater molasse, contain conglomerates with the first pebbles from the Mesozoic of the Helvetic nappes (Leupold *et al.*, 1942) and of sandstones from the Sub-alpine molasse thrust sheets (Tanner, 1944).

The Kronberg thrust sheet is shown to be emplaced on the Gäbris gravel fan, the basal thrust reaching down into the basement (solution 2 in the discussion above). This geometry is responsible for the high relief in the reconstruction. Also due to the choice of solution 2, no more movement of the Glarus thrust is allowed for. Geochronological data (Hunziker, personal communication) give ages maybe as young as 14 Myr to be correlated with Ruchi phase movements on the Glarus thrust. These dates are valid for the internal part of the Glarus thrust above the Aar massif and, using solution 1, these displacements could be transferred to thrust faults within the Sub-alpine molasse.

Erosional stage (5 Myr and present)

Since the Upper Freshwater molasse already contains pebbles derived from the Sub-alpine molasse, local erosion of the foreland basin infill must have begun by Mid-Miocene times.

For the *Early Pliocene (5 Myr)* no sedimentation is known to have occurred. In Fig. 4 the Gäbris gravel fan is shown to be deformed into a thrust sheet bordered in the north by an anticlinal 'triangle zone'. The importance of the antiform is underlined by the juxtaposition of reference points S (Scheftenau; base of the lower Lower Freshwater molasse) and W (Wattwil; base of the upper Lower Freshwater molasse). The formation of this complex 'triangle zone' includes a north-dipping thrust or back-thrust (Randunterschiebung). The basal thrust of the Gäbris thrust sheet is shown to merge rearwards with the Kronberg thrust before it reaches basement.

Between the Early Pliocene and the *Present (0 Myr)* vertical movements took place and still go on. For the reconstruction the present-day vertical uplift rates and gradients were used. This implies an increase from about 0.2 mm yr^{-1} at the northern end to 1.8 mm yr^{-1} at the southern end of the cross-section. This situation could reflect an isostatic response of the thickened crust: the Moho drops from 35 km at the northern to 50 km at the southern margin of the cross-section and is accompanied by a negative gravity anomaly (see, e.g. Müller, 1982). The situation may well be different in western Switzerland, where additional younger thrust faults linking the deformation of the Jura mountains to crustal shortening in the Alps could have caused uplift within the latter.

DISCUSSION AND CONCLUSION

The flysch–molasse foreland basin in eastern Switzerland is a peripheral foreland basin which formed at the close of a continent–continent collision. Loading

by tectonic thickening and subsequent sediment accumulation can be envisaged as a continuous process migrating towards the foreland with time. From the North Helvetic flysch trough, which initiated in Late Eocene times, to the end of molasse sedimentation in Middle Miocene times the foredeep migrated over at least 70 km, which corresponds to an average rate of 3 mm yr^{-1} for the flysch and Lower Marine molasse, and 2 mm yr^{-1} from the Lower on to the Upper Freshwater molasse. This migration is linked to deformation in the adjoining hinterland, where a complex succession of folding and thrusting buried the most internal part of the foreland basin infill to a depth of at least 10–12 km. The loading of the lithosphere was coeval with shortening and thickening of the upper crust. Downwarping of the lithosphere was accompanied by normal, syn-sedimentary faulting in the basin. The existence of these growth faults is indicated by abrupt facies and thickness changes or disappearance of individual stratigraphic units. Examples include the middle Eocene Nummulite limestones and Globigerina shales (Fig. 4), the termination of the Ruchi sandstone in the footwall of the Vorstegstock thrust (Fig. 3) and the termination of the Oligocene Lower Marine and Freshwater molasse (Fig. 4). Within the molasse foredeep a tectonic control of the downwarp is indicated by the fact that the thickness of the infill is independent of the sedimentary facies. In particular at the lateral limits of gravel fans, where very rapid changes from conglomeratic to shaly facies occur, stratigraphic thicknesses remain unchanged (Habicht, 1945). Sediment accumulation (or preservation) rates in the proximal parts were of the order of 0.65 mm yr^{-1} in the flysch, 0.4 mm yr^{-1} in the Lower Freshwater and about 0.3 mm yr^{-1} in the Upper Freshwater molasse.

Within the deformed parts of the foreland basin and the Helvetic zone thrusting seems to have migrated towards the foreland. However, some of the thrust faults are clearly out of sequence: some thrust sheets travelled over great distances and were subsequently sliced up by thrust faults developing in their footwall. This process is called 'Einwicklung' in Alpine literature and typical examples comprise the exotic units.

Loading in this particular example of foreland basin must be seen as both surface and subsurface loading (Karner & Watts, 1983). The specified subsurface load added in the course of the basin formation includes a piggyback stack of thrust sheets from the Helvetic, Penninic and Austroalpine zones of the Alps, comprising crustal rocks. This load is added over the entire length of the orogen due to the extension and dip of individual active thrust faults. Displacement on hinterland dipping thrust faults reaching deeply into the crust is interpreted as being responsible for loading and uplift. The model differs in this way from the one presented by Jordan (1981) for the Rocky Mountains in the Western U.S., where displacement occurs on flat lying detachment horizons. It resembles the model discussed by Beaumont (1981) with the difference that in the Alps the metamorphic core is involved in the loading history to a greater extent. The load added and the amount of uplift are directly dependent on the angle of dip of the active thrust faults. The thrust faults are interpreted to correlate with the formation of basement wedges (e.g. Aar, Tavetsch and Gotthard massifs), which would speak of a relatively high angle of dip for these faults.

Erosion replacing part of the specific load added is seen to be more or less evenly distributed over the entire uplifted region, but it seems that erosion could not keep pace with uplift at two stages. These two stages are characterized by the high elevation of the Alps in mid-Oligocene and Early Miocene times and thus also correspond to periods of extensive surface loading. In the model these stages coincide with phases of pronounced basement shortening, rapid orogenic contraction and rapid downwarp. The orogenic contraction, defined by the shortening of the horizontal distance between reference point 0 (in the flysch basin) and the toe of the Silvretta basement, is estimated to be of the order of 7–10 mm a^{-1} and must have been accommodated by the whole zone between these two points, i.e. individual thrust sheets may have travelled even more slowly. The rate of downwarp of point 0 during these two phases of seemingly increased activity are of the order of 0·5–1·3 mm a^{-1}, i.e. somewhat higher than the sediment preservation rate. This is to be expected since the basin receiving sediments had already migrated to a more external position than point 0 at this time.

The classical approach of unveiling the unroofing history of the orogen by the analysis of pebble and heavy mineral spectra in the foreland sediments turns out to be of limited use. This stems from the fact that many of the thrust sheets which yield characteristic spectra were bound by surfaces at very low angle to the palaeoland surface. It is therefore difficult to assess the geometry of distributive provinces.

As for the correlation between deposition of specific foreland sediments and the timing of specific phases of deformation in the hinterland the following remarks (to be taken with caution) can be made. The formation and filling of the most internal part of the foreland basin, the flysch trough, is contemporaneous with the emplacement of exotic units with their piggyback sequence of Penninic and Austroalpine nappes and the emplacement of the internal Helvetic nappes (Säntis to Mürtschen thrust sheets). The next following Lower Marine molasse foredeep can be related to the formation of the more external Glarus thrust sheet s.s. of the Helvetic nappes and the Vorstegstock thrust sheet (flysch) of the Infrahelvetic complex, and extensive basement thrusting between the Gotthard and Tavetsch massifs. The Lower Freshwater molasse coincides with further transport of the Helvetic nappes along the Glarus thrust, extensive internal, Calanda phase deformation in the Infrahelvetic complex (essentially the Aar massif and its cover) and basement shortening between Gotthard and Tavetsch, as well as Tavetsch and Aar massifs. During the Upper Marine and Freshwater molasse the foredeep had reached its most external position and the deformation associated with it includes thrusting within the internal, Sub-alpine molasse and the basement shortening related to that. In the segment of the north Alpine foreland basin considered here, uplift and erosion seemed to have been the next following dominant process, still prevailing at present.

REFERENCES

BEAUMONT, C. (1981) Foreland basins. *Geophys. J. R. astr. Soc.* **65**, 291–329.

BRÜCKNER, W. (1946) Neue Konglomeratfunde in den Schiefermergeln des jüngeren helvetischen Eocäns der Zentral- und Ostschweiz. *Eclog. geol. Helv.* **38**, 315–328.

BÜCHI, U.P., WIENER, G. & HOFMANN, F. (1960) Neue Erkenntnisse im Molassebecken auf Grund von Erdöltiefbohrungen in der Zentral- und Ostschweiz. *Eclog. geol. Helv.* **58**, 87–108.

BÜRGISSER, H.M. (1980) Zur Mittel-Miozänen Sedimentation im Nordalpinen Molassebecken: das "Appenzellergranit"-Leitniveau des Hörnli-Schuttfächers (Obere Süsswassermolasse, Nordostschweiz). *Mitt. geol. Inst. ETH u. Univ. Zürich*, N.F. **232**, 196 pp.

FREI, H.-P. (1979) Stratigraphische Untersuchungen in der subalpinen Molasse der Nordost-Schweiz, zwischen Wägitaler Aa und Urnäsch. *Mitt. geol. Inst. ETH u. Univ. Zürich*, N.F. **233**, 217 pp.

FREY, M. (1978) Progressive low-grade metamorphism of a black shale formation, Central Swiss Alps, with special reference to pyrophyllite and margarite bearing assemblages. *J. Petrol.* **19**, 93–135.

FREY, M., HUNZIKER, J.C., FRANK, W., BOCQUET, J. DAL PIAZ, G.V., JÄGER, E. & NIGGLI, E. (1974) Alpine metamorphism of the Alps: a review. *Schweiz. miner. petrol. Mitt.* **54**, 247–290.

FUNK, H.P., LABHART, T., MILNES, A.G., PFIFFNER, O.A., SCHALTEGGER, U., SCHINDLER, C., SCHMID, S.M. & TRÜMPY, R. (1983) Bericht über die Jubiläumsexkursion "Mechanismus der Gebirgsbildung" der Schweizerischen Geologischen Gesellschaft in das ost- und zentralschweizerische Helvetikum und in das nördliche Aarmassiv vom 12. bis 17. September 1982. *Eclog. geol. Helv.* **76**, 91–123.

GROSHONG, R.H., PFIFFNER, O.A. & PRINGLE, L.R. (1984) Strain partitioning in the Helvetic thrust belt of eastern Switzerland from the leading edge to the internal zone. *J. struct. Geol.* **6**(1 + 2), 5–18.

HABICHT, K. (1945) Geologische Untersuchungen im südlichen sanktgallisch-appenzellischen Molassegebiet. *Beitr. geol. Karte Schweiz*, N.F. **83**, 166 pp.

HEIM, ARN (1910–1917) Monographie der Churfirsten-Mattstock-Gruppe. *Beitr. geol. Karte Schweiz*, N.F. **20** (1–4), 662 pp.

HEIM, ARN (1932) Zum Problem des Alpen-Molasse Kontaktes. *Eclog. geol. Helv.* **25**, 223–231.

HUNZIKER, J.C., FREY, M., CLAUER, N., DALLMEYER, R.D., FRIEDRICHSEN, H., ROGGWILER, P. & SCHWANDER, H. (1986) The evolution of illite to muscovite: mineralogical and isotopic data from the Glarus Alps, Switzerland. *Contr. Miner. Petrol.* **92**, 157–180.

JORDAN, T.E. (1981) Thrust loads and foreland basin evolution, Cretaceous, Western United States. *Bull. Am. Ass. Petrol. Geol.* **65**(12), 2506–2520.

KARNER, G.D. & WATTS, A.B. (1983) Gravity anomalies and flexure of the lithosphere at mountain ranges. *J. geophys. Res.* **88** (B12) 10 449–10 477.

LAUBSCHER, H.P. (1983) Detachment, shear, and compression in the Central Alps. *Mem. geol. Soc. Am.* **158**, 191–211.

LAUBSCHER, H.P. & BERNOULLI, D. (1982) History and deformation of the Alps. In: *Mountain Building Processes* (Ed. by K. J. Hsü), pp. 169–180.

LEMCKE, K., BÜCHI, U.P. & WIENER, G. (1968) Einige Ergebnisse der Erdölexploration auf die mittelländische Molasse in der Zentralschweiz. *Bull. Ver. Schweiz. Petrol.-Geol. -Ing.* **35**(87), 15–34.

LEUPOLD, W. (1942) Neue Beobachtungen zur Gliederung der Flyschbildungen der Alpen zwischen Reuss und Rhein. *Eclog. geol. Helv.* **33**, 247–291.

LEUPOLD, W., TANNER, H. & SPECK, J. (1942) Neue Geröllstudien in der Molasse. *Eclog. geol. Helv.* **35**, 235–246.

MILNES, A.G. & PFIFFNER, O.A. (1977) Structural development of the Infrahelvetic Complex, eastern Switzerland. *Eclog. geol. Helv.* **70**, 83–95.

MILNES, A.G. & PFIFFNER, O.A. (1980) Tectonic evolution of the Central Alps in the cross section St. Gallen—Como. *Eclog. geol. Helv.* **73**, 619–633.

MÜLLER, St (1982) Deep structure and recent dynamics in the Alps. In: *Mountain Building Processes* (Ed. by K. J. Hsü), pp. 181–199.

PFIFFNER, O.A. (1977) Tektonische Untersuchungen im Infrahelvetikum der Ostschweiz. *Mitt. geol. Inst. ETH u. Univ. Zürich*, N.F. **217**, 432 pp.

PFIFFNER, O.A. (1978) Der Falten- und Kleindeckenbau im Infrahelvetikum der Ostschweiz. *Eclog. geol. Helv.* **71**, 61–84.

PFIFFNER, O.A. (1981) Fold-and-thrust tectonics in the Helvetic nappes (E. Switzerland). In: *Thrust and Nappe Tectonics* (Ed. by K. R. McClay & N. J. Price), pp. 319–327. *Spec. Publ. geol. Soc. London*, **9**. Blackwell Scientific Publications, Oxford.

PFIFFNER, O.A. (1982) Deformation mechanisms and flow regimes in limestones from the Helvetic zone of the Swiss Alps. *J. struct. Geol.* **4**, 429–442.

PFIFFNER, O.A. (1985) Displacements on thrust faults. *Eclog. geol. Helv.* **78**, 313–333.

RESCH, W. (1976) Bericht über geologische Aufnahmen im Grenzbereich Molasse-Helvetikum bei Dornbirn auf Blatt 111, Dornbirn. *Verh. geol. B.-A. Wien* 1976/1, A122–A126.

RÜEFLI, W.H. (1959) Stratigraphie und Tektonik des eingeschlossenen Glarner Flysches im Weisstannental (St. Galler Oberland). *Mitt. geol. Inst. ETH u. Univ. Zürich*, C/75, 194 pp.

SIEGENTHALER, C. (1974) *Die Nordhelvetische Flysch-Gruppe im Sernftal (Kt. Glarus)*. Ph.D. Thesis, University of Zürich, 83 pp.

STYGER, G.A. (1961) Bau und Stratigraphie der nordhelvetischen Tertiärbildungen in der Hausstock- und westlichen Kärpfgruppe. *Mitt. geol. Inst. ETH u. Univ. Zürich*, C/77, 151 pp.

TANNER, H. (1944) Beitrag zur Geologie der Molasse zwischen Ricken und Hörnli. *Mitt. geol. Inst. ETH u. Univ. Zürich*, C/22, 108 pp.

TRÜMPY, R. (1969) Die helvetischen Decken der Ostschweiz. Versuch einer palinspastischen Korrelation und Ansätze zu einer kinematischen Analyse. *Eclog. geol. Helv.* **62**, 105–142.

TRÜMPY, R. (1973) The timing of orogenic events in the Central Alps. In: *Gravity and Tectonics* (Ed. by K. A. DeJong & R. Scholten), pp. 229–251. Wiley, London.

TRÜMPY, R. (1980) *Geology of Switzerland—a Guide Book. Part A: an Outline of the Geology of Switzerland*. Wepf & Co. 104 pp.

WEGMANN, R. (1961) Zur Geologie der Flyschgebiete südlich Elm. *Mitt. geol. Inst. ETH u. Univ. Zürich*, C/76, 256 pp.

Spec. Publs int. Ass. Sediment. (1986) **8**, 229–246

Thrust belt development in the eastern Pyrenees and related depositional sequences in the southern foreland basin

C. PUIGDEFÀBREGAS*, J. A. MUÑOZ* *and* M. MARZO†

* Servei Geològic de Catalunya, Travessera de Gràcia, 56.08006 Barcelona, Spain; † Departament d'Estratigrafia, Universitat de Barcelona, Granvia, 585.08007 Barcelona, Spain

ABSTRACT

Nine depositional sequences have been distinguished in the South Pyrenean foreland basin, east of the central South Pyrenean thrust sheets. Their characteristics and evolution and the southward migration of depocentres closely correlate with the emplacement of the south Pyrenean thrust sheets. These thrust sheets have been grouped into three main structural units: Upper Thrust Sheets, Middle Thrust Sheets and Lower Thrust Sheets. The thrusts developed in a piggyback sequence, the upper units being displaced southwards by the lower thrusts. The foreland basin was progressively incorporated into the younger thrust sheets. During the submarine emplacement of the Upper Thrust Sheets (early Eocene), tectonic loading caused a deepening of the northern margin of the basin, and a southwards migrating trough formed ahead of the thrust sheets. Contemporaneously, a shallow platform developed to the south. At the final stages of the emplacement of the Upper Thrust Sheets, the basin was plugged by evaporites. The emplacement of the Middle Thrust Sheets and the development of the antiformal stack underneath (Lower Thrust Sheet) during the middle and late Eocene resulted in uplift of cover and basement units which fed southward prograding deltas and fan-deltas. At this time, in the Catalan Coastal Range (southern margin), north directed thrusts and strike-slip faults induced fan-delta progradation, contributing to the final restriction of the basin and deposition of the second evaporite plug (late Eocene), the last marine episode in the southern foreland basin. During the Oligocene, the sole of the South Pyrenean thrusts emerged at the synorogenic surface producing alluvial fans and progressive unconformities. The nine depositional sequences have been grouped in three sedimentary cycles separated by evaporitic horizons. The three cycles correlate with different thrusting styles. The evaporitic sequences seem to coincide with the motion of thrusts along flats, whilst clastic wedges would coincide with thrust ramping.

INTRODUCTION

The Pyrenean orogen extends from Provence in the east, to the Cantabrian platform in the west with an elongation of about 1500 km and an average width of 200 km. The Pyrenean chain is of Alpine age but its structures do not show continuity with those of the Alpine Mediterranean orogenic belt. During the main tectonic event of the Pyrenees a thrust-and-fold belt developed. Thrusting occurred from late Cretaceous to Miocene times. The Pyrenees can be considered as an asymmetrical chain, with a prevailing south displacement of the thrust sheets, although to the north of the axial zone (the biggest outcrop of the Hercynian basement of the Pyrenees) northwards

directed thrusts occur (Fig. 1). The total calculated shortening across the western Pyrenees is approximately 120 km (Hossack, Deramond & Graham, 1984); however, in the eastern Pyrenees no calculation of total shortening has as yet been made, but balanced cross-sections across the southern side of the belt (Muñoz, 1985), show at least 50% shortening, indicating that a similar displacement of about 100 km is possible.

The Southern Pyrenees can be divided into a series of southerly transported thrust sheets (Muñoz, Martinez & Verges, 1986). The Upper Thrust Sheets consist of a mainly Mesozoic cover (e.g. the Central-South

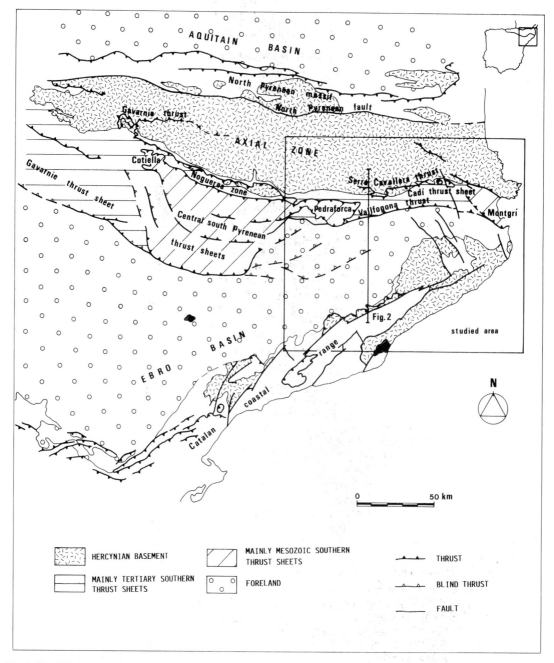

Fig. 1. Simplified map of the north-eastern Iberian Peninsula showing the main structural units of the Pyrenees, their related foreland basins and the location of the studied area and the cross-section of Fig. 2.

Pyrenean thrust sheets, Pedraforca and Montgrí nappes) and were emplaced during the late Cretaceous and early Eocene times. The Middle Thrust Sheets were developed below and after the former ones. They involve both cover sediments and Hercynian basement (e.g. Gavarnie and Cadí nappes). The cover includes an incomplete and reduced Mesozoic sequence overlain by Palaeogene foreland basin sequences deposited forward (to the south) of the previous emplaced Upper Thrust Sheets. The basement mainly consists of post-Silurian rocks. The lowermost outcropping thrust sheets (Lower Thrust Sheets) were emplaced during the late Eocene–Oligocene and they are formed by mainly pre-Silurian basement rocks with a very reduced Mesozoic cover.

This paper is concerned with the analysis of the South Pyrenean Tertiary foreland basin, situated between the Central South Pyrenean thrust sheets, and the Mediterranean sea (Fig. 1). It aims to provide an insight into the relationships between thrust development and the sequential evolution of the foreland basin.

STRUCTURE

The southward displacement of the South Pyrenean thrust sheets over a stable, lower Eocene carbonate platform resulted in the development of the South Pyrenean foreland basin. The thrusts developed in a piggyback sequence, the Upper Thrust Sheets being the first to be emplaced. In the studied area, these Upper Thrust Sheets are represented by the Pedraforca and Montgrí nappes (Fig. 1).

The Pedraforca nappe, mainly consisting of Upper Cretaceous limestones, is overlapped by Lutetian sediments. Its internal structure is characterized by an imbricate thrust fan with hangingwall folds. The north-westernmost thrust sheet of the Pedraforca nappe is characterized by a more complete Mesozoic sequence and could be equivalent to the western thrust sheets (e.g. Boixols), emplaced during the Maastrichtian to early Eocene. The Montgrí nappe consists of Cretaceous sediments mainly covered by Neogene deposits. Its internal structure is poorly understood.

The majority of the studied area is occupied by the Cadí Middle Thrust Sheet, which underlies the Pedraforca and Montgrí nappes. The Cadí thrust sheet is comprised of Devonian and Lower Carboniferous basement, overlain by a thick sequence of uppermost Cretaceous to Palaeocene continental sediments (the Garumnian facies), and a thick Eocene sequence representing the northern parts of the South Pyrenean foreland basin. The foreland basin was later displaced to the south by the floor thrust of the Cadí thrust sheet. This floor thrust, called the Serra Cavallera thrust (Fig. 1) occurs in the weak black Silurian shales and climbs up section southwards into the lower Eocene marls. It branches at its leading edge into the Vallfogona thrust which represents the sole

thrust of the eastern South Pyrenean thrust sheets (Fig. 2). The Vallfogona thrust flattens downwards to the north of its present outcrop and can be recognized on seismic sections. The Pedraforca and Montgrí nappes and the peripheral foreland basin formed to the south of these nappes were carried in a piggyback fashion during the displacement of the Serra Cavallera and Vallfogona thrusts in Lutetian to Oligocene times.

The Serra Cavallera thrust represents the roof thrust of antiformal stacks comprising horses of Hercynian basement rocks and their incomplete cover series (Permian volcanics and Garumnian deposits). The Freser Valley antiformal stack is the most obvious of these duplexes and forms a kilometric-scale culmination (Fig. 2). It developed by the piggyback thrusting of numerous pre-Silurian basement horses (Muñoz, 1985) and it is cut to the north by a northward dipping out-of-sequence thrust. This thrust extends all along the northern part of the studied area in the axial zone and represents the last stage in the development of the antiformal stacks when the floor thrust stuck, producing renewed deformation hindwards of the sticking point.

The progressive piling up of the horses in the Freser Valley antiformal stack produced a space problem in the Eocene beds which was overcome by the development of tight folds. In addition the emplacement of the antiformal stacks along the southern border of the axial zone also initiated the formation of a large syncline which folds the Cadí and Pedraforca thrust sheets (Fig. 2).

The southern margin of the South Pyrenean foreland basin is occupied by the Catalan Coastal Range (Fig. 1). During Alpine tectonism this area was characterized by sinistral strike-slip faulting (Guimerá, 1984). Synchronous with these faults, some northward-directed thrusts developed. As a result of this faulting the foreland basin's southern margin was active during this time, although its activity is of minor importance when contrasted with the activity of the northern margin during the same period of time (Anadón *et al.*, 1986).

DEPOSITIONAL SEQUENCES

The studied sector of the South Pyrenean foreland basin contains 3000 m of Palaeocene to Oligocene sediments. The most complete successions are exposed in the northern thrust sheets and along the borders of the Catalan Coastal Range. In the central parts of the

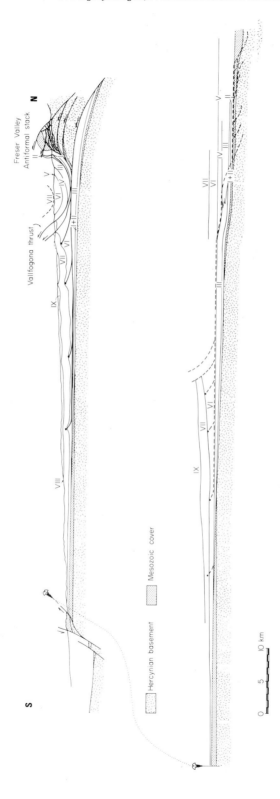

Fig. 2. Balanced and restored cross-sections of the Eastern Pyrenees and South Pyrenean Foreland Basin. Roman numbers refer to depositional sequences (see Fig. 3).

Ebro foreland basin the lower sequences are not exposed and are only known from oil well data.

In order to establish the tectonosedimentary evolution of this area, several depositional sequences (in the sense of Vail, Mitchum & Thompson, 1977a) have been differentiated. Each sequence comprises a body of genetically related strata bounded by unconformities, by their correlative conformities, or by abrupt vertical changes in regional facies distribution. The defined sequences are related to distinct events in the development of the thrust belt. A similar approach has been used by several authors to differentiate sequences in the Tertiary and Mesozoic of the Pyrenees (e.g. Soler & Puigdefàbregas, 1970; Garrido Megias, 1973; Puigdefàbregas, 1975; Puigdefàbregas & Souquet, 1985; Mutti *et al.*, 1985; Simó, 1985).

The sequential subdivision here proposed and its chronostratigraphic significance is shown in Fig. 3. The subdivision is based in part on field data, available sub-surface information and from published studies by various authors (e.g. Anadón, 1978; Busquests, 1981; Colombo, 1980; Ferrer, 1971; Gich, 1972; Luterbacher, Ferrer & Rosell, 1973; Pallí, 1972; Reguant, 1967; Riba, 1975; Serra-Kiel, 1971; Taberner, 1983). The names given to sequences, and to some of their related formations are intended to avoid the confusion which originated by some previous definitions.

In the facies distribution maps of each sequence, the displacement of the South Pyrenean thrust sheets has been taken into account. The minimum displacement values for the Vallfogona thrust ('a' vector in Figs 4–10) have been calculated from balanced cross-sections (Fig. 2). The displacements of the Pedraforca and Montgrí nappes ('b' vector in Figs 4–10) have been approximated using data from previous authors (Seguret, 1972).

Cadí sequence

The Cadí sequence is characterized by transgressive sediments over all of the studied area, that gently thicken to the north, attaining a maximum thickness of 300 m in the northernmost outcrops (isopach map, Fig. 4).

The sequence consists of Alveolina limestones named the Cadí Formation by Mey, Nagtegaal & Hartevelt (1968), and the Orpí Formation by Ferrer (1971). It represents a shallow marine environment grading to a slightly deeper marly environment (the Sagnari Formation, Gich, 1972) containing a fauna of

Nummulitids in the north. The southernmost extension of this carbonate platform is laterally transitional to red beds, and delineates the southern shoreline.

According to Ferrer (1971) the sequence is of Ilerdian age. It overlies the Garumnian facies, and is in turn overlain by the regressive deposits of the Corones sequence.

Corones sequence

The sequence represents a regressive event. The isopach map shows a gradual northward thickening of the sediments to a maximum observable thickness of 250 m (Fig. 5) with the depocentre probably located north of the present-day northern outcrops (Fig. 11). This sequence includes deltaic sediments which prograded southwards in the north and a thin cover of terrigenous red beds extending over the carbonate platform of the Cadí sequence in the south. The lower part of the sequence includes part of the Sagnari Formation (Gich, 1972), and the upper part, the Corones Formation (Gich, 1972).

The base of the deltaic succession is composed of coarsening-up sequences that were deposited in a restricted marine environment. The restricted character of the environment is indicated by the presence of faunal assemblages consisting almost exclusively of Alveolina and Miliolidae. These grade upwards into shallow marine sandstones, and finally into fluvial sandstones, with southerly palaeocurrents.

The top of the sequence is characterized by folded algal carbonates containing surficial hydrocarbon, that extends all along the northern outcrops.

The Corones sequence directly overlies the Cadí sequence, but due to the poor faunal content, its exact age is uncertain, coming between the upper Ilerdian and lower Cuisian.

Armàncies sequence

The Armàncies sequence records a sudden deepening indicated by the presence of slope deposits directly overlying the shallow algal limestone of the Corones sequence. The sediments of the Armàncies sequence, represented by slope shales, carbonate turbidites and slope breccias (megaturbidites), filled a deep northern trough (Fig. 6). The breccias contain Alveolina limestones clasts from the Cadí sequence, in a Nummulitid-rich matrix of Cuisian age resedimented from the contemporaneous shallower areas. The depocentre of this trough migrated a minimum of 15 km to the south relative to the depocentre of the

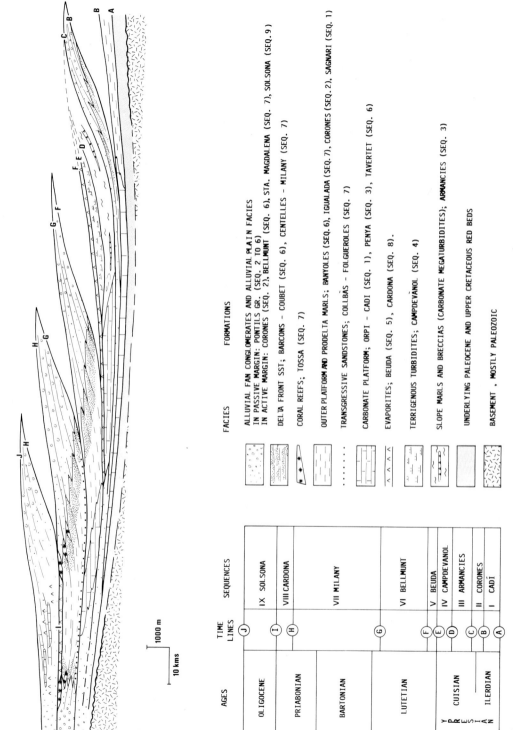

Fig. 3. Idealized cross-section of the studied area of the foreland basin showing depositional sequences and their constituent sedimentary facies.

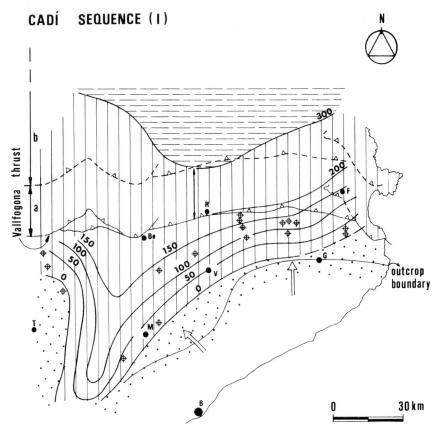

Fig. 4. Facies distribution map of Cadí sequence (Ilerdian). Isopach lines refer only to the carbonate platform facies.

Corones sequence (Fig. 11). At the southern margin of the trough, a shallow carbonate platform developed (Penya Formation of Estevez, 1973) with an abundant fauna of Nummulitids indicating a Cuisian age. This platform probably had an abrupt northern slope and a gradual transition southwards to the red beds of the Pontils Group (Anadón, 1978). The northern margin of the Armàncies trough was defined by the relief associated to the moving Pedraforca and Montgrí nappes. There is no evidence of a northern carbonate platform as a source for the megaturbidite breccias. These could be derived both from the southern platform and from northern thrust sheets.

Campdevànol sequence

The Campdevànol sequence represents a sudden clastic influx of terrigenous turbidites above the previous Armàncies carbonate slope deposits. This turbidite sequence is up to 900 m thick and has westerly orientated palaeocurrents suggesting a north-eastern source (Fig. 7).

The deep turbidite trough originated in front of the moving Pedraforca and Montgrí nappes. The emplacement of the Pedraforca thrust sheet is documented by the presence of Triassic olistoliths within the turbidites, and also by palaeomagnetic data, which shows a counter-clockwise rotation of the magnetic poles of the Campdevànol turbidites in the vicinity of the thrust sheet (Burbank & Puigdefàbregas, 1985). The poor faunal content and other palaeosalinity indicators suggest a hypersaline environment. This has been related to a relative fall in sea-level which would have created shallow evaporite ponds on the southern carbonate platform, and produced a regression favouring the input of terrigenous turbidites.

CORONES   SEQUENCE   (II)

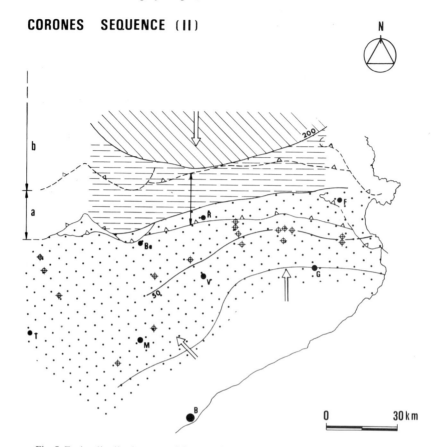

Fig. 5. Facies distribution map of Corones Sequence (upper Ilerdian–lower Cuisian).

Beuda sequence

This sequence is represented by an evaporite formation up to 70 m in thickness which extends over the previously marine area, including the Campdevànol turbiditic trough (Fig. 8). Orti, Busquets & Rosell (1985) suggest the evaporites originated during a progressive shallowing leading to eventual emersion, and point out the possibility of a tectonic control. We interpret the Beuda sequence as representing an evaporite plug deposited during basin constriction at the final stages of the emplacement of the Pedraforca–Montgrí thrust sheets. This evaporite event coincides with the generalized sea-level fall at the end of supercycle Ta of Vail, Mitchum & Thompson (1977b).

Bellmunt sequence

The Bellmunt sequence is characterized by a coarse-grained deltaic complex prograding southwards and a southerly retreating carbonate platform, which in turn overlies and transgresses passive margin alluvial fan sediments.

The northern deltaic complex includes three main facies belts: prodelta marls (Banyoles Formation: Rios & Masachs, 1953), delta-front sandstones (Barcons Formation: Gich, 1972) and delta plain to alluvial conglomerates and sandstones (Bellmunt Formation: Gich, 1972). The estimated maximum thickness is about 1000 m in the depocentre, the axis of which is located between 5 and 15 km south of the present Vallfogona thrust outcrop line and 30 km south of the depocentres of the Armàncies and Campdevànol sequences. The southern carbonate platform (Tavertet Formation: Reguant, 1967) represents a high energy Nummulites nearshore belt which grades southwards into, and eventually transgresses over, the coastal alluvial fan deposits of the passive margin (Romagats Formation: Colombo, 1980).

The southward migration of depocentres and facies

ARMANCIES SEQUENCE (III)

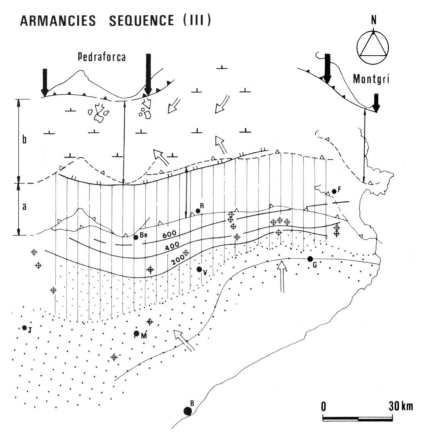

Fig. 6. Facies distribution map of Armàncies Sequence (Cuisian). Isopach lines refer only to the carbonate platform facies.

belts results in an abrupt vertical change in the facies distribution pattern with respect to the underlying sequences.

Milany sequence

This consists of a deltaic complex and represents a repetition of the scheme described for the previous Bellmunt sequence, with a renewed displacement of the depocentre axis about 10–20 km to the south (Figs 10 and 11). Maximum thickness is up to 1200 m. This sequence starts with a transgressive event over the previous Bellmunt sequence indicated by either glauconitic sandstones (Folgueroles Formation, Reguant, 1967) or by a transgressive nearshore complex of carbonate and sandstones (Collbàs Formation, Ferrer, 1971). This southward prograding coarse-grained delta complex includes three main facies

belts: prodelta marls (Igualada Formation, Ferrer, 1971), delta front sandstones (Milany Formation, Gich, 1972, and the upper part of the Rocacorba Formation of Pallí, 1972), and delta plain to alluvial fan conglomerates characterized by an increase of basement pebbles in contrast to the previous sequence (Sta. Magdalena Formation, Gich, 1972). In the Catalan Coastal Range (the southern passive margin) some uplift and subsequent northwards progradation of fan deltas occurred during the deposition of this sequence (Montserrat Formation, Anádon, Marzo & Puigdefàbregas, 1985). At the final stages of both northern and southern deltaic progradations, a coral reef belt (Tossa Formation, Ferrer, 1971) developed over the outer edge of the delta front sandstones. This reef formation is preceded by development of minor reefs over inactive deltaic lobes (St Bartomeu, Barnolas, Busquets & Serra-Kiel, 1981 and Centelles,

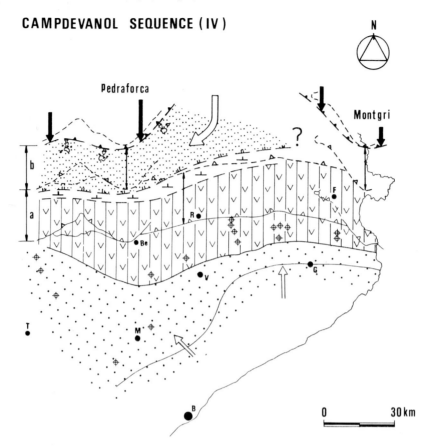

Fig. 7. Facies distribution map of Campdevànol Sequence (upper Cuisian).

Taberner, 1983). The reef stage immediately precedes the final constriction of the basin represented by the next Cardona sequence. The detailed relations between the reef stage and the Cardona evaporites are discussed in Busquets *et al.* (1985).

Cardona sequence

It consists of a distinct evaporite cycle which has been recently studied, documented, and summarized by Busquets *et al.* (1985). It developed in the centre of the residual basin at the end of the Milany sequence, mostly overlying the Igualada marls and surrounded by the Tossa reef belt (Fig. 10). A complete section of the Cardona Formation would include (Busquets *et al.*, 1985) a 5 m thick basal sulphate member, a lower salt member up to 200 m thick, a potassic member including a lower 5–20 m thick sylvinite, an upper 40–

80 m thick carnalite, and a post-potassic member consisting of grey lutites with gypsum and halite intercalations. This sequence grades upwards to the lacustrine deposits of the next Solsona sequence. Busquet *et al.* conclude the existence of a single depositional episode (saline macrocycle) which would involve a continuous concentration up to the potash member. The analogy with the same evaporite event in the western South Pyrenean foreland basin (Ortí *et al.* 1985) and the geometrical possibility or their subsurface continuity (Puigdefàbregas, 1975) accounts for the interpretation of the Cardona sequence in relation to the final stages of the emplacement of the Cadí thrust sheet and their westerly equivalents, which would result in the confinement of the last marine foreland basin. This important tectonic event coincides with the general sea-level fall indicated by Vail *et al.* (1977b) at Priabonian times.

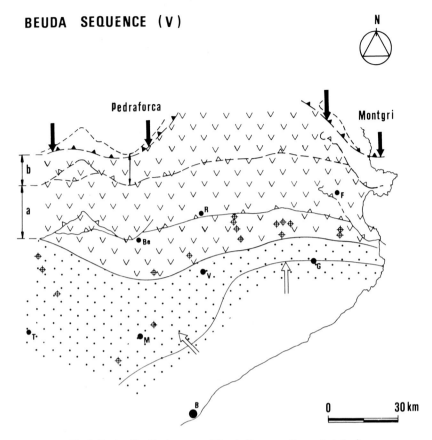

Fig. 8. Facies distribution map of Beuda Sequence (lower Lutetian).

Solsona sequence

The Solsona sequence consists of a succession of fluvial and lacustrine deposits filling, what can be considered as the last stage of the South Pyrenean foreland basin. Alluvial fans were active all around the basin margins, and progressive unconformities developed along the northern active margin (Riba, 1976), interpreted as the synsedimentary deformation due to the displacement of the Vallfogona thrust. In the centre of the basin, the area formerly occupied by the Cardona evaporite was flooded with fluvial waters coming from the active alluvial fans. These new lacustrine areas were first filled by thin turbidites (Suria Formation, Busquests *et al.*, 1985) and later on by carbonates, evaporites and intercalated alluvial deposits including lignites.

FORELAND BASIN EVOLUTION

On the basis of the combined structural, sedimentological and stratigraphical observations previously described, we propose the following sequential evolution for the South Pyrenean foreland basin in the studied area.

During Ilerdian times (Cadí sequence, Fig. 12) a basin-wide Atlantic transgression covered most of the area and resulted in a shallow carbonate platform overlying the Garumnian red beds of late Cretaceous to Palaeocene age. At this time there is no evidence of any tectonic event in the studied area, although it should be remembered that, in the Central Pyrenees, the first thrust sheets had been emplaced during the Maastrichtian (Bóixols thrust, Simó, 1985). According

BELLMUNT SEQUENCE (VI)

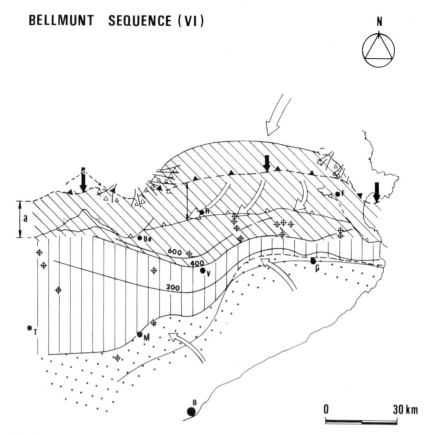

Fig. 9. Facies distribution map of Bellmunt Sequence (Lutetian). Isopach lines refer only to marine facies.

to this fact, the Garumnian facies could represent the earliest stages of the foreland basin development in the Pyrenees (Puigdefàbregas & Souquet, 1985).

The first thrust movements can be inferred from the first northerly supplied deltaic system during the late Ilerdian–early Cuisian (Corones sequence, Fig. 12). Tilting related to this thrust emplacement would induce uplift in the passive southern margin with subsequent deposition of southerly supplied clastics overlying the previous carbonate platform. This resulted in a generalized regressive episode with possible development of evaporites in the central part of the basin.

The sudden deepening of the basin recorded during the Cuisian (Armàncies sequence, Fig. 12) results from the loading effect induced by the initial stages of the emplacement of the Pedraforca and Montgrí nappes. Due to the relative sea-level rise, a well developed carbonate platform extended over the

southern clastic rimmed passive margin. At this stage, the former foreland sequences had already been incorporated into the thrust sheets.

Further thrusting and loading of the Upper Thrust Sheets (Pedraforca and Montgrí) during the late Cuisian resulted in the development of a terrigenous turbidite trough in front of them (Campdevànol sequence, Fig. 12). The subsequent bulge effect at the southern margin resulted in the deposition of shallow carbonate-evaporite sequences.

At the final stages of the Upper Thrust Sheets emplacement a general shallowing and basin restriction occurred allowing the deposition of an evaporite plug (Beuda sequence, Fig. 12) overlying both the turbidite trough and the southern platform.

During the development of the Middle Thrust Sheets (Lutetian), the Upper Thrust Sheets (Pedraforca–Montgrí) were eroded, feeding a series of southward prograding coarse-grained deltas (Bell-

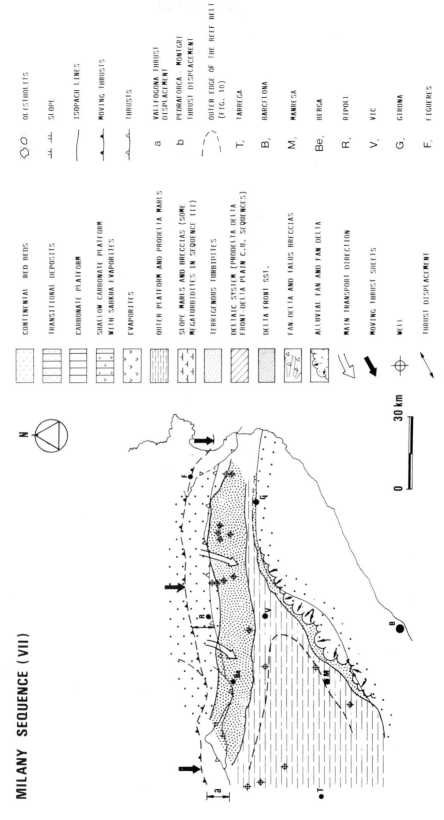

MILANY SEQUENCE (VII)

LEGEND   FIGS. 4-10

CONTINENTAL RED BEDS	
TRANSITIONAL DEPOSITS	
CARBONATE PLATFORM	
SHALLOW CARBONATE PLATFORM WITH SABKHA EVAPORITES	
EVAPORITES	
OUTER PLATFORM AND PRODELTA MARLS	
SLOPE MARLS AND BRECCIAS (SOME MEGATURBIDITIES IN SEQUENCE III)	
TERRIGENOUS TURBIDITES	
DELTAIC SYSTEM (PRODELTA-DELTA FRONT-DELTA PLAIN C.U. SEQUENCES)	
DELTA FRONT SST.	
FAN-DELTA AND TALUS BRECCIAS	
ALLUVIAL FAN AND FAN DELTA	
MAIN TRANSPORT DIRECTION	
MOVING THRUST SHEETS	
WELL	
THRUST DISPLACEMENT	

OLISTHOLITS

⊥ SLOPE

ISOPACH LINES

MOVING THRUSTS

THRUSTS

a VALLFOGONA THRUST DISPLACEMENT

b PEDRAFORCA - MONTGRI THRUST DISPLACEMENT

OUTER EDGE OF THE REEF BELT (FIG. 10)

T, TARREGA

B, BARCELONA

M, MANRESA

Be, BERGA

R, RIPOLL

V, VIC

G, GIRONA

F, FIGUERES

0   30 km

Fig. 10. Facies distribution map of Milany Sequence (upper Lutetian–Bartonian). The evaporites of the Cardona sequence (Priabonian) were deposited in the area bounded by the outer edge of the reef belt.

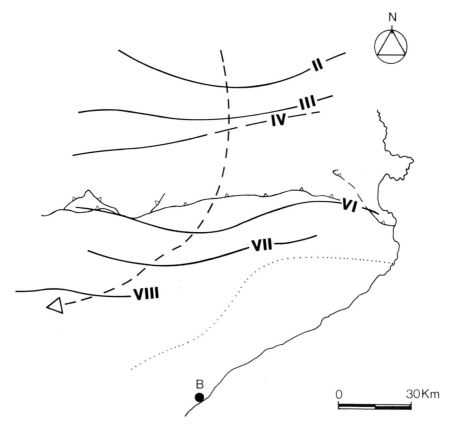

Fig. 11. Pattern of migration of the depocentres of the successive depositional sequences (II to VIII) during the evolution of the foreland basin.

munt sequence, Fig. 12). Deltaic progradation was accelerated by increased clastic input caused by the development of the Freser Valley antiformal stack below the Cadí thrust sheet. Deltaic southwards progradation caused the contemporaneous platform to retreat southwards.

During the late Lutetian and Bartonian, the floor thrust of the antiformal stack stuck and, as a result, out-of-sequence thrusts formed hindwards of its leading edge, bringing basement rocks to the surface. The sedimentary responses to this situation were: (i) an important transgression at the base of the Milany sequence, (ii) the southward shift of its depocentre, and (iii) a renewed delta progradation from the north with high percentage of basement clasts. The development of north-westerly vergent thrusts in the passive margin during this stage resulted in active alluvial fan and fan-delta progradation. This,

coupled with the northern progradation, led to the gradual restriction of the basin.

After the development of the Freser Valley antiformal stack, the Vallfogona thrust moved along a flat (Fig. 12). As this thrust was not yet emerging in the foreland basin, the clastic input diminished. This fact, possibly combined with a relative fall of the sea-level, led to the deposition of the Priabonian evaporites (Cardona sequence). This evaporites represent the last marine deposits of the basin.

During the Oligocene the Vallfogona thrust moved along a ramp emerging at the synorogenic surface and defining the northern margin of the youngest South Pyrenean foreland basin (Ebro basin). Contemporaneously the non-marine Solsona molasse filled this basin, the depocentre of which experienced a new migration to the SW (Figs 11 and 12). Synsedimentary movements of the Vallfogona thrust are recorded in

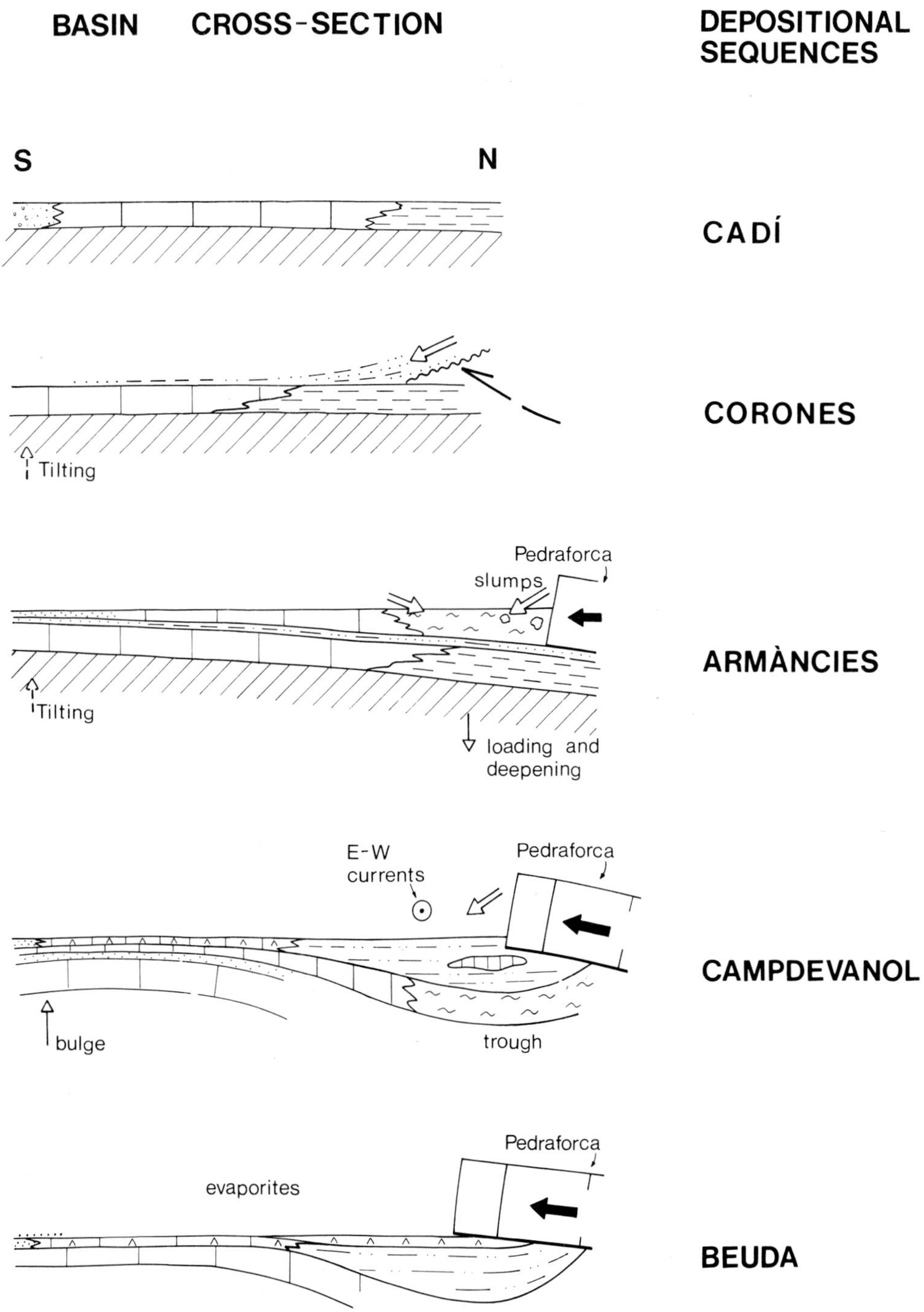

BASIN   CROSS-SECTION

DEPOSITIONAL SEQUENCES

S   N

CADÍ

Tilting

CORONES

Pedraforca
slumps

Tilting

loading and
deepening

ARMÀNCIES

E-W
currents

Pedraforca

bulge

trough

CAMPDEVANOL

evaporites

Pedraforca

BEUDA

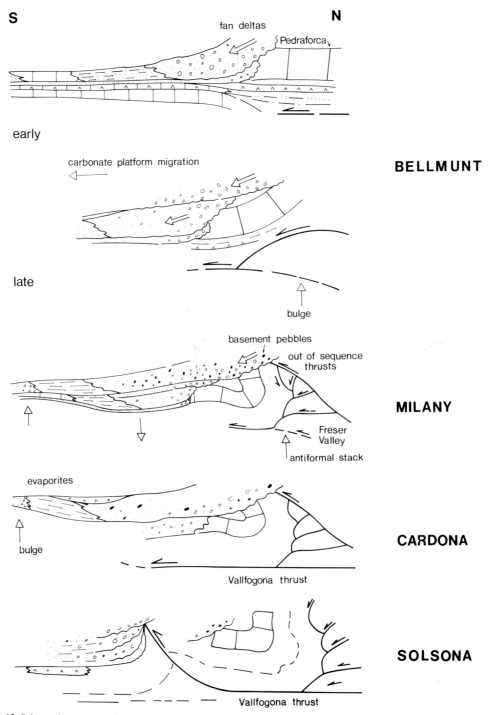

Fig. 12. Schematic representation of the tectonosedimentary evolution of the foreland basin. Explanation in the text.

the alluvial conglomerates by the development of progressive unconformities.

CONCLUSIONS

In the Southern Pyrenees, since the earlier stages of thrust movement in the Maastrichtian, a foreland basin developed ahead of the moving thrust sheets. As the deformation progressed to the south, the foreland basin also migrated southwards being progressively incorporated into the younger thrust sheets.

The foreland basin-fill includes nine depositional sequences whose characteristics were controlled principally by thrusting events along the northern margin of the basin.

The thrust sequence in the south-eastern Pyrenees displays the following cyclicity in the type of thrust system: (i) first foreland imbricate stack (Pedraforca, Montgrí thrust sheets), (ii) hinterland duplexes (Freser Valley antiformal stack) and (iii) second foreland imbricate stack (Vallfogona thrust). Changes in the type of thrust system correlate with evaporite deposition (Beuda and Cardona sequences). These evaporite events allow the differentiation of the basin-fill into three main depositional cycles related to the above-mentioned thrust cyclicity.

The first cycle, which includes the Cadí, Corones, Armancies and Campdevanol sequences, coincides with the submarine emplacement of the Upper Thrust Sheets. It is characterized by the progressive deepening of the northern margin of the basin due to tectonic loading. The resulting slope and turbidite trough experienced a southward migration.

The second cycle includes the Bellmunt and Milany sequences and coincides with the development of the antiformal stack, hindwards of the leading edge of the previously emplaced thrust sheets. Related relief rejuvenation induced deltaic progradation, at the same time that the basin depocentre migrated still farther to the south.

During the third cycle, which includes the Solsona and younger Ebro basin sequences, important southward thrusting carried piggyback all the other thrust sheets and incorporated the northern parts of the former foreland basin sequences. These thrust sheets, displaced by the emergent Vallfogona thrust, supplied clastics to the alluvial fans which were laterally connected with lacustrine areas in the centre of the basin.

Evaporitic events seem to coincide with the development of flat thrust trajectories whilst basinward progradation of clastic systems correlates with ramping episodes. In the Southern Pyrenees the striking coincidence between changes in type of thrust system (marked by evaporites) and general sea-level falls indicated by Vail *et al.* (1977b) suggests a tectonic control on the sea-level fluctuations.

REFERENCES

ANADÓN, P. (1978) *El Paleógeno continental anterior a la transgresión Biarritziense (Eoceno Medio) entre los rios Gaiá y Ripoll (provincias de Tarragona y Barcelona).* Tesis Doctoral, Universitat de Barcelona, 267 pp.

ANADÓN, P., MARZO, M. & PUIGDEFÀBREGAS, C. (1985) The Eocene fan-delta of Montserrat (Southeastern Ebro Basin, Spain), *6th European Regional Meeting Excursion Guidebook*, pp. 111–146.

ANADÓN, P., CABRERA, L., GUIMERA, J. & SANTANACH, P. (1986) Paleogene strike-slip deformation and sedimentation along the Southeastern margin of the Ebro Basin. In: *Strike-Slip Deformation, Basin Formation and Sedimentation* (Ed. by K. T. Biddle & N. Christie Blick). *Spec. Publ. Soc. econ. Paleont. Mineral.* **37**, 303–318.

BARNOLAS, A., BUSQUETS, P. & SERRA-KIEL, J. (1981) Características sedimentológicas de la terminación del ciclo marino del Eoceno superior en el sector oriental de la Depresión del Ebro (Catalunya, NE España). *Acta Geol. Hisp.* **16**, 215–221.

BUSQUETS, P. (1981) *Estratigrafia i Sedimentologia del Terciari pre-pirinenc entre els rius Llobregat i Freser-Ter.* Tesi Doctoral, Universitat de Barcelona, 543 pp.

BUSQUETS, P., ORTI, F., PUEYO, J.J., RIBA, O., ROSELL-ORTIZ, L., SAEZ, A.., SALAS, R. & TAVERNER, C. (1985) Evaporite deposition and diagenesis in the saline (potash) Catalan Basin, Upper Eocene. *6th European Regional Meeting. Excursion Guidebook*, pp. 13–59.

BURBANK, D.W. & PUIGDEFÀBREGAS, C. (1985) Chronologic investigations of the south Pyrenean basins: preliminary magnetostratigraphic results from the Ripoll basin. *6th European Regional Meeting, Abstr.*, pp. 66–69.

COLOMBO, F. (1980) *Estratigrafia y sedimentología del Terciario inferior continental de los Catalánides.* Tesis Doctoral, Universitat de Barcelona, 609 pp.

ESTEVEZ, A. (1973) *La vertiente meridional del Pirineo Catalan al N del curso medio del rio Fluvià.* Tesis Doctoral, Universidad de Granada, 514 pp.

FERRER, J. (1971) El Paleoceno y Eoceno del Borde suroriental de la Depressión del Ebro. *Mem. Suisses Paléont.* **90**, 70 pp.

GARRIDO MEGIAS, A. (1973) *Estudio geológico y relación entre tectónica y sedimentación del Secundario y Terciario de la vertiente meridional pirenaica en su zona central (provincias de Huesca y Lérida).* Tesis Doctoral, Universidad de Granada, 395 pp.

GICH, M. (1972) *Estudio geológico del Eoceno prepirenacio del Ripollés oriental.* Tesis Doctoral, Universitat de Barcelona, 477 pp.

GUIMERÀ, J. (1984) Paleogene evolution and deformation in the northeastern Iberian Peninsula. *Geol. Mag.* **121**, 413–420.

HOSSACK, J.R., DERAMOND, J. & GRAHAM, R.H. (1984) The geological structure and development of the Pyrenees. *Coll. Chevauchement et deformation, Toulouse. Abstr.*, pp. 46–47.

LUTERBACHER, H., FERRER, J. & ROSELL, J. (1973) El paleógeno marino del Noroeste de España. *XIII Coloquio Europeo de micropaleontología. Zona Pirenacia, C.N.G.*, Enadimsa, pp. 28–62.

MEY, P.H.W., NAGTEGAAL, P.J.C. & HARTEVELT, J.A. (1968) Lithostratigraphic subdivision of post-hercynian deposits in the South-Central Pyrenees. *Leidse. Geol. M.* **41**, 221–228.

MUÑOZ, J.A. (1985) *Estructura alpina i herciniana a la vora sud de la zona axial del Pirineu Oriental.* Tesi Doctoral, Universitat de Barcelona, 305 pp.

MUÑOZ, J.A., MARTINEZ, A. & VERGES, J. (1986) Thrust sequences in the Spanish Eastern Pyrenees. *J. struct. Geol.* **8**, 399–405.

MUTTI, E., REMACHA, E., SGAVETTI, M., ROSELL, J., VALLONI, R. & ZAMORANO, M. (1985) Stratigraphy and facies characteristics of the Eocene Hecho Group Turbidite systems, South-Central Pyrenees. *6th European Regional Meeting. Excursion Guidebook*, pp. 521–576.

ORTI, F., BUSQUETS, P. & ROSELL, L. (1985) *Estudi petrològic, sedimentòlogic i estratigràfic de la conca evaporítica catalana de l'Eocè Mitjà (Lutecià).* Memòria de l'Ajut d'investigació (1983), Universitat de Barcelona, 34 pp.

PALLÍ, L. (1972) Estratigrafia del Paleògeno del Empordà y zonas limitrofes. *Publ. Geol. Univ. Autònoma Barcelona*, 338 pp.

PUIGDEFÀBREGAS, C. (1975) La sedimentación molásica en la cuenca de Jaca. *Pirineos*, **104**, 1–188.

PUIGDEFÀBREGAS, C. & SOUQUET, P. (1985) Cyclicity, basin evolution and tectonic control in the Mesozoic and Cenozoic of the Pyrenees. *Terra cognita*, **5**, (2–3), 119.

REGUANT, S. (1967) El Eoceno marino de Vic (Barcelona). *Mem. Inst. Geol. Min. Esp.* **68**, 330 pp.

RIBA, O. (1975) Introduction. Le Bassin Tertiaire Catalan espagnol et les gisements de Potasse. *Livret de l'Excursion n. 20. IXe Congrès int. Sediment.* Nice, pp. 9–13.

RIBA, O. (1976) Syntectonic unconformities of the Alto Cardener, Spanish Pyrenees: a genetic interpretation. *Sedim. Geol.* **15**, 213–233.

RIOS, J.M. & MASACHS, V. (1953) *Hoja 295 (Bañolas) del Mapa Geológico de España, escala 1:50.000. IGME.*

SEGURET, M. (1972) *Etude tectonique des nappes et séries decollèes de la partie centrale du versant sud des Pyrenées.* Publ. USTL. Série Géol. Struct. **2**, 155 pp., Montpellier.

SERRA-KIEL, J. (1981) *Estudi sobre la sistemàtica, filogènia, bioestratigrafia i paleobiologia dels Nummulites del grup N. pernotus, N. perforatus (conca aquitana, catalana i balear).* Tesi Doctoral, Universitat de Barcelona, 543 pp.

SIMÓ, A. (1985) *Secuencias deposicionales del Cretácico superior de la Unidad del Montsec.* Tesis Doctoral, Universitat de Barcelona, 310 pp.

SOLER, M. & PUIGDEFÀBREGAS, C. (1970) Lineas generales de la geología del Alto Aragón Occidental. *Pirineos*, **96**, 5–20.

TABERNER, C. (1983) *Evolución ambiental y diagenética de los dépositos del Terciario inferior (Paleoceno y Eoceno) de la Cuenca de Vic.* Tesis Doctoral, Universitat de Barcelona, 1400 pp.

VAIL, P.R., MITCHUM, R.M. & THOMPSON, III, S. (1977a) Seismic stratigraphy and global changes of sea level, part 3: relative change of sea level from coastal onlap. In: *Seismic Stratigraphy. Applications to Hydrocarbon Exploration. Mem. Am. Ass. Petrol. Geol. Bull.* **26**, 63–81.

VAIL, P.R., MITCHUM, R.M. & THOMPSON, III, S. (1977b) Seismic stratigraphy and global changes of sea level, part 4: Global cycles of relative changes of sea level. In: *Seismic Stratigraphy. Applications to Hydrocarbon Exploration. Mem. Am. Ass. Petrol. Bull.* **26**, 83–97.

Spec. Publs int. Ass. Sediment. (1986) **8**, 247–258

Thrust tectonic controls on Miocene alluvial distribution patterns, southern Pyrenees

J. P. P. HIRST* *and* G. J. NICHOLS†

* B.P. Petroleum Development Ltd, Kirklington Road, Eakring, Newark NG22 0DA, U.K.; † Department of Geological Sciences, University of Liverpool, P.O. Box 147, Liverpool L69 3BX, U.K.

ABSTRACT

Alluvial dispersal patterns in the northern Ebro Basin, the southern foredeep of the Pyrenees, can be shown to be controlled by the positions of frontal and lateral ramps in the adjacent southern Pyrenean thrust belt. The apices of two early Miocene fluvial distributary systems in the north-western part of the Ebro Basin lie in front of structural lows. Detrital mineral suites in these distributary systems indicate that the drainage basins of these systems included the axial zone of the Pyrenees. Along the 100 km between the apices of the two systems the frontal ramp of the thrust sheet was developed as a structural salient, forming a barrier to the headward migration of streams during the early Miocene. A series of relatively small alluvial fans formed along this part of the basin margin at the topographic break between the frontal ramp and the Ebro Basin: analysis of the clast types present in these fans shows that their drainage basins lay within the topographic high of the thrust front.

It is suggested that the patterns observed in the northern Ebro Basin—fan-shaped fluvial distributary systems, the apices of which lie at structural lows in the basin margin, plus smaller alluvial fans along the higher parts of the margin—may be characteristic of foreland basins with internal drainage. The geometry of the thrust sheets in the adjacent thrust belt controls these alluvial distribution patterns.

INTRODUCTION

From late Eocene to early Miocene times, the Ebro Basin in northern Spain (Fig. 1) was an enclosed area of continental sedimentation (Quirantes Puertas, 1969; Turner, Hirst & Friend, 1984). Most of the clastic detritus was derived from the Pyrenees, which lie to the north of the basin. Clastic input from the Catalan ranges to the east and the Iberian ranges to the SW was volumetrically less important (Turner *et al.*, 1984). In cross-section the basin is asymmetric, there being up to 4000 m thickness of sediments along the northern margin, thinning towards the southern part of the basin where the maximum thickness is only a few hundred metres (Riba, 1971). The development of the basin was similarly asymmetric with the older basin filling sediments restricted to the northern side (Riba, 1971).

The principal source area, the Pyrenean chain, was a zone of orogenic activity resulting from the collision of the Iberian and European plates during the Mid-Tertiary (Boillot, 1984). The axial zone of the Pyrenees was an area of uplift and erosion during this time. In the southern Pyrenees a series of thrusts developed during the Eocene and Oligocene forming a thrust belt up to 50 km from north to south stretching along the entire length of the mountain chain (Mattauer & Henry, 1974). Frontal and oblique ramps to these thrusts formed topographically high areas which can be shown to have acted as orographic barriers in the Tertiary southern Pyrenean drainage system (see later).

The development of the southern Pyrenean thrust belt was contemporaneous with sedimentation in the northern Ebro Basin (Puigdefabregas & Soler, 1973; Mattauer & Henry, 1974). An integrated study of the structure of the thrust belt in part of the southern Pyrenees, and the sediments in the adjacent parts of the Ebro Basin, allows the relationship between the geometry of the thrusts and the facies distribution patterns to be analysed in what may be regarded as a typical foreland basin.

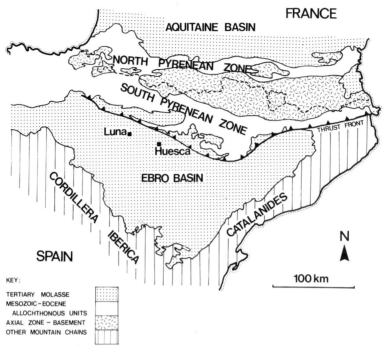

Fig. 1. Location map. This study concerns the area around the towns of Luna and Huesca in the district of Aragon. The major tectonic divisions are those of Choukroune & Seguret (1973).

STRUCTURAL SETTING

The structure of the area covered by this study is summarized in Fig. 2. The structural units used here are those of Garrido-Megias and Camara (Camara, personal communication, 1983) who modified the major structural divisions of Seguret (1972) using seismic data. The three major thrust units—the Guarga, Gavarnie–Boltaña and Cotiella–Monsec units—all have a Triassic décollement horizon. Each thrust sheet consists mainly of Mesozoic to early Eocene limestones overlain by terrigenous clastics (Seguret, 1972; Mattauer & Henry, 1974). The clastic units are of late Eocene to Oligocene age and were deposited in the Jaca, Ainsa and Tremp–Graus Basins (Nijman & Puigdefabregas, 1974). These basins were connected during the Early Tertiary and formed the southern foredeep of the Pyrenees. Thrusting in the Oligocene brought about south to south-westwards movement of the main thrust sheets and these three basins became largely disconnected and allochthonous. These thrust movements brought about a change in the areas of erosion and sedimentation in the southern Pyrenees. The older molasse basins became largely erosional, although there is some evidence of

continued sedimentation in the southern part of the Graus Basin (Ori & Friend, 1984) and the southern Pyrenean foredeep shifted south to the northern side of the Ebro Basin.

The uplift boundary (Eisbacher, Carrigy & Campbell, 1974) between the erosional realm of the southern Pyrenees and the Ebro Basin was formed by folding and thrusting at the leading edges of the Guarga Unit to the west and the Cotiella–Monsec Unit to the east. The leading edge of the Guarga Unit lies on a frontal ramp and is at present exposed as folded and thrust-faulted Mesozoic, Eocene and Oligocene beds in the External Sierras. The ramp structure is not so clearly developed at the leading edge of the Cotiella–Monsec Unit and the marginal Ebro Basin sediments have been folded into the Barbastro Anticline, partly by the southward movement of thrusts, but largely by Late Tertiary diapirism. Nevertheless, there is a distinct break between the folded southern Pyrenean zone and the Ebro Basin in this area. On the western side of the Cotiella–Monsec Unit an oblique ramp structure forms a surface culmination, the Mediano Anticline, which partly separates the Ainsa and Tremp–Graus Basins (Deramond et al., 1984). To the west lies a second oblique ramp structure at the south-

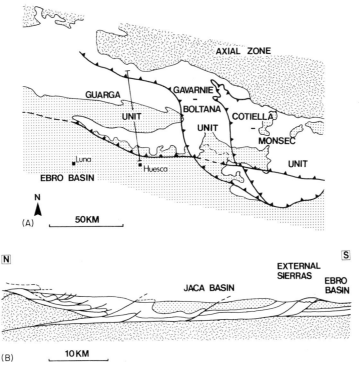

Fig. 2. (A) An outline of the structure of the western part of the southern Pyrenees; the structural units are those of Garrido-Megias & Camara (personal communication, 1985). (B) Cross-section through the southern Pyrenean zone along the line north of Huesca on (A) (from Deramond *et al.*, 1984, p. 6). (Key as in Fig. 1.)

western margin of the Gavarnie–Boltaña Unit. A culmination above this forms a major north–south feature, the Boltaña Anticline (Deramond *et al.*, 1984) which completely separates the Ainsa and Tremp–Graus Basins.

In the western part of the area, the nature of the tip line of the Guarga thrust changes from emergent to buried, and the leading edge no longer lies on an emergent frontal ramp (Nichols, 1984). At this point the ridge of limestones which form the backbone of the External Sierras becomes buried and there is no orographic barrier between the Jaca Basin and the Ebro Basin. The position where the leading edge becomes emergent is considered to have been an important factor in influencing the early Miocene drainage patterns.

In the Oligocene and Miocene the Pyrenean axial zone was a region of uplift and the Mesozoic cover was eroded from the Hercynian basement of metamorphic and acid plutonic rocks. This unrooting occurred in an asymmetric fashion as higher grade metamorphic rocks are presently exposed only in the

eastern part of the chain. These metamorphic and igneous basement rocks, the allochthonous carbonates of the southern Pyrenean thrust belt and the clastic fill of the Tremp–Graus, Ainsa and Jaca Basins formed the erosional realm from which detritus was shed into the Ebro Basin in the late Oligocene and Miocene.

SEDIMENTATION PATTERNS

Throughout the Oligocene and most of the Miocene the Ebro Basin was a basin of internal drainage. Clastic sediments were deposited at the margins. Basinwards, these passed into lacustrine limestones and marls, with evaporites being deposited in and around ephemeral lakes (Quirantes Puertas, 1969). Clastic input was greatest along the northern margin and a conglomeratic fringe, made up of a series of more or less discrete bodies, was developed in the study area. Several features indicate that these clastic deposits are continental sediments; for example the remains and ichofossils of terrestrial vertebrates, the presence of mudcracks and palaeosols. The unidirectional and mainly radial palaeocurrent patterns are indicative of fluvial deposition.

Within the study area (Fig. 1) several sediment dispersal patterns have been recognized. This has been achieved using palaeocurrent data, mapping the facies distributions and by analysing the clast types and sand petrographies. Two major facies associations may be defined: (a) alluvial fans, (b) fluvial distributary systems. These two facies associations are described in turn below.

Alluvial fans

The remnants of Tertiary alluvial fans are to be found all along the northern margin of the Ebro Basin (Fig. 3). In most cases they appear to have been discrete alluvial fan bodies which formed at the break in slope between the topographic high of the contemporary External Sierras and the adjacent parts of the Ebro Basin; parts of some of these fans overlapped on to the External Sierras outcrop, filling in structurally lower areas (e.g. the Vadiello fan—Fig. 3). The coarser, more proximal parts of these fans are often well exposed, forming towering masses of conglomerate, e.g. Roldan, Riglos (Figs 3 and 4). These conglomerates were deposited in the proximal parts of the fans as sheets of material decimetres to metres thick. Some beds fine upwards and individual beds are moderately well sorted. Amalgamation of conglomeratic units is common. Sandstones also occur as sheet-like bodies, with their upper surfaces frequently scoured and filled by the overlying conglomeratic unit. The thickness and proportion of conglomerate units decreases away from the apex of the fan body, the more distal material being centimetre to decimetre bedded sheets of gravelly sandstones, siltstones and mudstones. Where the direction of flow can be ascertained, an approximately radial palaeocurrent pattern is observed.

The sheet form of these units, the absence of cut banks and minimal evidence of bar formation suggests that the dominant mode of deposition was from sheet flows which spread over large areas of the fan surface. No evidence of mass flow was found in the conglomerates.

Petrographically the fanglomerate bodies fall into two groups: those which are made up of material almost entirely derived from the units which make up the External Sierras and those which also contain clasts derived from the Jaca Basin. The former type predominate in the western half of the area studied

(e.g. the Agüero, Murillo, Riglos and Roldan fanglomerate bodies—Fig. 3). Most of the clasts are Mesozoic and Eocene limestones and are angular to rounded. The proportion of sandy material is generally low. These were relatively small fans, with original areas of 2–6 km², with similarly restricted drainage basins which would have lain within the contemporary limestone outcrop of the External Sierras. To the east, some of the fan bodies (e.g. the Vadiello fanglomerate body—see Fig. 3) also contain rounded black cherts and other clasts which are considered to have been derived from the older alluvial deposits in the Jaca basin. These fans were somewhat larger (5–25 km²) and consequently had larger drainage basins which included the southern part of the Jaca Basin.

Fluvial distributary systems

In contrast to the alluvial fan deposits, which are dominantly red-brown, the majority of the early Miocene sediments in this part of the Ebro Basin are buff in colour. These buff coloured deposits are mainly sandstones, siltstones and mudstones, although conglomerates are significant near the basin margin. The sandstone units are the fill of fluvial channels which incised into a floodplain of horizontally bedded siltstones and mudstones.

Cross-sections of sandstone bodies, perpendicular to palaeocurrent, have width/height ratios ranging from 5 to over 200. On the basis of their width/height ratios, Friend, Slater & Williams (1979) subdivided such sand bodies into ribbons and sheets, ribbons having width/height ratio less than 15. It was inferred that ribbons were the products of laterally stable channels while sheets resulted from laterally unstable channels. This is a simplified division, particularly concerning the sheet units which may also result from unconfined flow (flood deposits, crevasse splays, dechannelization) as well as being the products of laterally migrating and braided channels (Hirst, 1983; Friend, Hirst & Nichols, 1986). In general, the presence of ribbon sand bodies is indicative of laterally stable channels and ribbons are a distinctive component of the fluvial distributary systems described below. A typical ribbon sandstone body from the area is illustrated in Fig. 5.

Palaeocurrent data

Palaeocurrent data from early Miocene sand bodies in the north-western part of the Ebro Basin are

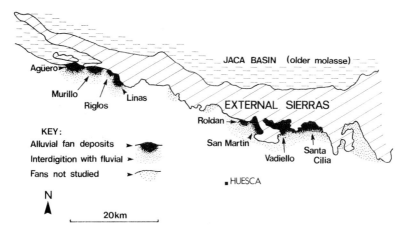

Fig. 3. The principal Lower Miocene alluvial fan deposits along the northern margin of the Ebro Basin in the study area.

Fig. 4. Los Mallos de Riglos. Amalgamated conglomerate beds deposited in the proximal parts of an alluvial fan have been partly eroded to leave towering masses of conglomerate.

Fig. 5. An exposure of the fluvial facies which makes up the bulk of the Luna and Huesca System deposits: a sandstone filled fluvial channel with a sharply erosional base cutting into overbank muds, silts and thin sheets sandstones. The sandstone body is 2·5 m thick.

summarized in Fig. 6. Each arrow represents the vector mean of a number of individual readings from the surrounding area. Each of these readings is in itself a summary of the palaeocurrent information from an individual sand body and represents the interpreted direction of flow of a short reach of a stream channel. This represents a certain rank of palaeoflow (Allen, 1966) and according to the hierarchy of Miall (1974) would be considered to be rank 3. Extensive plan exposures (the mean orientation of which represents the trend of an individual channel) were divided into 100 m stretches, a direction recorded for each stretch and treated as a rank 3 direction. Data from individual bars or large-scale ripple-form structures which could not be related to a channel axis have not been included.

The distribution of the mean palaeoflow orientations in Fig. 6 clearly defines two systems, one to the east, which has dominant westerly and southerly flow directions, and a second to the west which shows an almost complete radial pattern. These have been termed the *Huesca System* and *Luna System* respectively, and both were evidently radial, distributary fluvial systems. Back plotting of the palaeocurrent means allows the apices of these distributary systems to be estimated (see Fig. 6). The apex of the Luna System can be particularly well defined and is probably correct to within a kilometre. The apex of the Huesca System is less precisely defined due to tectonic deformation of the proximal deposits and the data suggest that the point of entry into the Ebro Basin was much broader than for the Luna System.

Correlation

The early Miocene sediments are well exposed over much of the area but the exposures tend to occur in isolated patches. Biostratigraphic correlation is not possible at present because of the lack of data on the sparse fauna and flora in the early Miocene sediments

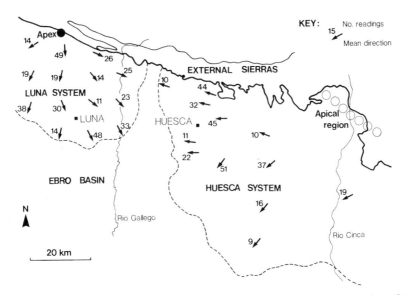

Fig. 6. Summary of the palaeocurrent data from the Luna and Huesca Systems. Palaeocurrent directions of 'rank 3' (see text) have been grouped geographically and a mean for each set calculated. Each arrow represents the mean direction of the number of data points (indicated alongside the arrow) in that area. Back-plotting of all the individual data points for each system indicates a tightly constrained apex for the Luna System and a broader apical region along the basin margin for the Huesca System.

of the area. A stratigraphy based on the remains of small mammals is currently being established in the Ebro Basin (Daams & Van der Meulen 1983) and a magnetic reversal stratigraphy programme is under way (Turner *et al.*, 1984; Friend, personal communication) but to date neither has been extensively applied to the area. No distinctive marker horizons (widespread volcanic layers, marine incursions, extensive palaeosols, etc.) have been identified in the area.

Over much of the studied area, the sediments are more or less horizontal. Dips rarely exceed 2° and it is possible to correlate over several kilometres by following sedimentary horizons. This is particularly so in parts of the Luna System where topographic relief may be several hundred metres. The regional dip is more of a problem in the Huesca System where the most distal sediments may be as much as 400 m stratigraphically higher than the medial deposits. Palaeocurrent data suggest that all the early Miocene fluvial sediments in the area were deposited by either the Huesca or the Luna river systems; interdigitation of the most distal deposits of these two systems near to the basin margin indicates that they were at least partly contemporaneous. Vertical exposures of up to 90 m in the Huesca System deposits show little or no evidence of any change of sedimentary style through

time (Hirst, 1983). However, vertical changes in the size and abundance of ribbon bodies have been recognized on a scale of hundreds of metres in the Luna System (Nichols, 1984). The data available do not allow quantification of either lateral or vertical variations in the area as a whole and the facies belts indicated in Fig. 7 are only considered to be approximate.

The major features of the Luna and Huesca Systems are briefly described below.

The Luna System

The deposits of the Luna System are spread over an approximately semicircular area with a radius of just over 40 km (see Fig. 6). The proximal part of the system is made up of conglomeratic sediments which are interpreted as being deposited by proximal braided streams. In exposures where the dimensions of the braided channels can be determined they can be seen to have been a hundred metres or more wide and up to 5 m deep. Deposition occurred mainly on mid-channel bars and in minor channel fills. Braided stream deposition of medium to very coarse sandstone is important in the sandy medial part of the system. In addition, sand body geometries indicating more

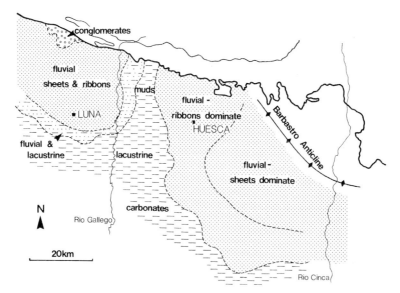

Fig. 7. Facies belts in the Early Miocene sediments in the study area (see text for discussion).

confined flow occur. There are examples of simple ribbons, complex multistorey and multilateral ribbons and sheets of sandstone produced by lateral migration of a single channel as well as braided stream deposits throughout the medial zone of the system. Towards the distal margins, sand bodies become smaller and sheet sandstones deposited by poorly channelized or unconfined flow become more abundant; the proportion of fine alluvium becomes more significant, channel fill bodies are less abundant and interdigitation with lacustrine deposits occurs. There is also a down-system decrease in grain size. These trends represent a transition from channelized flow with sediment accumulation in channel plugs and from overbank floods to poorly channelized flow with the overbank flood deposits passing into lacustrine deposits.

The Huesca System

The Huesca System, with a radius of over 60 km, is considerably larger than the Luna System. The proximal deposits are tectonically deformed, the major structural feature being the Barbastro Anticline (Fig. 7). To the south and west of this structure, the Miocene sediments are relatively undisturbed and are considered to be the medial and distal deposits of the Huesca System (Hirst, 1983). The deposits consist of channel sand bodies (both sheet and ribbon forms) enclosed within floodplain fines. In the distal part of the system the proportion of sediment deposited in the channels

is low (less than 5%) and the fluvial deposits interdigitate with lacustrine sediments. In the medial areas the sand bodies form a greater proportion of the deposits and are often amalgamated. Other radial changes include a decrease in grain size and sand body thickness downcurrent. Thus the overall sedimentary style is similar to the Luna System.

In the Huesca System there are also radial variations in the type of sand body. In the medial deposits, sheet sand bodies dominate with ribbon forms subordinate. The sheet bodies are mostly the products of laterally migrating rivers although a few were deposited by braided rivers. This is in contrast to the ribbon bodies which were deposited by laterally stable rivers, even though plan exposures indicate that the rivers were often sinuous. In the distal deposits, ribbon sand bodies dominate, indicating a greater lateral stability of the channels in this area. If the medial and distal deposits were contemporaneous it would appear that the lateral stability of the channels increased down-system. Several possible reasons may be invoked to explain this. A decrease in gradient of the system as it approached the lacustrine area would have caused a decrease in flow velocity and hence scouring power of the channel. This would be compounded by the greater cohesiveness of the channel banks distally as a result of the higher proportion of fine sediment deposited in the distal parts of the system. It has also been demonstrated empirically that lateral migration is favoured by intermediate rather than low gradients

(Schumm & Khan, 1972). Frequent channel avulsion has been proposed by Stokes (1961) and Wells (1983) to account for the lateral stability of palaeochannels described by these authors. The radial increase in the absolute number of channels on the present day Markanda Fan (Mukerji 1976) suggests more frequent avulsion downsystem. However, as mentioned above, it is possible that the distal deposits are somewhat younger than the medial deposits and the differences between these two areas may represent an increase in channel stability through time.

The Lacustrine Zone

In the lacustrine zone which interdigitates with both of these fluvial systems, fine to medium sandstones occur, but without erosional bases and with wave-ripple bedforms. The finer lacustrine deposits are largely oxidized muds near to the northern margin of the Ebro Basin where there was some input of material derived from the External Sierras. Near the basin centre the fine lacustrine deposits are grey marls and limestones which contain horizons rich in freshwater gastropods.

PROVENANCE

Analysis of the mineral suites in sandstones from the Luna System, the Huesca System and five marginal alluvial fan bodies has shown that certain minerals can be used to characterize the alluvial fans and the two fluvial systems (see Fig. 8).

Glauconite is present only in sandy material associated with the marginal alluvial fan bodies. It is a relatively unstable mineral and it may be inferred that the sands containing detrital glauconite have had a short transport history. Grains of this mineral were most probably derived from a glauconitic sand horizon exposed in the External Sierras at the top of the Middle Eocene limestone sequence.

Biotite is rare to absent in all the deposits associated with the marginal alluvial fans, reflecting the scarcity of this mineral in their restricted drainage basins. However, the Huesca System sandstones can be characterized by the presence of green chloritized biotite in significant quantities. Chloritized biotite occurs in the Luna System sandstones, but normally only in small amounts. This suggests that there was a more abundant source of biotite in the central and eastern Pyrenees than in the western Pyrenees, a

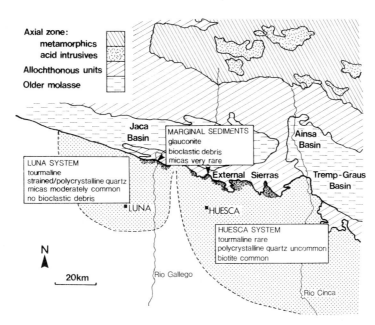

Fig. 8. The minerals and clast types indicated are those considered to be characteristic in the three groups of sediments. The 'marginal sediments' analysed were 12 samples of fine to medium sand from the five most westerly alluvial fan bodies (see Fig. 3). Thirty-four samples of fine to medium sand were analysed from the Luna and Huesca Systems. The source areas for the fluvial deposits are the meta-sediments and granodiorites in the axial zone, Mesozoic to Eocene allochthonous units (mainly limestones) and older molasse basin clastics.

difference which may be attributed to the extent of outcrop of acidic plutons (the most likely source of the biotite) which are more common in the east.

The presence of tourmaline in the Luna System sediments distinguishes them from Huesca System deposits. The occurrence of tourmaline in the Luna System samples suggests that this mineral must have been present in exposed axial zone rocks (possibly pegmatites) in the western part of the Pyrenees in the Miocene and was less common towards the east.

The rock types in the source areas of the fluvial systems can be deduced from the detrital mineral suites present. Minerals such as garnet, andalusite, kyanite and metamorphic (strained) quartz are all common in both the Huesca and Luna System deposits. They can be considered to be characteristic of a high-rank metamorphic source area (Pettijohn, 1975), such as the axial zone of the Pyrenees. Other minerals occurring frequently in both these systems are apatite, biotite and euhedral zircon, and these are characteristic of acid igneous source areas. However, as much of the material in the Luna System is derived from the Jaca Basin, and some of the Huesca System detritus was derived from the Tremp–Graus and Ainsa Basins, these interpretations of source rock types should be used with care: they may be more representative of the drainage basins of the older molasse deposits in the Jaca, Ainsa and Tremp–Graus Basins than the Miocene molasse in the Ebro Basin. Nevertheless, the characteristic mineral suites in the various alluvial deposits in the north-western Ebro Basin allow the deposits of the two fluvial systems and the marginal alluvial fans to be distinguished on the basis of their petrographies.

STRUCTURAL CONTROLS ON SEDIMENTATION PATTERNS

The frontal ramp of one of the south Pyrenean thrust sheets, the Guarga Unit, produced a structural high along most of the northern edge of the Ebro Basin in the late Oligocene (Puigdefabregas & Soler, 1973). This high extended as far as the western end of the External Sierras, where the tip line becomes buried. At this point the frontal ramp is no longer present and hence there is a structurally lower part of the leading edge of the thrust sheet. The frontal ramp of the Guarga Unit formed a barrier along the northern margin of the Ebro Basin as far as the Rio Cinca, 100 km to the east, where there is no surface evidence for a ramp structure between the Mediano

and Boltaña Anticlines. Palaeocurrent directions and facies distribution patterns show that the apex of the Luna System occurs at the structural low at the western end of the External Sierras. The apex of the Huesca System formed in a broader structural low between the Mediano and Boltaña Anticlines. It is suggested that runoff from the southern Pyrenean zone entered the Ebro Basin at these two points. Coarser clastic load was deposited near to the basin margin and the finer sediments were carried out by the rivers of the fluvial distributary systems. Rivers from the axial zone and the southern Pyrenean zone did not cross the structurally higher line of the frontal ramp of the thrust sheet at any other point during this period. This is because at the time of the emergence of a frontal thrust system the structurally lower areas offered the easiest route for headward stream capture, whereas the higher areas acted as barriers to headward capture. Restricted runoff from this emergent frontal thrust system gave rise to a series of small terminal alluvial fans along the basin margin. The volume of sediment deposited by these alluvial fans was minor relative to the Luna and Huesca Systems.

The relationship between the positions of frontal thrust systems and fluvial distributary systems in the north-western part of the Ebro Basin is summarized in Fig. 9. Within the thrust belt the Boltaña Anticline, a culmination above an oblique ramp (Deramond *et al.*, 1984), acted as an east–west barrier to drainage. This effectively separated the drainage basins of the Luna and Huesca fluvial systems and this would account for the distinct petrographies of the sands deposited by the two fluvial systems. In a consideration of the palaeodrainage pattern in late orogenic basins, Eisbacher *et al.* (1974) established a series of structural features which control drainage basin evolution. An *erosional realm* (in this case the southern Pyrenean Zone) is separated from the *aggradational basin* (the Ebro Basin) by a zone of reverse faulting defining the *uplift boundary* of the mountain belt (the frontal thrusts of the south Pyrenean thrust sheets). Rivers enter the molasse basin through *structural re-entrants* where they deposit their load (the apices of the Luna and Huesca Systems) and the re-entrants are separated by *structural salients* from which only short, vigorous streams issue into the molasse basin (the marginal alluvial fans along the south side of the External Sierras). A longitudinal drainage pattern is predicted in the mountains, changing course on reaching a structural re-entrant. It is suggested here that this is likely to have been the case in the drainage basin of the Luna System, with flow along the axis of the old

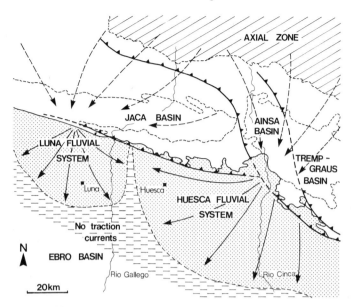

Fig. 9. Summary map. Sediment laden run-off from the western half of the axial zone and the Jaca Basin entered the Ebro Basin at the western end of the External Sierras and spread out to form the Luna Fluvial System. The source area for the Huesca System included granodiorite masses in the axial zone (see Fig. 8), the Ainsa Basin and the Tremp–Graus Basin. This fluvial distributary system had a less well defined apical area between the oblique ramps of the Gavarnie–Boltaña and the Cotiella–Monsec thrust units (see Fig. 2A). Both systems terminated in a lacustrine zone.

Jaca Basin (probably controlled by east–west folding and faulting) as far as the structural low at the western end of the External Sierras.

The main difference between the Miocene southern Pyrenees described here and the parts of the modern Himalayas and the Mesozoic–Tertiary Canadian Cordillera described by Eisbacher *et al.* (1974) is that the palaeoflow in the Ebro Basin was not longitudinal with respect to the trend of the Pyrenees. Palaeoflow in the Ebro Basin seems to have been dominantly radial from each structural re-entrant, with each fluvial system terminating in a lacustrine zone in the basin centre. It is this internal drainage pattern of the Miocene Ebro Basin which makes it different from many other foreland basins such as the southern Himalaya and the Oligo–Miocene Swiss Molasse Basin: the major Himalayan rivers (e.g. the Ganges and Brahmaputra) drain longitudinally, and at times the palaeoflow in the Swiss Molasse Basin was longitudinal into an open marine environment (Trümpy, 1980).

The occurrence of discrete points of input from the erosional domain into the aggradational basin appears to be a common feature of many foreland zones—in parts of the Himalaya and the Canadian Cordillera (Eisbacher *et al.*, 1974), the Swiss Molasse Basin

(Trümpy, 1980) and the Ebro Basin described here. However, terminal, radial distributary systems are restricted to areas of internal drainage, such as the Ebro Basin in the Mid-Tertiary. Since the late Miocene the Ebro Basin has drained into the Mediterranean by the River Ebro (which flows sub-parallel to the Pyrenean trend) and these radial distributary patterns no longer exist.

ACKNOWLEDGMENTS

The authors would like to thank Dr P. F. Friend for his help and advice at all stages in this work. The work was undertaken whilst J. P. P. Hirst held a Shell research studentship and G. J. Nichols held a Natural Environment Research Council studentship at the University of Cambridge.

REFERENCES

ALLEN, J.R.L. (1966) On bedforms and palaeocurrents. *Sedimentology*, **6**, 153–190.

BOILLOT, G. (1984) Some remarks on the continental margins in the Aquitaine Basin and French Pyrenees. *Geol. Mag.* **121**, 407–412.

CHOUKROUNE, P. & SEGURET, M. (1973) Tectonics of the Pyrenees: role of compression and gravity. In: *Gravity and Tectonics* (Ed. by De Jong & Scholten), pp. 141–156. Wiley, New York.

DAAMS, R. & VAN DER MEULEN, A.J. (1983) Palaeoecological interpretation of micromammal faunal successions in the Upper Oligocene and Miocene of Spain. *R.C.M.N.S. Interim-colloquium*, Montpellier, p. 17.

DERAMOND, J., FISCHER, M., HOSSACK, J., LABAUME, P., SEGURET, M., SOULA, J.-C., VIALLARD, P. & WILLIAMS, G.D. (1984) *Field guide of conference trip to the Pyrenees: Chevauchement et deformation conference*, pp. 1–28. Paul Sabatier University, Toulouse.

EISBACHER, G.H., CARRIGY, M.A. & CAMPBELL, R.B. (1974) Palaeodrainage patterns and late-orogenic basins of the Canadian Cordillera. In: *Tectonics and Sedimentation* (Ed. by W. N. Dickinson). *Spec. Publ. Soc. econ. Palaeont. Mineral., Tulsa*, **22**, 143–166.

FRIEND, P.F., HIRST, J.P.P. & NICHOLS, G.J. (1986) Sandstone-body structure and river processes in the Ebro Basin of Aragon, Spain. *Cuadernos de Geologia: Fluvial Sedimentation in Spain*, Madrid.

FRIEND, P.F., SLATER, M.J. & WILLIAMS, R.C. (1979) Vertical and lateral building of river sandstone bodies, Ebro Basin, Spain. *J. geol. Soc. London*, **136**, 39–46.

HIRST, J.P.P. (1983) *Oligo–Miocene alluvial systems in the northern Ebro Basin, Huesca Province, Spain*. Unpublished Ph.D. Thesis, University of Cambridge, 247 pp.

MATTAUER, M. & HENRY, J. (1974) The Pyrenees. In: *Mesozoic–Cenozoic Orogenic Belts; data for orogenic studies* (Ed. by A. M. Spencer). *Spec. Publ. geol. Soc. London*, Scottish Academic Press, 814 pp.

MIALL, A.D. (1974) Palaeocurrent analysis of alluvial sediments: a discussion of directional variance and vector magnitude. *J. sedim. Petrol.* **44**, 1174–1185.

MUKERJI, A.B. (1976) Terminal fans of inland streams in Sutlej-Yamuna Plain, India. *Z. Geomorph. Neue Folge.* **20**, 190–204.

NICHOLS, G.J. (1984) *Thrust tectonics and alluvial sedimentation, Aragon, Spain*. Unpublished Ph.D. Thesis, University of Cambridge, 243 pp.

NIJMAN, W. & PUIGDEFABREGAS, C. (1974) In: *Field guides to modern and ancient fluvial systems in Britain and Spain* (Ed. by T. Elliott). Part of the proceedings of the Second International Conference on Fluvial Sediments, University of Keele, U.K. 1981 (fig. 4.30).

ORI, G.G. & FRIEND, P.F. (1984) Sedimentary basins formed and carried piggyback on active thrust sheets. *Geology*, **12**, 475–478.

PETTIJOHN, F.J. (1975) *Sedimentary Rocks*. Harper & Row, New York, 628 pp.

PUIGDEFÀBREGAS, C. & SOLER, M. (1973) Estructura de las Sierras Exteriores Pirenaicos en el corte del Rio Gallego (Provincia de Huesca). *Pireneos*, **109**, 5–15.

QUIRANTES PUERTAS, J. (1969) *Estudio sedimentologico y estratigrafico de Terciaro continental de los Monegros*. Tesis (C.S.I.C.), Zaragoza, 101 pp.

RIBA, O. (1971) *Mapa Geologico de España, E.1:200000. Sintesis de la cartografia existente Lerida*, primer edition. Departamiento de publicaciones del Instituto de Geologico y Mineralogico de España, Madrid.

SCHUMM, S.A. & KHAN, H.R. (1972) Experimental study of channel patterns. *Bull. geol. Soc. Am.* **85**, 1755–1770.

SEGURET, M. (1972) *Etude tectonique des nappes et series decollees de la partie centrale du versant sud des Pyrenees*. Thesis, Montpellier. Publications USTECA, 1972, Montpellier, 155 pp.

STOKES, W.L. (1961) Fluvial and aeolian sandstone bodies in the Colorado Plateau. In: *Geometry of Sandstone Bodies* (Ed. by J. A. Peterson & J. C. Osmond), pp. 151–178. American Association of Petroleum Geology, Tulsa.

TRÜMPY, R. (1980) *Geology of Switzerland. A guide book. Part A. An outline of the geology of Switzerland*, pp. 24–30. Schweiz. Geol. Komm. Basel.

TURNER, P., HIRST, J.P.P. & FRIEND, P.F. (1984) A palaeomagnetic analysis of Miocene sediments at Pertusa, near Huesca, Ebro Basin, Spain. *Geol. Mag.* **121** (4), 279–290.

WELLS, N.A. (1983) Transient streams in sand-poor redbeds: early–Middle Eocene Kuldana Formation of Northern Pakistan. In: *Modern and Ancient Fluvial Systems* (Ed. by J. D. Collinson & J. Lewin). *Spec. Publs int. Ass. Sediment.* **6**, 393–403. Blackwell Scientific Publications, Oxford.

Spec. Publs int. Ass. Sediment. (1986) **8**, 259–271

Syntectonic intraformational unconformities in alluvial fan deposits, eastern Ebro Basin margins (NE Spain)

PERE ANADÓN

Institut Jaume Almera, C.S.I.C. (Barcelona), Martí Franques s/n. 08028 Barcelona, Spain

LLUÍS CABRERA, FERRAN COLOMBO, MARIANO MARZO *and* ORIOL RIBA

Departament d'Estratigrafia, Universitat de Barcelona, Gran Via, 585, 08007 Barcelona, Spain

ABSTRACT

Several types of syntectonic intraformational unconformities can be observed in the Palaeogene alluvial fan conglomerates deposited in the eastern Ebro Basin margins. Progressive unconformities are the most common but unconformities displayed by supratenuous folds are also observed. We describe syntectonic intraformational unconformities related to the following structural settings: (a) Nappe fronts (i.e. Cadí Nappe in the southern Pyrenees). Development and deformation of the progressive unconformities are related to successive movements during the late phases of floor thrust emergence. (b) Uplifting structures related to the basement-involved strike-slip fault system of the Catalan Coastal Range: monoclinal folds linked to uplifting blocks and stretched cover folds. The successive motion of basement faults located more and more basinward resulted in a shifting of the basin boundary in some sectors. Syntectonic intraformational unconformities developed in the alluvial fan sequences which accumulated along each new basin boundary. (c) Folds affecting a thick Mesozoic cover (Linking Zone). In this case, folding during sedimentation produced progressive unconformities in the Tertiary strata at the fold limbs. Supratenuous synclines affecting the Tertiary syntectonic deposits developed between the anticlines.

Progressive syntectonic unconformities recognized in the structural settings mentioned above show broadly similar geometric features. The syntectonic intraformational unconformities are related to uplifting structures such as nappe fronts, monoclinal folds, overfolds and limited overthrusts. The spatial and temporal distribution of the syntectonic intraformational unconformities record the diachronous development of these structures along the margins of the eastern Ebro Basin.

INTRODUCTION

Geological-structural setting

During the Palaeogene a N–S shortening affected the NE part of the Iberian Peninsula, as a consequence of a likely limited collision between the European (French) and the Iberian plates with the French plate overriding that of Iberia. Resulting from this collision the Pyrenees arose as an E–W asymmetric thrust belt with most of the movement directed towards the south. While several southern Pyrenean nappes displaced southward several tens of kilometres (Seguret, 1970; Solé-Sugrañes, 1978) other north-moving thrust-sheets involved noticeably less displacement (Deramond *et al.*, 1984).

The Tertiary Ebro Basin, NE Spain (Fig. 1) is the intact southern foreland basin of the Pyrenees. This basin formed during the Palaeogene as an asymmetric sedimentary trough and its origin was closely linked to the southward emplacement of Pyrenean thrust sheets together with their related piggy-back basins over the northern sedimentary areas of the Ebro Basin.

Apart from the Pyrenees thrust belt, the Ebro Basin is also bounded by other structural units (Fig. 1). The southern margins of the Ebro basin are formed by the Iberian Range to the SW, the Catalan Coastal Range

259

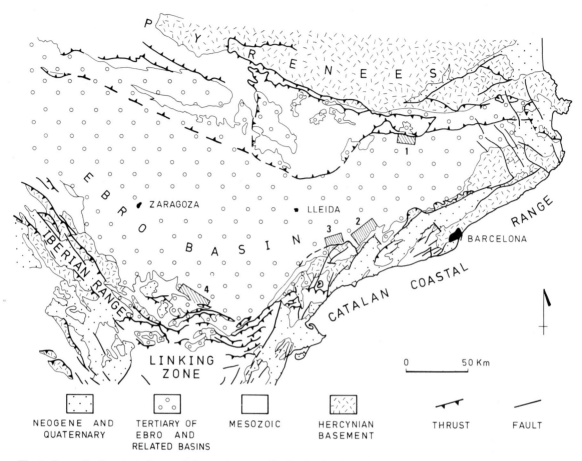

Fig. 1. Generalized geological map of the north-eastern Iberian Peninsula showing the location of the studied Palaeogene alluvial fan deposits: (1) Sant Llorenç del Morunys, (2) Sant Miquel del Montclar, (3) Montsant–Serra de La Llena, (4) Andorra–Alloza.

to the SE and a Linking Zone between these ranges (Guimerà, 1984). Each one of these units displays characteristic structural features.

The Iberian Range has been termed an Intermediate Range (Julivert *et al.*, 1974), where a typical basement-cover tectonics took place during the Palaeogene. This chain consists of an array of NW–SE trending cover folds and limited thrusts strongly determined by major basement faults (Guimerà, 1984).

The Catalan Coastal Range is located along the SE margin of the Ebro Basin. This structural unit is a Palaeogene strike-slip system which consists of a set of NE–SW right stepping, *en échelon* left lateral strike-slip faults which extend for about 200 km (Guimerà, 1984; Anadón *et al.*, 1986).

The Linking Zone (Guimerà, 1984) consists of an array of folds and thrusts of E–W direction. This unit

differs from the Iberian Range and the Catalan Coastal Range on the basis of a thicker Mesozoic cover affected by low angle, larger thrusts. A more extensive folding and thrusting of the cover took place in this unit during the Palaeogene.

Tectonics and sedimentation during the Palaeogene at the eastern Ebro Basin margins

The Palaeogene evolution of deformation in the NE Iberian Peninsula can be explained as a consequence of the regional compressions resulting from the global processes of plate collision which formed the Pyrenees and the Betic Chain (Guimerà, 1984).

The diverse features of the structural units which surround the eastern Ebro Basin and their location in relation to the active plate margins of the Iberian plate (Pyrenean to the north, Betic to the south) caused distinct Palaeogene tectonics at each basin margin. Thrust sheet emplacement closely linked to strong crustal shortening took place in the central and eastern Pyrenees (Muñoz, Martínez & Verges, 1986). The setting in motion of the NE–SW strike-slip basement faults of the Catalan Coastal Range gave rise to upthrust structures of basement rocks and push-up blocks or folds and flexures where a thick Mesozoic cover existed. The motion on the NW–SE basement faults of the Iberian Range caused the interference of these faults with those of the Catalan Coastal Range in the Linking Zone. As a result a widespread development of cover folds and thrusts took place in this zone where a thick Mesozoic cover with a well developed décollement level existed (Guimerà, 1984).

Well-developed alluvial fan systems formed all along the margins of the eastern Ebro Basin affected by the rising of the above mentioned tectonic structures. Proximal alluvial fan conglomerates stacked up along the basin margins undergoing rapid uplift and erosion. Most of these alluvial fan conglomerates were syntectonic and, resulting from this, the interdependence of sedimentation and tectonics gave rise to complex structural and stratigraphic relationships. The facies distribution of the alluvial fan systems in the eastern Ebro Basin shows a complex temporal and spatial distribution because of the influence exerted on the formation and evolution of the alluvial systems by the tectonic evolution of the basin margins where diverse structures developed at different times during sedimentation.

Apart from the coarse grained alluvial facies distribution, the relationship between tectonics and sedimentation along the eastern Ebro Basin margins is also recorded by the structural arrangement displayed by these facies. Along these margins internal unconformities indicating the occurrence of tectonic pulses of uplift contemporaneously with sedimentation are common. Riba (1976a) made a first general review describing several types of internal unconformities in the northern margins of the Ebro Basin which he termed 'progressive unconformities' and 'syntectonic unconformities'. Later, this kind of tectono-sedimentary feature has attracted a different terminology from other authors (Miall, 1978).

In this paper, the study area is extended to the Catalan Coastal Range and the Linking Zone margins. Description and comparison of the main features of several case-studies of internal unconformities from diverse tectonic settings enable us to develop a general discussion on their interpretation.

Progressive and angular syntectonic unconformities

Riba (1976a, b) suggested that a single progressive unconformity is a syntectonic cumulative wedge system attached to a tilted depositional surface, in turn linked to an uplifting structure (anticline flanks, over fold flanks, high angle faults, nappe fronts, diapiric flanks). This cumulative wedge system results directly from the uplifting process and may be developed either as a rotative offlap or a rotative onlap (Fig. 2). A rotative offlap would develop under a constant rate of sedimentation during a period of accelerating uplift (A in Fig. 2). On the contrary a rotative onlap would record a decelerating rate of uplift (B in Fig. 2). In both cases the structure of the syntectonic cumulative wedge system would result from the tilting of the beds around a rotation axis.

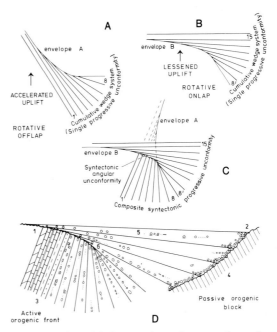

Fig. 2. Genetic model of progressive and syntectonic angular unconformities (after Riba, 1976a). (A) Rotative offlap envelope developed during accelerated uplift of a block adjacent to the basin. (B) Rotative onlap recording a decrease in the rate of uplift. (C) Combination of (A) and (B) envelopes: a composite syntectonic progressive unconformity results in the inner zones of the basin, and a syntectonic angular unconformity develops near the uplifted block. (D) Synoptic model of syntectonic unconformities: (1) syntectonic angular unconformity of active orogenic front; (2) syntectonic climbing angular unconformity (progressive overlap of passive orogenic block); (3) preorogenic sequence; (4) passive orogenic block with buried topography; (5) syntectonic sequence; (6) articulate axis of envelope surfaces.

Very often a cumulative wedge system developed as a rotative offlap is overlain by a wedge system arranged as a rotative onlap. In this case the upper part of the underlying wedge system is partially eroded and an obvious angular unconformity develops (composite progressive unconformity, C in Fig. 2). The angular unconformity, which may locally show a strong discordance, is called a syntectonic angular unconformity (Riba, 1976a, b) and tends to be laterally extensive along strike for up to a few kilometres before dying out. On the contrary it disappears rather abruptly perpendicular to strike towards the inner zones of the basin where sedimentation was not interrupted by erosion.

In the genetic model of progressive and angular syntectonic unconformities proposed by Riba (1976a, b) several elements are involved: the offlap-onlap development of cumulative wedge systems, the single progressive unconformities and their relationship to accelerated-decelerated uplifting episodes. The mechanism of development of these structures can have a clear polarity in some cases. Sometimes an 'active' quickly rising block can exist opposite another 'passive' block. In the latter, because of the spreading of alluvial deposits, the wedge systems can overlap the pre-uplift surface at some angle, giving rise to a progressive downlap-onlap (D in Fig. 2, climbing unconformity after Riba, 1976a). On the contrary, a composite progressive unconformity will develop over the 'active' block. These unconformable structures can develop in confined synclinal basins of varying size.

In broad terms the shape of the progressive and angular syntectonic unconformities results from the relative rates of sedimentation-erosion and uplift-basin subsidence. Moreover, the presence of several internal progressive to angular syntectonic unconformities affecting a sedimentary sequence records the tectonic evolution of a certain area of a basin margin. Progressive and angular syntectonic unconformities may have in some cases a regional tectonic significance, but very often they just record the occurrence of rapid, localized uplift contemporaneously with sedimentation along the basin margin.

Severe tilting and deformation of progressive unconformities at basin margins can take place where tectonic evolution goes on for a long time span. As a consequence they can be totally obscured. Depending on the structural setting and processes where they develop, progressive unconformities formed along active basin margins will have diverse potentials of development and preservation.

SYNTECTONIC UNCONFORMITIES RELATED TO THRUST SHEET EMPLACEMENT (SOUTH-EASTERN PYRENEAN FRONT)

East of the Segre river (sector 1 in Fig. 1) two main thrust sheets were emplaced during the middle–late Cuisian to early Lutetian (Pedraforca thrust sheet: Seguret, 1970; Solé-Sugrañes, 1978; Megías & Posadas, 1981) and during early to middle Oligocene (Cadí thrust sheet: Puigdefàbregas et al., this volume).

Related with the emergent floor thrust of the Cadí thrust sheet, alluvial fan conglomerates (Berga Formation) accumulated at the margins of the newly created northern margin of the foreland basin (Fig. 3).

The syntectonic Berga conglomerates, up to 2500 m thick, lie conformably on the underlying blue marine marls of the Sant Llorenç dels Morunys Formation (upper Biarritzian–lower Priabonian). These conglomerates display spectacular syntectonic unconformities which have been extensively described by Riba (1976a, b) and will not be treated again here.

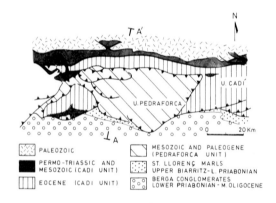

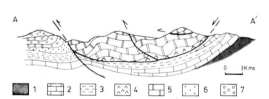

Fig. 3. (A) Simplified geological map of South Pyrenean thrust sheets to the east of the Segre river. (B) Cross-section along line A–A' in the geological map: (1) Permo–Triassic; (2) Lower Eocene Alveolina limestones; (3) Eocene marls; (4) Keuper facies; (5) Mesozoic and Tertiary of the Pedraforca nappe; (6) St Llorenç dels Morunys Formation; (7) Berga Conglomerates. Modified from Megías & Posadas (1981).

Basically the structural geometry shown by the Berga conglomerates (Fig. 4) can be described as repetitive composite syntectonic progressive unconformities with the older related angular unconformities tectonically deformed. This pattern suggests a combination of variable rates of uplift with horizontal shortening, which induced a basinward migration of the foreland margin. Lateral transitions from a syntectonic angular unconformity to a progressive unconformity and finally to a conformable sequence have been observed in the Berga conglomerates.

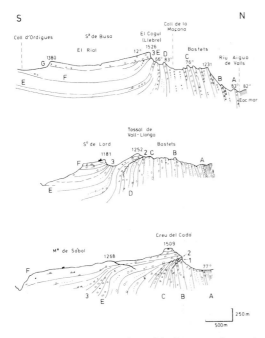

Fig. 4. Sequence of cross-sections of the Berga conglomerates, showing repetitive composite progressive unconformities. A to G, photo-geologic key beds. 1, 2 and 3 are angular unconformities. After Riba (1976a).

SYNTECTONIC UNCONFORMITIES RELATED TO UPLIFTING BLOCKS (CATALAN COASTAL RANGE)

The Catalan Range is located along the SE margin of the Ebro Basin, and is composed of Hercynian basement and an unconformably overlying Mesozoic cover (Fig. 1). The upper part of the cover (mainly Jurassic and Cretaceous carbonates) is detached from the basement at a main décollement level within Keuper (uppermost Triassic) mudstones and evaporites (Llopis, 1947). The lower part of the cover, however, remains attached to the basement.

The Catalan Coastal Range is characterized by an array of nearly vertical, strike-slip basement faults, arranged as a set of right-stepping *en échelon* faults (Guimerà, 1984; Anadón *et al.*, 1986). They developed probably during the Late Palaeozoic and were active as normal faults during Mesozoic times as can be inferred from the study of the Mesozoic cover rocks (Anadón *et al.*, 1979).

During the Alpine orogeny the most important and intense deformation took place along these basement-involved faults. In the basement the motion along the nearly vertical faults gave rise to the development of fault gouges as well as basement slices which overthrust the Mesozoic cover and even the Palaeogene deposits of the Ebro Basin (Julià & Santanach, 1984). Where a thick Mesozoic cover is present the basement faults may not be obvious at the surface, and movement is indicated by the presence of flexures and folds (Fig. 5).

During Palaeogene time, large amounts of terrigenous sediments accumulated in alluvial systems developed along the basin boundaries defined by the active basement faults. Generally, in this strike-slip system the SE fault blocks experienced an uplift relative to those in the NW. The slip along the diverse faults started at different times as was recorded by the diachroneity of the alluvial fan conglomerates deposited along the fault-defined basin boundaries (Anadón *et al.*, 1986). The beginning of development of these alluvial fan system along the SE Ebro Basin margin (Montserrat, Sant Miquel del Montclar, Montsant, Fig. 5) spans at least from the early to the late Eocene (Anadón *et al.*, 1979).

Several syntectonic unconformities are found in the proximal conglomeratic facies of the alluvial fan systems developed along the Catalan Coast Range during the Palaeogene. They demonstrate the synchrony between the slip on the faults and the sedimentation as well as the local shifting of the southeastern boundaries of the Ebro Basin during the Palaeogene. These features are well developed in the Sant Miquel del Montclar area and Serra de la Llena area (Figs 1 and 5).

Sant Miquel del Montclar area

The Sant Miquel Montclar area is located in the SE Ebro Basin margin close to the central part of the Catalan Coastal Range (number 2 on Fig. 1 and Fig. 5). In this area the lower part of the Palaeogene sequence reflects no activity along the nearest strike-

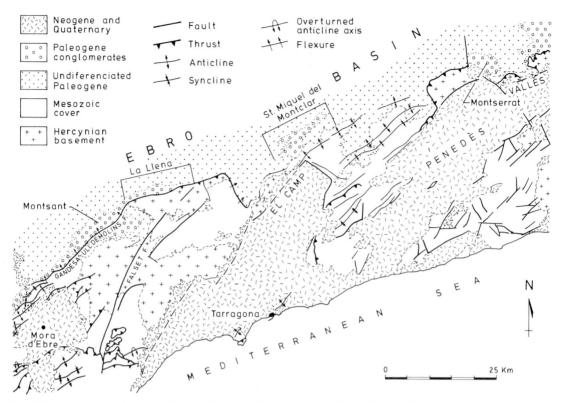

Fig. 5. Generalized geological map of the Catalan Coastal Range and the adjacent Ebro Basin showing the structural pattern of the Palaeogene strike-slip system. Location of Sant Miquel del Montclar and La Llena conglomerates is indicated.

slip fault (El Camp Fault). This sequence is formed from base to top of the following units:

Mediona formation (uppermost Palaeocene). This unit is up to 50 m thick and consists of red sandy mudstones with interbedded sandstones and calcareous palaeosols. It overlies the Triassic cover unconformably.

Orpí Formation (lowermost Eocene). Formed by up to 100 m of foraminiferal grainstones and wackestones (*Alveolina* limestones). The Orpí Formation represents shallow water platform deposits formed during the Ilerdian (early Eocene) transgression (Ferrer, 1971).

Pontils Group (lower–middle Eocene). The thickness of this unit is about 800 m and consists mainly of thick mudstones with thin interbeds of sandstones and minor carbonates and evaporites. The Pontils Group was deposited in a complex of closely related depositional environments including flood plains, dry and ponded mud flats and paludine-lacustrine areas. These non-marine environments were located far from the basin margin during the early and middle Eocene (Anadón, 1978).

Santa Maria Group. This Group consists mainly of fine grained terrigenous material and some subordinate carbonate deposits of marine origin (Ferrer, 1971). In the NE part of the studied area it is up to 700 m thick, but it is not present in the SW part due to a change of facies (Anadón *et al.*, 1979; Allen *et al.*, 1983, p. 135). It is overlain by Sant Miquel del Montclar conglomerates.

The upper part of the Palaeogene sequence consists of alluvial deposits. The more proximal deposits are represented by the Sant Miquel del Montclar conglomerates and the distal facies constitute the Montblanc Formation (upper levels).

The syntectonic Sant Miquel del Montclar conglomerates (uppermost Eocene) consists of more than 500 m of massive conglomerates with interbedded red sandstones and mudstones lenses (Colombo, 1980). The map pattern of this unit (Fig. 6) shows its fan-shaped geometry as well as the basinward lateral passage from massive conglomerates to red sandstone, mudstone and minor conglomerates.

The conglomerates lie by means of an angular

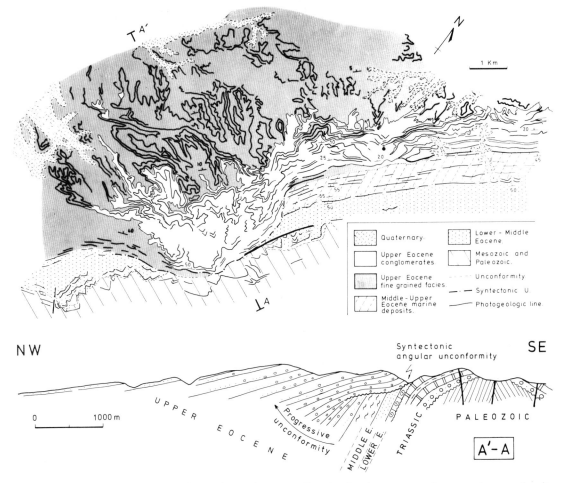

Fig. 6. Map pattern of the Sant Miquel del Montclar conglomerates. Cross-section shows a composite progressive unconformity.

unconformity over the lower to upper Eocene deposits (Fig. 6). This unconformity is related to the previous development of a single progressive unconformity affecting both the proximal Sant Miquel del Montclar conglomerates and more distal facies of the Montblanc Formation. The structural disposition of the Tertiary and Mesozoic rocks at this basin margin (cross-section in Fig. 6) suggests a rotative offlap followed by a rotative onlap caused by the uplift of the SE block of the El Camp Fault during the latest Eocene. This pattern can be termed composite progressive unconformity and can be ascribed to acceleration in tectonic uplift followed by a deceleration under a constant rate of sedimentation. In terms of sedimentary facies the result is a coarsening and thickening upwards sequence followed by a fining upward sequence in the zones of the rotation axis of the envelope surfaces.

Another important feature can be deduced from this pattern. In this area the lower parts of the Palaeogene sequence reflect no activity along the nearest El Camp basement fault (Fig. 5). On the other hand the coarse conglomerate deposits of Sant Miquel del Montclar that display a composite progressive unconformity would be linked to the uplifting of the SE block of the El Camp Fault and the formation of a new basin margin closer to the basin centre.

Serra de La Llena area

In the Palaeogene depositional sequence of the Serra de la Llena (Fig. 5) and the nearby Montsant area a slightly different evolution occurred. The lower part of the Palaeogene depositional sequence does not record tectonic activity along the nearby basement faults (Gandesa–Ulldemolins and Falset faults, Fig. 5). This lower sequence constitutes the Cornudella Group (Colombo, 1980) that can be correlated with the Pontils Group in the Sant Miquel del Montclar area (Anadón *et al.*, 1986). The Cornudella Group crops out along the SW part of the Catalan. Coastal Range. This unit, up to 400 m thick, consists mainly of fine grained terrigenous rocks with minor carbonate and evaporite horizons. In this area the marine Eocene sequences observed in the Sant Miquel del Montclar area are not present. The Cornudella Group was deposited in alluvial flood plains including paludine-lacustrine areas. These environments were located far from the basin margin during the early and middle Eocene (Colombo, 1980).

Overlying the Cornudella Group a thick succession (up to 1000 m) of coarse grained deposits constitutes the upper part of the Palaeogene sequence (Scala Dei Group). This Scala Dei Group of late Eocene–Oligocene age, consists of proximal alluvial fan, massive stacked-up conglomerates (Montsant Formation) that pass laterally basinwards (Allen *et al.*, 1983) into more distal red sandstones, mudstones and conglomerates (Margalef Formation, Fig. 7). In the Serra de La Llena area the lower part of the Palaeogene sequence does not crop out due to the overthrust of the Falset–Prades block (Fig. 5). Only in some places close to the Prades block are the uppermost levels of the Cornudella Group exposed.

In the studied area several syntectonic unconform-ities have been observed in the Scala Dei Group sequence. The lower part of the Montsant conglom-erates show a local syntectonic angular unconformity which has been tilted to vertical (no. 1 in cross-section B, Fig. 7) and a single progressive unconformity (no. 2 in cross-section B, Fig. 7). The upper levels of the Montsant conglomerates display a composite progres-sive unconformity (no. 3 in cross-section A, Fig. 7) that can be followed several kilometres along strike and constitutes the wedge system associated with the Montsant flexure (Fig. 5). The lower unconformity is restricted in space, laterally disappearing and being substituted by a conformable succession.

The lower part of the Montsant conglomerates in the Serra de La Llena area records the uplift linked to the activity along the Falset Fault (syntectonic unconformities 1 and 2). Later the Ulldemolins–Gandesa strike-slip basement fault also became active. In the cover this fault formed the Montsant flexure and gave rise to the well developed progressive unconformity which affects the upper levels of the

Montsant Conglomerates (no. 3 on cross-section A, Fig. 7). The diachronous initiation of motion on the faults and their *en échelon* right-stepping pattern results in a diachronous uplift of blocks. In a transverse section in the Serra de La Llena area there was a NW-ward (basinward) migration of tectonic deformation causing the formation of new sedimentary boundaries marked by successive progressive and angular syntec-tonic unconformities.

SUPRATENUOUS SYNSEDIMENTARY SYNCLINES AND RELATED SYNTECTONIC UNCONFORMITIES (LINKING ZONE)

While a NE–SW structural fabric dominates in the Catalan Coastal Range, a NW–SE trend is dominant in the Iberian Range (Fig. 1). The Linking Zone between these ranges is characterized by an E–W array of folds and thrust developed on a thick mainly Mesozoic cover. In this zone the dominant structural fabric is approximately E–W but there are two important virgations where they take a NE–SW direction. These virgations are probably due to the action of NE–SW basement faults with a sinistral movement that acted simultaneously to the formation of E–W folds and thrusts (Guimerà, 1984). The general vergence is to the north, but there are some reverse faults with a southerly vergence.

The E–W folds and thrusts formed later than those with NE–SW and NW–SE orientations (Simón Gómez, 1980; Guimerà, 1984). Nevertheless, locally, during a certain period the NW–SE and the E–W structural directions developed simultaneously (Gui-merà, 1984).

While the southern part of the Linking Zone (Maestrat area) is formed mainly by a thick, subhori-zontal pile of Mesozoic rocks, the northern part, close to the Ebro Basin, is characterized by folds and thrusts affecting Mesozoic, Palaeogene and lower Miocene rocks. The Palaeogene and lower Miocene deposits show supratenuous synclines and related syntectonic unconformities in the limbs of the anticlines. These features are common along the northern part of the Linking Zone from Alcorisa to Fuentespalda. The Alloza–Andorra area provides a good example of these synsedimentary folds and syntectonic uncon-formities developed in alluvial deposits (Fig. 8).

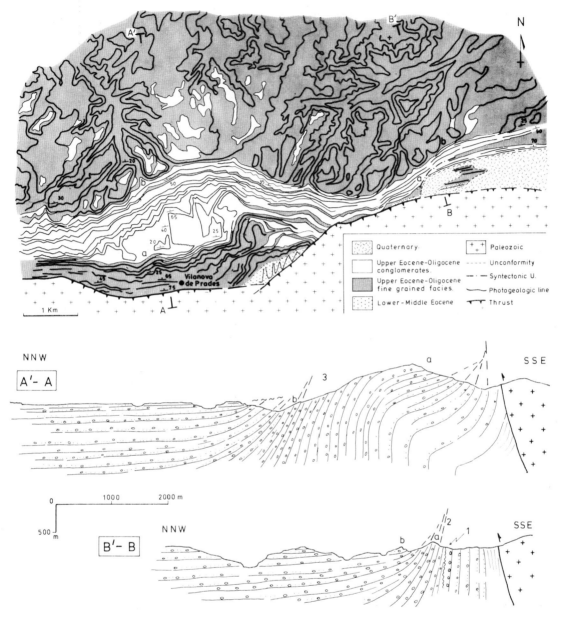

Fig. 7. Geological map and cross-sections of the Serra de La Llena conglomerates. 1, 2 and 3 indicate syntectonic unconformities; a and b are key beds. See text for further explanation.

Stratigraphy and structure of the Tertiary deposits in the Alloza–Andorra area

In this area the Hercynian basement does not crop out and Mesozoic (Keuper to Cenomanian) cover is up to 1000 m thick although this thickness varies strongly.

The stratigraphy in the Alloza–Andorra area has been established by González *et al.* (1984) although formal lithostratigraphic units have not been defined. The dating of the Tertiary sequences is difficult due to the scarcity of fossil remains.

For our purposes three main informal lithostratigraphic units can be distinguished: Lower, Interme-

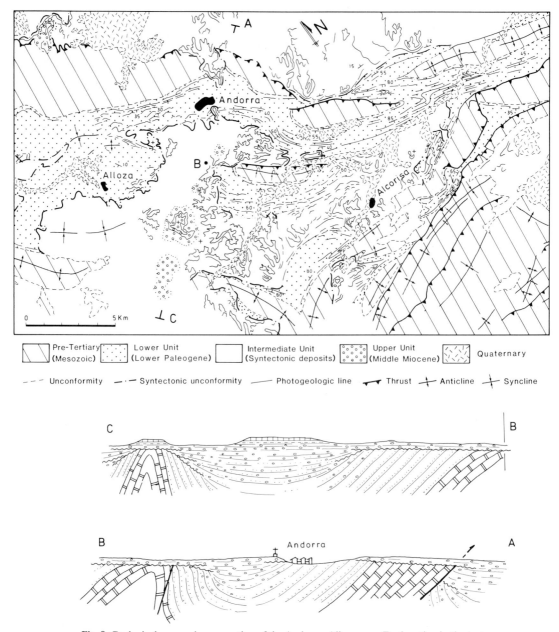

Fig. 8. Geological map and cross-section of the Andorra–Alloza area. Explanation in the text.

diate and Upper. The Lower unit (lower Palaeogene) ranges from upper Palaeocene to lower Oligocene. The base of this unit is formed by 10 m of red mudstones and calcareous crusts. The bulk of the sequence, up to 600 m thick, is formed mainly of fluvial red to yellow mudstones with interbedded sandstones and conglomerates. The Intermediate unit

(upper Palaeogene to lower Miocene) consists mainly of red sandstones and conglomerates with minor interbedded red mudstones of alluvial fan origin. The thickness of this unit is difficult to evaluate because of its structural arrangement but in some places it is at least several hundred metres thick.

The Upper unit (middle Miocene) is formed, in its

lower part, by 60 m of distal alluvial mudstones with minor interbedded sheet sandstones. The upper part, 50 m thick, consists of lacustrine limestones with interbedded marls.

The Lower unit (lower Palaeogene) lies unconformably on the Mesozoic rocks and it has been folded together with the Mesozoic strata (Fig. 8).

The Intermediate unit (upper Palaeogene–lower Miocene) displays a different cartographic pattern and show in its lower parts a supratenuous fold disposition. In some places (see cross-section in Fig. 8) the crests of the anticlines have been eroded. Only the upper parts of the intermediate unit are represented on these crests onlapping the Mesozoic and Lower Palaeogene rocks. In the limbs of the synsedimentary folds several syntectonic angular unconformities related to an overall composite progressive unconformity are found. The rotation axis for the progressive unconformities is nearly horizontal and parallel to the fold axis. All of these facts indicate a synsedimentary folding of this unit reflected in the distribution of the sediment thickness, the geometry of the structure and the related syntectonic unconformities.

Thus the Intermediate unit has a syntectonic origin and the upper onlapping beds have been deposited in the latest stages of deformation, after decrease of the uplift due to folding.

The Upper unit (middle Miocene) was deposited after cessation of the main tectonic phase. Similar syntectonic unconformities to these described in the Andorra–Alloza area have been observed in other places of the Linking Zone (Ashauer & Teichmuler, 1935).

DISCUSSION AND CONCLUDING REMARKS

During the Palaeogene the N–S shortening which affected the NE part of the Iberian Peninsula gave rise to the Pyrenean thrust belt and controlled the evolution of the Catalan Coastal Range strike-slip system, as well as that of the arrays of cover folds and thrusts which make up the Iberian Range and the Linking Zone (Fig. 1).

Several types of syntectonic intraformational unconformities can be observed in the Palaeogene alluvial fan conglomerates deposited in the eastern Ebro Basin margins formed by the above mentioned structural units (Figs 4, 6, 7 and 8). These syntectonic unconformities show broadly similar geometric features since all of them are arranged as cumulative wedges with strata rotating around slightly plunging to nearly horizontal rotations axes. This can be explained because, despite these unconformities developed in rather diverse tectonic settings and resulting from different tectonic processes, they are always linked to uplifting movements, i.e. those linked to the emergent floor thrust in the Pyrenees and to the synsedimentary cover folds and thrusts in the Linking Zone. Even in the case of the Catalan Coastal Range strike-slip system, where noticeable horizontal slip was involved, sedimentation was strongly influenced by uplifting structures. It must be emphasized that in the Catalan Coastal Range, the Tertiary syntectonic coarse grained alluvial facies are hardly ever attached to the planes of major strike-slip basement faults. Between these faults and the Tertiary sediments, the Mesozoic cover is usually present. As a consequence the horizontal slip on the basement faults resulted in basement upthrusts, stretched cover folds and flexures with horizontal axes, which caused vertical movements to affect the active basin margins. The resulting syntectonic unconformities will differ from those which are formed when horizontal movements on strike-slip faults directly affect the syntectonic alluvial fan deposits. In such cases, syntectonic facies are affected by drag folds and progressive to angular syntectonic unconformities with strongly plunging to nearly-vertical rotation axes developed (as in the case-study described by Duée *et al.*, 1979).

The progressive unconformities observed in the eastern Ebro Basin are restricted to the marginal areas of the basin and they die out in the not very distant inner zones of the basin, where the sequences are continuous and not affected by synsedimentary deformation and erosion. The time span during which an unconformity was developed could be very short. According to the palaeontological data provided by the Tertiary sequences around the Sant Miquel del Montclar area (Fig. 6) the progressive unconformity observed there was developed during the latest Priabonian, that is in about 2 Myr. This quick development of the structures suggests that progressive unconformities may record strong, quick local uplifts.

Some general inferences can be established from the local arrangement and regional distribution of progressive unconformities recorded in the eastern Ebro Basin margins. Single or isolated composite progressive unconformities occur at basin margin sectors which have experienced a relatively simple tectonic evolution, in general clearly dominated by uplifting (Figs 9-II, III and IV). On the other hand, assembled composite progressive unconformities, lo-

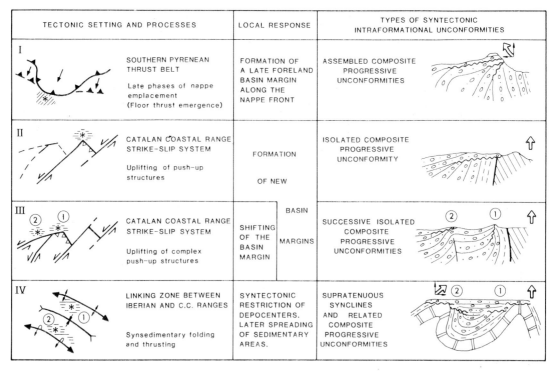

TECTONIC SETTING AND PROCESSES	LOCAL RESPONSE	TYPES OF SYNTECTONIC INTRAFORMATIONAL UNCONFORMITIES	
I SOUTHERN PYRENEAN THRUST BELT Late phases of nappe emplacement (Floor thrust emergence)	FORMATION OF A LATE FORELAND BASIN MARGIN ALONG THE NAPPE FRONT	ASSEMBLED COMPOSITE PROGRESSIVE UNCONFORMITIES	
II CATALAN COASTAL RANGE STRIKE–SLIP SYSTEM Uplifting of push–up structures	FORMATION OF NEW	ISOLATED COMPOSITE PROGRESSIVE UNCONFORMITY	
III CATALAN COASTAL RANGE STRIKE–SLIP SYSTEM Uplifting of complex push–up structures	SHIFTING OF THE BASIN MARGIN	BASIN MARGINS	SUCCESSIVE ISOLATED COMPOSITE PROGRESSIVE UNCONFORMITIES
IV LINKING ZONE BETWEEN IBERIAN AND C.C. RANGES Synsedimentary folding and thrusting	SYNTECTONIC RESTRICTION OF DEPOCENTERS. LATER SPREADING OF SEDIMENTARY AREAS.	SUPRATENUOUS SYNCLINES AND RELATED COMPOSITE PROGRESSIVE UNCONFORMITIES	

Fig. 9. Summary of main conclusions.

cally stacked and deformed occur at basin margin sectors where successive, well marked, tectonic pulses have taken place involving a slight horizontal component of movement (Fig. 9-I and III-2). The abrupt lateral transitions from angular to progressive syntectonic unconformities, observed in some basin margin sectors, emphasize the local character of the tectonic movements recorded by progressive unconformities.

In basin margins related with active thrust belts (as the Pyrenees) only the progressive unconformities linked to the outer, late foreland basin margins (developed during floor thrust emergence and the rising of related uplifting structures) may have a noticeable potential of preservation (Fig. 3). Earlier foreland basin margins will be usually concealed by main thrust sheet emplacements and the progressive unconformities linked to these earlier margins would not be preserved or severely obscured (Fig. 3).

Diachroneity in the development of several progressive unconformities along the sole thrust tip line may enable one to evaluate nappe movements during late emplacement phases.

The diachronous formation of new basin margins, or the limited shifting of pre-existing ones, both related with the movement on strike-slip faults (as in the Catalan Coastal Range), can also be recorded by the development of progressive unconformities (Fig. 9-II and III). The overall linear distribution of the syntectonic unconformities along the strike-slip system can be modified by the general arrangement of the main faults.

In basin margins made up by zones of cover folding and limited thrusting (as the Iberian Range and the Linking Zone) synsedimentary folding results in a progressive restriction of the sedimentary areas followed by later spreading after cessation of deformation. Where there is not a simple linear arrangement of the involved folds and thrusts a complex spatial distribution of the progressive unconformities may result. This can make the accurate establishment of basin margin evolution difficult.

In short, spatial and temporal interrelationships between the progressive unconformities recorded along any basin margin will reveal the overall dynamics and timing of its tectonic evolution. The accuracy of this kind of analysis will depend on the potential of preservation of these tectonosedimentary features and their dating.

ACKNOWLEDGMENTS

We would like to thank Dr Muñoz for helpful comments on Pyrenean Geology and Dr P. A. Allen for critically reading the manuscript and providing helpful suggestions.

REFERENCES

ALLEN, P.A., CABRERA, L., COLOMBO, F. & MATTER, A. (1983) Variations in fluvial style on the Eocene-Oligocene alluvial fan of the Scala Dei Group, SE Ebro Basin, Spain. *J. geol. Soc. London*, **140**, 133–146.

ANADÓN, P. (1978) El Paleógeno continental anterior a la transgresión Biarritziense (Eoceno medio) entre los rios Gaià y Ripoll (Provs. Tarragona y Barcelona). *Est. Geol.* **84**, 431–440.

ANADÓN, P., COLOMBO, F., ESTEBAN, M., MARZO, M., SANTANACH, P. & SOLÉ-SUGRAÑES, L. (1979) Evolución tectonoestratigráfica de los Catalánides. *Acta Geol. Hispánica*, **14**, 242–270.

ANADÓN, P., CABRERA, Ll., GUIMERÀ, J. & SANTANACH, P.F. (1986) Paleogene strike-slip deformation and sedimentation along the Southeastern margin of the Ebro Basin. In: *Strike-Slip Deformation, Basin Formation and Sedimentation* (Ed. by K. T. Biddle & N. Christie-Blick). *Spec. Publ. Soc. econ. Paleont. Mineral.* **37**, 303–318.

ASHAUER, H. & TEICHMULER, R. (1935) Die variscische und alpidische gebirgsbildung cataloniens. *Abb. Gassells, Wiss. Gottingen. Math-Phys.* III. **16**, 78 pp.

COLOMBO, F. (1980) *Estratigrafía y sedimentología del Terciario inferior continental de los Catalánides.* Unpublished Ph.D. Thesis, University of Barcelona.

DERAMOND, J., FISHER, M., HOSSACK, J., LABAUME, P., SEGURET, M., SOULA, J.C., VIALLARD, P. & WILLIAMS, G. (1984) Thrusting and deformation. *Field Guide of conference trip to the Pyrenees.* Paul Sabatier University, Toulouse, France.

DUÉE, G., HERVOUET, Y., LAVILLE, E., LUCA, P. & ROBILLARD, D. (1979) L'accident nord moyen-atlasique dans la région de Boulemane (Maroc) une zone de coulissement synsédimentaire. *Ann. Soc. Geol. Nord. Lille,* **98**, 145–162.

FERRER, J. (1971) El Paleoceno y Eoceno del Borde suroriental de la depresión del Ebro (Cataluña). *Mem. Suisses Paléont.* **90**, 1–70.

GONZÁLEZ, A., PARDO, G., VILLENA, J. & PÉREZ, A. (1984) Estratigrafía y Sedimentología del Terciario de la Cubeta de Alloza (Prov. de Teruel). *Bol. Geol. Min.* **95**, 407–428.

GUIMERÀ, J. (1984) Paleogene evolution of deformation in the north-eastern Iberian peninsula. *Geol. Mag.* **121**, 413–420.

JULIÀ, R. & SANTANACH, P. (1984) Estructuras en la salbanda de falla paleogena de la falla del Vallès-Penedès (Cadenas Costeras Catalanas): su relación con el deslizamiento de la falla. *Primer Congr. Esp. Geol.* **1**, 47–59.

JULIVERT, M., FONTBOTÉ, J.M., RIBEIRO, A. & CONDE, L. (1974) *Mapa Tectónico de la Península Ibérica y Baleares.* IGME.

LLOPIS, N. (1947) *Contribución al conocimiento de la Morfoestructura de los Catalánides.* Pub. CSIC.

MEGIAS, A.G. & POSADAS, M. (1981) Precisiones sobre la colocación del Manto del Pedraforca (Pirineo Oriental, España). *Est. Geol.* **37**, 221–225.

MIALL, A.D. (1978) Tectonic setting and syndepositional deformation of molasse and other non marine-paralic sedimentary basins. *Can. J. Earth Sci.* **15**, 1613–1632.

MUÑOZ, J.A., MARTÍNEZ, A. & VERGES, J. (1986) Thrust sequences in the Spanish Eastern Pyrenees. *J. struct. Geol.* **8**, 399–405.

RIBA, O. (1976a) Syntectonic unconformities of the Alto Cardener, Spanish Pyrenees: a genetic interpretation. *Sediment. Geol.* **15**, 213–233.

RIBA, O. (1976b) Tectogenèse et Sédimentation: deux modèles de discordances syntectoniques pyrénéennes. *Bull. Bur. Rech Géol. min.* 2ème sér. Sect. I, **4**, 387–405.

SEGURET, M. (1970) *Etude tectonique des nappes et séries décollées de la partie centrale du versant Sud des Pyrénées. Caractère synsédimentaire, rôle de la compression et de la gravité.* Thesis, University of Montpellier.

SIMÓN GÓMEZ, J.L. (1980) Estructuras de superposición de plegamientos en el borde NE de la Cadena Ibéria. *Acta Geol. Hisp.* **15**, 137–140.

SOLÉ-SUGRAÑES, L. (1978) Gravity and compressive nappes in the Central Southern Pyrenees (Spain). *Am. J. Sci.* **278**, 609–637.

Spec. Publs int. Ass. Sediment. (1986) **8**, 273–291

Late Cenozoic tectonics and sedimentation in the north-western Himalayan foredeep: I. Thrust ramping and associated deformation in the Potwar region

GARY D. JOHNSON*, ROBERT G. H. RAYNOLDS†
and DOUGLAS W. BURBANK‡

* *Department of Earth Sciences, Fairchild Science Center, Dartmouth College, Hanover, New Hampshire 03755, U.S.A.; † AMOCO Production Company, P.O. Box 3092, Houston, Texas 77253, U.S.A.; ‡ Department of Geological Sciences, University of Southern California, Los Angeles, California 90089-0741, U.S.A.*

ABSTRACT

The deformed proximal margin of the Himalayan foreland basin of northern India and Pakistan contains a sequence of over 3000 m of Neogene and Quaternary clastic rocks known as the Siwalik Group. These rocks record the influence of external tectonic controls exerted by the adjacent Himalayan orogenic belt together with internal, syndepositional structural controls emanating from within the foreland basin itself. The application of a magnetic polarity stratigraphy to constrain the chronology of sediment accumulation allows for the documentation of the onset, duration, and termination of various structural events preserved in the sedimentary record with a precision which may approach ~20,000 to ~50,000 yr. Analysis of magnetostratigraphies from more than 20 sites across the Potwar Plateau and adjacent regions of the north-western Himalayan foredeep define two intervals of major deformation, both of which are associated with minor precursor and post-deformational sedimentologic changes. The first event initiates in the Miocene and culminates in the mid-Pliocene between 4·5 and 3·5 Myr BP. This event is manifested by strong folding, uplift and rotation in areas adjacent to the present Salt Range. The second deformational interval spans 2·1–1·6 Myr BP and preserves a record of massive deformation throughout much of the Potwar region. This disruption is interpreted to be a response to progressive stress accumulation and paroxysmal release due to the overriding of a major basement fault by the Salt Range detachment.

INTRODUCTION

The record of NW Himalayan tectonism is preserved in an extensive apron of late Palaeogene to Quaternary sedimentary succession exposed in the outer Himalayan foothills and adjacent Indo–Gangetic Plain of northern Pakistan and India (Gansser, 1964). The onset of molasse sedimentation is a response to initial orogenic activity in the Himalaya, and is represented by the mixed fluvial/deltaic and near-shore marine facies of the Murree/Dharamsala/ Dagshai/Kasauli Formations of the lesser Himalaya. These Eocene to Oligocene and early Miocene facies represent the initial clastic record of the development of intracontinental subduction which began in the Mid-Eocene due to plate collision. Subsequently, progressive outward (southward) displacement of the zone of Himalayan deformation has resulted in motion within the Main Boundary Thrust (MBT) zone and

the resulting confinement of a zone of late (syn-) orogenic fluvial facies, the Siwalik Group, to a deformed foredeep margin and evolving foreland basin primarily to the south of the zone of deformation (Fig. 1).

Seismic reflection data from south of the MBT zone in Pakistan has confirmed the existence of salt-constrained detachment development under the greater part of the Potwar Plateau and the presence of sub-thrust basement faults (Lillie *et al.*, 1985). The latter appear to have acted as stress concentrators during the late Neogene and Quaternary. This paper addresses the implications of this evidence in conjunction with a detailed consideration of the magnetic polarity stratigraphy of a number of localities exposing the Siwalik Group. It explores their bearing on the interpretation of deformational events and sedimen-

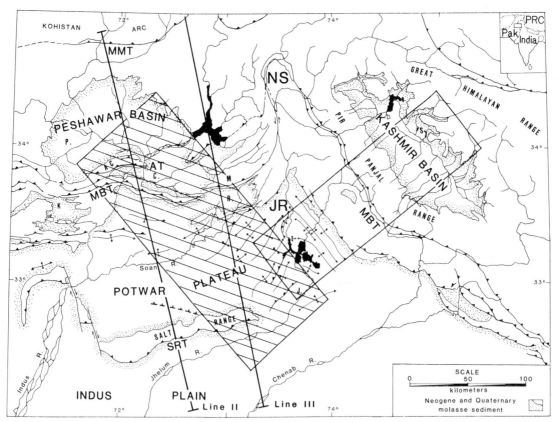

Fig. 1. Schematic outline of major structural features of the Himalayan foreland basin, northern Pakistan and India. Area covered in this report, outlined and cross-hatched; companion paper (Burbank *et al.*, this volume) outlined. Location of schematic structure sections (Fig. 2) are shown. Symbols as follows: AC, Attock–Cherat Range; AT, Attock Thrust; C, Campbellpore (now Attock); H, Hirpur; J, Jhelum; JR, Jhelum Re-entrant; K, Kohat; M, Margala Hills; MBT, Main Boundary Thrust; MMT, Main Mantle Thrust; NS, NW Syntaxis; P, Peshawar; R, Rawalpindi; S, Srinagar; SRT, Salt Range Thrust.

tologic responses during the latest Neogene and Quaternary within the proximal Himalayan foredeep and adjacent intermontane basins lying primarily to the west of the Jhelum River. In a companion paper (Burbank *et al.*, this volume), the region to the east of the Jhelum River and Northwest Syntaxis (Fig. 1) are analysed and the data from the entire region are synthesized.

STRUCTURAL ELEMENTS OF THE NORTHERN PUNJAB AND ADJACENT AREAS

Faulting in the Punjab foreland

It has long been recognized that exposures of the Late Tertiary molasse sequence of the Himalaya are restricted almost entirely to the first rugged foothills rising to the north of the Indo–Gangetic plain. In India, this Tertiary Belt is often quite narrow, ranging from 50 to less than 5 km in width. In the Potwar Plateau and related Salt Range of Pakistan, the deformed belt attains unusual widths of over 100 km. The belt is terminated on its northern side by the MBT zone which carries early Tertiary or Mesozoic rocks of the Hazara Himalaya (Gansser, 1964; Fuchs, 1975) over the younger Siwalik molasse sequence. The southern limit of molasse exposure is defined by the outer limit of compressional deformation induced by the plate convergence in the vicinity of the Salt Range.

The Salt Range and Potwar Plateau (Fig. 2) are the result of a shallow dipping thrust that has carried an entire post-Cambrian stratigraphic sequence southwards along a detachment localized within the

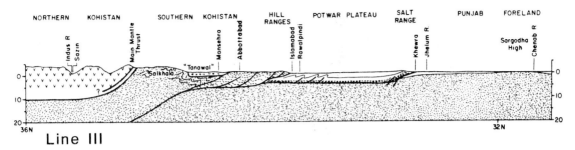

Line III

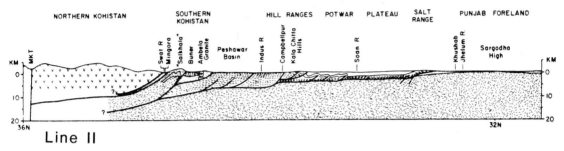

Line II

Fig. 2. Schematic cross-sections of the foreland margin of northern Pakistan. See Fig. 1 for location of sections (modified from Yeats & Lawrence, 1984).

Cambrian-aged Salt Range Formation. This structural interpretation is founded on early work in the Salt Range by Wynne (1878), Cotter (1933), Wadia (1945a, b) and Gee (1945, 1947, 1980). That the entire Potwar Plateau is allochthonous was first postulated in a cross-section by Cotter (1933) and later discussed by Voskresenskiy (1978). Cotter's views have been supported by recent data which allow for the interpretation that active deformation is taking place with very little seismic response (Seeber & Armbruster, 1979). This aseismic deformation appears due to aseismic slippage within a ductile detachment surface primarily constrained within the salt horizon. Recent evaluation of reflection seismic data from the central Potwar Plateau has further confirmed the presence of an extensive salt basin (the Salt Range Formation) at the stratigraphic base of the Phanerozoic section underlying large portions of the central and southern Potwar, as well as occurring slightly south of the Salt Range itself (Lillie & Yousuf, 1986; Lillie *et al.*, 1985).

The Salt Range and Potwar Plateau are bounded on the east by the Jhelum River, where thrusting diminishes, and an increasing proportion of the compressional deformation is taken up in a series of frontal monoclines, simple folds, blind thrust-cored folds and thrust-cored folds. Whereas in the central and western reaches of the Plateau, the Soan syncline

broadens and occupies much of the allochthonous sheet (see, e.g. cross-section by Pilgrim, 1913 and Fig. 2), to the east, the Potwar Plateau sediments become more intensely folded until at the Grand Trunk Road in the easternmost Potwar Plateau, the entire Siwalik outcrop is thrown into folds (Figs 3 and 4). This phenomenon, likely a response to thinning of the Salt Range Formation at depth, is an expression of the repeatedly upward cutting of the detachment slip surface, resulting in the frequent thrust-cored anticlines encountered along this trend. Similar relationships constrained by salt basin-boundary conditions have been reported elsewhere and have been shown to grossly affect the style of deformation in the supra-thrust sheet (Davis & Engelder, 1985).

There is also a systematic change in structural style as one proceeds from north to south across the Potwar Plateau. Several writers (Pinfold, 1919; Cotter, 1933; Gill, 1951b) have described the transition from steeply faulted, isoclinally folded beds in the north (generally north of the Soan River), through a zone of less steep faulting and tight folds to a realm of gentle folds in the south. In general, examination of the variation in and style of structural development of the Potwar Plateau (Gill, 1951a, b; Martin, 1962) has been concerned with the analysis of folding patterns defined by concentric folding. Martin (1962), in a series of

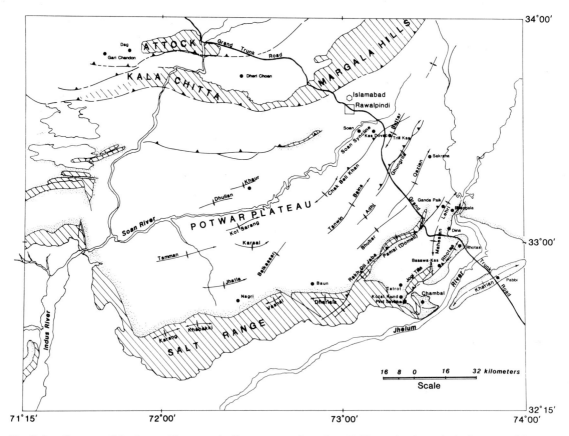

Fig. 3. Locality map of the Potwar Plateau and adjacent areas of northern Pakistan showing principal structural features discussed in text. Location of palaeomagnetic sections utilized as data base (see Fig. 5) are indicated (closed circles).

sections across several structures of the Potwar including the Lehri–Mahesian folds (Fig. 3), points out the interrelationship of folds and thrust faults in describing faults that are warped by subsequent folding. This may explain the anomalous, steeply dipping faults reported in various structures of the eastern Potwar (Gill, 1951b). Martin sees this deformation to be the result of a continuous application of compressive stress, with the rock response varying according to the stress trajectory and the stage of deformation attained by the structure.

Basement faulting

Voskresenskiy (1978) pointed out the likelihood of pre-existing basement faulting playing an active role in controlling the behaviour of the Potwar detachment. At the time, however, the nature of the presumed ramping was not defined. Reflection seismic data recently released by the Pakistan Oil and Gas Development Corporation (OGDC) (Lillie & Yousuf, 1986; Lillie *et al.*, 1985) now demonstrate clearly the character of this thrust geometry. Several north-dipping normal faults affect the sub-thrust basement in the areas of the central Potwar Plateau and Salt Range south of the Soan River (Lillie, 1986). At least one of these (Figs 1 and 2) has exerted a major control on detachment behaviour, and it perhaps ultimately constrained the greater Potwar Plateau during initial compressive deformation until such time as a ramping episode occurred resulting in the Salt Range over-thrust. The influence of basement faults on the fold/thrust geometry in the eastern Potwar Plateau has not been fully evaluated. We hypothesize that due to a diminution in the thickness of the lubricating Salt Range Formation, an increase in the frictional component of the thrust along the slip surface results in the frequent upthrusts and the observed tightening of deformational style (Fig. 4).

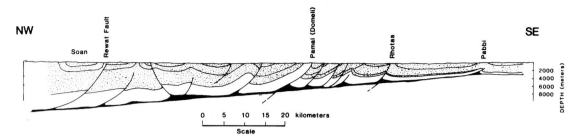

Fig. 4. Schematic structural cross-section along the Grand Trunk Road traverse, eastern Potwar Plateau. See Fig. 3 for location of traverse. Salt Range Formation (black); pre-orogenic platform sediments (white); syn-orogenic clastics, including Murree and Siwalik Groups (stippled).

The nature of basement control on foreland sequences elsewhere (Wiltscko & Eastman, 1983) and the influence of associated salt basin geometry on fold/thrust geometry (Davis & Engelder, 1985; Jenyon, 1985) provide evidence that the greater Potwar Plateau/Salt Range detachment represents an interplay between two rather complex mechanisms. Further evaluation of this relationship is warranted in so far as a genetic relationship may be shown to exist between the normal faulting in the foreland basement and the distribution of the Salt Range Formation in the late Proterozoic/early Cambrian.

A discussion of the timing of events associated with the deformation of the Himalayan foreland basin and the Salt Range ramping event(s) is developed herein. In it we explore the time-constrained stratigraphic record and use it to analyse both the general pattern of structural disruption, as well as the precise timing and consequences of a ramping event.

SEDIMENTARY DYNAMICS OF THE NORTHERN PAKISTANI FOREDEEP MARGIN

Sedimentary dynamics of molasse deposition

Peshawar Basin

To the north of the Main Boundary Thrust, a series of intermontane basins are developed on allochthonous bedrock terranes (Figs 1 and 2). From the Peshawar Basin in the NW across the Campbellpore Basin to the overthrusted northern margin of the Potwar Plateau, these intermontane basins become progressively smaller and more abbreviated in their stratigraphic record to the south. Magnetostratigraphic data in combination with the physical stratigraphy at a number of dated locations permit us to delineate details in the sequence of basin development. Although we cannot at present specify the timing of the initiation of thrusting in areas immediately adjacent to the Peshawar Basin, we can nevertheless use magnetostratigraphic data from synorogenic clastic facies within the basin to provide a chonologic constraint on the cessation of thrusting along the southern margin of the basin. This serves to constrain the history of movement along the Main Boundary Thrust zone in the Trans–Indus regions.

Two sites within the Peshawar basin provide these data (Burbank & Tahirkheli, 1985): the Dag section (Fig. 3) lies adjacent to a thrust ridge(?) of severely deformed Palaeozoic(?) slates and early Tertiary carbonates (Tahirkheli, 1970), whereas the Garhi Chandan sequence rests unconformably on strongly folded and erosionally truncated Murree strata of Late Oligocene to Early Miocene age. The ~3 Myr age of onset of synorogenic clastic deposition in both stratigraphic sections (Fig. 5) indicates that major structural deformation had shifted out of the basin interior at this time. In each section, initial low-relief floodbasin or lacustrine facies are displaced basinward by an influx of fanglomerates shed centripetally off the basin margin. Between 2·6 and 1·6 Myr these fan sediments prograded northwards across the southern basin margin at an average rate of 2 cm yr^{-1}. We interpret these chronologic and stratigraphic data as indicating that: (1) thrusting and major uplift of the Attock–Cherat Range (and segments of the Main Boundary Thrust zone in this area) occurred just prior to 3 Myr ago; (2) this uplift impounded the pre-existing southerly flowing river systems; and (3) the resultant low-energy sedimentation was displaced to the north by the gradual influx of fanglomerates from the uplifted southern basin margin.

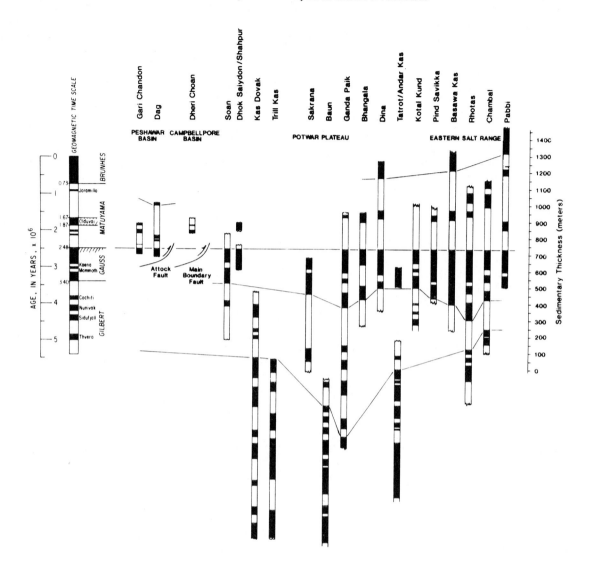

Fig. 5. Correlation of local magnetic polarity sections from the Peshawar and Campbellpore Basins, the Potwar Plateau and Eastern Salt Range regions of northern Pakistan with the magnetic polarity time-scale. See Fig. 3 for location of sections. Normal polarity intervals (black); reversed polarity intervals (white). Data from numerous sources (see text).

Attock Range

In the Attock–Cherat Range (Fig. 3) along the southern margin of the Peshawar basin and in isolated nearby inliers, exposures of the Oligocene and Early Miocene-aged Murree Formation are well developed

(Meissener *et al.*, 1974). These terrestrial redbeds represent an early stage of molasse deposition following the initiation of intracontinental subduction along the Main Mantle Thrust of the Hindu Kush and Kohistan Himalaya in the earlier Cenozoic. In the Cherat Range, the Murrees and the underlying carbonates of Eocene age have been complexly deformed by a series of faults and low-angle thrusts (Burbank, 1983; Burbank & Tahirkheli, 1985). This deformation in combination with the absence within the Peshawar basin of molasse facies equivalent to Lower and Middle Siwalik strata of middle and late Miocene age suggest that major tectonic disruption of this terrane occurred in the vicinity of the Peshawar basin between 5 and 10 Myr, a timing which may help to explain the relatively source-proximal, sandy-bedded, braided stream environments characteristic of the type Nagri Formation *sensu strictu* of the Siwalik Group in the central and western Potwar Plateau further to the south.

Campbellpore Basin and Kala Chitta Range

Major movement along the Main Boundary Thrust zone and the development of the Campbellpore basin (Figs 1 and 2) to the south are well constrained by chronologic data. The local basement of Eocene Nummulitic limestones is strongly deformed, and there are no known exposures of Miocene and early Pliocene-aged Murree and Siwalik molasse within the basin. On the basis of two fission-track dates (G. D. Johnson *et al.*, 1982; Burbank, 1982) and the local magnetic polarity stratigraphy, the oldest exposed intermontane sediments in the Campbellpore basin (Dheri Choan, Fig. 5) date from ~1·8 Myr (Burbank, 1982, G. D. Johnson *et al.*, 1982). Although no large, stable lakes appear to have existed within the basin, the sedimentation pattern is interpreted as resulting from a sluggish, meandering fluvial system (Bloch, 1981) with transient floodbasin lacustrine deposits several kilometers in diameter that were frequently interstratified with fluvial deposits. The intermontane nature of deposition is emphasized by this depositional pattern which is typical of ponded systems found in basins with actively rising flanks (Miall, 1979), rather than of the braided systems that characterized the external molasse of the Himalaya. The record of uplift and deformation of the Kala Chitta Range (Fig. 3), the eastern extension of the Cherat Range, is essentially a record of the history of movement of the Main Boundary Thrust in this region bordering the northern Potwar Plateau.

Khairi Murat Range and associated terrain

The area of the northern Potwar Plateau to the south of the Kala Chitta Range and the MBT zone

and north of the Soan Syncline is dominated by complexly folded and faulted rocks of Eocene, Oligocene and Early Miocene age (the Nummulitic limestones, Murree and Lower Siwalik formations). Characterized by often vertical dips, thrust-bounded ridges and tight isoclinal folding, the terrane represents a deeper level of structural exposure than is found to the south in the Potwar. Seismic data (Lillie *et al.*, 1985) shows poor reflector characterization making difficult the resolution of thrust/detachment geometry in this region. Because no well developed late Neogene and Quaternary molasse is exposed in this structural block in the central and eastern Potwar (with the exception of the Lei Conglomerate—see below), our interpretation of the timing of structural deformation in this zone must be inferred from adjacent terrain.

Soan Syncline

The Soan Syncline is the first major structural flexure encountered south of Rawalpindi. This syncline broadens and flattens to the SW and is the major structural feature of the central Potwar Plateau (Figs 3 and 4). The syncline is a flat bottomed, SW plunging fold with a steep north-western flank and a more shallow south-eastern flank. Moragne (1979) measured and interpreted the magnetic polarity stratigraphy of a steeply dipping 650 m section near the north-eastern nose of the syncline in the vicinity of the Grand Trunk Road (Soan section), and he determined that the exposed rocks range in age from the middle Gilbert to lower Matuyama Chron (Fig. 5). Moragne's interpretation of the magnetic polarity stratigraphy is supported by a fission track (zircon) date of 1·9 ± 0·4 Myr on a volcanic ash near the polarity reversal interpreted as the Gauss/Matuyama boundary (Fig. 6).

Further SW, on the same flank of the syncline, an unconformable and flat-lying ash has been dated by the fission track method (zircon) at 1·6 ± 0·22 Myr (G. D. Johnson *et al.*, 1982). Exposed in portions of the central Soan Syncline and overlapping on to its north-western margin are the Lei conglomerates, a Siwalik formation which overlies the flat-lying ash and which is consanguinous with deformation of the Main Boundary Thrust zone to the north and of the bulk of the foredeep margin of the northern Potwar Plateau (Fig. 6). The two ash beds discussed above and their associated magnetic polarity stratigraphies constrain the timing of deformation of this fold to within the interval from 2·1 to 1·9 Myr BP (Raynolds & Johnson, 1985). Evidence from additional palaeo-

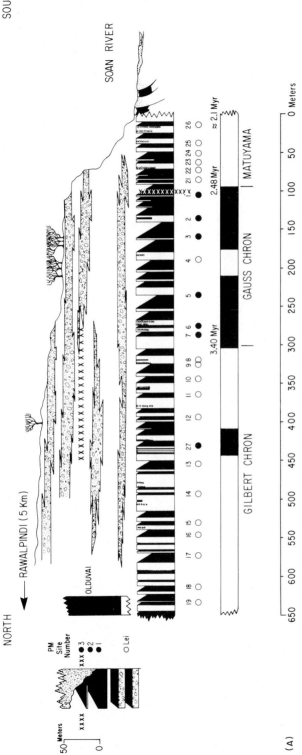

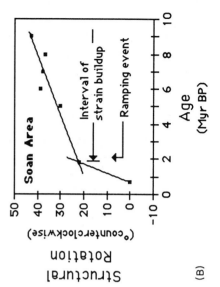

Fig. 6. (A) Schematic representation of the Soan Syncline sequence 2 km south of Rawalpindi showing evidence of termination of deposition, folding, truncation, and renewed deposition of upper Siwalik Group sediments between approximately 2·1 and 1·8 Myr BP. Modified from Raynolds & Johnson (1985). (B) Amount of structural rotation of Siwalik Group sediments determined for rocks of known age from the Soan area (Kas Dovak; Dhok Sayidan; Soan localities). Data from Raynolds (1980).

magnetic considerations suggests that truncation of the Soan sequence and deposition of the Lei Conglomerate was followed by minimal deformation subsequent to the Olduvai magnetic polarity subchron.

Chak Beli Khan Anticline

The Chak Beli Khan anticline (Fig. 3) is a small fold that defines the gently dipping southern margin of the Soan Syncline in the central Potwar Plateau. Good exposures of sandstones which characterize the ancestral Indus depositional system in the Middle Siwalik formations (Raynolds, 1981) occur in the core of this fold. The south-eastern flank of this structure, steepened and in places fractured and overturned, is expressed along strike to the NE as the Riwat Fault. In the SW, the structure does not expose rocks of Pliocene–Pleistocene age and does not provide palaeomagnetic data useful to the interpretation of Late Neogene/Quaternary deformational events.

Along the NW flank of the Chak Beli Khan structure, Late Neogene rocks are well displayed along Kas Dovac where over 3000 m of Siwalik strata are exposed (Raynolds, 1980). This structural flank is truncated by the Riwat Fault (Wadia, 1928), which places Lower Siwalik sediments over Middle Siwalik formations near the town of Riwat (Fig. 3). As a classic example of a thrust fault dwindling in throw and merging into an overturned fold, the Riwat Fault dies out to the NE near the apex of the Soan syncline, and to the SW along the north-eastern flank of the Chak Beli Khan anticline.

The data from Kas Dovac (Raynolds, 1980) suggest that the Riwat fault and the uplift associated with its overthrusting block was initially expressed between the upper Gilbert and upper Gauss chrons yielding depositional hiatus of nearly 0·7 Myr. The abrupt exclusion of limestone clasts from the supra-unconformable sequence at Kas Dovac implies that axial drainage had become established in the basin which prevented the deposition of NW-derived clasts on the SE flank of the Soan Syncline. A scenario of deformation can be evolved whereby the Riwat Fault uplifted the SE flank of the syncline sometime between 3·4 and 2·7 Myr. By 2·6 Myr the Riwat Fault had largely ceased to move, inducing only a very slight dip to the post-2·6 Myr section at Kas Dovac. Deposition continued as consequent drainage within the newly developed structure became established resulting in unusual thicknesses of fine-grained facies being deposited until about 2·1 Myr.

Tanwin/Bains Anticline

The Bains fold trends to the NE, where it is kinked into a more northerly trend (Fig. 3). This fold is dissected by tributaries of the Soan River and exposes additional ancestral Indus River facies of various Middle and Lower Siwalik formations in its core. Sandstones which characterize the ancestral Jhelum River system in various Upper Siwalik formations in the eastern Potwar Plateau (Raynolds, 1981) are only sparsely preserved on the flanks of the fold due to a relatively deep level of erosion across this structure. The synclines separating the Bain fold from the Chak Beli Khan and the Adhi–Gungrilla structures do not appear to contain extensive sequences of Upper Siwalik-aged rocks except in the structural saddle near the Grand Trunk Road. The Riwat fault truncates the fold axes of both the Bain Anticline and that of the syncline to the west. These cross-cutting relationships, which are very clear on aerial photographs, form one of the most decisive pieces of evidence supporting evidence for extensive thrust faulting within the eastern portion of the Potwar thrust sheet prior to 2·6 Myr ago.

Buttar (Mankiala) Anticline

North-east of the Bains structure and across the Mandra structural saddle, the *en echelon* Buttar fold (Fig. 3) is developed. This fold was investigated by Small (1980) and preserves on its western flank over 1000 m of ancestral Jhelum River sandstones and mudstones above the characteristic ancestral Indus River sandstones of the Middle Siwaliks. The Buttar fold trends NE and is truncated by the Kahuta Fault. East of the Buttar fold, in the syncline separating it from the Gungrila fold, a thick lens of quartzite conglomerate is preserved. This conglomerate facies forms gently rounded resistant hills and is a characteristic facies found associated with many of the synclines in the eastern Potwar. Magnetostratigraphic studies by Small (1980) indicate that the Buttar fold had not yet undergone deformation by 5·5 Myr ago as this time is characterized by rivers coursing freely across the site of the structure. Rocks younger than 5·5 Myr are not preserved along the west flank of this structure.

Adhi–Ghungrila Anticline

The Adhi structure (Fig. 3) is developed entirely in Middle and Lower Siwalik formations, and no ancestral Jhelum River sandstones of Upper Siwalik age are found on either of its flanks. This structure has proven hydrocarbon reserves (determined by PPL and AMOCO). To the north, the Ghungrila structure is flanked by a northward thickening sequence of Upper Siwalik rocks. A 600 m section was measured and sampled near the village of Pandori on the western flank of the Ghungrila fold (Fig. 3) (Raynolds, 1980). In this area, over 1300 m of ancestral Jhelum River sandstones of Upper Siwalik age overlie ancestral Indus River sandstones of Middle Siwalik age. On the northern end of the Ghungrila anticline (at about 33·5° latitude) a distinct change in sandstone coloration takes place. Brown coloured, Jhelum facies sandstones of the Upper Siwalik sequence are gradually replaced along strike by pale grey-green coloured sandstones that resemble more closely white sandstones of Indus-facies type common in Middle Siwalik formations. On the eastern flank of the structure, in the vicinity of the Grand Trunk Road, a small thickness of quartzite conglomerates is preserved, dipping towards the town of Gujar Khan. These conglomerates are approximately 1000 m above the Indus facies sandstones exposed in the anticlinal core.

Qazian Anticline

East of Gujar Khan, the poorly exposed Qazian fold occurs. This fold is largely covered by the late Pleistocene Potwar Silts on its southern end, while to the north it is flanked by massive quartzite conglomerates. This structure was recently drilled by Gulf Oil with hydrocarbon shows having been reported. We have no precise chronostratigraphic data on this structure.

Jabbar Anticline

The Jabbar structure is a short anticline with a box-fold termination on its southern edge. To the north, this fold is truncated by the Jhelum Thrust (Fig. 3) (Wadia, 1928). An 800 m section was measured and sampled near the village of Sakrana on the western flank of the fold (Raynolds, 1980) (Fig. 5). The top of this section encounters thick quartzite conglomerates which are dated as occurring in the middle portion of the Gauss Chron. These conglomerates are inferred to have been transported across the site of the future structure at that time. The surface relief of this structure thus dates from less than about 3 Myr. Below the conglomerates, the sandstones are transitional between the typical Jhelum-facies sandstones of the Upper Siwalik formations and the paler white

sandtones of Indus-facies type described from the northern end of the Ghungrila structure. The Jhelum Thrust carries gently dipping Middle Siwalik rocks westward over the Jabbar structure. This is the locus of the change in structural strike associated with the Jhelum structural re-entrant (Visser & Johnson, 1978).

Baun Basin

A 1700 m section of Siwalik sediments exposed along the Sauj Kas, 10 km south of the central Potwar town of Chakwal and along the north flank of the Salt Range has been studied, and a palaeomagnetic stratigraphy developed which spans a nearly 5 Myr record from *c.* 9·5 to 4·0 Myr BP (Opdyke *et al.*, 1979, 1982; N. M. Johnson *et al.*, 1982) (Figs 3 and 5). This site, near the eastern termination of one of the principal sub-thrust normal faults under the Potwar detachment (Lillie *et al.*, 1985; Lillie & Yousuf, 1986), records a differential rotation of pre-Gilbert Chron Dhok Pathan Formation and Gilbert chron-aged Upper Siwalik sediments (Opdyke *et al.*, 1982). In addition to containing coarse clastics representing both Eocene-aged carbonates and late Palaeozoic Talchir Formation clasts of Salt Range-derived fanglomerates beginning about 4·5 Myr BP, the entire earlier Siwalik section has been rotated up to 35° counter-clockwise. Younger sediments of apparent Gilbert Chron age from this same site are essentially unrotated. As a result, it can be interpreted that the rotation of the detachment in the Sauj Kas region was completed by the Gilbert Chron (Opdyke *et al.*, 1982). The rocks of this sequence can be traced well to the east to just north of the western Pamal (Domeli) Ridge.

Pamal (Domeli) Anticline

To the east of the Baun Basin exposures and to the south of the Jabbar fold, the Pamal (Domeli) Ridge (Fig. 3) rises as a major asymmetrical thrust-faulted anticline overturned towards the south. This fold exposes Eocene-aged carbonates in its core. Northwards, the fold is offset by cross-cutting faults of small throw and finally merges with the Jabbar fold. In its southern extension, the Domeli ridge overthrusts the adjacent syncline to the SE. Within the syncline, beds become more conglomeratic along strike as the thrust trace is approached. Further south, the thrust overlies its own debris. This structure was examined in the Ganda Paik section where a 1500 m section was measured (Raynolds, 1980) (Fig. 5). The section displays a marked facies change with the exclusion of thick channel sandstones and the advent of conglomeratic stringers. Those conglomerates containing derived Nummulitic debris occur at about 2·5 Myr. This is interpreted to be the time at which Eocene strata began to be unroofed along the nearby portion of the Domeli Ridge, and this detritus was shed into the adjacent syncline, a process continuing today. The structural setting was such that the large rivers draining the northern mountains no longer flowed across this site after 2·5 Myr. This exclusion of the ancestral Jhelum River system is a manifestation of the confinement of the river course into the Jhelum re-entrant axis to the east of Ganda Paik.

Kotal Kund Syncline and Mount Jogi Tilla

One of the most significant records of Pliocene–Pleistocene sedimentation events in the eastern Salt Range occurs in the Kotal Kund syncline, occupied in its greater extent by the Bunha River (Fig. 3). Here three sections have been studied (Opdyke *et al.*, 1979; Frost, 1979; G. D. Johnson *et al.*, 1982; N. M. Johnson *et al.*, 1982) which provide evidence of uplift and truncation of a portion of the Middle Siwalik Dhok Pathan Formation sometime after 4·6 Myr BP and before the deposition of the Upper Siwalik Tatrot beds of Gauss age (Tatrot–Andar Kas locality) on the northern flanks of the Mount Jogi Tilla structure. Jogi Tilla is a complex of four recumbent, south-verging thrust folds which expose most of the Phanerozoic section present in the eastern Salt Range (Wynne, 1870; Pascoe, 1930; Fermor, 1931; Gee, 1980).

To the west of the Jogi Tilla lies the Kotal Kund syncline which continued to receive sediments until about 1·6 Myr BP (Frost, 1979) at which time the major centre of deposition shifted out and to the SE to the vicinity of the syncline now occupied by the combined Bunha/Jhelum Rivers. The Pind Savikka and Kotal Kund localities collectively record this event. It appears that continued deformation has taken place in the form of tilting and rotation, affecting all sites within the basin.

Mahesian–Lehri Anticline

The Lehri and Mahesian structures (Fig. 3) are a tightly joined pair of anticlines which have been described by Martin (1962). These two concentric folds are *en echelon* features that have been compressed together along a series of thrust faults. Lower Siwalik facies rocks are exposed in the core of the Mahesian fold. This fold pair is surrounded by a mantle of Upper Siwalik sediments. A section was measured on each flank: to the west at Bhangala and to the east near the town of Dina (Reynolds, 1980). The northern end of the Lehri structure is terminated in a dramatic box fold that is well exposed along the shore of Mangla Reservoir. Forming the top of the measured section near Dina on the western side of these structures, is a thick sequence of conformable quartzitic conglomerates evincing subsequent activity of the ancestral Jhelum River across the region. The section sampled at Bhangala illustrates that channel sandstones persisted at this site until just after 2·4 Myr. After this time, thin and laterally discontinuous sandstones carry angular Siwalik clasts. The implication is that this locality began being influenced by deformation about 2·3 Myr. Although separated from Ganda Paik by only a few kilometres, the channel sandstones persist about 75 m higher in the section at Bhangala. At the prevailing rate of sediment accumulation (30 cm/1000 yr), this thickness represents a lag of 250,000 yr between the onset of the deformation which affected these adjacent folds.

Rhotas Anticline

The Rhotas anticline occurs east of the Mahesian/Lehri pair of folds (Fig. 3). Although it is a simple fold on its northern end, the Rhotas structure is complex and broken by faults to the south in the vicinity of Mount Jogi Tilla (see above). A well exposed 1350 m sequence of Upper Siwalik rocks was sampled along the northern termination of this fold SW of the Grand Trunk Road (Reynolds, 1980). This sequence also preserves quartzitic conglomerates in the top part of the section (preserved on the western flank of the fold, just north of the Kahan River). The Rhotas fold is overlain by a distinctly unconformable quartzite conglomerate which sweeps across the anticlinal axis and is preserved in patches mantling the crest of the fold. The section measured on the apex of the Rhotas fold ends in rocks that are about 1·4 Myr old (Reynolds & Johnson, 1985). At that time, the Jhelum River flowed across the structure. The central part of the fold has been investigated by Opdyke *et al.* (1979) and G. D. Johnson *et al.* (1979) who describe a section measured from Basawa Kas, 12 km SW (Figs 3 and 5). This section preserves a younger sequence of rocks, allowing for a maximum age of surface expression of the Rhotas fold of 0·4 Myr to be derived.

Kharian (Pabbi) Anticline

East of the Rhotas structure, on the east side of the Jhelum River (Fig. 3), the outermost NE-trending anticline is developing. The Kharian or Pabbi fold (Wynne, 1877, 1879), mapped and studied by palaeomagnetic criteria by Keller (1975) and Keller *et al.* (1977), is a gently asymmetric anticline probably representing draping of the Tertiary section over a blind thrust. The age of surface expression for the Pabbi Hills is less than 0·4 Myr (Johnson *et al.*, 1979). While structural models imply that this fold is underlain at depth by a north-westwardly dipping blind thrust, a lateral component of compression, perhaps related to stratigraphic thinning of the Salt Range Formation at depth, has resulted in the structure being offset into three distinct segments to the SW.

SALT RANGE THRUST RAMPING

Time resolution and interpretation of magnetostratigraphies

The dynamics of sediment accumulation in an actively evolving foreland basin can be assessed utilizing time-constrained measures of synorogenic sedimentary facies behaviour (Raynolds & Johnson, 1985; Burbank & Raynolds, 1984). Whereas most palaeontologically based determinations of time in stratigraphic successions involve resolution limits of 10^6 yr, magnetic polarity stratigraphy, determined from rocks as young as the late Neogene and Quaternary, may be useful at resolution limits 10^4–10^5 yr, yielding a significant improvement in our ability to characterize the rates of geological processes (Raynolds & Johnson, 1985). When a local magnetic polarity stratigraphy is successfully correlated to the global magnetic polarity time-scale, the resultant ability to define chronostratigraphically the reversal succession at a locality is a powerful tool in describing sedimentary dynamics. We have evaluated more than 20 localities exposing Siwalik Group rocks and their palaeomagnetic stratigraphy in this report (Fig. 5) and have been able to recognize a number of relationships which allow for precise statements to be made regarding the timing of certain sedimentation-controlling tectonic events in the subsiding Himalayan foredeep. The reader is referred elsewhere to the details of the magnetic polarity analysis upon which some of our interpretations have been based (Burbank, 1982; G. D. Johnson *et al.*, 1979; N. M. Johnson *et*

al., 1982, 1985; Opdyke *et al.*, 1979, 1982; Raynolds, 1980).

The net sediment accumulation at a given depositional site has a tendency to define a sigmoidal curve when accumulated sediment thickness is plotted against time (Fig. 7). This response can be interpreted as representing the transition from low sediment-accumulation-rate conditions in the distal foredeep position to passage into maximum sediment-accumulation-rate conditions in the foredeep depocentre followed by attenuation of sediment accumulation rates in the deforming proximal foredeep position. Important to the issues raised in this paper, the record of sediment accumulation at a given site may illustrate several inflection points along the graphic plot discussed above which represent significant tectonic events affecting the site (Fig. 7) (see also Johnson *et al.*, 1979; Raynolds & Johnson, 1985). In particular, the initiation of uplift and deformation and cessation of the depositional record at a given site can be used collectively with similar data from other sites to effect an interpretation of a more regional nature. We have attempted this for the northern Pakistani foredeep.

Initial early to middle Pliocene deformational events in the southern Potwar

The previously described chronostratigraphic data serve to define the nature and timing of compressional events in the Potwar Plateau and adjacent areas of northern Pakistan. Arguments presented by N. M. Johnson *et al.* (1982) and Opdyke *et al.* (1979, 1982) based on palaeomagnetic criteria allow for the recognition of a period of deformation occurring in the southern Potwar Plateau–eastern Salt Range in latest Miocene (*c.* 8–4 Myr BP).

Within the Kotal Kund Syncline localities in the eastern Salt Range, sediment-accumulation rates diminished about 7·9 Myr BP (N. M. Johnson *et al.*, 1982) (Fig. 7) and the depositional record is truncated at about 4·6 Myr BP. Approximately 32° of counter-clockwise rotation occurred after this time. Subsequently, deposition recommenced during the mid-Gauss Chron at approximately 3·0 Myr BP (Opdyke *et al.*, 1979) (Fig. 8). We interpret this to represent the initial response to the ramping thrust of the Potwar detachment which resulted in a complex arrangement of frontal anticlines and the tip-stick thrust fronts of the Pamal (Domeli) and Jogi Tilla structures.

In the Baun Basin locality at Sauj Kas, in the central Salt Range, diminished sediment-accumulation rates began in post-Chron 9 time (∼8·5 Myr BP) (N. M.

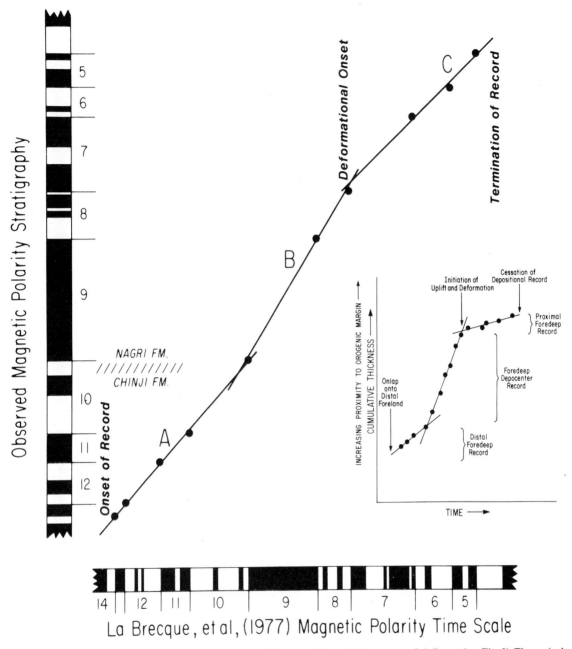

Fig. 7. Sediment accumulation record of Siwalik Group sediments at Tatrot-Andar, eastern Salt Range (see Fig. 3). Theoretical sediment accumulation record of a typical external molasse. Inflection points represent tectonically important events influencing the sedimentation record. Data from N. M. Johnson *et al.* (1982).

Johnson *et al.*, 1982) and a transition from typical Middle Siwalik facies (Dhok Pathan Formation (?)) to Upper Siwalik facies containing identifiable clasts derived from the Salt Range occurs in the medial Gilbert Chron (~4·5 Myr BP) (Opdyke *et al.*, 1979). Additionally, the Dhok Pathan beds exhibit a significant tectonic rotation of nearly 30° not present in the overlying Upper Siwaliks (Fig. 8).

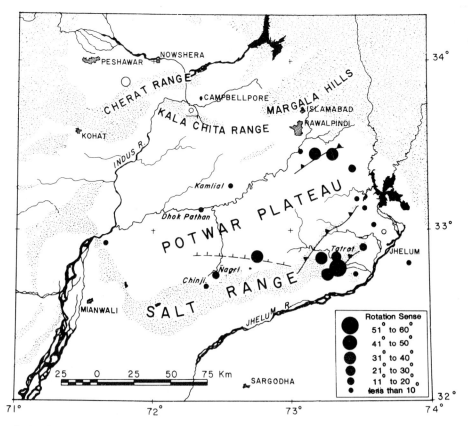

Fig. 8. Magnitude of structural rotation determined from mean of normally magnetized sites from various localities in the Potwar Plateau and adjacent areas of northern Pakistan. See Fig. 3 for locality names. Counter-clockwise rotation, solid circles; clockwise rotation, open circles.

The combined evidence from the various late Miocene events described above provides reasonable constraints on the initial motion on the Potwar–Salt Range detachment. Although now appearing to be constrained by faulting in the sub-thrust basement, this motion may be viewed as one of slippage of the Potwar detachment in the late Miocene until fault-bounded basement highs or steep ramps in the Indian foreland were encountered at some time less than 6·0 Myr ago. The main phase of this early deformation leading to both major uplift and rotation appears to be confined to an interval of 4·5–3·0 Myr ago.

The initial structural response to the encounter of the Potwar detachment with the basement highs of the foreland appears to be the buckling of portions upper thrust sheet, structural rotation of certain segments and the shedding of some of the upper stratigraphic section along the uplifted dorsal flank

and leading frontal edge of the thrust sheet. The data from the Kotal Kund area, as well as other sections bounding the Pamal (Domeli) Ridge and Jogi Tilla suggest that these features, having experienced rapid deformation in the early Pliocene, exerted some control on sedimentation patterns in the eastern Salt Range basins during latest Noegene and Quaternary. It appears that the break-up of the Salt Range in the east into a number of smaller, sub-parallel, digitate ranges (the Bakrala, Domeli and Joji Tilla, including the Chak Beli Khan Anticline/Riwat Fault further to the north) may reflect the cutting upwards of the Potwar detachment at a number of locations coinciding with proximity to the eastern edge of the Salt Range Formation salt basin.

Just to the west of the Kotal Kund area, the position of the eastern limit of one of the major sub-thrust faults of the Potwar basement (Figs 2 and 3) can be defined. Whereas this fault may have restricted motion of the detachment in the central Potwar for a period of time, its absence, coupled with the thinning of the Salt Formation of the eastern Potwar, resulted in a more properly defined imbricate thrust system as seen along the Grand Trunk Road traverse (Fig. 4).

Late Pliocene to early Pleistocene deformational events

The development of numerous folds affecting Siwalik Group rocks within the central and eastern Potwar is a record of approximately 20% shortening along the breadth of the detachment which has occurred from approximately 3·0 Myr ago to the present (Fig. 4). A south-eastward younging of the onset of deformation and the initiation of 'surface expression' (Johnson *et al.*, 1979; Burbank & Raynolds, 1984) can be generally demonstrated based on palaeomagnetically constrained data from Siwalik Group outcrops located on the structures encountered in the Grand Trunk Road traverse (Figs 5 and 6). These data suggest that significant deformation occurred in the middle Matuyama (from approximately 2·2 to 1·5 Myr BP). It is suggested here that this 700,000 yr episode was primarily characterized by the Potwar detachment having successfully overridden the basement warp or fault with the subsequent development of the Salt Range proper. In fact, we view the major movement on the MBT, the 'snapping shut' of the Soan Syncline, and the related deformation across the Potwar to be a direct response to the stress transmitted across the breadth of the Potwar from the impinging basement ramp near the Salt Range.

Truncation of the Soan Syncline sequence and deposition of the Lei Conglomerate

The timing of thrusting along the Main Boundary Thrust zone that is inferred from Campbellpore sediments receive spectacular confirmation from the stratigraphy and chronology of the strata exposed in the Soan Syncline, south of the Main Boundary Thrust zone (Fig. 4). Immediately adjacent to the fault zone, Siwalik molasse strata have been severely folded and deeply eroded. Farther south, the Soan Syncline provides the first good exposures of the youngest pre-thrusting strata. Here, in the vertical northern limit of an asymmetric syncline, at least 3000 m of Siwalik sediments are preserved. According to the local magnetic polarity stratigraphy (Raynolds, 1980; Moragne, 1979), these range in age from greater than 9 to

2·1 Myr (Fig. 6). These upturned strata are overlain by the flat-lying Lei conglomerate which has a basal age (G. D. Johnson *et al.*, 1982; Raynolds, 1980) of ~1·9 Myr. The conglomerates are polymictic and reflect lithologies exposed in the uplifted bedrock terrain along the Main Boundary Thrust zone.

These chronologies indicate that, prior to 2·1 Myr, typical external molasse deposition was occurring at the site of the present Soan Syncline. During the succeeding 200,000 yr, as over 3,000 m of uplift and erosion took place at a mean minimum rate of 15 m Myr^{-1}, the syncline snapped shut. Subsequently, the Lei conglomerate prograded across the erosionally truncated syncline. We interpret this succession of events as a direct response to thrusting along the Main Boundary Thrust zone, deformation of the proximal strata adjacent to the thrust, and erosion of the newly uplifted regions along the boundary fault area. In agreement with the basal age (~1·8 Myr) of the Campbellpore sediments in the piggy-back basin, chronologic data from the Soan Syncline define a period between 2·1 and 1·9 Myr of rapid deformation within this region in response to movement along the MBT.

The subsequent termination of significant deformation in the northern Potwar plateau beginning after 1·9 Myr (Figs 5 and 6) reflects the transference of deformational events to the region of the Salt Range and its eastern extensions. The entire northern portion of the Potwar Plateau thus became a passenger on the detachment.

Salt Range ramping

The Salt Range ramping event was initiated approximately 2·2 Myr ago. In addition to generating the large-scale deformation in the northern Potwar, it significantly altered the character of sediment accumulation in the eastern extension of the Salt Range at a number of localities. Probably the most affected were the most proximal sites in the Bunha River basin localities of Tatrot, Kotal Kund and Pind Savikka, which experienced an abrupt termination of sediment accumulation about 1·5 Myr ago (Frost, 1980; G. D. Johnson *et al.*, 1982; Raynolds & Johnson, 1985). The onset of uplift of the Salt Range and a presumed further outward displacement of proximal compressive effects can be seen in the Rhotas and Pabbi Hills data, sites which lie possibly at the edge of the Salt Range Formation salt basin and which, therefore, are likely related to the periodic upward cutting of the Salt Range fault discussed above. The attainment of surface expression in the Rhotas anticline, based upon the magnetic polarity record, is less than 0·4 Myr BP. The same data for the Pabbi Hills anticline is a little less certain, but the attainment of surface expression is also about 0·4 Myr BP (G. D. Johnson et al., 1979).

SUMMARY

The timing of events associated with the development of the Potwar Plateau/Salt Range detachment and related basins to the NW can be interpreted by means of local magnetic polarity stratigraphies. From the evidence discussed above, it appears that the Himalayan foreland basin of northern Pakistan, which is dominated by the Peshawar and Campbellpore basins and the Potwar Plateau and associated thrust-related structures, has a history of deformation which extends from latest Miocene (c. 8·0 Myr) to the present.

Initial compressive events in the foreland margin in late Miocene are evinced by a depositional hiatus of mid-Miocene to mid-Pliocene duration in areas to the north of the Main Boundary Thrust zone (the Peshawar and Campbellpore Basins). To the south, in the vicinity of what is now the northern margin of the Salt Range, attenuation of sediment accumulation rates is noted at a number of localities, particularly along the eastern portions of the range (Kotal Kund and Baun localities).

Strong deformation of the area north of the Main Boundary Thrust zone continued through the Pliocene and culminated with the structural development of the Peshawar Basin ∼ 3 Myr ago. Prior to the major deformation along the MBT, two intervals of earlier deformation can be discerned in the Potwar. Within the southern Potwar and eastern Salt Range, data from Kotal Kund and Baun indicate that slowing sediment-accumulation rates preceded an interval of folding, uplift, and rotation that culminated between about 4·5 and 3·5 Myr BP. Subsequently, major movement along the Riwat fault sometime between about 3·4 and 2·7 Myr BP and initial uplift in the Ganda Paik area ∼ 2·5 Myr BP are seen as precursor events presenting an early response to increasing stress accumulation within the Potwar as the basement ramp was initially expressed. Deformation is expressed in several styles depending upon the geographical limits of the Salt Range Formation (the detachment slip surface). In the central Potwar, little evidence of early Pliocene deformation in the form of thrusting exists. Rotation of the detachment sheet, in

some localities of up to 30°, is recorded (Fig. 8). To the east, the upward forcing of the detachment surface resulted in the early Pliocene expression of the Chak Beli Khan/Riwat, and Jogi Tilla structures.

The open folds developed in the central and eastern Potwar during the Pliocene (the Soan, Bain, Buttar, Gungrilla/Adhi, Jabbar, and Ganda Paik areas) exhibit attenuation of sediment accumulation in post-3 Myr time (Fig. 9), suggesting that these sites were actively deforming at that time. This appears to be an expression of increasing strain within the Potwar detachment.

The onset of late-orogenic sedimentation beginning in the Peshawar and Campbellpore basins and evidence of significant detrital input from an uplifting Main Boundary Thrust zone terrane (perhaps the Attock Thrust) beginning about 2·7 Myr ago, and the deformation of the Soan Syncline commencing about 2·1 Myr, suggests significant motion along the fault zone at that time.

The abrupt initiation of folding of the Soan sequence about 2·1 Myr ago, with an associated progradation of locally derived clastics across the basin, the termination of sedimentation in the Campbellpore basin and basin-flanking sites in the Peshawar basin about 1·5 Myr ago, the termination of sedimentation over many of the folds in the eastern Potwar (listed above) and the cessation of sedimentation in the Kotal Kund region at 1·6 Myr serves to fix the timing of development of the Salt Range ramping event. This major structural episode, resulting in the Potwar detachment overriding a major sub-thrust basement fault having a throw in excess of 1 km (Lillie et al., 1985), can be inferred to have taken place between 2·1 and 1·6 Myr ago. The resultant termination of significant deformational events in the northern Potwar at that time is an expression of the transference of strain release to the frontal portion of the detachment.

Since 1·6 Myr, the central and northern Potwar, as inferred from our analysis of the palaeomagnetic stratigraphy, has ridden passively upon the detachment. Deformation along the leading edge of the detachment continues as in the case of the Lilla structure of the Jhelum plain south of the central Salt Range and in the thrust-cored Rhotas and Pabbi structures of the eastern extension of the Salt Range (Fig. 3).

Although complex, and not representing a case of a simple migrating foredeep, the data from the northern Pakistani foreland basin serves as an excellent example of the intimate interaction between tectonic

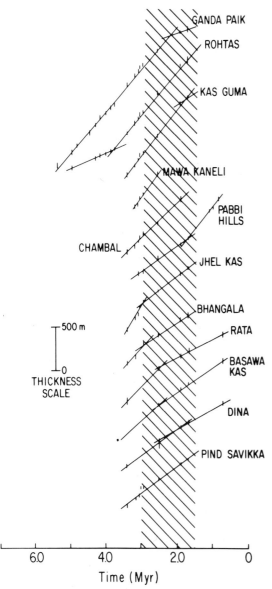

Fig. 9. Sediment accumulation records for a number of sites in the eastern Potwar Plateau, northern Pakistan. An attenuation of the rate of sediment accumulation is generally noted between 3·0 and 1·6 Myr BP in these localities evincing the onset of structural deformation. See Fig. 3 for locations. Selected sites from the Pir Panjal foothill belt (Kas Guma, Mawa Kaneli, Jhel Kas, and Rata) (see Burbank *et al.*, this volume) are included for comparison.

cause and sedimentological effect. We hope this example can serve as a model to illustrate the surprising abruptness and out-of-sequence thrust chronologies which may be observed in certain types of foreland basins.

ACKNOWLEDGMENTS

This research was supported by NSF grants EAR 8018779, EAR 8407052, INT 8019373, INT 8308069, and INT 8517353 to Dartmouth College and the R. E. Stoiber Field Fund at Dartmouth College. R. A. K. Tahirkheli, N. M. Johnson, R. J. Lillie and N. D. Opdyke contributed at various stages of this study. We thank the personnel of the University of Peshawar and the Geological Survey of Pakistan for their continued interest in and support of this project.

REFERENCES

BLOCH, R. (1981) *Stratigraphy of and facies variation within subvolcanic lithosomes, Siwalik Group, Campbellpore Basin, Pakistan.* Unpublished Honours Thesis, Kresge Library, Dartmouth College, 74 pp.

BURBANK, D.W. (1982) *The chronologic and stratigraphic evolution of the Kashmir and Peshawar intermontane basins, northwestern Himalaya.* Unpublished Ph.D. Thesis, Kresge Library, Dartmouth College, 291 pp.

BURBANK, D.W. (1983) The chronology of intermontane-basin development in the northwestern Himalaya and the evolution of the Northwest Syntaxis. *Earth planet. Sci. Lett.* **64**, 77–92.

BURBANK, D.W. & RAYNOLDS, R.G.H. (1984) Sequential late Cenozoic disruption of the northern Himalayan foredeep. *Nature*, **311**, 114–118.

BURBANK, D.W. & TAHIRKHELI, R.A.K. (1985) The magnetostratigraphy, fission-track dating, and stratigraphic evolution of the Peshawar intermontane basin, northern Pakistan. *Bull. geol. Soc. Am.* **96**, 539–552.

COTTER, G. DE P. (1933) The geology of the part of the Attock District west of longitude 72° 45′ E. *Mem. geol. Surv. India*, **55**, 63–161.

COWARD, M.P. & BUTLER, R.W.H. (1985) Thrust tectonics and the deep structure of the Pakistan Himalaya. *Geology*, **13**, 417–420.

DAVIS, D.M. & ENGELDER, T. (1985) The role of salt in fold-and-thrust belts. *Tectonophys.* **119**, 67–88.

FERMOR, L.L. (1931) General report for 1930—Eastern Salt Range, Punjab, etc. *Rec. geol. Surv. India*, **65**, 118–125.

FROST, C.D. (1979) *Geochronology and depositional environment of a late Pliocene age Siwalik sequence enclosing several volcanic tuff horizons, Pind Savikka area, eastern Salt Range, Parkistan.* Unpublished Thesis, Kresge Library, Dartmouth College, Hanover, N.H., 54 pp.

FUCHS, G. (1975) Contributions to the geology of the northwestern Himalayas. *Abh. Geol. Bund.-A*, **32**, 1–59.

GANSSER, A. (1964) *Geology of the Himalaya.* Wiley (Interscience), London, 289 pp.

GEE, E.R. (1945) The age of the Saline Series of the Punjab and Kohat. *Proc. natn. Acad. Sci. India*, **14**, 269–312.

GEE, E.R. (1947) Further note on the age of the Saline Series of the Punjab and of Kohat. *Proc. natn. Acad. Sci. India*, **16**, 95–116.

GEE, E.R. (1980) *Pakistan Geological: Salt Range Series Maps (1:500,000)*, Geological Survey of Pakistan, Quetta.

GILL, W.D. (1951a) Stratigraphy of the Siwalik Series in the Northern Potwar, Punjab, Pakistan. *Q. Jl geol. Soc. London,* **107**, 375–394.

GILL, W.D. (1951b) The tectonics of the sub-Himalayan fault zone in the northern Potwar region and in the Kangra District of the Punjab. *Q. Jl geol. Soc. London,* **107**, 395–421.

JENYON, M.K. (1985) Fault-associated salt flow and mass movement. *J. geol. Soc. London,* **142**, 547–553.

JOHNSON, G.D., JOHNSON, N.M., OPDYKE, N.D. & TAHIRKHELI, R.A.K. (1979) Magnetic reversal stratigraphy and sedimentary tectonic history of the Upper Siwalik Group, Eastern Salt Range and southwestern Kashmir. In: *Geodynamics of Pakistan* (Ed. by A. Farah & K. A. De Jong), pp. 149–165. Geological Survey of Pakistan, Quetta.

JOHNSON, G.D., ZEITLER, P., NAESER, C.W., JOHNSON, N.M., SUMMERS, D.M., FROST, C.D., OPDYKE, N.D. & TAHIRKHELI, R.A.K. (1982) The occurrence and fission-track ages of late Neogene and Quaternary volcanic sediments, Siwalik Group, northern Pakistan. *Pakistan Palaeogeogr. Palaeoclim. Palaeoecol.* **37**, 63–93.

JOHNSON, N.M., OPDYKE, N.D., JOHNSON, G.D., LINDSAY, E.H. & TAHIRKHELI, R.A.K. (1982) Magnetic polarity stratigraphy and ages of Siwalik Group rocks of the Potwar Plateau. *Pakistan Palaeogeogr. Palaeoclim. Palaeoecol.* **37**, 17–42.

JOHNSON, N.M., STIX, J., TAUXE, L., CERVENY, P.F. & TAHIRKHELI, R.A.K. (1985) Paleomagnetic chronology, fluvial processes and tectonic implications of the Siwalik deposits near Chinji Village, Pakistan. *J. Geol.* **93**, 27–40.

KELLER, H.M. (1975) *The magnetic polarity stratigraphy of an Upper Siwalik sequence in the Pabbi Hills of Pakistan.* Unpublished M.A. Thesis, Kresge Library, Dartmouth College.

KELLER, H.M., TAHIRKHELI, R.A.K., MIRZA, M.A., JOHNSON, G.D., JOHNSON, N.M. & OPDYKE, N.D. (1977) Magnetic polarity stratigraphy of the Upper Siwalik deposits, Pabbi Hills, Pakistan. *Earth planet. Sci. Lett.* **36**, 187–201.

LILLIE, R.J., YEATS, R.S., LEATHERS, M., BAKER, D.M., YOUSUF, M. & JAUME, S.C. (1985) Interpretation of seismic reflection data across the Himalayan foreland thrust belt in Pakistan (abstract no. 75980). *Abstr. Prog. geol. Soc. Am. Ann. Mtg.* **17**, 644.

LILLIE, R.J. & YOUSUF, M. (1986) Modern analogs for some midcrustal reflections observed beneath collisional mountain belts. In: *Deep Structure of the Continental Crust: Results from Reflection Seismology* (Ed. by M. Marazangi & L. Brown) 25 pp. (in press).

MARTIN, N.R. (1962) Tectonic style in the Potwar, West Pakistan. *Geol. Bull. Panjab Univ.* **2**, 39–50.

MEISSENER, C.R., MASTER, J.M., RASHID, M.A. & HUSSAIN, M. (1974) Stratigraphy of the Kohat Quadrangle, Pakistan. *Prof. Pap. U.S. geol Surv.* **716-D**, 1–29.

MIALL, A.D. (1979) Tertiary fluvial sediments in the Lake Hazen intermontane basin, Ellesmere Island, Arctic Canada. *Pap. geol. Surv. Can.* **79-9**, 1–25.

MORAGNE, J.H. (1979) *Magnetic polarity stratigraphy and timing of deformation for an Upper Siwalik sedimentary sequence, Soan syncline, Pakistan.* Unpublished Thesis, Kresge Library, Dartmouth College, Hanover, N.H., 31 pp.

OPDYKE, N.D., JOHNSON, N.M., JOHNSON, G.D., LINDSAY,

E. & TAHIRKHELI, R.A.K. (1982) Paleomagnetism of the middle Siwalik formations of northern Pakistan and rotation of the Salt Range detachment. *Palaeogeogr. Palaeoclim. Palaeocol.* **37**, 1–15.

OPDYKE, N.D., LINDSAY, E., JOHNSON, G.D., JOHNSON, N.M., TAHIRKHELI, R.A.K. & MIRZA, M.A. (1979) Magnetic polarity stratigraphy and vertebrate paleontology of the Upper Siwalik Subgroup of northern Pakistan. *Palaeogeogr. Palaeoclim. Palaeoecol.* **27**, 1–34.

PASCOE, E. (1930) General report for 1929—Potwar Plateau, etc. *Rec. geol. Surv. India,* **63**, 125–141.

PILGRIM, G.E. (1913) The correlation of the Siwaliks with mammal horizons of Europe. *Geol. Surv. India Rec.,* **43**, 264–326.

PINFOLD, E.S. (1919) Notes on structure and stratigraphy in the north-west Punjab. *Rec. geol. Surv. India,* **49**, 137–159.

RAYNOLDS, R.G.H. (1980) *The Plio–Pleistocene structural and stratigraphic evolution of the eastern Potwar Plateau, Pakistan.* Unpublished Ph.D. Thesis, Kresge Library, Dartmouth College. 265 pp.

RAYNOLDS, R.G.H. (1981) Did the ancestral Indus flow into the Ganges drainage? *Univ. Peshawar geol. Bull.* **14**, 141–150.

RAYNOLDS, R.G.H. & JOHNSON, G.D. (1985) Rates of Neogene depositional and deformational processes, northwest Himalayan foredeep margin, Pakistan. In: *The Chronology of the Geological Record* (Ed. by N. J. Snelling) *Mem. geol. Soc. London,* **10**, 297–311.

SEEBER, L. & ARMBRUSTER, J. (1979) Seismicity of the Hazara arc in northern Pakistan: decollement vs. basement faulting. In: *Geodynamics of Pakistan* (Ed. by A. Farah and K. A. De Jong), pp. 131–142. Survey of Pakistan, Quetta.

SMALL, P.A., III (1980) *The magnetic polarity stratigraphy of the Buttar Anticline and an estimated age for the Mankiala Bentonite, Eastern Potwar Plateau, Pakistan.* Unpublished B.A. thesis. Kresge Library, Dartmouth College, 41 pp.

TAHIRKHELI, R.A.K. (1970) Geology of the Attock–Cherat Range. *Geol. Bull. Peshawar Univ.* **5**, 1–26.

VISSER, C.F. & JOHNSON, G.D. (1978) Tectonic control of Late Pliocene Molasse sedimentation in a portion of the Jhelum Re-entrant, Pakistan. *Geol. Rdsh.* **67**, 15–37.

VOSKRESENSKIY, I.A. (1978) Structure of the Potwar Highlands and the Salt Range as an indication of horizontal movement of the Hindustan Platform. *Geotectonics,* **12**, 70–75.

WADIA, D.N. (1928) The geology of Poonch State (Kashmir) and adjacent portions of the Punjab. *Mem. geol. Surv. India,* **51**, 185–370.

WADIA, D.N. (1945a) A note on the repeated overthrusts of the Cambrian rocks on the Eocene in the north-eastern part of the Salt Range. *Proc. natn. Acad. Sci. India, A,* **14**, 214–221.

WADIA, D.N. (1945b) The significance of thrust structure of the Salt Range. *Proc. natn. Acad. Sci. India, B,* **16**, 249–252.

WILTSCHKO, D. & EASTMAN, D. (1983) Role of basement warps and faults in localizing thrust fault ramps. *Mem. geol. Soc. Am.* **158**, 177–190.

WYNNE, A.B. (1870) On the geology of Mount Tilla in the Punjab. *Rec. geol. Surv. India,* **3**, 79–86.

WYNNE, A.B. (1875) Geological notes on the Khareean Hills in the upper Punjab. *Rec. geol. Surv. India,* **8**, 46–49.

WYNNE, A.B. (1877) Note on the Tertiary zone and underlying rocks in the North-west Panjab. *Rec. geol. Surv. India*, **10**, 107–132.

WYNNE, A.B. (1878) On the geology of the Salt Range in the Punjab. *Mem. geol. Surv. India*, **14**, 1–313.

WYNNE, A.B. (1879) Further notes on the geology of the Upper Punjab. *Rec. geol. Surv. India*, **12**, 114–133.

YEATS, R.S. & LAWRENCE, R.D. (1984) Tectonics of the Himalayan thrust belt in northern Pakistan. In: *Marine geology and oceanography of Arabian Sea and coastal Pakistan* (Ed. by B. U. Haq & J. D. Milliman), pp. 177–198. Van Nostrand Reinhold.

Spec. Publs int. Ass. Sediment. (1986) **8**, 293–306

Late Cenozoic tectonics and sedimentation in the north-western Himalayan foredeep: II. Eastern limb of the Northwest Syntaxis and regional synthesis

DOUGLAS W. BURBANK*, ROBERT G. H. RAYNOLDS†
and GARY D. JOHNSON‡

**Department of Geological Sciences, University of Southern California, Los Angeles, CA 90089–0741, U.S.A.;
†AMOCO Production Company, P.O. Box 3092, Houston, TX 77253, U.S.A.; ‡Department of Earth Sciences,
Dartmouth College, Hanover, New Hampshire 03755, U.S.A.*

ABSTRACT

In order to help delineate the succession of late Cenozoic tectonic and stratigraphic events in the north-western Himalayan foredeep and adjacent ranges, a geological transect is described which extends from the north-eastern margin of the Kashmir Basin to the axis of the Jhelum Re-entrant along the eastern boundary of the Potwar Plateau. When combined with previous bedrock mapping, chronologic and stratigraphic studies of nine sections in the intermontane basin and the bounding foredeep define three primary pulses of late Cenozoic uplift affecting the Pir Panjal Range (5–4, 1·9–1·5, and 0·4–0 Myr ago). Many of the changes in facies, provenance, and palaeocurrents observed in the sedimentary rocks along the transect can be related to these deformational episodes.

When these data are combined with those from the Potwar Plateau and adjacent intermontane basins in Pakistan (Johnson *et al.*, this volume), a synthesis emerges illustrating a complex evolution of the foredeep during the past 5 Myr. Early episodes of uplift and rotation in the vicinity of the Salt Range are shown to be synchronous with initial uplift of the Pir Panjal Range. Extensive deformation between 2·1 and 1·6 Myr ago across much of the Potwar Plateau and, perhaps, along the bounding thrusts of the Pir Panjal appears causally related to a thrust ramping event in the Salt Range. In addition to providing a history of sedimentation and deformation that is more temporally constrained than has previously been possible, this study suggests a synchrony of several sets of structural events across a broad portion of the foredeep. This widespread synchrony may represent diverse responses to a common cause: stress accumulation and release due to interactions between irregular basement topography on the underthrusting Indian plate and the basal detachment of the overriding foredeep.

INTRODUCTION

As a consequence of the past and continuing collision of the Indian subcontinent with Eurasia, propagating faults and folds have disrupted the proximal margin of the Himalayan foredeep. Although previous geological studies have served to delineate the structural style of this disruption, the timing and tempo of these deformational events have been only loosely constrained. During the past decade, numerous chronologies based on magnetic polarity stratigraphy and fission-track dating have been developed for the terrestrial sediments of the Himalayan foredeep and the adjacent intermontane basins in the vicinity of the Northwest Syntaxis (Fig. 1). This spectacular bend in the collisional ranges bordering peninsular India occurs at a plexus of mountains where the Pamirs, Hindu Kush and Himalaya meet. These latter two ranges lie on the west and east flanks, respectively, of the Northwest Syntaxis, and it is the deformation along their southern margins that has disrupted molasse deposition in the adjoining Indo–Gangetic foredeep and has created the Peshawar, Campbellpore and Kashmir intermontane basins. A detailed transect of the western limb of the Northwest Syntaxis from the Peshawar Basin to the Jhelum River (Fig. 1) has been described in an accompanying paper (Johnson *et al.*, this volume). The objectives of this paper are two-fold: first, to describe a similar transect through the eastern arm of the Syntaxis from Kashmir

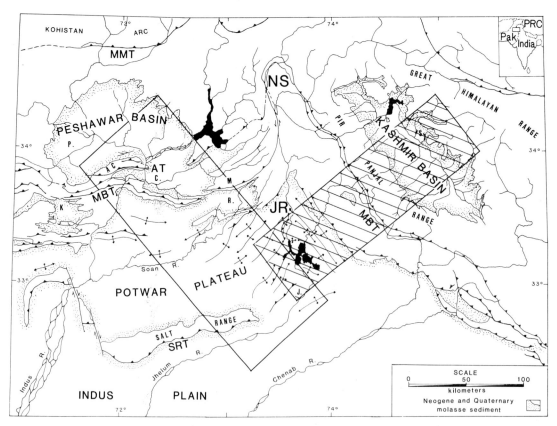

Fig. 1. Map of the north-western Himalayan foredeep, the southern margin of the Hindu-Kush, the south-western margin of the Himalaya, and the major intermontane basins in the vicinity of the Northwest Syntaxis (NS). The major anticlinal axes in the deformed molasse sediments, as well as major thrust faults (barbed lines) and strike-slip faults in the region surrounding the Jhelum Re-entrant (JR) are delineated. The hachured box delineates the area described in detail in this study. The box on the left, running from the Jhelum River to the Peshawar Basin, delineates the area described in a companion paper (Johnson *et al.*, this volume). Data from both areas are synthesized in this paper. Major thrust faults: Attock Thrust (AT); Main Boundary Thrust (MBT); Main Mantle Thrust (MMT); Salt Range Thrust (SRT). Other localities: Campbellpore (C); Jhelum (J); Kohat (K); Peshawar (P); Rawalpindi (R); Srinagar (S); Attock-Cherat Range (AC); Margala Hills (M).

to Jhelum, and, second, to synthesize the data from both limbs to generate a tectonic history for this region. This eastern transect is of considerable interest, because it spans a major late Cenozoic basin in which over 1 km of intermontane sediments have accumulated, it traverses a fault-bounded mountain range of 4,000 m high peaks which has experienced rapid and recent uplift, and after crossing deformed strata of the proximal molasse, it terminates in a zone of continuing molasse deposition.

Structural applications of chronologic data

When a local magnetostratigraphy is successfully correlated with the magnetic polarity time-scale, a time plane is introduced into the local stratigraphy at each of the identified reversal boundaries. This succession of chronologic data permits both delineation of the timing of specific events recorded by the sediments and examination of rates of processes and changes in these rates through time. In terms of developing a history of structural disruption, four aspects of these chronostratigraphies are particularly useful. First, the timing of changes in palaeocurrents, provenance and depositional style can be specified. Second, the limiting ages of sediments above and below unconformities can be used to bracket deformational episodes. Third, because rates of sediment accumulation often respond to tectonic controls, we can interpret decreases in the long-term sediment-

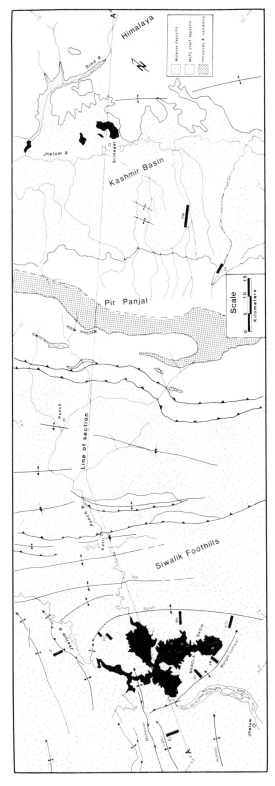

Fig. 2. Geologic map extending from the Jhelum River to the north-eastern margin of the Kashmir Basin. The Cenozoic molasse sediments of the Himalaya are shown by the stippled pattern. The Siwaliks (Neogene) and the Murrees (mid-to late Palaeogene) are not differentiated, because they represent a depositional continuum. The molasse is overthrusted along its north-eastern margin by an imbricated stack of bedrock slices composed of Eocene limestones and pre-Cenozoic bedrock. The Panjal Traps and associated gabbroic intrusions form much of the crestal area of the Pir Panjal Range. The gentle north flank of the range is mantled with glacial deposits which obscure most of the bedrock. The Kashmir Basin is filled with 1·3 km of Plio–Pleistocene Karewa sediments and is bounded on its north-eastern margin by folded and faulted Palaeozoic and Mesozoic bedrock. Geology is based in part on previous mapping by Middlemiss (1919), Wadia (1928, 1931, 1934), Shah (1968, 1978, 1980) and Fuchs (1975). Locations of measured and dated stratigraphical sections are shown by black boxes: Dina (DI); Hirpur (H); Jari (JA); Jhel Kas (JK); Kas Guma (KG); Mawa Kaneli (MK); Rata-Dadial (RD); Romushi (RM); Sakrana (S). Line AA' indicates the position of the cross-section (Fig. 3).

accumulation rate as reflecting the initiation of local deformation. Fourth, the timing of the initial surficial expression of a structure and a minimum amount of subsequent uplift can be estimated. When combined with stratigraphic and structural information, these chronologic data facilitate a more detailed tectonic synthesis than has previously been possible.

The Kashmir to Jhelum Transect

This transect comprises essentially four structural and stratigraphic elements (Figs 2 and 3): the Kashmir basin; the bedrock terrain of the Pir Panjal Range; the deformed molasse of the proximal foredeep margin; and the largely undisturbed foredeep. This transect begins in the north in the Kashmir basin and ends to the south in the modern plain of the Jhelum River. Along the transect, we have compiled the presently available structural (Fig. 3), chronologic (Fig. 4), and stratigraphic (Figs 5 and 6) data.

The Kashmir Basin

The south-western margin of the Kashmir basin is topographically defined by the Pir Panjal Range (Fig. 1) which has been uplifted along a series of northward-dipping thrust faults (see next section). Flowing down the axis of the basin for much of its course, the Jhelum River exits the basin near Baramula, after which it occupies successive strike valleys as it obliquely traverses the Pir Panjal Range. The structural nature of the north-eastern margin of the basin, as delineated by the Great Himalaya (Fig. 1), is not well known. The mountains rise abruptly for 2,000 m from the valley floor. Both open and tight folds in the Palaeozoic bedrock verge to the SW, and Middlemiss (1910) mapped a series of small north-dipping thrusts along the northern basin margin. A similar style of folding, thrusting, and vergence is also shown by the Palaeozoic sequence 30 km farther towards the interior of the range (Fuchs, 1975). We infer that major, northward-dipping thrusts have modulated uplift along this segment of the range and that these thrusts are largely buried by the steep alluvial cones mantling the northern flank of the Kashmir basin (Burbank, 1982; Burbank & Johnson, 1983; Burbank & Reynolds, 1984).

The Palaeozoic to Mesozoic bedrock within the basin (Wadia, 1934; Shah, 1972, 1978) comprises primarily volcanic and marine rocks that were uplifted and deeply weathered during the late Cretaceous and early Cenozoic. This bedrock sequence was trans-

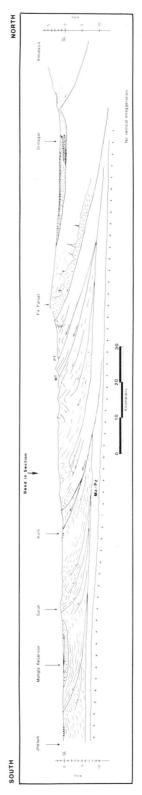

Fig. 3. Cross-section from the Jhelum plain to the Himalaya. Line of section (AA') is shown in Fig. 2. The surficial distribution and bedding attitudes of lithologic units and the location of faults are based on our mapping and on previously published maps of Wadia (1928, 1931, 1934); Shah (1968, 1978, 1980) and Fuchs (1975). With the exception of some scattered well-log data and some low-resolution seismic data (Kaila *et al.*, 1978), few subsurface data are available. Whereas the vergence of thrusting and major structures is generally unambiguous throughout the area, the actual geometry in vertical section is unknown. The thrust fault shown bordering the northeastern margin of the Kashmir Basin is undocumented, but is inferred from mapping by Middlemiss (1910), the geological history of the Kashmir Basin (Burbank & Johnson, 1983), and analagous geometries associated with nearby ranges. Murree Thrust (MT); Panjal Thrust (PT).

gressed by a shallow Eocene-aged sea from which the 'nummulitic' limestone (Godwin-Austin, 1959; Wadia, 1928) was deposited. Remnants of this limestone are found both in Ladakh and the Greater Himalaya to the north and in the south flank of the Pir Panjal Range. However, within the basin itself, uplift has caused complete erosion of these limestones and any additional pre-Pliocene sediments which may have overlain these strata.

Above this extensive unconformity, the intermontane sediments of the Karewa Formation (Lydekker, 1876, 1883) were deposited (Fig. 2). The depositional contact of the Karewas with the underlying bedrock is only rarely exposed within the Kashmir basin. Where it does crop out, however, such as in the river valleys at the southern end of the basin, deeply weathered palaeosols 10 m or more thick attest to a long interval of non-deposition and soil development that preceded initiation of intermontane sedimentation.

The asymmetry of uplift along the margins of the Kashmir basin and the continuing deposition in its northern portions have restricted extensive exposures of the Karewa Formation to eroded valleys along the north-western flank of the Pir Panjal Range (Fig. 2). Exposures here indicate that, in sharp contrast to the fluvial deposition that characterizes the Indo–Gangetic external molasse basin, the approximately 1300 m thick Karewa succession (Karunakaram & Rao, 1976; Burbank, 1982) is dominated by lacustrine mudstones, lignites, and deltaic siltstones and sandstones (Bhatt, 1975; Singh, 1982; Burbank & Johnson, 1983). These low-energy deposits occur immediately above the exposed basal unconformity and suggest that an early pulse of uplift of the ancestral Pir Panjal Range caused extensive ponding of the pre-existing fluvial systems within the newly defined Kashmir basin. Lacustrine sedimentation persisted throughout the interval of intermontane sedimentation and continues today in the shallow lakes along the northern basin margin (Fig. 2).

Within this predominantly mudstone Karewa sequence, coarse conglomerates punctuate the record of quiet-water deposition. Palaeocurrent measurements taken from imbricated and cross-bedded conglomerates (Fig. 5) in the lower two-thirds of the Karewas exposed on the flanks of the Pir Panjal Range indicate derivation from a north-easterly source, presumably the Great Himalaya. We infer that pulses of uplift along the northern basin margin caused thick wedges of conglomerates to be shed south-westwards across nearly the entire low-relief basin. In the upper third

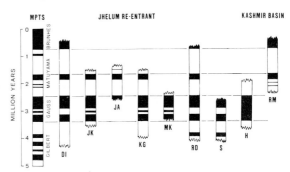

Fig. 4. Correlation of local magnetic polarity stratigraphy with the magnetic polarity time-scale (Mankinen & Dalrymple, 1979). Both fossil occurrences and the frequent presence of prominent ashes associated with the Gauss–Matuyama boundary assist in the recognition of identifiable chrons and subchrons at the local scale. Volcanic ashes in Kashmir and further west in northern Pakistan have been fission-track dated (Burbank, 1982; Johnson *et al.*, 1982) and serve to reinforce the correlations shown here. For a more detailed discussion of the sections shown here, see Johnson *et al.* (1979), Raynolds (1980) and Burbank (1982). The sections are arranged in the spatial sequence in which they would be encountered on a traverse starting from the Jhelum River and ending in Kashmir. Dina (DI); Hirpur (H); Jari (JA); Jhel Kas (JK); Kas Guma (KG); Mawa Kaneli (MK); Rata-Dadial (RD); Romushi (RM); Sakrana (S).

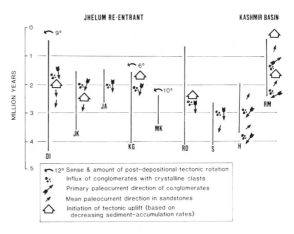

Fig. 5. Schematic representation of tectonic and sedimentologic events for the local sequences aligned along the cross-sectional traverse. The vertical line represents the temporal duration of each section as interpreted from the magnetostratigraphic data. The amount of post-depositional rotation is determined from the magnetostratigraphic data from each section. Those sections showing no rotation are those whose mean magnetic orientation is not significantly different from geographical north and south. Each of the events (influx of conglomerates, initiation of tectonic uplift, or palaeocurrent direction) is shown in its proper temporal position in the local sequence. Section abbreviations are the same as those for the previous figure.

of the Karewa succession, palaeocurrents on the Pir Panjal flank (Fig. 5) indicate a centripetal drainage pattern and suggest a shift of the basin depocentre to the NE.

The conglomerate facies were used by both de Terra & Paterson (1939) and Bhatt (1976) to correlate between exposures and to develop a stratigraphic framework for deposition of the Karewas. Recent studies based on magnetic polarity stratigraphy and fission-track dating (Burbank, 1982; Agrawal, 1985) have created a new chronology spanning the interval from about 3·5–0·4 Myr ago (Fig. 4). Because the exposures of the oldest Karewa sediments are either limited in extent or faulted and deformed, they cannot be directly dated. Their age can be estimated by downwards extrapolation of sediment-accumulation rates derived from the overlying, dated strata. Such estimates place the initiation of Karewa deposition at between 4 and 5 Myr ago. The conglomerates derived from inferred uplift events within the Great Himalaya occur sporadically between around 3·8 and 2·0 Myr ago, whereas between 1·7 and 0·4 Myr (Fig. 5), the primarily centripetal drainage persisted. The change in palaeocurrent direction between 2·0 and 1·7 Myr ago coincides with a decrease in sediment-accumulation rates (from 32 to 16 cm kyr^{-1}) (Fig. 6) which we infer to indicate enhanced uplift (Fig. 5) within the Pir Panjal Range and diminished subsidence rates in the Kashmir basin. Termination of thrusting within the Great Himalaya may have decreased both the loading and isostatic depression of the basin margin, with a resultant reduction in the rate of clastic influx.

Widespread Karewa sedimentation persisted until about 0·4 Myr (Burbank & Johnson, 1983), after which a major upheaval of the south-western half of the basin (Figs 3 and 5) caused Karewa lacustrine beds to be uplifted 1500 m or more along the flanks of the Pir Panjal Range at a mean rate of ∼4 mm yr^{-1}. This uplift generated a series of range-parallel gentle folds and minor thrusts in the Karewas that are inferred to have developed above a gravity-driven detachment of small displacement at the base of the Karewas. Erosion of the tilted and uplifted strata and progradation of a conglomerate across this newly bevelled surface created the prominant, planar, northward-sloping 'Karewa' surfaces that head in the Pir Panjal and project most of the way across the present basin (Fig. 3). The palaeomagnetic data (Burbank, 1982) show negligible amounts of tectonic rotation within the Karewa strata, and they indicate that the 45° of mid and late Tertiary clockwise rotation determined for this region by Klootwijk *et al.* (1983)

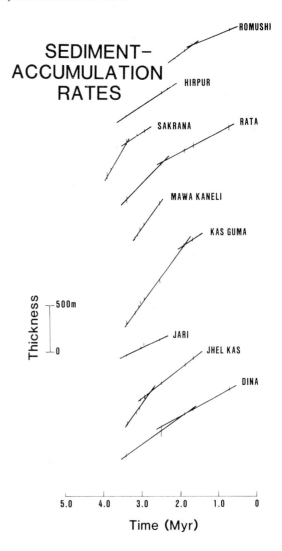

Fig. 6. Sediment-accumulation rates for sections along the transect. The control points in the time domain are the magnetozone boundaries according to their correlation with the magnetic polarity time-scale (see Fig. 4). The vertical thickness scale is the same for each of the sections, and the vertical length of each control point represents the uncertainty associated with the stratigraphic position of the reversal boundary, i.e. it is the stratigraphic distance between the two adjacent magnetic sites of opposite polarity. The sediment-accumulation rate curves are drawn as best-fit line segments through the data points. The slope of the line is directly proportional to the sediment-accumulation rate. Inflections towards slower sediment-accumulation rates are inferred to represent the encroachment of tectonic deformation and initiation of relative uplift at that particular site.

occurred prior to the last 4 Myr of active tectonism and deposition.

Pir Panjal Range

On the NE slopes of the Pir Panjal Range, glacial deposits and the extensive 'Karewa' surfaces obscure most of the bedrock. However, from the range crest for nearly 25 km to the SW, surficial outcrops are abundant and reveal a complex sequence of folded and faulted Palaeozoic to early Tertiary rocks that lie to the north of the deformed sediments of the foredeep (Figs 2 and 3). Because political considerations have led to restricted access to this region, most of our interpretations are based on previous mapping by Middlemiss (1919), Wadia (1928, 1931, 1934), Shah (1968, 1972, 1978, 1980) and Fuchs (1975).

The bedrock units in the Pir Panjal generally strike NNW to SSE, and, although there is lateral variation along the range, the structure depicted along the transect (Figs 2 and 3) is largely representative of the style of deformation in the remainder of the range. Along the transect, four major stratigraphic and structural units are apparent (Wadia, 1928). Upper Carboniferous 'Agglomeritic Slates', Carboniferous to early Triassic Panjal Traps, and associated gabbroic intrusives form the crestal region of the Pir Panjal Range. These are overlain by, but may be in fault contact with, the largely terrestrial Gondwana strata. Both of these units have been tightly folded and moderately overturned to the SW (Fig. 3). They, in turn, rest unconformably above the early Palaeozoic (?) Dogra slates (Wadia, 1928; Shah, 1968; Fuchs, 1975) which are also overturned and verging to the SW. This entire sequence has been thrusted along the Panjal Thrust and over the fourth stratigraphic/structural unit which is dominated by Eocene limestones and shales. This thrust may be structurally equivalent to the 'Main Central Thrust' of the central and eastern Himalaya (Seeber, Armbruster & Quittmeyer, 1981). Additional north-dipping thrusts are present within this zone (Shah, 1980), in each case bringing Palaeozoic rocks on to overturned Eocene strata. The southern margin of the Pir Panjal zone is delineated by the 'Main Boundary Thrust' (MBT) or 'Murree Thrust' which carries the Eocene and older rocks above the proximal molasse sediments of the Murree Formation.

The timing of deformation within this Pir Panjal sequence is not well constrained, although at least the outermost zone was deformed in post-Eocene times. The singularity of south-westward vergence throughout the range suggests that, despite evidence of multiple cleavages resulting from earlier deformations (Wadia, 1928), the Cenozoic Himalayan compression has overprinted the entire sequence. Our chronologic data from Kashmir for the initiation of intermontane sedimentation, as well as palaeocurrent and provenance data from the molasse to the south (described in the next section), suggest to us that much of the thrusting commenced in the early Pliocene (Burbank, 1983). On the other hand, the absence of external molasse strata north of the MBT and the presence of thick palaeosols below the Karewa beds suggest that the region of the present Pir Panjal may have represented an area of positive, but low relief that stood slightly above the molasse basin during the late Miocene.

Data from the Murree molasse strata to the SE of the transect indicate that these rocks, like those in the Kashmir basin, have undergone ~45° of clockwise rotation (Klootwijk *et al.*, 1983). In the Punjab Re-entrant 200 km farther SE, Siwalik strata as young as 7 Myr have been clockwise rotated 7–10° (Johnson *et al.*, 1983). Because the Plio–Pleistocene Karewa beds are unrotated (Burbank, 1982), the interval of rotation is most likely to have occurred between post-middle Miocene and pre-middle Pliocene (~10–4.5 Myr) and perhaps can be constrained to an even narrower time slice between 7 and 5 Myr. Because of the regional nature of this late Tertiary rotation, it is unlikely to be due to movement along a specific thrust. We interpret it as resulting from a broad-scale rotation related to the initial structural development of the Jhelum Re-entrant (Fig. 1). Apparently once the thrusts bounding the Pir Panjal Range developed around 4–5 Myr, differential rotation of the allochthonous rocks of the thrust sheets ceased, despite continued motion along the thrusts and continuing Indo-Asian convergence.

The deformed foredeep

A 30 km wide zone dominated by the Oligocene–Miocene molasse strata of the Murree Formation extends south from the MBT (Wadia, 1928; Rao & Rao, 1976; Shah, 1980). Near the southern margin of this zone in the vicinity of the transect (Figs 2 and 3), a major thrust fault brings both Eocene limestones and Permo–Triassic marine sediments to the surface. Near Kotli, 5 km farther south, a second thrust (Middlemiss, 1919) carries the Murrees above the Middle Siwalik strata of late Miocene age. Broad, open folds characterize most of the 'Murree Zone', and within one gentle syncline just south of the transect, a small inlier of lower Siwalik molasse strata is preserved (Wadia, 1928; Feibel, 1980). There are few chronologic or stratigraphic data within this region to constrain the timing of deformation of any of the structures described above. Deposition of the Murrees and the Lower Siwaliks pre-dated the deformation in this region, and the Plio–Pleistocene Siwalik rocks that would record the deformational events have been largely eroded away.

Within the Siwalik strata south of the fault at Kotli, however, recent magnetostratigraphic studies (Raynolds, 1980; Johnson et al., 1979) have both provided new insights into the chronology of local deformational events and permitted inferences to be drawn concerning tectonic movements on the faults bordering the Pir Panjal Range. In the northern half of this region, four major anticlines trending NW–SE and including the Nar and Suruh anticlines (Fig. 2) provide a straightforward structural picture. To the south, however, a more complicated pattern results from interference between the E–W trending Mangla–Samwal anticline, NE–SW trending Rohtas, Mahesian, and Lehri anticlines, and several sets of thrust faults that parallel these folded structures (Figs 2 and 3).

Although many of the above named structures have been studied magnetostratigraphically, only the E–W and NW–SE trending anticlines will be discussed here, because only these are subparallel to the major thrusts of the south-western Pir Panjal (Fig. 2) and are clearly related to the pattern of deformation along this transect. (Several of the other structures are discussed in the companion paper by Johnson et al. (this volume).) Both structural and stratigraphic approaches can be used to constrain the timing of tectonic movements in this area. There are deformed sediments immediately adjacent to the thrust at Kotli, and the age of these could help to delimit the interval of thrusting. However, these coarse conglomerates have not been amenable to palaeomagnetic dating, and consequently there are no available chronological data on the structures south of Kotli until the Suruh anticline is encountered.

Magnetostratigraphic studies have been completed at three localities (Rata-Dadial, Mawa Kaneli and Kas Guma) along this 50 km long anticline (Raynolds, 1980) (Figs 2 and 4). In the northernmost site at Rata-Dadial, decelerating sediment-accumulation rates, interpreted by us to represent the onset of deformation, begin at ~2·5 Myr ago (Fig. 6). This anticline and many others in the foredeep appear to be thrust-cored, and consequently, we regard their initial growth as an

indication of thrust fault propagation into these localities. The rate of growth of the Suruh structure was certainly slow, because this area continued to accumulate sediments for an additional 1·5 Myr until less than 1 Myr ago. Near the SE end of the Suruh fold at Kas Guma, deformation appears to have begun by 1·8 Myr ago (Figs 5 and 6). These data suggest that anticlinal growth began in the NW and slowly migrated to the SE along the Suruh structure.

Sections at Jari (Raynolds, 1980) and Jhel Kas (Johnson et al., 1979) have been dated along the Mangla–Samwal anticline (Figs 2 and 4). Sediment-accumulation rates here begin to decrease between 2·9 and 2·4 Myr ago and suggest the commencement of deformation (Fig. 6). Surficial expression of this fold as an element of positive relief occurred not more than 1·5 Myr ago. In sum, the chronologic data from the Suruh and Mangla–Samwal anticlines suggest that deformation began in the north around 2·5 Myr along the Suruh structure and that this folding slowly propagated to the SE along the incipient anticline. Concurrently, much of the deformation was apparently transferred to the southern margin of the modern Mangla Basin, where the Mangla–Samwal anticline was experiencing its initial phase of growth.

From a stratigraphic viewpoint, we can date the first appearance of a high-energy conglomeratic facies as it transgressed across this region. This conglomeratic facies is significant, because it represents the encroachment of a source area whose clasts reflect lithologies present in the Pir Panjal Range. Thus, we view this facies as a distinctive product of the thrusting and uplift of the Pir Panjal and progressive denudation of the pre-Tertiary bedrock. The rate of south-westward progradation of this facies along the transect (Fig. 7) averaged about 3 cm yr^{-1}. If this rate is extrapolated back in time and north-eastwards towards the Pir Panjal, it suggests that conglomeratic sedimentation and, hence, the initiation of major thrusting along the faults bounding the SW margin of the Pir Panjal commenced around 4–5 Myr.

None of the dated sections along this transect extend back this far in time. West of the Jhelum River, however, in many of the older successions, two distinctive sandstone complexes are present. The lower, older 'white sandstones' (Raynolds, 1980) contain abundant hornblende, as well as white quartzitic and granitic pebbles. These sandstones display current structures indicating deposition from rivers flowing to the ENE. The younger 'brown sandstones' lack hornblende and granitic clasts, but carry brown quartzitic and volcanic clasts. These

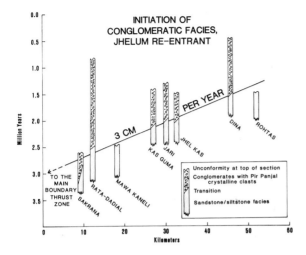

Fig. 7. Initiation of conglomeratic facies in the Pir Panjal foothills and the easternmost Potwar Plateau. The relative position of each section is derived from an orthogonal projection of the geographical location of the section on to the line of the cross-section (see Fig. 2). The age of the facies boundary represents the onset of conglomeratic sedimentation at each location. The magnetostratigraphic age of this stratigraphic horizon and a regression through these facies boundaries indicates that the south-westward onslaught on conglomerates proceeded at a mean rate of about 3 cm yr^{-1}. If this mean rate is extrapolated back in time and space towards the Main Boundary Thrust zone, it suggests that these conglomerates would first have appeared between 4 and 5 Myr ago in response to thrusting and early uplift of the Pir Panjal Range.

sandstones preserve south-directed palaeocurrents. We interpret the white sandstones to represent the ancestral Indus River flowing longitudinally along the axis of the foredeep and draining into the Ganges River (Raynolds, 1982). The brown sandstones, which are also the dominant sandstone type found east of the Jhelum River along the transect described here, are interpreted to reflect the ancestral Jhelum River and to be a response to the structural development of the Jhelum Re-entrant and uplift of the Pir Panjal Range. This fundamental re-arrangement of the drainage systems occurs in several dated sequences between 4 and 5 Myr. Thus, the timing of the initiation of Karewa sedimentation in Kashmir, the onslaught of conglomerates, and the re-arrangement of drainage patterns converge on a date of 4–5 Myr ago for initial uplift of the Pir Panjal and concurrent development of the Jhelum Re-entrant.

SYNTHESIS AND SUMMARY

When the chronologic, stratigraphic, and structural data from both west (Johnson *et al.*, this volume) and east of the Jhelum River are considered, it is possible to develop a tectonic history for the north-western Himalayan foredeep that is well constrained in time and space over the past 5 Myr (Fig. 8). Clearly, there are many limitations to such a synthesis. Although data from over 18 dated locations are considered in this reconstruction, there are many faults and folded structures that have not been studied. This makes it necessary to interpolate between dated structures. The techniques used in this study do not permit us to date certain structures due to the absence of syndeformation strata in outcrop. For example, although the onset of sedimentation in the Peshawar Basin provides an upper age limit for the time of deformation, the long history of deformation of foredeep strata in the vicinity of the Peshawar basin (Burbank, 1983; Burbank & Tahirkheli, 1985) and the thick alluvial cover south of the Attock Range seem to preclude the direct determination of a reliable date for the onset of major movement along the Attock thrust. Similar restrictions apply to the thrusts bordering the south-western margin of the Pir Panjal Range. In addition, within the Potwar Plateau (Figs 1 and 3), there are a number of undated thrusts that are considered to be of only secondary importance to the overall deformational pattern, but which may, in fact, have played a more intrinsic role. Despite these caveats, the reconstruction presented here provides one of the most tightly constrained syntheses of foredeep deformation that is presently available. This synthesis builds upon an earlier analysis (Burbank & Raynolds, 1984) and incorporates additional data from stratigraphic, chronologic, and geophysical studies.

In this north-western portion of the Himalayan foredeep, we can delineate three discrete regions that are discriminated on the basis of the style and causes of the deformation experienced by each. One region lies to the east of the Jhelum River and encompasses the structures along the transect described here, all of which are oriented subparallel to the strike of the Himalaya and appear to reflect a NE–SW oriented compressional regime. The second region comprises the eastern Potwar Plateau west of the Jhelum River and extends north-westwards to the Peshawar basin (Figs 3 and 4, Johnson *et al.*, this volume). Here, the major structural features trend NE–SW, subparallel to the strike of Attock–Cherat, Margala, and Hindu Kush Ranges. The western margin of this region is

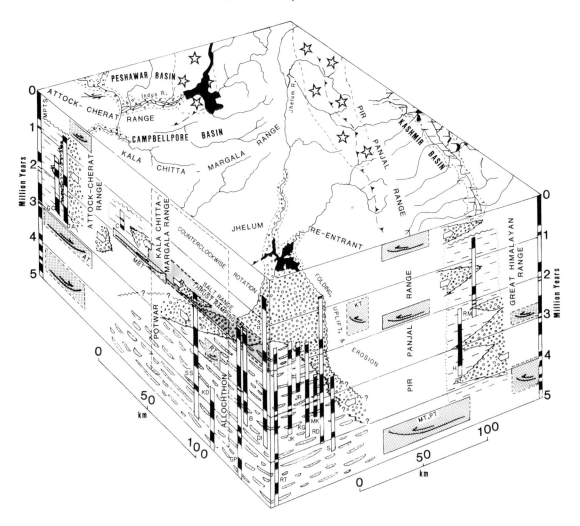

Fig. 8. Time–space block diagram of the north-western Himalayan foredeep and adjacent ranges in the vicinity of the Northwest Syntaxis. The vertical dimension extends back to 5 Myr, and the magnetic polarity time-scale is displayed along that dimension. The dated sequences described in this study (Fig. 4) and in Johnson *et al.* (this volume) are depicted in their chronological range and are projected from their geographical locations on to the transects defining the sides of the block. The right panel includes columns representing Romushi (RM); Hirpir (H); Sakrana (S); Rata-Dadial (RD); Mawa Kaneli (MK); Kas Guma (KG); Jari (JA); Jhel Kas (JK) and Rohtas (RH). The columns shown on the left panel include: Bangala (B); Campbellpore (C); Dag (DA); Dina (DI); Garhi Chandan (GC); Ganda Paik (GP); Kas Dovac (KD); Lei (L), and Soan Syncline (SS). For each of the stratigraphic sections, generalized facies and depositional relationships are shown in their proper chronological position (for example, the north-westwardly derived conglomerates at 3·8 and 3·0 Myr ago in Kashmir and the southward progradation of conglomerates from Sakrana across the Jhelum Re-entrant after 3 Myr). The intermontane basins are dominated by low-energy, largely lacustrine facies that are punctuated by conglomeratic influxes. Two broad phases of molasse sedimentation across the Potwar Plateau and Jhelum Re-entrant are shown: a channel-and-floodplain interval succeeded by late-stage polymictic conglomerates. Large, open arrows indicate prevailing palaeocurrent directions at different times. The stippled boxes enclosing a thrust-fault symbol delineate both the spatial and the chronological intervals over which the thrusts are interpreted to have been active. Dashed borders indicate less well constrained periods of thrusting. Nearly all of the sequences in the Jhelum Re-entrant terminate in the conglomeratic facies. Many are presently overlain by thick, undated conglomeratic successions. Although Quaternary tectonism has generated the complex series of anticlines on which the stratigraphic sections were measured (Figs 1 and 2), many of the intervening basins, such as the Mangla Basin (Figs 2 and

transitional to the third area (Fig. 4, Johnson *et al.*, this volume) which encompasses the remainder of the Potwar Plateau extending from the Salt Range to the Kala Chitta Range.

Deformation in the easternmost area is quite clearly related to tectonic history of the NW–SE trending Pir Panjal and Great Himalayan Ranges. The Potwar Plateau to the west appears to be underlain by a very efficient, salt-lubricated detachment (Seeber *et al.*, 1981; Burbank, 1983; Lillie *et al.*, 1985; Johnson *et al.*, this volume). Much of the latest Cenozoic Indo–Asian convergence in this area appears to be accommodated along this detachment with a resultant decrease in the magnitude of deformation in the strata north of the Salt Range. The intermediate region between these two areas contains structures similar to those of the central and western Potwar Plateau inasmuch as they are parallel to the structural trends of the crystalline ranges to the north. However, rather than passively rafting the entire Phanerozoic section above a detachment as occurs to the west, the distal thrusts cut up section, and the structural deformation is considerably more intense. We interpret these changes as reflecting palaeogeographic control, whereby the Eocambrian salt deposits are attenuated or absent to the east, and, consequently, no single widespread detachment developed in this region.

Within the window of time (0–5 Myr) under consideration, the earliest major phase of deformation between 4 and 5 Myr ago is related to initial uplift of the Pir Panjal Range along its bounding thrusts and structural definition of the Kashmir intermontane basin (Fig. 8). The re-arrangement of pre-existing drainage patterns as the south-flowing ancestral

Jhelum River supplanted the east-flowing ancestral Indus River is viewed as a response to this uplift and to the incipient development of the Jhelum Re-entrant. The clockwise rotation of the region east of the syntaxial axis, at least within the newly formed Kashmir basin, ceased at this time. From the newly uplifted fault front of the Pir Panjal, a conglomeratic wedge bearing volcanic clasts prograded to the south and transgressed sequentially across the dated localities in the eastern Potwar Plateau (Figs 7 and 8).

The folding, erosion, and rotation that occurred in both the Baun and the Kotal Kund /Tatrot/Andar area between ~4·5 and 3·5 Myr ago (Johnson *et al.*, this volume) suggests that the Salt Range detachment may have become active at this time and that, as a result, the 'Potwar allochthon' was structurally defined for the first time (Fig. 8). The coincidence in time (Fig. 8) of this initial deformational episode near the Salt Range with the first interval of major uplift of the Pir Panjal Range suggests these events may be causally linked. Whether or not this deformation resulted from encounters between irregular basement topography with existing thrust ramps (Burbank, 1983) is a matter of speculation. It is interesting to note that, if these events are indeed related, stress release occurred synchronously at localities more than 100 km apart with little or no apparent deformation in the intervening area where molasse sedimentation continued unabated (Fig. 8).

Within the Kashmir basin itself, the predominantly lacustrine Karewa sediments began to aggrade by 4 Myr ago. Sporadic, thrust-modulated pulses of uplift along the north-western margin of the basin shed coarse conglomerates south-westward across the basin

3) have continued to accumulate coarse sediments until the present time. The large dashed box on the left panel labelled 'Potwar Allochthon' depicts that area riding above the eastern edge of the Salt Range detachment surface. On the right panel, three intervals of uplift in the vicinity of the Pir Panjal Range have defined the Kashmir Basin, controlled palaeocurrent patterns, and deformed the adjacent foredeep to the SW. The Kotli Thrust (KT) is likely to have been active in the early to middle Pleistocene, but it is not well dated. The earliest interval of deformation along the Panjal (PT) and Murree (MT) Thrusts in the Pir Panjal (4–5 Myr ago) is contemporaneous with the deformation inferred from the records at Baun and Kotal Kund (Johnson *et al.*, this volume). Although these latter sites lie to the south of the left panel, the 'Potwar Allochthon' is inferred to have become structurally defined at this time. Early Pleistocene deformation in the Attock–Cherat Range along the Attock Thrust (AT) is succeeded by younger deformation to the south and east. Out-of-sequence thrusting is indicated by movement along the Riwat Thrust (RT) between 3·0 and 2·5 Myr ago and by later, major movement on the MBT between 2·1 and 1·9 Myr ago. This rapid motion on the MBT, the strong deformation of the Soan Syncline (SS), and related deformation that rippled across the proximal foredeep are all interpreted as responses to a major thrust ramping event in the vicinity of the Salt Range (Johnson *et al.*, this volume). Following this event, counter-clockwise rotation of portions of the Potwar Allochthon and adjacent regions occurred. The surface of the block illustrates active present-day processes. Deposition is largely restricted to the axial portions of the intermontane basins and narrow floodplains. Regions of high seismicity within 10–15 km of the surface (Seeber *et al.*, 1981) are shown by stars. Presently or recently active faults that break the surface seem to be associated with shallow seismicity.

during the succeeding 2 Myr. By 2·5 Myr ago, the effects of the Pir Panjal deformation were beginning to be expressed in the Potwar Plateau east of the Jhelum River. Not only were the uplift-related conglomerates transgressing across the region (Figs 5 and 8), but sediment-accumulation rates (Fig. 6) were beginning to slow as uplift began to replace subsidence in the proximal foredeep (Fig. 5). These growing anticlines attained surficial expression sometime after 1·5 Myr ago. However, the Mangla Basin north of the Mangla–Samwal anticline and south of the Suruh anticline continued to receive coarse-grained sediments until the present time (Fig. 8). After about 2 Myr ago, most of the active thrusting to the NE of the Kashmir basin apparently ceased. The transfer of major deformation to the south appears to have instigated renewed uplift of the Pir Panjal Range during the early Pleistocene. Around 0·4 Myr ago, a dramatic pulse of uplift, undoubtedly caused by thrusting along the boundary thrusts of the Pir Panjal raised the range more than 1 km and focused Karewa sedimentation on to the NE side of the Kashmir Basin, where it continues to this day.

To the west of the Jhelum River, movement along the Attock thrust (Fig. 1) partially ponded the course of the ancestral Indus River and subsequently intermontane sediments began accumulating in the Peshawar basin around 3 Myr ago (Fig. 8). Less than 0·5 Myr later, movement along the Riwat fault folded the Kas Dovac sequence and presaged the major deformation in the Soan River region (Raynolds, 1980). About 2·1 Myr ago, rapid movement along the Main Boundary Thrust bounding the southern margin of the Kala Chitta–Margala Ranges initiated sedimentation in the newly formed Campbellpore basin and caused the spectacular folding of the Soan syncline south of the fault zone (Fig. 8). As described by Johnson *et al.* (this volume), these two deformational events were synchronous with or slightly antedated a pulse of uplift in the Salt Range some 100 km to the south. As the ancestral Salt Range terrane became buttressed against a down-to-the-north bedrock high (Voskresenskiy, 1978; Lillie *et al.*, 1985) in the northward-converging Indian subcontinent, rising stresses were apparently transmitted across the breadth of the Potwar Plateau. Compression along the Riwat fault was likely the initial response to this enhanced compression (Fig. 8). Subsequently, these stresses either precipitated or enhanced movement along the Main Boundary Thrust (causing the Soan deformation) and were probably responsible both for both the slowing of sedimentation rates and the initial

folding in the previously undisturbed foredeep of the eastern Potwar Plateau and for the resultant wave of intraformational conglomerates that spread across the region (Fig. 8). Progradation of the Lei Conglomerate from the uplifted terrane across the Soan syncline at 1·9 Myr signals the end of major uplift in this area. The presently undeformed nature of this conglomerate suggests that subsequent deformation was transferred to the south at this time and was concurrent with the ramping of the Potwar decollement and the development of the present Salt Range.

Chronologically constrained palaeomagnetic data indicate that up to 25° of counter-clockwise rotation within the central and eastern Potwar Plateau (Raynolds, 1980) and in the Salt Range itself (Opdyke *et al.*, 1982) occurred above this detachment during the Pleistocene as a result of sinistral shearing along the western margin of the indenting Indian subcontinent. Along the length of the Salt Range, essentially the entire Phanerozoic succession is being carried southwards as a detached mass. Further to the east, however, in the vicinity of the Jhelum River and the eastern Potwar Plateau, no single stratigraphically controlled detachment surface appears to exist, and instead, it may be replaced by low-angle thrusts cutting progressively upsection as hypothesized by Coward & Butler (1985). In both cases, the edge of the actively deforming foredeep lies at the latitude of the southern margin of the Salt Range. To the south of this zone, active foredeep deposition is occurring along the modern Jhelum and Indus Rivers.

The overall picture that emerges from this analysis is one suggesting that the Indo–Asian collision has caused deformation to ripple across the proximal foredeep. As convergence brings uplifted source areas progressively closer to the molasse depocentre, initial deformation in the form of waning sediment-accumulation rates encroaches through the nearby foredeep. Whereas this deformation by folding seems to occur nearly continuously, episodes of faulting appear to be constrained to narrow time slices. Taking the MBT and the sediments of the Soan Syncline as an example, the thrusting event occurs as a brief pulse of rapid movement with a discrete termination after which little or no additional deformation occurs. Within this context, the thrusting, where not confined to an efficient detachment zone, can be seen to have stepped out across the foredeep in distinct increments and limited time intervals during the Plio–Pleistocene deformation of this portion of the Himalayan external molasse basin. However, if our interpretation of the history of the Pliocene deformation of the Salt Range is correct (Johnson *et al.*, this volume), out-of-sequence deformation also occurs, such that initial buttressing and uplift of the Salt Range ultimately results in severe deformation 100 km farther north in a more proximal position within the Himalayan foredeep. Similarly, the initial deformation episodes 4–5 Myr ago in the Pir Panjal and Kotal Kund suggest a causal linkage between two widely separated areas.

Finally, the style of Plio–Pleistocene structural disruption in the north-western Indo–Gangetic foredeep appears strongly dependent on the distribution of the incompetent (particularly evaporitic) horizons in the Phanerozoic succession. Although the orientation of the bedrock ranges, bounding thrusts, and fold axes strike nearly perpendicular to each other in the eastern Potwar Plateau on either side of the Jhelum Re-entrant, the style of deformation appears very similar in both areas. Deformation is limited to a narrow zone and is seen to have spread sequentially across these areas. In contrast, in the central Potwar Plateau above the salt-lubricated detachment, deformation occurred in widely separated areas with minimal involvement of the intervening strata. Moreover, except for brief pulses of stress release along the trailing edge of the detached mass, nearly all subsequent convergence in this portion of the foredeep appears to be accommodated along the efficient, largely aseismic Salt Range detachment.

ACKNOWLEDGMENTS

This research was supported by grants from the Geological Society of America, Sigma Xi, the Richard E. Stoiber Field Fund, and the Amoco, Marathon, and Shell Oil Companies to D. W. Burbank, and through NSF grants EAR 8018779, EAR 8407052, INT 8019373, and INT 8308069 to Dartmouth College. Our research in Pakistan was made possible through the close collaboration of the Centre for Excellence in Geology, Peshawar University and Dr R. A. K. Tahirkheli.

REFERENCES

AGRAWAL, D.P. (1985) Cenozoic climatic changes in Kashmir: the multidisciplinary data. *In: Climate and Geology of Kashmir: the Last 4 Million Years* (Ed. by D. P. Agrawal, S. Kusumgar & R. V. Krishnamurthy). *Current Trends in Geology*, **6**, 1–12. Today and Tomorrow's, New Delhi.

BHATT, D.K. (1975) On the Quaternary geology of the Kashmir Valley with special reference to stratigraphy and sedimentation. *Misc. Publ. geol. Surv. India*, **24**, 188–204.

BHATT, D.K. (1976) Stratigraphic status of the Karewa Group of Kashmir, India. *Himalayan Geol.* **6**, 197–208.

BURBANK, D.W. (1982) *The chronologic and stratigraphic evolution of the Kashmir and Peshawar intermontane basins, northwestern Himalaya.* (Unpublished) Thesis, Dartmouth College, Hanover, N.H., 291 pp.

BURBANK, D.W. (1983) The chronology of intermontane-basin development in the northwestern Himalayan and evolution of the Northwest Syntaxis. *Earth planet. Sci. Lett.* **64**, 77–92.

BURBANK, D.W. & JOHNSON, G.D. (1983) The late Cenozoic chronologic and stratigraphic development of the Kashmir intermontane basin, northwestern Himalaya. *Palaeogeogr. Palaeoclim. Palaeoecol.* **43**, 205–235.

BURBANK, D.W. & RAYNOLDS, R.G.H. (1984) Sequential Late Cenozoic structural disruption of the northern Himalayan foredeep. *Nature*, **311**, 114–118.

BURBANK, D.W. & TAHIRKHELI, R.A.K. (1985) The magnetostratigraphy, fission-track dating, and stratigraphic evolution of the Peshawar intermontane basin, northern Pakistan. *Bull. geol. Soc. Am.* **96**, 539–552.

COWARD, M.K. & BUTLER, R.W.H. (1985) Thrust tectonics and the deep structure of the Pakistan Himalaya. *Geology*, **13**, 417–420.

DE TERRA, H. & PATERSON, T. (1939) Studies on the ice age in India and associated human cultures. *Publ. Carnegie Inst. Wash.* **493**, 354 pp.

FEIBEL, C.S. (1980) *Geological mapping of the Himalayan foothills using LANDSAT imagery.* Unpublished B.S. Thesis, Dartmouth College, Hanover, N.H. 15 pp.

FUCHS, G. (1975) Contributions to the geology of the northwestern Himalayas: *Abh. geol. Bundesanstalt.* **32**, 3–59.

GODWIN-AUSTIN, H.H. (1959) On the lacustrine Karewa deposits of Kashmir, *Q. Jl geol. Soc. Lond.* **15**, 221–229.

JOHNSON, G.D., JOHNSON, N.M., OPDYKE, N.D. & TAHIRKHELI, R.A.K. (1979) Magnetic reversal stratigraphy and sedimentary tectonics of the Upper Siwalik Group, eastern Salt Range and southwestern Kashmir. In: *Geodynamics of Pakistan* (Ed. by A. Farah & K. de Jong), pp. 149–165. Geological Survey of Pakistan, Quetta.

JOHNSON, G.D., RAYNOLDS, R.G.H. & BURBANK, D.W. (1986) Late Cenozoic tectonics and sedimentation in the northwestern Himalayan foredeep: I. Thrust ramping and associated deformation in the Potwar region. In: *Foreland Basins* (Ed. by P. Allen & P. Homewood). *Spec. Publs int. Ass. Sediment.* **8**, 273–291. Blackwell Scientific Publications, Oxford.

JOHNSON, G.D., OPDYKE, N.D., TANDON, S.K. & NANDA, A.C. (1983) The magnetic polarity stratigraphy of the Siwalik Group at Haritalyangar (India) and a new last appearance datum for *Ramapithecus* and *Sivapithecus* in Asia. *Palaeogeogr. Palaeoclim. Palaeoecol.* **44**, 223–249.

KAILA, K.L., KRISHNA, V.G., CHOWDHURY, K.R. & NARAIN, H. (1978) Structure of the Kashmir Himalaya from deep seismic soundings. *J. geol. Soc. India*, **19**, 1–22.

KARUNAKARAN, C. & RAO, A.R. (1976) Status of exploration for hydrocarbons in the Himalayan region—contribution to stratigraphy and structure. *Geol. Surv. India Himalayan Geology Seminar*, 1976, New Delhi, 72 pp.

KLOOTWIJK, C.T., SHAH, S.K., GERGAN, J., SHARMA, M.L., TIRKEY, B. & GUPTA, B.K. (1983) A palaeomagnetic reconnaissance of Kashmir, northwestern Himalaya, India. *Earth planet. Sci. Lett.* **63**, 305–324.

LILLIE, R.J., YEATES, R.S., LEATHERS, M., BAKER, D.M., YOUSUF, M. & JAYME, S.L. (1985) Interpretation of seismic reflection data across the Himalayan foreland thrust belt in Pakistan. *Abstr. geol. Soc. Prog.* **17**, 644.

LYDEKKER, R. (1876) Notes on the geology of the Pir Panjal and neighboring districts. *Rec. geol. Surv. India,* **9**, 155–183.

LYDEKKER, R. (1883) The geology of Kashmir, Chamba, and Khayam. *Mem. geol. Surv. India,* **22**, 1–344.

MANKINEN, E.A. & DALRYMPLE, G.B. (1979) Revised geomagnetic polarity time scale for the interval 0–5 m.y. B.P. *J. geophys Res.,* **84**, 615–626.

MIDDLEMISS, C.S. (1910) The Silurian–Triassic sequence in Kashmir. *Rec. geol. Surv. India,* **40**, 206–260.

MIDDLEMISS, C.S. (1919) The inclination of the thrust plane between the Siwalik and Murree zones near Kotli, Jammu Province. *Rec. geol. Surv. India,* **50**, 122–125.

OPDYKE, N.D., JOHNSON, N.M., JOHNSON, G.D., LINDSAY, E.H. & TAHINKHELI, R.A.K. (1982) Paleomagnetism of the middle Siwalik formations of northern Pakistan and rotation of the Salt Range decollement. *Palaeogeogr, Palaeoclim. Palaeoecol.* **37**, 1–15.

RAO, V.V.K. & RAO, R.P. (1976) Geology of Tertiary belt of Northwest Himalaya, Jammu and Kashmir states, India. *Himalayan Geology Seminar,* September 1976, Geological Survey of India, 26 pp.

RAYNOLDS, R.G.H. (1980) *The Plio–Pleistocene structural and stratigraphic evolution of the eastern Potwar Plateau.* Unpublished Ph.D. Thesis, Dartmouth College, Hanover, N.H., 265 pp.

RAYNOLDS, R.G.H. (1982) Did the ancestral Indus flow into the Ganges drainage? *Geol. Bull. Univ. Peshawar,* **14**, 141–150.

SEEBER, L. ARMBRUSTER, J.C. & QUITTMEYER, R.C. (1981) Seismicity and continental subduction in the Himalayan arc. In: *Zagros-Hindu Kush-Himalaya Geodynamic Evolution* (Ed. by H. K. Gupta & F. M. Delaney). *Am. Geophys. Un. Geodyn. Ser.* **3**, 215–242.

SHAH, S.K. (1968) A lithostratigraphic classification of the lower Palaeozoic Slate Group of Pohru Valley. *Un. Rev.* **32**, 10–17.

SHAH, S.K. (1972) Stratigraphic studies on lower Palaeozoic sequence of Anantnag district, Kashmir. *Himalayan Geol.* **2**, 468–480.

SHAH, S.K. (1978) Facies pattern of Kashmir within the tectonic framework of the Himalaya. In: *Tectonic Geology of the Himalaya* (Ed. by P. S. Saklani), pp. 63–78. Today and Tomorrow's, New Delhi.

SHAH, S.K. (1980) Stratigraphy and tectonic setting of the Lesser Himalayan Belt of Jammu. In: *Stratigraphy and Correlations of Lesser Himalayan Formations* (Ed. by K. S. Valdiya & S. B. Bhatia), pp. 152–160. Hindustan Publications, Delhi.

SINGH, I.B. (1982) Sedimentation pattern in the Karewa basin, Kashmir Valley, India and its geological significance. *J. Paleont. Soc. India,* **27**, 71–110.

VOSKRESENSKIY, I.A. (1978) Structure of the Potwar highlands and the Salt Range as an indication of horizontal movement of the Hindustan Platform, *Geotectonics,* **12**, 70–75.

WADIA, D.N. (1928) The geology of Poonch State (Kashmir) and adjacent portions of the Punjab. *Mem. geol. Surv. India,* **65**, 2–155.

WADIA, D.N. (1931) The syntaxis of the Northwest Himalaya: its rocks, tectonics, and orogeny. *Rec. geol. Surv. India,* **65**, 189–220.

WADIA, D.N. (1934) The Cambrian–Trias sequence of Northwestern Kashmir. *Rec. geol. Surv. India,* **68**, 121–176.

Palaeozoic of North America

Spec. Publs int. Ass. Sediment. (1986) **8**, 309–325

Progressive filling of a confined Middle Ordovician foreland basin associated with the Taconic Orogeny, Quebec, Canada

R. N. HISCOTT,* K. T. PICKERING† *and* D. R. BEEDEN‡

*Centre for Earth Resources Research, Department of Earth Sciences, Memorial University of Newfoundland, St John's, Newfoundland A1B 3X5, Canada ; †Department of Geology, University of Leicester, University Road, Leicester LE1 7RH, U.K.; ‡Reservoir Geology Division, Poroperm-Geochem Limited, Chester Street, Saltney, Chester CH4 8RD, U.K.

ABSTRACT

The Ordovician Taconic Orogeny involved collision of the eastern continental margin of North America with a volcanic arc(s) situated above a SE-dipping subduction zone. Loading of the North American plate by thrust sheets of the thickened accretionary prism produced a complex foreland basin, with depositional remnants extending from Alabama to Newfoundland. In eastern Quebec, the oldest flysch deposits associated with the Taconic Orogeny are Arenig sand-rich submarine fans of the allochthonous Tourelle, Métis and St Modeste formations. These fans are characterized by packets of coarse-grained sandstones separated by interchannel or abandonment-facies siltstones and mudstones. The Tourelle Formation and coeval correlatives were deposited on uplifted continental rise deposits of the earlier passive-margin, suggesting deposition in small slope basins on the orogenic side of the foreland basin.

The main Caradoc–Ashgill fill of the foreland basin consists of (a) basin-floor deposits of the 2·3 km thick allochthonous Deslandes Formation, (b) basin-floor, lobe-fringe and sandstone lobe deposits of the 4·0 km thick parautochthonous Cloridorme Formation, and (c), on the craton side of the basin, black shales and silty carbonates of the autochthonous *c.* 100 m thick Macasty Formation and the *c.* 1·0 km thick Vauréal Formation. These units rest, or are inferred to rest, on foundered carbonates like those exposed north of Anticosti Island along the north shore of the St Lawrence River. The lower 1·5–2·0 km of the Cloridorme Formation consists of basin floor deposits with megaturbidites up to 10 m thick that resemble beds described from the modern Mediterranean basins and from the Italian Apennines. The upper 2·0 km consists of alternating coarse sandy lobe deposits like those of the Tourelle Formation, and muddy fan-fringe or lobe-fringe facies to about 500 m thick. The thicker mud intervals appear to represent times of elevated sea-level during filling of the foreland basin. Source-area denudation was essentially complete before the central part of the basin ever reached shelf depths, so that no molasse-type deposits overlie the foreland-basin flysch of eastern Quebec.

INTRODUCTION

The Appalachian Orogen experienced three discrete episodes of deformation: the Middle Ordovician Taconic Orogeny; the Devonian Acadian Orogeny; and the Carboniferous Alleghenian (Hercynian) Orogeny (Williams & Hatcher, 1983). The Taconic Orogeny involved collision of the passive continental margin of eastern North America with a belt of oceanic island arcs situated above a south-eastward-dipping subduction zone; remnants of the island arcs are found in the Dunnage tectonostratigraphic zone

of the Appalachian Orogen (Williams, 1979). This kinematic model is substantiated by data from all segments of the orogen (Malpas & Stevens, 1977; Hiscott, 1978; Robinson & Hall, 1980; Rowley, 1981; Rowley & Kidd, 1981; Shanmugam & Lash, 1982).

As the North American continental margin approached the subduction zone, the flourishing carbonate platform foundered, presumably due to normal faulting controlled by plate flexure with downbending, and was covered by (1) argillaceous limestone with slide sheets and slide scars, (2) black, graptolitic shale, and (3) flysch derived from the orogenic belt. Detailed discussion of local stratigraphic relationships of the flysch throughout the Appalachian Orogen is available elsewhere (Stevens, 1970; Rickard & Fisher, 1973; Shanmugam & Walker, 1980; Belt & Bussières, 1981; Rowley & Kidd, 1981; Shanmugam & Lash, 1982). The downwarped continental margin, bounded seaward by the emergent orogenic belt with its thick pile of thrust sheets, including the accretionary prism and remnants of volcanic arcs (St Julien & Hubert, 1975; Williams & Hatcher, 1983), defined an elongate foreland basin (Quinlan & Beaumont, 1984) that extended from Newfoundland in the north to Alabama in the south (Fig. 1). This type of foreland basin, formed by load-induced subsidence of a continental margin as it is driven toward an active or recently active oceanward-dipping subduction zone, has no good modern analogy, except perhaps for the northern margin of Australia, depressed beneath the Banda Arc (Hamilton, 1979). In the case of the Taconic orogenic event, thrust sheets on the upper plate eventually became extremely thick (Quinlan & Beaumont, 1984), and produced (a) a major fold-thrust belt facing the interior of the craton, and (b) a flexural depression on the landward side of the orogen that was geometrically, even if not mechanically, like

classic foreland basins in the Alps and western North America.

Deposits oceanward of, and contemporaneous with, the carbonate platform include platform-derived deep sea fan and slope sediments with turbidites, but these are compositionally distinct and easily separated from the Taconic-age flysch that locally overlies them (Hiscott, 1978, 1984). On a regional scale, the oldest flysch overlies passive-margin slope and rise deposits and is now allochthonous. The youngest flysch overlies the foundered carbonate platform and is autochthonous.

Timing of crustal downwarping and the introduction of flysch (Fig. 2) indicates that foreland-basin development was broadly diachronous along the length of the orogen. The earliest foundering of the carbonate platform occurred in Tennessee and Newfoundland. In Newfoundland, the early collision occurred at the St Lawrence Promontory (Thomas, 1977), whereas in Tennessee the early collision occurred between a microcontinental block (Piedmont Terrane; Williams & Hatcher, 1983), and the eastern margin of North America. Subsequent closure of the sea between the island arc belt and the continental margin was complicated, due to the irregular shape of the North American margin with its promontories and re-entrants (Thomas, 1977), and the unknown,

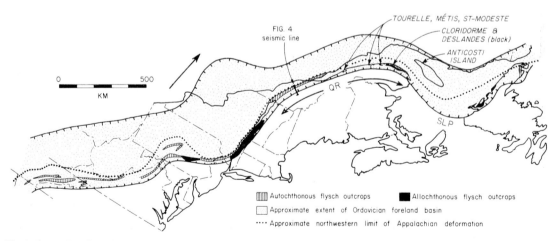

Fig. 1. General outline of Taconic-age foreland basin in the Appalachian Orogen. Development of the basin was diachronous (Fig. 2), so this sketch is really an integration of data throughout the Middle and Late Ordovician. The westward limit of the basin is arbitrary, but coincides roughly with the inferred most westward position of the peripheral bulge formed by loading of the continental crust (Quinlan & Beaumont, 1984). Thrust sheets were thinnest in the Canadian Appalachians, resulting in a relatively narrow basin. Also shown are outcrop belts of allochthonous and autochthonous (including parautochthonous) flysch (Enos, 1969a), approximate westward limit of Appalachian deformation, location of units in Quebec discussed in the text, Quebec Re-entrant (QR), St Lawrence Promontory (SLP), and location of Fig. 4 seismic line.

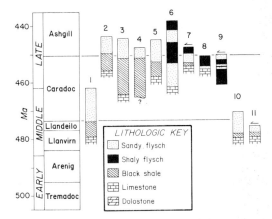

Fig. 2. Ages of autochthonous and parautochthonous flysch, adapted from Barnes *et al.* (1981), Ross *et al.* (1982) and numerous primary references. Faunal zones specified in primary references were related to European series using sheet 1 in Ross *et al.* (1982). 1 = Tellico (Tennessee) and Knobs (Virginia) Formations; 2 = Reedsville Formation (Pennsylvania); 3 = Martinsburg Formation (Tennessee, Virginia, West Virginia, Maryland, Pennsylvania, New Jersey); 4 = Shochary Ridge sequence (Pennsylvania); 5 = Schenectady Formation (New York); 6 = Nicolet River Formation (western Quebec); 7 = Beaupré (sandy) and Lotbinière (shaly) Formations (Quebec); 8 = St Irénée (sandy) and Lotbinière (shaly) Formations (Quebec); 9 = Cloridorme Formation (eastern Quebec); 10 = Mainland sandstone (southwest Newfoundland); 11 = Goose Tickle Formation (northwest Newfoundland). Units 7, 9 and 11 are overlain by thrust sheets.

but probably complex, spatial distribution of island arc segments. In general, however, collision occurred first at promontories in the ancient continental margin, substantiated by the observation that palaeoflow in flysch units is generally directed away from promontories and toward adjacent re-entrants, or embayments (Fig. 3).

The deepest axial zone of the foreland basin was close to the orogenic front, and was covered by deep-water facies deposited on basin plains and submarine fans. Cratonward parts of the basin generally had sufficiently low subsidence rates for shallow-water carbonate sedimentation to continue during development of the basin. These carbonate sections are thicker than time-equivalent strata on the stable craton (Quinlan & Beaumont, 1984), and thicken toward the orogenic belt before passing laterally into slope mudstones and then axial flysch.

This paper considers foreland-basin evolution in the eastern part of the Quebec Re-entrant (Thomas, 1977), where sediments derived from the orogenic belt consist entirely of deep-water black shales and flysch

facies. Elsewhere, in western Newfoundland and SW of Quebec City, the flysch is overlain by shallow-marine and terrestrial molasse deposits.

The most comprehensive synthesis of the geology of the Quebec Re-entrant is that of St Julien & Hubert (1975), with additional details on allochthonous flysch units provided by Hiscott (1978, 1980). In general, continental slope and rise sediments of the Cambro–Ordovician passive margin succession, often assigned to the Cap-des-Rosiers Group in eastern Quebec (St Julien & Hubert, 1975), were thrust over the foundered carbonate platform during the collisional event, and acted as the major source of detritus for (a) allochthon-

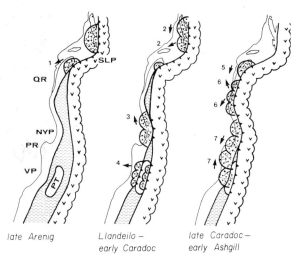

Fig. 3. Simplified model for collision of the submerged margin of eastern North America with an offshore belt of volcanic arcs. Initial collision was in the north. Arrows summarize palaeoflow data, derived from the following sources: (1) Tourelle Formation (Hiscott, 1978); (2) Mainland sandstone and Goose Tickle Formation (R. K. Stevens, unpublished data); (3) Austin Glen Member of Normanskill Formation (Middleton, 1965 and Fagan & Edwards, 1969); (4) Tellico and Knobs Formations (Walker & Keller, 1977 and Shanmugam & Walker, 1980); (5) Cloridorme Formation (Enos, 1969a); (6) Beaupré and St Irénée Formations (Belt & Bussières, 1981); (7) Martinsburg Formation (McBride, 1962). Palaeoflow was generally away from promontories and into re-entrants. For the geographic positions of the promontories and re-entrants, refer to Fig. 1, dotted line. VP, Virginia Promontory; PR, Pennsylvania Re-entrant; NYP, New York Promontory; QR, Quebec Re-entrant; SLP, St Lawrence Promontory (Thomas, 1977). PT = Piedmont Terrane microcontinent (Williams & Hatcher, 1983) responsible for production of flysch at two different times in the southern Appalachians (Rodgers, 1971; Quinlan & Beaumont, 1984). The volcanic arc terrane was probably much more complex than this diagram suggests, perhaps resembling the complex collage of arcs in the modern western Pacific Ocean.

ous flysch units deposited above the thrust slices, probably in small slope basins, and (b) younger autochthonous and parautochthonous flysch deposited in front of the thrust system in the foreland basin *sensu stricto*. Other minor sources of detritus were the volcanic arcs, and obducted ophiolites (Enos, 1969a; Hiscott, 1978, 1984; Belt & Bussières, 1981). The relationship of the autochthonous and parautochthonous flysch to the Taconic-age nappes and thrust sheets, and the general asymmetric shape of the foreland-basin fill, are shown by a deep seismic-reflection profile from the Quebec City area (Fig. 4) (Laroche, 1983). The structural configuration of the Taconic-age foreland basin was modified by later Acadian (Silurian–Devonian) folding and thrusting. These younger structures are seen most clearly on a seismic line shot from the north side of Anticosti Island in the Gulf of St Lawrence to within 5 km of outcrops of the

Cloridorme and Deslandes formations on the north coast of the Gaspé Peninsula (Fig. 5) (Roksandic & Granger, 1981).

STRATIGRAPHY AND FACIES

The Taconic-age flysch units in the Quebec Re-entrant are the allochthonous Tourelle, Métis, St Modeste and Deslandes formations, the parautochthonous Cloridorme Formation, and the autochthonous lowest silty unit (English Head 'facies', Roliff, 1968) of the Vauréal Formation in the Anticosti Basin. Unit 1 of the Vauréal Formation overlies black bituminous shales of the Macasty Formation, and has been compared lithologically to the Nicolet River Formation between Montreal and Quebec City (INRS–Pétrole, 1976). The Nicolet River Formation

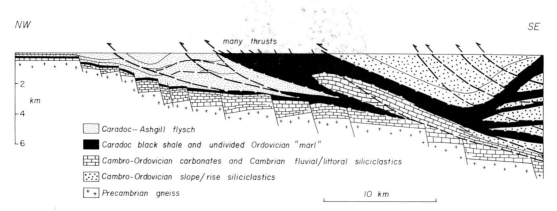

Fig. 4. Simplified interpretation of part of a SOQUIP seismic line fully described by Laroche (1983). This segment of the line is located on Fig. 1. Foreland-basin flysch lies stratigraphically above Caradoc black shales, which lie above Cambro–Ordovician carbonates of a passive-margin sequence. The flysch was subsequently overthrust by nappes containing older rocks originally deposited farther southeast. Unit 6 in Fig. 2 (Nicolet River and Lotbinière Formations) forms the flysch in this section.

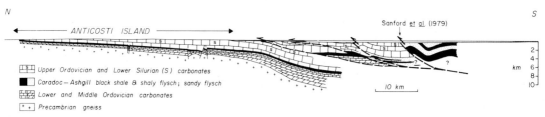

Fig. 5. Simplified interpretation of seismic line across Anticosti Island and toward the Gaspé Peninsula coast (located in Fig. 6). The imbricate thrust faults involve Lower Silurian rocks, and are therefore of Acadian age. The southern part of the line crosses a part of the Gulf of St Lawrence for which there are no bore-hole data, so that interpretations of facies changes at depth are speculative. Maps prepared by Sanford *et al.* (1979) indicate a major change from fault blocks containing Silurian carbonates to parautochthonous Cloridorme Formation flysch near the southern end of the line.

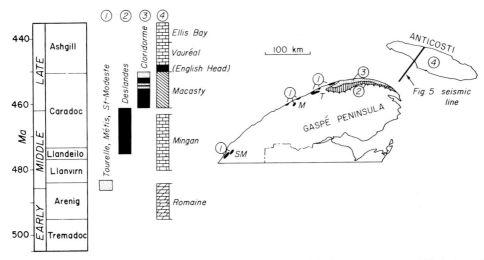

Fig. 6. Ages of Ordovician flysch and platform deposits in the eastern part of the Quebec Re-entrant. Lithologic symbols are given in Fig. 2, but do not apply to the map. On the map, units are distinguished by different patterns and are keyed to stratigraphic columns 1–4. Ages are taken from Barnes *et al.* (1981) (columns 1–3), and SOQUIP (1983) (column 4).

has been described by Beaulieu, Lajoie & Hubert (1980), and is a shaly flysch deposit with deep-sea fan characteristics. Ages of these units and other Ordovician strata on Anticosti Island are given in Fig. 6; thickness estimates appear in Table 1. The Tourelle, Métis and St Modeste formations are equivalent in age and in facies. As noted by Biron (1972, 1973), the Deslandes Formation is equivalent in age and facies to the bulk of the α block of the Cloridorme Formation

as defined by Enos (1969a); therefore, we will also apply the name Deslandes to strata of the α block. The constituent members of the Cloridorme Formation (Fig. 7, Table 1) are informal, and are not those recognized by Enos (1969a). Enos' members were designated by Greek letters and Arabic numbers, but we have found serious overlaps in the definitions of, in particular, Enos' β and lowermost γ units. Furthermore, we feel that some of Enos' members are defined

Table 1. Thickness estimates of units in Figs 6 and 7

Unit	Thickness (m)	Source
1. Tourelle Formation and equivalents	300–500	Hiscott (1977)
2. Deslandes Formation (α)	2300	Enos (1969a)
3. Cloridorme Formation	4000	Our work & Enos (1969a)
Marsoui (M) member	1000	Beeden (1983)
Mont-St-Pierre (MSP) member	475	Enos (1969a)
Petite–Vallée (PV) member	835	Enos (1969a)
Pointe-à-la-Frégate (PF) member	580	Our work & Enos (1969a)
Manche d'Epée (ME) member*	550	Enos (1969a)
St Hélier (SH) member	1140	Our work & Enos (1969a)
4. Romaine Formation	440–800	SOQUIP (1983)
5. Mingan Formation	440–550	SOQUIP (1983)
6. Macasty Formation	30–170	SOQUIP (1983)
7. Vauréal Formation	925–1160	SOQUIP (1983)
Unit 1 'flysch'	290–700	SOQUIP (1983)
8. Ellis Bay Formation	100–275	SOQUIP (1983)

*The ME member is a lateral equivalent of the PF member, and is omitted in calculating total thickness of the formation.

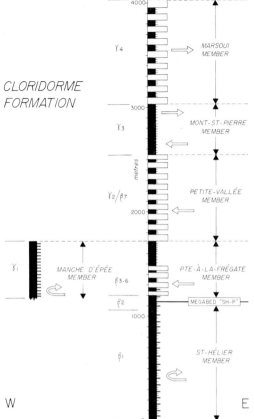

Table 2. Facies classes, groups, and selected facies

Class		
	Group	
		Facies
A	Gravel, muddy gravel, gravelly mud, pebbly sand, $\geq 5\%$ gravel	
B	Sand, $\geq 80\%$ sand grade, $<5\%$ pebble grade	
	B1	Disorganized sands
	B2	Organized sands
C	Sand-mud couplets and muddy sands, 20–80% sand grade, $<80\%$ mud grade (mostly silt)	
	C1	Disorganized muddy sands
	C2	Organized sand-mud couplets
		C2.4 Very thick/thick-bedded, mud-dominated, sand-mud couplets (megaturbidites)
D	Silts, silty muds, and silt-mud couplets, $>80\%$ mud, $\geq 40\%$ silt, 0–20% sand	
	D2	Organized silts and muddy silts
E	$\geq 95\%$ mud grade, $<40\%$ silt grade, $<5\%$ sand and coarser grade	
	E1	Disorganized muds and clays
	E2	Organized muds
F	Chaotic deposits	
G	Biogenic oozes, hemipelagites and chemogenic sediments, $<5\%$ terrigenous sand and gravel	

Fig. 7. Informal members of the Cloridorme Formation, and correlation with members of Enos (1969a). The top of the St Hélier member is the top of a distinctive thick megaturbidite (Fig. 8) described by Pickering & Hiscott (1985). Sandstone packets (stipple) are schematic. Black = shale, thin ticks = thin-bedded turbidites, and thick ticks = megaturbidites (Pickering & Hiscott, 1985). Large arrows crudely summarize palaeoflow. In the St Hélier and Manche d'Epée members, megaturbidites show flow reversals due to reflection from basin margins.

on too fine a scale given the apparently abrupt lateral facies changes that characterize parts of the Cloridorme Formation.

Detailed descriptions of facies present in the Tourelle, Métis, St Modeste, Deslandes, and Cloridorme formations are available elsewhere (Enos, 1969a; Skipper & Middleton, 1975; Hiscott & Middleton, 1979; Hiscott, 1980; Beeden, 1983; Pickering & Hiscott, 1985). In this paper, we refer to facies using the classification scheme of Pickering *et al.* (1986) (Table 2), modified from the Mutti & Ricci Lucchi (1972) scheme. A brief summary of facies present in the foreland-basin fill is given below.

Pebble conglomerates are rare or absent in these flysch units, with the coarsest facies being granule-rich sandstones of Facies Class B. Disorganized sands of Facies Group B1 are restricted to the Tourelle, Métis and St Modeste formations, and consist of poorly graded, coarse-grained sandstone beds that range in thickness from about 1 to 20 m. Organized sands of Facies Group B2 are associated with sands of Facies Group B1, but also occur without Group B1 in the Petite–Vallée and Marsoui members of the Cloridorme Formation as medium- to coarse-grained, parallel-stratified and cross-stratified sandstones from about 1 to 2 m thick.

Disorganized muddy sands of Facies Group C1 have been described from the Tourelle Formation (slurry-sandstone facies of Hiscott, 1980), the Marsoui member of the Cloridorme Formation ('type 2 greywackes' of Enos, 1969a), and the St Hélier member of the Cloridorme Formation (Skipper & Middleton, 1975; 'type 3 greywackes' of Enos, 1969a). In the latter case, beds are about 20–50% muddy sand, with the remainder being an homogeneous mudstone cap. The beds are generally >1 m thick, and range up to 8·4 m, including the mudstone cap. The disorganized muddy sands represent the deposits of mud-rich, very high density turbidity currents or fluid sand-mud debris flows.

Facies Group C2 includes (a) classic turbidites that can be described using the Bouma scheme, and (b) thick-bedded and very thick bedded, mud-dominated, sand-mud couplets (Facies C2.4), that occur as megaturbidites 1–10 m thick. The classic turbidites occur in all flysch formations, both within sandstone packets and in shale-dominated units. The megaturbidites (Fig. 8) have thick mud caps, are up to approximately 10 m thick, and have sole and internal structures, generally flutes and ripple lamination, testifying to multiple reversals of flow direction during deposition of single, step-wise-graded beds (Hiscott & Pickering, 1984; Pickering & Hiscott, 1985). Such megaturbidites are now widely interpreted as basin

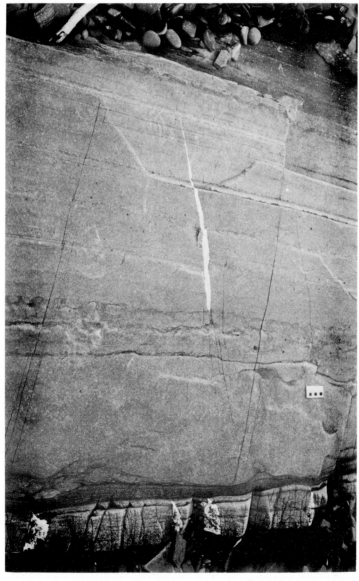

Fig. 8. Lower sandy part of megaturbidite SH-P in the St Hélier member of the Cloridorme Formation (Pickering & Hiscott, 1985), with a parallel-laminated to cross-stratified coarse sandstone base, about 10 cm thick, followed abruptly by alternating divisions of pseudonodules and laminated fine sandstone. The total thickness of this bed, with its mudstone cap (starting at beach gravel), is 554 cm. Scale divisions are in centimetres.

plain deposits of confined basins (Ricci Lucchi & Valmori, 1980; Cita *et al.*, 1984; Hiecke, 1984; Mutti *et al.*, 1984; Seguret, Labaume & Madariaga, 1984).

Most fine-grained intervals in all flysch units consist primarily of Facies Group D2, in the form of thin, graded, laminated siltstone beds interlayered with laminated shale. These are siltstone turbidites, interpreted as deposits of (a) mid-fan (*sensu* Normark, 1970) interchannel and channel levee complexes, (b) interlobe mud blankets on the middle fan, and (c) lobe-fringe and perhaps more distal lower-fan settings.

No other facies classes or groups are significant components of the Quebec foreland-basin fill. Rare chaotic slide deposits of Facies Class F occur in several formations.

ARENIG–LLANVIRN SLOPE BASINS

The Tourelle, Métis and St Modeste formations were deposited on uplifted and probably tilted slope and rise sediments of the earlier passive continental-margin succession. These slope and rise deposits comprise varicoloured shales, carbonate turbidites and conglomerates, quartzose turbidites, and bedded cherts, all so chaotically deformed as to have been termed a mélange by Biron (1974). The timing of this deformation is conjectural, but the occurrence of folded slope and rise blocks in the Tourelle Formation (Hiscott, 1977) and evidence of injection of mud into sandstone blocks in the mélange (R. K. Stevens, personal communication), suggesting water-rich, poorly-consolidated muds, both support early deformation before, during or shortly after deposition of the Tourelle Formation and equivalent units. That the older slope and rise deposits underlay the NW-ward-dipping slope down which the flysch detritus was transported into the basin is shown by the presence of large slabs (olistoliths) of carbonate turbidites and bedded chert in sandy sediment gravity flow deposits (Facies Group B1) of the Arenig-Llanvirn flysch (Hiscott, 1980). Also, the transition from passive-margin slope and rise deposits into coarse-grained, thick-bedded sandstones of Facies Groups B1, B2 and C2 is abrupt, with slides, slide scars and sandstone injections (Hiscott, 1979) characterizing the contact. These features are believed to be more consistent with (a) rapid introduction of sands into a small slope basin, or onlap of sheet or channelled sands on to a pre-existing slope, than with (b) progradation of a turbidite system into a larger foreland basin.

The only Arenig–Llanvirn unit that has been studied in detail is the Tourelle Formation (Hiscott, 1979, 1980; Hiscott & Middleton, 1979, 1980). The Métis and St Modeste formations have been mapped and described briefly by Liard (1973) and Vallières (1977), respectively. Amalgamated medium- to coarse-grained sandstones and sandstone beds (>2 m thick (Facies Groups B1 and B2) comprise about 50% of the Tourelle Formation: these deep-water systems were relatively sand-rich. The thickest continuous sections are on the order of 250 m thick; the formation may have originally been up to twice that thick. Some sections contain very little shale, whereas others are characterized by an alternation of (a) sand-rich packets of Facies Groups B1, B2 and C2 and (b) shaly units of Facies Groups D2 and E2, each on the order of 10–50 m thick. Channels of outcrop scale are present but not common. Four channels with minimum depths of 7, 13, 9 and 10 m are described by Hiscott (1980).

The Tourelle formation, and by comparison the Métis and St Modeste formations, is interpreted as the deposit of several overlapping, small, sand-rich submarine fans (radius about 20 km) derived from a south-eastern source (Hiscott, 1980). The fans are believed to have been confined to small slope basins, low on a slope formed by stacking of Taconic-age thrust sheets. The region in front of these thrusts was, at this time, an incipient foreland basin that deepened and widened later as the thrust prism became thicker.

LLANDEILO–MID-CARADOC BASIN PLAIN

The Deslandes Formation and the St Hélier (SH) member of the Cloridorme Formation are generally fine-grained, shale-rich sequences dominated by Facies Group D2, with spaced megaturbidite beds up to 5·5 m thick (Facies C2·4) and thick, disorganized muddy sandstones up to 8·4 m thick (Facies Group C1) (Skipper & Middleton, 1975; Pickering & Hiscott, 1985). The Manche d'Epée (ME) member (γ1 member of Enos, 1969a), exposed 50 km west of St Helier, consists almost entirely of megaturbidites, the thickest being 9·5 m thick. Although the ME member is similar in facies to the Deslandes Formation and SH member, it is younger than both (Fig. 7), and probably is the basinward equivalent of westward-building lobes of the Point-à-la-Frégate member discussed below.

The fine grain size of the sections mentioned above, and the megaturbidites, are consistent with a basin plain or basin floor setting for these sediments. The basin floor was essentially flat to allow flow of megaturbidity currents in either direction along the foreland-basin axis, but must have been confined by relatively steep basin margins and intrabasinal highs to force flow reflections (or deflections) without the very large energy and sediment losses that would result from relatively gradual deceleration on gentle basin slopes. Steep slopes are also consistent with an origin for the megaturbidity currents involving major slope failures, perhaps triggered by seismic shocks (cf. Mutti *et al.*, 1984) or by increased load as a result of a sudden sea-level rise (see Discussion).

It is possible to calculate the approximate transit distances of these large flows between successive passes by a given station, based on grain-size differences across grading breaks, estimates of flow velocities from sedimentary structures, and rudimentary models for the energy budget of turbidites (Komar, 1977). Preliminary results suggest that the effective length of this bathymetrically confined segment of the foreland basin was on the order of 100 km. This length compares favourably with the spacing between intrabasinal structural highs, about 75 km, in the Tertiary Periadriatic foreland basins of Italy (Ricci Lucchi, 1975), and is 3–4 times greater than the actual distance that individual megaturbidites can be traced along strike in the SH member (Pickering & Hiscott, 1985). We believe that the PF and ME members were deposited synchronously in the same segment of the basin, with the ME member being the basin-floor equivalent of sandstone lobes of

the PF member (Fig. 9). The present distance between the most widely separated outcrops of these two members is 62 km, well within our 100 km estimate.

The shales and thin fine-grained sand and silt turbidites of the Deslandes Formation and SH member (Facies Group D2) were probably derived from multiple feeder channels at the south-eastern basin margin, and formed small isolated depositional lobes in the basin.

MID-CARADOC SANDY LOBES

The Pointe-à-la-Frégate (PF) and Petite-Vallée (PV) members of the Cloridorme Formation have a total thickness of about 1400 m. Except for a 255 m interval of predominantly shale with siltstone turbidites (Facies Group D2, member $\beta6$ of Enos, 1969a) forming the upper half of the PF member, the 1400 m is characterized by an alternation of (a) 10–30 m thick sandstone packets of Facies Group B2 and C2, and (b) 30–80 m thick units of shale and siltstone turbidites (Facies Group D2), with rare <5 m thick megaturbidites (Facies C2.4) in the lower 200 m of the PF member. The medium- to coarse-grained sandstone packets show no trends in bed thickness or grain size, and only rarely have metre-scale channelling at their base. Correlated sections in the PF member (Fig. 10) and in the PV member (Fig. 11) (Enos, 1969a, b) show that sandstone packets thin, and in some cases are replaced entirely by siltstone/shale facies, over dis-

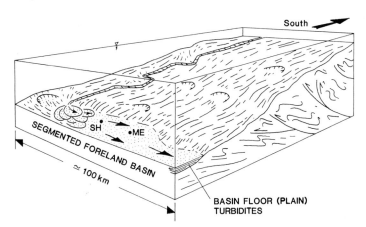

Fig. 9. Palaeogeographic sketch of foreland-basin during Llandeilo–mid-Carodoc time. The basin floor was confined by steep side slopes, and by intrabasinal highs with spacing of about 100 km. Large-volume turbidity currents were multiply deflected and reflected within the confined basin centre, while small sandstone lobes prograded from east to west. SH = St Hélier, ME = Manche d'Epée.

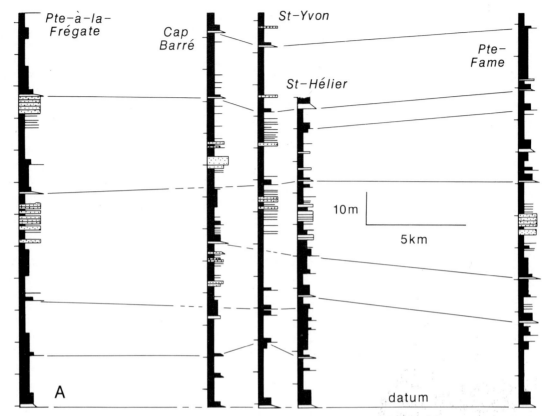

Fig. 10. Correlated sections through the Pointe-à-la-Frégate member, from coastal localities up to about 25 km apart. The datum (A) is the base of megaturbidite SH-P (Fig. 8), the top of which defines the base of the PF member. (B) is positioned immediately above (A). Inspection shows that sandstone packets (stippled) have limited extent, and in several cases cannot be traced between adjacent sections. Megaturbidites are draughted with a pointed base (sandstone part) and a black mudstone cap. These caps project further to the right than black bars representing Facies Group D2 with >40% siltstone turbidites, which in turn project further right than bars representing Facies Group D2 and E2 with <40% siltstone turbidites. Unpatterned beds without pointed bases are thicker siltstone turbidites.

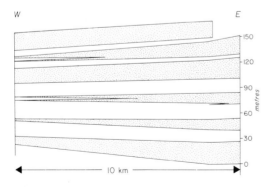

Fig. 11. Geometry of sandstone packets (stippled) in the Petite–Vallée member, correlated over a distance of 10 km by Enos (1969a). Mean current directions are from the east (right). Un-patterned intervals are mud-rich packets.

tances of 5–10 km. In the case of packets in the PF member, these lateral changes represent the combined effects of downcurrent and cross-current variability, whereas in the PV member, the thinning observed by Enos (1969a, b) is entirely in the downcurrent direction.

We interpret the sandstone packets as small lobes that were fed from slope channels to the east, closer to the St Lawrence Promontory. The lobes had relief above the general level of the seafloor, as some megaturbidites did not deposit in the vicinity of lobes, probably because of deflection by the lobe topography, while other megaturbidites thin over the tops of the lobes. Extremely high sediment accumulation rates led to rapid lobe switching as the system built westward across the former basin plain (Fig. 12). Fine-grained

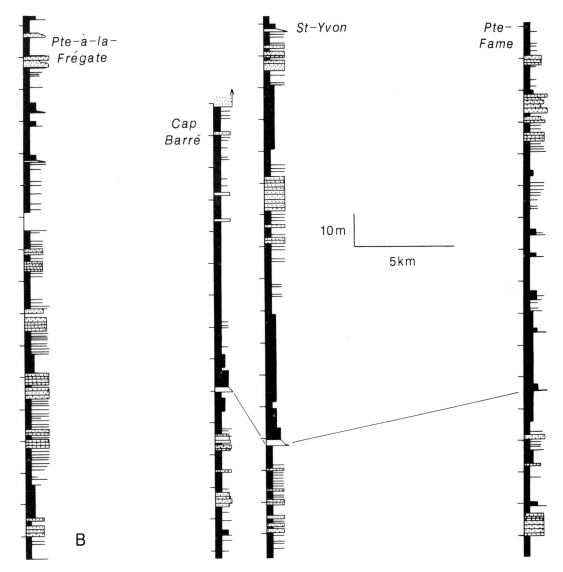

deposits of Facies Group D2 represent the lateral and distal margins of lobes, where rapid deposition from suspension often led to the formation of climbing-ripple lamination.

We hesitate to use the word 'fan' for the system containing the sandy lobes, primarily because (a) we do not know the shape of the depositional system (Pickering, 1982), (b) we do not know if there was one or several feeder channels, and (c) palaeoflow is consistently toward about 280°, along the axis of the foreland basin. If the term 'fan' were to be used, then it would be essential to specify that the system was grossly oversupplied, with deposit shape determined not by depositional processes, but rather by basin shape.

The alternation of packets of Facies Groups B2 and C2 with units of Facies Group D2 (Figs 10 and 11) probably records the dynamics of rapidly shifting lobes, an autocyclic control. High sediment accumulation rates, with the entire 1400 m succession of the PF and PV members deposited within the span of one or two graptolite zones, makes it impossible to assess the importance and effects of short-duration local or global eustatic sea-level changes, which are not documented in sufficient detail for this purpose (e.g. Fortey, 1984).

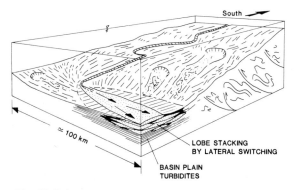

Fig. 12. Palaeogeographic sketch of foreland basin during mid-Caradoc time. Sandstone lobes were constrained by the east–west orientation of the basin margins, and were stacked vertically and laterally as the basin filled. The lobes pass laterally and distally into interbedded shales and thin-bedded turbidites (Facies Group D2). Because of relatively high sedimentation rates, megaturbidites generated by large slides are rare, and are only found in the lower 200 m of the PF member.

LATE CARADOC MUD BLANKET

The Mont-St-Pierre (MSP) member of the Cloridorme Formation (or γ3 member of Enos, 1969a) consists of about 475 m of laminated shale (70%), siltstone turbidites (13%) and minor sandstone turbidites and dolomite concretions. The major Facies Groups represented are D2, E2 and C2.

Not only is this stratigraphic interval finer grained than the vertically adjacent units, but palaeoflow for 80% of the siltstone turbidites is opposite from that of underlying members, indicating a mean flow toward 070° (Enos, 1969a).

The fine grain size and the demise of the former eastern source area probably are genetically related, and indicate either (a) construction of an intrabasinal barrier to exclude most flows from the east, or (b) rapid decline in the relative relief of the eastern source coupled with progressive tectonically-induced rise of a new source to the west, so that the sediment contribution from the eastern source was considerably reduced and eventually eliminated. We favour the latter alternative, because an intrabasinal high could only be expected to hold back temporarily the contribution of a vigorous source. Instead, the 20% contribution from the east, restricted to the base of the MSP member, is the last input from that direction. Also, the transition from sandstone packets of the PV member to Facies Group D2 deposits of the MSP member is abrupt (Enos, 1969a), a feature that we

would not expect from relatively slow uplift of intrabasinal highs.

The mechanism that we find most attractive to account for the rapid reduction in the relief of the eastern source, and therefore in the quantity of coarse detritus, is a eustatic rise in sea-level. The sea-level curve of Fortey (1984) indicates a mid-Caradoc high stand of sea-level, consistent with this hypothesis.

LATE CARADOC–EARLY ASHGILL EAST-DIRECTED FAN

The Marsoui (M) member of the Cloridorme Formation (γ4 member of Enos, 1969a) is approximately 1000 m thick, and is characterized by mean palaeoflow toward 070° (Enos, 1969a; Beeden, 1983). Like the lower part of the PF member and the PV member, the section consists of alternating sandstone packets, 3–45 m thick, dominated by Facies Group C2 Bouma Ta turbidites (Facies C2.1), and units of shale and siltstone turbidites (Facies Groups D2 and E2). A single observed channel-fill sequence, 10 m thick, does not show a well-developed layer-thickness trend. In general, sandstone packets are characterized by gradational bases, multiple or single thickening-upward sequences internally, and in some cases by a rather rapid thinning-upward trend at the top (Beeden, 1983).

The Marsoui member is interpreted as a large, eastward-building deep-sea fan, overlying finer grained muds of the MSP member that were also transported by eastward-flowing turbidity currents. Like the mid-Caradoc sand packets, those of the Marsoui member were formed by rapidly shifting lobes on an oversupplied, elongate 'fan'. The shale/siltstone units represent finer grained lobe-fringe sediments.

CARADOC–ASHGILL MUDS ON THE PLATFORM

Both the Macasty Formation and unit 1 of the Vauréal Formation are not exposed in outcrop on Anticosti Island. They have been sampled in several bore holes (SOQUIP, 1983). The Macasty is a bituminous black shale that resembles the Utica Shale exposed between Quebec City and New York State (Belt & Bussières, 1981). These shale units probably represent hemipelagic muds, and thin distal mud turbidites emplaced by flows that periodically reached beyond the northern limit of the foreland basin and out on to the adjacent craton. Unit 1 of the Vauréal Formation consists of terrigenous siltstones and marls that resemble flysch of the Nicolet River Formation west of Quebec City (Beaulieu *et al.* 1980). Both the Macasty and Vauréal formations thicken toward the axis of the foreland basin (Roliff, 1968), but the nature of the transition into facies like those present in the Cloridorme Formation is unknown.

POST-CARADOC FORELAND-BASIN EVOLUTION

The top of the Cloridorme Formation is truncated by modern erosion, and there is no indication of shallowing at the top of the Marsoui member. The section on Anticosti Island indicates that sediment supply from the orogenic belt diminished to the extent that the foreland basin did not receive any molasse-type deposits like those present near the New York Promontory (Queenston 'delta'): siltstones and marls of the lower unit of the Vauréal Formation are overlain by increasingly shallow-water carbonates (Fig. 6). For this segment of the Taconic-age foreland basin, the volume of sediment eroded from the orogenic highlands must have been equal to or somewhat less than the submarine volume of the basin. What began as an oversupplied basin with high rates of sediment accumulation ended its evolution as an essentially full, load-induced downwarp, starved of additional clastic detritus. Additional evidence that the source terrane was virtually eroded to base level is the local presence of Ashgillian shelf carbonates on top of the thrust belt in the Gaspé Peninsula (White Head Formation; Lespérence, 1985)—shallow seas now overlay parts of the former source terrane.

DISCUSSION

The segment of the foreland basin that we have described was located in a re-entrant in the continental margin, and therefore was protected from the most intense effects of plate collision. Nevertheless, proximity to intensely deformed and uplifted rocks at the St Lawrence Promontory resulted in high sediment accumulation rates. Mean sediment accumulation rate can be calculated using the estimates of (a) stratigraphic thickness (Table 1), and (b) duration of sedimentation (Fig. 6). The latter quantity involves the greatest uncertainties. The age of the Cloridorme

Formation (Riva, 1968, 1974; Barnes, Norford & Skevington, 1981) extends from the base of the *Corynoides americanus* Zone to about the middle of the *Climacograptus spiniferus* Zone. According to a consensus view outlined in charts of Ross *et al.* (1982), this is a time interval of about 10 Myr. The thickness of the Cloridorme Formation is about 4000 m (500 m more than estimate of Pickering & Hiscott, 1985, based on new data of Beeden, 1983), which gives an average sediment accumulation rate of 400 m Myr^{-1}. This rate is similar to those reported for (a) the Astoria Fan during the Pleistocene of 360–600 m Myr^{-1} (Nelson, 1976), and (b) the Miocene Marnoso-arenacea of 150–750 m Myr^{-1} (Ricci Lucchi, 1981, p. 233), but is considerably lower than (c) accumulation rates of up to 11 m 1000 yr^{-1} (11,000 m Myr^{-1}) determined for the Pleistocene succession on the Mississippi Fan (Kohl *et al.*, 1985). The Mississippi Fan rate is unusually high due to augmented low-stand sediment yields, and due to the large drainage-basin area of the Mississippi River, including the ice front of a large continental ice sheet. In examples (a) and (b), rates are still considered to be high, due to either low sea-level (Astoria Fan) or active tectonics feeding sediments into confined basins (Marnoso-arenacea). The calculated mean rate for the Cloridorme Formation is probably a minimum, because there may be unrecognized breaks in sedimentation in such a tectonically active area. Certainly rates for the sand-rich members of the formation must have been considerably more than the mean value. Also, the Cloridorme rate was calculated without allowing for post-depositional compaction, which has not been so severe in the Tertiary and Quaternary examples presented above.

We have estimated the approximate length, about 100 km, of the segment of the foreland basin in which basin-plain turbidity currents were reflected and deflected. We have no data on the original width of the axial part of the basin. Given that turbidity currents were effectively constrained to flow along the axis of the foreland basin, the basin shape in its deepest parts must have been narrow and elongate, suggesting an effective width in the deepest axial part of the basin of up to only a few tens of kilometres. From this axial trough, the seafloor then rose gently to the north, eventually reaching the effective limit of the foreland basin tens of kilometres to the north of Anticosti Island (Fig. 1).

We are unaware of any previous suggestion of structural highs within the Taconic-age foreland basin at high angles to its axis. Such intrabasin highs have been noted elsewhere in younger foreland basins (Ricci Lucchi, 1975), and may be the result of reactivation of basement faults originally formed by rifting of continental crust during formation of the older passive margin. Alternatively, uneven distribution of structural loads, or compensation effects required by curvature of the axis of the foreland basin in re-entrants, may have been responsible for generation of cross-basin uplifts. Subdivision of the longer foreland basin into short segments by intrabasinal highs is a necessary condition for the confinement and reflection of large turbidity currents needed to produce reversed palaeocurrents and thick mudstone caps in single beds of Facies C2.4 that occur in the Deslandes and Cloridorme Formations.

Despite apparently high sediment accumulation rates throughout its deposition, the Cloridorme Formation is characterized by alternating sand-rich and mud-rich members; the PF member also has a thick mudstone sequence at its top. Given the short time-scale involved, we are reluctant to suggest alternating periods of tectonic uplift and denudation to explain these alternations. Instead, we favour a model in which most sand-rich parts of the section were deposited during times of normal or depressed sea-level, and mud-rich parts of the section were deposited during times of elevated sea-level. This explanation of alternating sand-rich and sand-poor deep-sea sequences is identical to that proposed by Mutti (1985), and finds support in the Ordovician sea-level curve of Fortey (1984) (Fig. 13). In particular, sand-rich fans of late Arenig and late Caradoc–early Ashgill age

coincide in time with inferred low stands of sea-level. Basin-plain fine-grained sediments were deposited during a Caradocian sea-level rise, which may have contributed to instability of basin-margin muds, causing massive slope failures and emplacement of megaturbidites: increased water depths augment load pressures on the sea-bed, and increase the probability of sliding. The late Caradocian mud blanket coincides with an extreme high in global Ordovician sea-level. The only anomaly in this general model is mid-Caradoc sand lobes that appear to have been emplaced when global sea-level was relatively high. This is only compatible with the sea-level curve if tectonic uplift was responsible for introduction of these particular sand bodies. These sands are not sheet-like, but rather interfinger laterally and distally with muds, a feature that may help distinguish extensive low-stand sands (Mutti, 1985) from high-stand, locally-derived sands in tectonically active areas.

In an earlier section, we discounted a sea-level control for alternating sandstone and shale packets *within* members of the Cloridorme Formation. These packets are 1–2 orders of magnitude thinner than the larger scale alternations shown in Fig. 7, and we see no reason to seek a similar explanation for both scales of cyclicity.

Inferred slope-basin fans of the Tourelle, Métis and St Modeste formations are more proximal in character and more sand-rich than lobe and lobe-fringe deposits in the Cloridorme Formation. If the Cloridorme deposits were fans *sensu stricto*, then the two types of fan correspond, respectively, to the low and high efficiency types of Mutti (1979). Rather than suggest significant changes in the population of grain sizes delivered to the marine environment during the course of the Taconic Orogeny, we believe that the higher proportion of sand in the slope-basin fans was the result of less effective entrapment of the finer sediment sizes by the slope basins. Instead, the fine-grained suspended load of turbidity currents and the bulk of hemipelagic sediments bypassed the slope basins and accumulated in the deeper parts of the developing foreland basin.

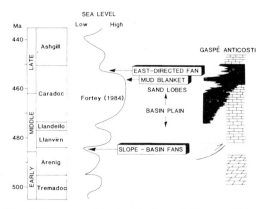

Fig. 13. Summary of foreland-basin evolution set within a framework of inferred variations in global sea-level (Fortey, 1984). Those deposits believed to have been most affected by eustatic variations in sea-level are boxed and referenced to the sea-level curve with arrows. The late Arenig sand-rich fans are allochthonous.

SUMMARY

During the Ordovician Taconic Orogeny, the Quebec Re-entrant was a site for the deposition of basin-plain and submarine-fan facies (a) in slope basins on a stack of thrust sheets and (b) in a large foreland basin that developed on top of foundered platform carbonates of the former passive continental margin of eastern North America. The preserved deposits of the slope basins are Arenig–Llanvirn sand-rich fans, about 300–500 m thick. The foreland basin *sensu stricto* was filled with (a) about 1150 m of basin-plain turbidites, megaturbidites and laminated shales; (b) about 1400 m of alternating fan-lobe and lobe-fringe deposits originating from an easterly source near the St Lawrence Promontory; (c) about 475 m of muds and thin-bedded turbidites probably deposited during a high stand of sea-level, and (d) about 1000 m of alternating fan-lobe and lobe-fringe deposits origi-nating from a westerly source. In all but one case, sand-rich intervals appear to have been deposited at times of low sea-level, and shaly intervals at times of high global sea-level.

The foreland basin appears to have been character-ized by intrabasinal highs forming saddles along the axis of the basin. The areas between these highs were bathymetric depressions that had steep enough mar-ginal slopes to cause reflections and deflections of large turbidity currents. Evidence from the Clori-dorme Formation suggests a bathymetric depression with length scale of about 100 km. Such internal highs may be common in foreland basins developed above former passive continental margins.

ACKNOWLEDGMENTS

This research was funded by Canadian N.S.E.R.C. Operating Grants to R. Hiscott, and to G. V. Middleton who supervised theses of Hiscott (1977) and Beeden (1983). Field support for K. Pickering was generously provided by the J. B. Tyrell Fund of the Geological Society of London. Technical assist-ance from L. Nolan and W. Marsh is gratefully acknowledged. An early version of the manuscript was improved by criticisms of C. R. Barnes. Reviewer D. Cant made several suggestions which improved the quality of the manuscript.

REFERENCES

BARNES, C.R., NORFORD, B.S. & SKEVINGTON, D. (1981) *The Ordovician System in Canada. Publ. int. Un. geol. Sci.* **8**, 27 pp.

BEAULIEU, J., LAJOIE, J. & HUBERT, C. (1980) Provenance et modèle de dépôt de la Formation de la Rivière Nicolet: flysch taconique du Domaine autochton et du Domaine externe des Appalaches du Québec. *Can. J. Earth Sci.* **17**, 855–865.

BEEDEN, D.R. (1983) *Sedimentology of some turbidites and related rocks from the Cloridorme Group, Ordovician, Quebec.* M.Sc. Thesis, McMaster University, Hamilton, Canada.

BELT, E.S. & BUSSIÈRES, L. (1981) Upper Middle Ordovician submarine fans and associated facies northeast of Quebec City. *Can. J. Earth Sci.* **18**, 981–994.

BIRON, S. (1972) Géologie de la région de Ste-Anne des Monts. *Min. Richesses Nat. Quebec, Service Cartographie, open file rep.*

BIRON, S. (1973) Géologie de la région de Marsoui. *Min. Richesses Nat. Quebec, Service Cartographie, open file rep.*

BIRON, S. (1974) Géologie de la région des Méchins. *Min. Richesses Nat. Quebec, Service Cartographie, open file rep.*, 12 pp.

CITA, M.B., BEGHI, C., CAMERLENGHI, A., KASTENS, K.A., McCOY, F.W., NOSETTO, A., PARISI, E., SCOLARI, F. & TOMADIN, L. (1984) Turbidites and megaturbidites from the Herodotus abyssal plain (eastern Mediterranean) unrelated to seismic events. *Mar. Geol.* **55**, 79–101.

ENOS, P. (1969a) Cloridorme Formation, Middle Ordovician flysch, northern Gaspé Peninsula, Quebec. *Spec. Pap. geol. Soc. Am.* **117**, 66 pp.

ENOS, P. (1969b) Anatomy of a flysch. *J. sedim. Petrol.* **39**, 680–723.

FAGAN, J.J. & EDWARDS, M. (1969) Paleocurrents in Normanskill Formation, south of Hudson and Catskill, New York. *Bull. geol. Soc. Am.* **80**, 121–124.

FORTEY, R.A. (1984) Global earlier Ordovician transgres-sions and regressions and their biological implications. In: *Aspects of the Ordovician System* (Ed. by D. L. Bruton), pp. 37–50. *Palaeont. Contr. Univ. Oslo 295.*

HAMILTON, W. (1979) Tectonics of the Indonesian Region. *Prof. Pap. U.S. geol. Surv. 1078,* 345 pp.

HIECKE, W. (1984) A thick Holocene homogenite from the Ionian Abyssal Plain (eastern Mediterranean). *Mar. Geol.* **55**, 63–78.

HISCOTT, R.N. (1977) *Sedimentology and regional implications of deep-water sandstones of the Tourelle Formation, Ordovi-cian, Quebec.* Ph.D. Thesis, McMaster University, Ham-ilton, Canada.

HISCOTT, R.N. (1978) Provenance of Ordovician deep-water sandstones, Tourelle Formation, Quebec, and implications for initiation of the Taconic orogeny. *Can. J. Earth Sci.* **15**, 1579–1597.

HISCOTT, R.N. (1979) Clastic sills and dikes associated with deep-water sandstones. Tourelle Formation, Ordovician, Quebec. *J. sedim. Petrol.* **49**, 1–10.

HISCOTT, R.N. (1980) Depositional framework of sandy mid-fan complexes of Tourelle Formation, Ordovician, Que-bec. *Am. Ass. Petrol. Geol.* **64**, 1052–1077.

HISCOTT, R.N. (1984) Ophiolitic source rocks for Taconic-age flysch: trace-element evidence. *Bull. geol. Soc. Am.* **95**, 1261–1267.

HISCOTT, R.N. & MIDDLETON, G.V. (1979) Depositional mechanics of thick-bedded sandstones at the base of a submarine slope, Tourelle Formation (Lower Ordovician), Quebec, Canada. In: *Geology of Continental Slopes* (Ed. by L. J. Doyle & O. H. Pilkey), pp. 307–326. *Spec. Publ. Soc. econ. Paleont. Mineral., Tulsa,* **27**.

HISCOTT, R.N. & MIDDLETON, G.V. (1980) Fabric of coarse deep-water sandstones, Tourelle Formation, Quebec, Canada. *J. sedim. Petrol.* **50**, 703–722.

HISCOTT, R.N. & PICKERING, K.T. (1984) Reflected turbidity currents on an Ordovician basin floor, Canadian Appalachians. *Nature*, **311**, 143–145.

INRS-PÉTROLE (1976) Stratigraphie, potential roche-mere, diagenèse minerale-organique du forage New Associated Consolidated Paper (N.A.C.P.) Anticosti no. 1. *Min. Richesses Nat. Quebec, Serv. Documentation Technique DP-360*, 78 pp.

KOHL, B. & LEG 96 SHIPBOARD SCIENTISTS (1985) Biostratigraphy and sedimentation rates of the Mississippi Fan. In: *Submarine Fans and Related Turbidite Systems* (Ed. by A. H. Bouma, W. R. Normark & N. E. Barnes), pp. 267–273. Springer-Verlag, New York.

KOMAR, P.D. (1977) Computer simulation of turbidity current flow and the study of deep-sea channels and fan sedimentation. In: *The Sea, vol. 6* (Ed. by E. D. Goldberg, I. N. McCave, J. J. O'Brien & J. H. Steele), pp. 603–621. Wiley, New York.

LAROCHE, P.J. (1983) Appalachians of southern Quebec seen through Seismic Line no. 2001. In: *Seismic Expression of Structural Styles* (Ed. by A. W. Bally), 3, 3.2.1–7 to 3.2.1–24. *Am. Ass. Petrol. Geol., Stud. Geol.* **15**.

LESPÉRENCE, P.J. (1985) Faunal distributions across the Ordovician–Silurian boundary, Anticosti Island and Percé, Québec, Canada. *Can. J. Earth Sci.* **22**, 838–849.

LIARD, P. (1973) Legend for map sheets of Mont Joli, Matane, Sayabec, Ste-Blandine E. *Min. Richesses Nat. Quebec, Service Cartographie DP-290*.

MALPAS, J. & STEVENS, R.K. (1977) The origin and emplacement of the ophiolite suite with examples from western Newfoundland [English translation]. *Geotectonics*, **11**, 453–466.

MCBRIDE, E.F. (1962) Flysch and associated beds of the Martinsburg Formation (Ordovician), Central Appalachians. *J. sedim. Petrol.* **32**, 39–91.

MUTTI, E. (1979) Turbidites et cônes sous-marins profonds. In: *Sédimentation Détritique (Fluviatile, Littorale et Marine)* (Ed. by P. Homewood), pp. 353–419. Institut Géologique, University of Fribourg, Switzerland.

MUTTI, E. (1985) Turbidite systems and their relations to depositional sequences. In: *Provenance of Arenites* (Ed. by G. G. Zuffa), pp. 65–93. NATO Advanced Scientific Institute, Reidel, Derdrecht.

MUTTI, E. & RICCI LUCCHI, F. (1972) Le torbiditi dell'Appennino settentrionale: introduzione all'analisi di facies. *Mem. Soc. geol. Ital.* **11**, 161–199.

MUTTI, E., RICCI LUCCHI, F., SEGURET, M. & ZANZUCCHI, G. (1984) Seismoturbidites: a new group of resedimented deposits. *Mar. Geol.* **55**, 103–116.

NELSON, C.H. (1976) Late Pleistocene and Holocene depositional trends, processes, and history of Astoria deep-sea fan, northeast Pacific. *Mar. Geol.* **20**, 129–173.

NORMARK, W.R. (1970) Growth patterns of deep-sea fans. *Bull. Am. Ass. Petrol. Geol.* **54**, 2170–2195.

PICKERING, K.T. (1982) The shape of deep-water siliciclastic systems: a discussion. *Geomar. Letts* **2**, 41–46.

PICKERING, K.T. & HISCOTT, R.N. (1985) Contained (reflected) turbidity currents from the Middle Ordovician Cloridorme Formation, Quebec, Canada: an alternative to the antidune hypothesis. *Sedimentology*, **32**, 373–394.

PICKERING, K.T., STOW, D.A.V., WATSON, M. & HISCOTT, R.N. (1986) Deep-water facies, processes and models: a review and classification scheme for modern and ancient sediments. *Earth Sci. Rev.* **22**, 75–174.

QUINLAN, G. & BEAUMONT, C. (1984) Appalachian overthrusting, lithospheric flexure and the development of Paleozoic stratigraphy in the eastern interior region, U.S.A. *Can. J. Earth Sci.* **21**, 973–996.

RICCI LUCCHI, F. (1975) Miocene palaeogeography and basin analysis in the Periadriatic Apennines. In: *Geology of Italy* (Ed. by C. Squyres), pp. 5–111. Petroleum Exploration Society of Libya, Tripoli.

RICCI LUCCHI, F. (1981) The Miocene Marnoso-Arenacea turbidites, Romagna and Umbria Apennines. In: *Excursion Guidebook with Contributions on Sedimentology of some Italian Basins* (Ed. by F. Ricci Lucchi), pp. 229–303. International Association of Sedimentologists, Second European Regional Meeting, Bologna, Italy.

RICCI LUCCHI, F. & VALMORI, E. (1980) Basin-wide turbidites in a Miocene, over-supplied deep-sea plain: a geometrical analysis. *Sedimentology*, **27**, 241–270.

RICKARD, L.V. & FISHER, D.W. (1973) Middle Ordovician Normanskill Formation, eastern New York: age, stratigraphic and structural position. *Am. J. Sci.* **273**, 580–590.

RIVA, J. (1968) Graptolite faunas from the Middle Ordovician of the Gaspé north shore. *Natur. Can.* **95**, 1379–1400.

RIVA, J. (1974) A revision of some Ordovician graptolites of eastern North America. *Paleontology*, **17**, 1–40.

ROBINSON, P. & HALL, L.M. (1980) Tectonic synthesis of southern New England. In: *The Caledonides in the U.S.A.* (Ed. by D. R. Wones), pp. 73–82. *Virginia Poly. Inst. State Univ., Dept geol. Sci., Mem.* 2.

RODGERS, J. (1971) The Taconic Orogeny. *Bull. geol. Soc. Am.* **82**, 1141–1178.

ROKSANDIC, M.M. & GRANGER, B. (1981) Structural styles of Anticosti Island, Gaspé Passage, and eastern Gaspé Peninsula inferred from reflection seismic data. In: *Field Meeting, Anticosti-Gaspé, Québec, 1981. Vol. II: Stratigraphy and Paleontology* (Ed. by P. J. Lespérance), pp. 211–221. International Union of Geological Sciences, Subcommission on Silurian Stratigraphy, Ordovician-Silurian Boundary Working Group. Département de Géologie, Université de Montréal.

ROLIFF, W.A. (1968) Oil and gas exploration—Anticosti Island, Quebec. *Proc. geol. Ass. Can.* **19**, 31–36.

ROSS, R.J., JR et al. (1982) The Ordovician System in the United States. *Publ. int. Un. geol. Sci.* **12**, 73 pp.

ROWLEY, D.B. (1981) Accretionary collage of terrains assembled against eastern North America during the medial Ordovician Taconic Orogeny. *Abstr. Progr. geol. Soc. Am.* **13**, 542.

ROWLEY, D.B. & KIDD, W.S.F. (1981) Stratigraphic relationships and detrital composition of the medial Ordovician flysch of western New England: implications for the tectonic evolution of the Taconic Orogeny. *J. Geol.* **89**, 199–218.

ST JULIEN, P. & HUBERT, C. (1975) Evolution of the Taconian Orogen in the Quebec Appalachians. *Am. J. Sci.* **275-A**, 337–362.

SANFORD, B.V., GRANT, A.C., WADE, J.A. & BARSS, M.S. (1979) *Geology of eastern Canada and adjacent areas*. Map *geol. Surv. Can. 1401A*.

SEGURET, M., LABAUME, P. & MADARIAGA, R. (1984) Eocene seismicity in the Pyrenees from megaturbidites of the South Pyrenean Basin (Spain). *Mar. Geol.* **55**, 117–131.

SHANMUGAM, G. & LASH, G.G. (1982) Analogous tectonic evolution of the Ordovician foredeeps, southern and central Appalachians. *Geology*, **10**, 562–566.

SHANMUGAM, G. & WALKER, K.R. (1980) Sedimentation, subsidence, and evolution of a foredeep basin in the Middle Ordovician, southern Appalachians. *Am. J. Sci.* **280**, 479–496.

SKIPPER, K. & MIDDLETON, G.V. (1975) The sedimentary structures and depositional mechanics of certain Ordovician turbidites, Cloridorme Formation, Gaspé Peninsula, Quebec. *Can. J. Earth Sci.* **12**, 1934–1952.

SOQUIP (1983) *Correlation structurale des puits de l'île d'Anticosti*. Société Québecoise d'Initiatives Pétrolières, Québec City, Canada, *chart C-3876*.

STEVENS, R.K. (1970) Cambro–Ordovician flysch sedimentation and tectonics in west Newfoundland and their possible bearing on a proto-Atlantic Ocean. In: *Flysch Sedimentology in North America* (Ed. by J. Lajoie), pp. 165–178. *Spec. Pap. geol. Ass. Can.* **7**.

THOMAS, W.A. (1977) Evolution of Appalachian–Ouachita salients and recesses from re-entrants and promontories in the continental margin. *Am. J. Sci.* **277**, 1233–1278.

VALLIÈRES, A. (1977) Géologie de la région de Cacouna à Saint-André-de-Kamouraska, Comté de Rivière-du-Loups et de Kamouraska. *Min. Richesses Nat. Québec, Service Cartographie, Prelim. Rep. DPV-513*, 31 pp.

WALKER, K.R. & KELLER, F.B. (1977) Tellico Formation: submarine fan, proximal to distal turbidite environments. In: *The Ecostratigraphy of the Middle Ordovician of the Southern Appalachians (Kentucky, Tennessee and Virginia)*, *U.S.A.* (Ed. by S. C. Ruppel & K. R. Walker), pp. 134–140. *Knoxville, Univ. Tennessee, Dept geol. Sci., Stud. Geol.* **77–1**.

WILLIAMS, H. (1979) Appalachian Orogen in Canada. *Can. J. Earth Sci.* **16**, 792–807.

WILLIAMS, H. & HATCHER, R.D., JR (1983) Appalachian suspect terranes. In: *Contributions to the Tectonics and Geophysics of Mountain Chains* (Ed. by R. D. Hatcher, Jr et al.), pp. 33–53. *Mem. geol. Soc. Am.* **158**.

Spec. Publs int. Ass. Sediment. (1986) **8**, 327–345

Evolution from passive margin to foreland basin: the Atoka Formation of the Arkoma Basin, south-central U.S.A.

DAVID W. HOUSEKNECHT

Department of Geology, University of Missouri, Columbia, Missouri 65211, U.S.A.

ABSTRACT

Atokan strata of the Arkoma basin record the transition from sedimentation on a passive, rifted margin to sedimentation in a foreland basin developed by convergent tectonic activity along the Ouachita orogenic belt. The basal Atoka Spiro sandstone was deposited along a tidally swept coastline on a tectonically stable shelf that had prevailed since the late Cambrian. The remainder of the Atoka Formation was deposited during the breakdown of that shelfal area by normal faults, apparently induced by obduction of the Ouachita accretionary prism on to the southern margin of North American crust. The resulting wedge of Atokan strata displays evidence of syndepositional fault movement that significantly influenced sediment dispersal patterns, distribution of certain facies, and thickness patterns within the basin. Deposition on a muddy slope dissected by tectonically localized slope channels characterized that part of the basin where active fault movement was occurring. By the end of Atokan time, syndepositional faulting had ceased, and flexural subsidence and deposition of coal-bearing molasse characterized the final phase of foreland basin evolution.

INTRODUCTION

Strata of the Arkoma basin and Ouachita orogenic belt reflect the opening and subsequent closing of a Palaeozoic ocean basin (Iapetus). Carboniferous strata of the region were deposited during the final phase of ocean basin closing, and record the transition from sedimentation on a passive, rifted margin to synorogenic sedimentation in a rapidly evolving foreland basin. The objectives of this paper are to review the Palaeozoic tectonic history of the southern margin of North America culminating with the Ouachita orogenic event, to document geometric characteristics of the foreland basin that formed as a consequence of that orogenic event, and to present an overview of the Pennsylvanian depositional history of the basin with emphasis on the active influence exerted by tectonic activity.

GEOLOGICAL SETTING

The Arkoma is one of a series of foreland basins that formed along the North American side of the Ouachita orogenic belt during the Carboniferous. The Ouachita Mountains, extending from central Arkansas into south-eastern Oklahoma (Fig. 1), represent the largest exposure of this orogenic belt that lies mostly buried beneath Mesozoic and Cenozoic strata of the Gulf coastal plain. The foreland region along the entire length of the Ouachita orogenic belt generally shared a common history of rifted margin sedimentation during the early to middle Palaeozoic followed by foreland basin development induced by convergent tectonism during the late Palaeozoic (Flawn *et al.*, 1961; Graham, Dickinson & Ingersoll, 1975; Thomas, 1985). Only during the final phases of Ouachita orogenesis was the foreland region segmented into individual structural basins, one of which is the Arkoma.

The Arkoma basin is an accurate synclinorium that extends from east-central Arkansas to south-eastern Oklahoma (Fig. 1). It lies immediately north of the Ouachita orogenic belt, to which it is intimately related. Although the southern margin of the basin is historically defined as the northern edge of the Ouachita frontal thrust belt, it is now clear that Atokan strata exposed in the frontal thrust belt were

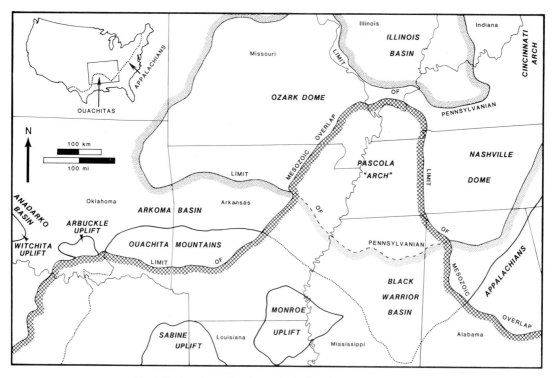

Fig. 1. Regional map showing geological setting of Arkoma basin.

deposited in continuity with Arkoma basin strata and that older Arkoma basin strata extend southwards for an unknown distance beneath the thrust belt.

Figure 2 schematically summarizes the Palaeozoic stratigraphy of the Arkoma basin and Ouachita Mountains. Upper Cambrian to basal Atokan strata within the basin comprise shallow marine carbonates, shales, and quartzose sandstones whose maximum aggregate thickness along the southern part of the basin is about 1·5 km. Because of their peripheral importance to this paper, they are grouped as 'undifferentiated, rifted-margin shelf strata' in Fig. 2. The overlying Atoka Formation comprises shales and sandstones that display complex stratigraphic and facies characteristics that are detailed in the following text. Deposited during a mere 5 Myr interval of time, the Atoka attains a maximum thickness of more than 5·5 km along the southern margin of the basin. Conformably overlying the Atoka are coal-bearing formations of Desmoinesian age that attain a maximum aggregate thickness of about 2·5 km. The Ouachita stratigraphic section is readily divisible into two parts. The lower part includes Ordovician to lowermost Mississippian strata that probably attain a

maximum thickness of about 3·7 km. These strata are predominantly deep marine shales with subordinate beds of limestone, quartzose sandstone, and bedded chert. Because of their peripheral importance to this paper, they are grouped as 'undifferentiated, rifted-margin, deep basin strata' in Fig. 2. Mississippian to middle Atokan strata of the Ouachitas comprise shales and sandstones (Stanley, Jackfork, Johns Valley, and Atoka Formations of Fig. 2) whose aggregate thickness exceeds 10 km. These Carboniferous strata constitute a flysch sequence whose deposition on submarine fans and in associated deep water environments is well documented (e.g. Morris, 1974; Moiola & Shanmugam, 1984).

Within the Arkoma basin, broad synclines separated by narrow anticlines dominate the structure at the surface (Fig. 3). The axes of these folds generally parallel the overall arcuate trend of the basin and that of the Ouachita frontal thrust belt. Listric thrust faults are known to underlie much of the folded section and ramp to the surface along the crests of many anticlines (Fig. 3). Beneath thrust fault horizons, the structural style is dominated by normal faults that offset Precambrian basement and the entire sub-Atoka

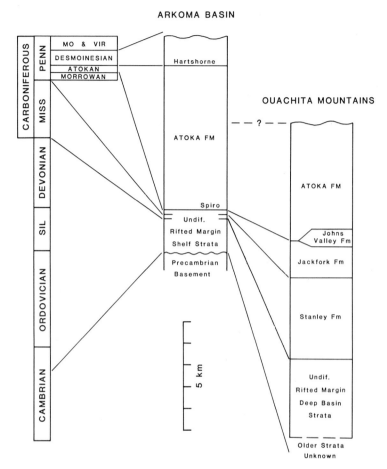

Fig. 2. Generalized stratigraphic columns of Arkoma basin and Ouachita Mountains. Wavy lines at top of columns represent present erosion surface.

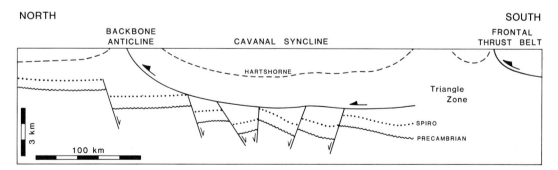

Fig. 3. Generalized tracing of a seismic line extending from north of Backbone anticline southward to Ouachita frontal thrust belt (see Fig. 5), illustrating the structural style of the Arkoma basin. Wavy line is approximate Precambrian–Cambrian unconformity; dotted line is approximate horizon of basal Atoka Spiro sandstone; dashed line is approximate horizon of basal Desmoinesian Hartshorne Sandstone and heavy solid lines are faults, with displacements indicated by arrows. Upper parts of syndepositional normal faults are displaced northwards by thrust fault, but are not shown on this figure.

sedimentary section (Fig. 3). These faults, most of which display southward dips, were active during deposition of lower to middle Atokan strata as indicated by significant thickening of those strata across the faults.

TECTONIC HISTORY

During the past two decades, numerous tectonic models have been proposed to explain the geological history of the Ouachita orogenic belt and the associated foreland. In recent years, most tectonic interpretations have converged on a scenario that involves consumption of oceanic crust and lithosphere via southward-dipping subduction and consequent collision between an Atlantic-type continental margin (the southern margin of North America) and either an island arc or continental plate (commonly called Llanoria) (see Houseknecht & Matthews, 1985 and Thomas, 1985 for citation of various models).

According to this scenario, a major episode of rifting resulted in the opening of an ocean basin during the latest Precambrian or earliest Palaeozoic (Fig. 4A) (e.g. Thomas, 1977). As a consequence of this rift opening, the southern margin of North America evolved into an Atlantic-type margin that persisted through the middle Palaeozoic (Fig. 4B). The prism of sediment that accumulated along this passive margin includes Cambrian to lowermost Mississipian strata deposited in shelf and off-the-shelf environments. Shelf facies are predominately carbonates with subordinate volumes of shale and quartzose sandstone. Although their original southward extent is unknown, they are present throughout the Arkoma basin and the lower Palaeozoic portion is believed to extend southwards beneath the highly deformed central uplift of the Ouachitas (Lillie *et al.*, 1983; Thomas, 1985). Off-the-shelf facies, exposed only in allochthonous positions within the central uplift of the Ouachitas, are predominately dark coloured shales with subordinate volumes of limestone, quartzose sandstone, and bedded chert. Known collectively as 'Ouachita facies', these strata are thought to represent slope, rise, and abyssal sediments deposited in a starved basin.

During the Devonian or early Mississippian, the ocean basin began to close, accommodated by southwards subduction beneath Llanoria (Fig. 4C). Although it is impossible to determine precisely when subduction began, it was clearly under way during the Mississippian, as suggested by detritus indicative of an orogenic provenance (Morris, 1974) and locally abundant volcanic detritus in the Stanley Formation of the Ouachitas (Fig. 2). Carboniferous volcanic rocks that have been encountered in the subsurface south of the Ouachitas, along the flanks of the Sabine uplift (Fig. 1) (Nicholas & Waddell, 1982), probably represent vestiges of a magmatic arc that developed along the northern margin of Llanoria (Fig. 4C). Within this convergent tectonic setting, the incipient Ouachita orogenic belt began to form as an accretionary prism associated with the subduction zone (Fig. 4C).

Throughout the Mississippian and into the earliest Atokan (culminating with deposition of the basal Atokan Spiro Sandstone, Fig. 2), the shelf along the southern margin of North America remained a site of relatively slow sedimentation in shallow marine through non-marine environments (Fig. 4C). Carbonates, shales and quartzose sandstones continued to be deposited in much the same realm that had characterized the area since the late Cambrian. However, the deep, remnant ocean basin (Dickinson, 1974) became the site of rapid deposition of flysch. Derived primarily from the east, where collision orogenesis had already resulted in uplift along the Ouachita trend (Thomas, 1985), sediment poured into the deep basin where it was dispersed longitudinally westwards and ultimately deposited on submarine fans and in associated abyssal environments. As a result, more than 5 km of flysch was deposited during Mississippian and Morrowan time (Stanley, Jackfork and Johns Valley Formations, Fig. 2).

By early Atokan time, the remnant ocean basin had been consumed by subduction and the northward advancing subduction complex was being obducted on to the rifted continental margin of North America (Fig. 4D). Partly as a result of attenuated continental crust being drawn into the subduction zone and partly because of vertical loading by the overriding accretionary prism (Dickinson, 1974, 1976), the southern margin of the North American continental crust was subjected to flexural bending. Apparently as a result of this flexural bending, widespread normal faulting occurred in the foreland. The normal faults generally strike parallel to the Ouachita trend, are mostly downthrown to the south, and offset both crystalline basement and overlying Cambrian to basal Atokan strata of the rifted margin prism. Subsurface and seismic evidence suggests that most of these faults broke previously undeformed continental crust, although it is likely that some of them may represent reactivation of faults formed during early Palaeozoic rifting (Fig. 4). The shelf-slope-rise geometry that had

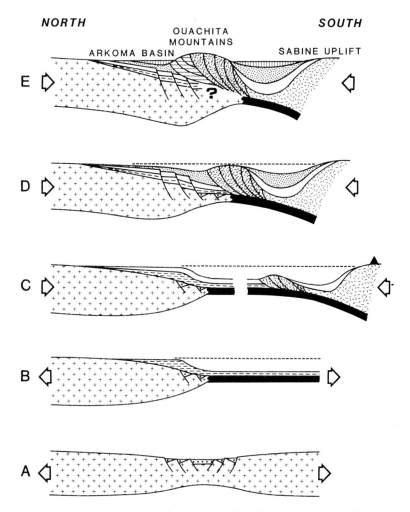

Fig. 4. Hypothetical cross-sections depicting tectonic evolution of southern margin of North America during (A) late Precambrian–earliest Palaeozoic, (B) late Cambrian–earliest Mississippian, (C) early Mississippian–earliest Atokan, (D) early–middle Atokan, and (E) late Atokan–Desmoinesian. Key to patterns: crosses = continental crust (undifferentiated in A, North American in B–E); straw hachures = 'Llanorian' crust; black = oceanic crust; heavy dots = basal Palaeozoic strata; horizontal hachures = upper Cambrian–basal Mississippian strata; white = Mississippian–basal Atokan strata; sand stippling = lower–middle Atoka strata; vertical lines = upper Atokan–Desmoinesian strata; mottled = Ouachita subduction complex; wavy lines = Ouachita foreland thrust belt; black trinagle = magmatic arc volcanoes. Figure modified from Houseknecht & Matthews (1985).

prevailed along the passive continental margin since the early Palaeozoic was broken down by the tectonically induced normal faults and rates of both subsidence and sedimentation increased markedly. Fault movement was contemporaneous with deposition of lower to middle Atokan shales and sandstones, resulting in abrupt thickness increases across faults (Fig. 4D). These Atokan strata represent a critical transition between passive margin sedimentation and

foreland basin sedimentation, and are the topic of most of the following discussion.

By late Atokan time (Fig. 4E), foreland-style thrusting became predominant as the subduction complex pushed northwards against strata deposited in settings illustrated by Fig. 4 (C, D). Resultant uplift along the frontal thrust belt of the Ouachitas completed the formation of a peripheral foreland basin (Dickinson, 1974) in which shallow marine, deltaic, and fluvial sedimentation prevailed. Upper Atokan and Desmoinesian strata of the Arkoma basin constitute a typical coal-bearing molasse. At that time, the gross structural configuration of the Arkoma–Ouachita system was essentially the same as at present, although relatively minor folding and thrusting continued after the Desmoinesian.

SYNDEPOSITIONAL NORMAL FAULTS

From a depositional edge along the northern margin of the basin, Atokan strata thicken to a maximum of more than 5·5 km along the northern edge of the Ouachita frontal thrust belt (Fig. 5). Most of this increase in thickness is accommodated by abrupt expansion of the lower to middle part of the Atoka section across syndepositional normal faults, the locations of which are shown in Fig. 5. Most of these faults have been mapped during the course of this research, although certain of the faults have been mapped locally by previous workers (Koinm & Dickey, 1967; Berry & Trumbly, 1968; Buchanan & Johnson, 1968; Lumsden, Pittman & Buchanan, 1971; Haley, 1982; Zachry, 1983). Several of the large displacement faults have acquired informal names that are widely used in the petroleum industry, and those names are shown on Fig. 5. Three faults which have not been reported in previous literature are labelled X, Y and Z, although the presence of fault Z is equivocal as discussed below.

Figure 6 is a north–south stratigraphic cross-section whose datum (labelled A) is a widespread log-response anomaly in the upper part of the Atoka Formation and whose lower key bed is the basal Atoka Spiro sandstone (labelled S). Constructed from wire-line logs of 41 wells, cross-section A–A' illustrates the fault-controlled geometry of Atokan strata in a part of the basin that is unaffected by thrust faults of large displacement (Fig. 5). Southwards along the line of section, the Atoka increases in thickness from 760 m to more than 3700 m, with most of the increase occurring in abrupt steps across a few faults with displacements of several hundred to more than 1000 m. Some of the large displacement faults appear to have a single plane of movement (e.g. Kinta fault) whereas others splay into two (San Bois fault) or more (Mulberry fault) discrete fault planes (Figs 5 and 6). In addition to those with large displacements, numerous syndepositional normal faults with smaller displacements (generally less than a few hundred metres)

are common in certain parts of the basin (e.g. between Kinta and San Bois faults in Fig. 6). All of the faults shown in Fig. 6 are downthrown to the south. Although this is true of the vast majority of syndepositional faults in the basin, north-dipping normal faults (i.e. downthrown to the north) are locally common (see Fig. 3). Where present, the north-dipping faults are locally paired with adjacent south-dipping faults to form syndepositional horst and graben geometries.

The trace of the basal Atoka Spiro sandstone on Fig. 6 further defines the geometry of individual syndepositional fault blocks. On some downthrown blocks, the Spiro displays apparent southward dip (e.g. downthrown to the Kinta fault). However, on most downthrown blocks, the Spiro displays apparent northwards dip (bear in mind that the cross-section is hung on a stratigraphic marker and does not depict true structure), indicating that most of the sydepositional normal faults had a rotational sense of movement. Whereas most faults in the Arkoma basin display this geometry, both the direction and magnitude of apparent dip on the Spiro, as well as the absolute magnitude of fault displacement, are variable along the strike of individual faults.

In contrast to the Atoka section, overlying Desmoinesian strata display no evidence of abrupt thickening across normal faults. For example, the interval between datum bed A and the Hartshorne Sandstone (H) shows a gradual, though asymmetrical increase in thickness to the south (Fig. 6), a characteristic shared by all of the Desmoinesian strata that are preserved over large areas of the basin. Such thickness patterns are common to foreland basins and probably result from flexural subsidence induced by foreland-style thrust loading along margins of basins adjacent to orogenic belts (Jordan, 1981).

Figure 7 is a NW–SE cross-section constructed from wire-line logs of five wells. This section is located in a part of the basin that has been subjected to thrust faulting and is necessarily more schematic than Fig. 6 because of fewer data points. Nevertheless, it can be used to illustrate important aspects of syndepositional normal faulting and Atokan stratigraphy. Significantly, cross-section B–B' extends southward into a deeper portion of the basin than cross-section A–A'. Notice that the San Bois fault is present near the south end of cross-section A–A' and near the north end of cross-section B–B'. As an example of variable fault displacement along strike, there is over 3300 m of section between bed A and the Spiro downthrown to the San Bois fault in cross-section A–A' (Fig. 6) whereas the same interval downthrown to the San

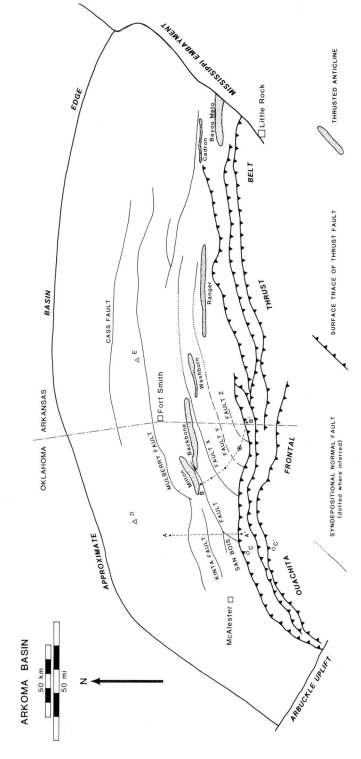

Fig. 5. Base map of Arkoma basin showing locations of known syndepositional normal faults, and lines of cross-section and other localities discussed in text.

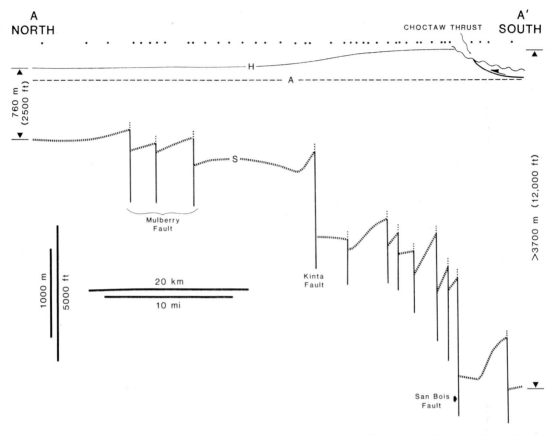

Fig. 6. Stratigraphic cross-section A–A′ constructed from wire-line logs from 41 wells, whose locations are indicated by black dots. S = Spiro sandstone, A = widespread upper Atoka key bed that serves as datum for cross-section, H = Hartshorne Sandstone.

Bois fault in cross-section B–B′ (Fig. 7) is only about 2400 m thick.

Southwards along the line of section, the Atoka thickens from less than 1200 m to more than 5500 m (estimated by projecting the horizon of the Hartshorne Sandstone), with syndepositional normal faults accounting for most of the increase. Moreover, the well at the south-eastern end of cross-section B–B′ did not penetrate the basal Atoka Spiro sandstone, suggesting that the estimate of 5500 m may be conservative. Alternatively, an undetected thrust fault may be present in the lower part of that well and bed E may be a subthrust equivalent of bed D (their log responses are similar); no other well has penetrated so deeply into the Atoka Formation and so this problem cannot be solved with available data. Nevertheless, the thickness of the Atoka section in the two south-easternmost wells approaches estimates of maximum

Atoka thickness within the Ouachita frontal thrust belt, suggesting that the Atoka Formation in the frontal Ouachitas is simply the distal equivalent of the Atoka in the Arkoma basin and that both were deposited within a single basin.

There are an insufficient number of Spiro penetrations in this deep portion of the basin to analyse fault block geometry as was done for cross-section A–A′ and so the Spiro is schematically shown to have no apparent dip in Fig. 7. However, the extreme thickness of the Atoka Formation and the presence of several widespread key beds within the section provide an opportunity to discuss the relative timing of fault movement. It is apparent from the distribution of key beds (Fig. 7) that fault movement was not synchronous. Expansion of the section below bed D is limited to fault Y (and perhaps Z); expansion of the sections immediately below bed C and immediately below bed

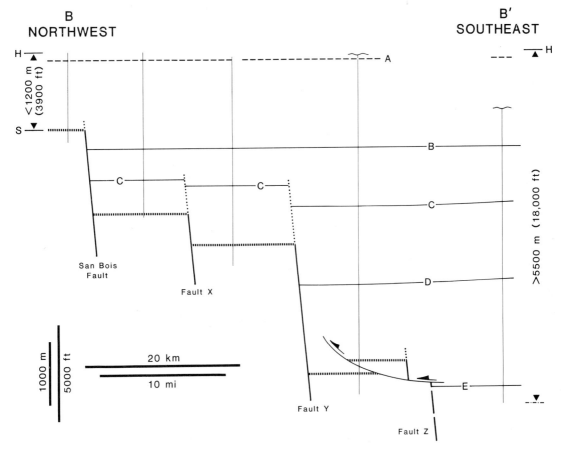

Fig. 7. Stratigraphic cross-section B–B′ constructed from wire-line logs from five wells, whose locations are indicated by vertical lines. Lines representing three wells on left extend from point of Hartshorne Sandstone penetration to total depth; lines representing two wells on right extend from present erosion surface (wavy lines) to total depth. S = Spiro sandstone; A = same bed labelled A in Fig. 6; B, C, D, and E = key beds within Atoka Formation; H = Hartshorne Sandstone. Bed A serves as datum for four wells on left; bed B serves as datum for well on right. Note thrust repeat of Spiro sandstone in fourth well from left.

B occurs across fault Y, fault X and the San Bois fault; expansion of the section immediately below bed A is limited to the San Bois fault. It appears that the southernmost syndepositional faults became active earliest and that active faulting migrated northwards with time. This general pattern is present throughout the basin.

One final aspect of syndepositional faulting merits description. Even though thickness increases in excess of 1000 m occur across many of these faults, there is absolutely no evidence of erosion of the Spiro sandstone or overlying strata on upthrown sides of the faults. There is typically a 100–200 m thick, dark grey shale immediately overlying the Spiro throughout the basin, and that shale is apparently the stratal equivalent of expanded sections downthrown to major faults.

To summarize, syndepositional normal faults fundamentally control the distribution and thickness of Atokan strata throughout the basin. Neither the basal Atoka Spiro sandstone nor the basal Demoinesian Hartshorne Sandstone displays evidence of syndepositional fault movement, thereby constraining normal faulting to an interval of time less than the duration of the Atokan age (approximately 5 Myr). Active normal faulting migrated northwards with time, resulting in a step-like onlap sequence within the

Atoka Formation. This northward migration of faulting was undoubtedly caused by the northward advance of the Ouachita orogenic front.

SPIRO SEDIMENTOLOGY

The basal Atoka Spiro sandstone represents the final phase of sedimentation on the stable shelf that had characterized the passive southern margin of North America since the late Cambrian (Fig. 4). In order to discuss the transition from passive margin to foreland basin sedimentation in the Arkoma basin properly, it is necessary to describe the nature of Spiro facies and inferred depositional environments. This summary is based on previously unpublished observations as well as the work of Lumsden *et al.* (1971).

In eastern Oklahoma, the Spiro is a fine to medium grained sandstone composed of subequant to equant, well rounded quartz that comprises well over 95% of the framework grains. Medium grained sand is concentrated in channel deposits that are 1–3 km wide and oriented NW–SE. Some of the channel facies contain only quartz sand and display unidirectional cross-bedding (Fig. 8A), and these are interpreted as fluvially dominated channel deposits. Others contain a mixture of quartz sand and carbonate grains (mostly echinoderm fragments) (Fig. 8B) and display bidirectional cross-bedding; these are interpreted as tidally dominated channel deposits. Interchannel deposits are composed of fine grained sand and shale organized into ripple bedded and thoroughly bioturbated sequences that were probably deposited in shallow subtidal through tidal flat environments (Fig. 8C, D). Overall, these characteristics suggest that rivers eroded sand from older, platform strata on the north side of the basin and delivered that sand to a broad, tidally swept, coastal environment. Full diameter cores from the southernmost part of the basin display similar sandstone facies as well as carbonate bank and interbank facies, indicating sedimentation on a marine shelf offshore from the tidal coastline.

As mentioned previously, the location of the original southern shelf edge is unknown. However, the southernmost known occurrences of Spiro coastal and shelfal facies are shown on Fig. 5. The open circle labelled C represents the location of an autochthonous (subthrust) cored interval and the one labelled C′ represents a palinspastically restored location of an allochthonous (thrust) cored interval in the same well; both display shoreline to shallow marine facies. Farther east, the circled location on the line of cross-

section B–B′ is the southernmost well that has penetrated the Spiro in that part of the basin. These occurrences indicate that stable shelfal conditions prevailed throughout the Oklahoma portion of the Arkoma basin, and probably considerably farther south, during Spiro deposition. Even though less is known about the Spiro in the Arkansas portion of the basin, it appears that the same is true there (Zachry, 1983), although deltaic facies may be volumetrically more important as a result of detrital influx from the Ozark dome to the north (Fig. 1).

ATOKAN (POST-SPIRO) SEDIMENTOLOGY

Provenance and sediment dispersal

Previous studies have demonstrated the value of palaeocurrent analysis in reconstructing the depositional history of Atoka submarine fan facies that were deposited in the deepest part of the sedimentary basin and which are now exposed in the Ouachita frontal thrust belt (e.g. Morris, 1974). Unfortunately, Atoka outcrops north of the frontal thrust belt are limited to the northern edge of the basin and to the thrusted anticlines shown on Fig. 5. The former display proximal facies that provide little information about deposition farther south in the basin, and the latter are so restricted in size and amount of section exposed that they are of limited utility. Neither provides sufficient palaeocurrent data to allow reconstruction of sediment dispersal patterns within the basin proper. However, the petrology of Atoka sandstones can be used to interpret sediment dispersal patterns, at least on a gross scale.

Palaeotectonic and palaeographic reconstructions of the Arkoma basin and surrounding region (Graham *et al.*, 1975; Graham, Ingersoll & Dickinson, 1976; Houseknecht & Kacena, 1983; Thomas, 1985) suggest that the basin received sediment from three major dispersal systems during Atokan sedimentation. Detritus continued to be eroded from older platform strata to the north of the basin (Ozark dome and beyond) and sand derived from this continental provenance was predominantly composed of mature quartz grains like those deposited in the Spiro. Sediment was also funnelled southwards through the Illinois basin into the eastern Arkoma basin (Fig. 1), and sand associated with that dispersal system was also composed predominantly of quartz (Potter & Glass, 1958). The third dispersal system derived sediment from the Ouachita

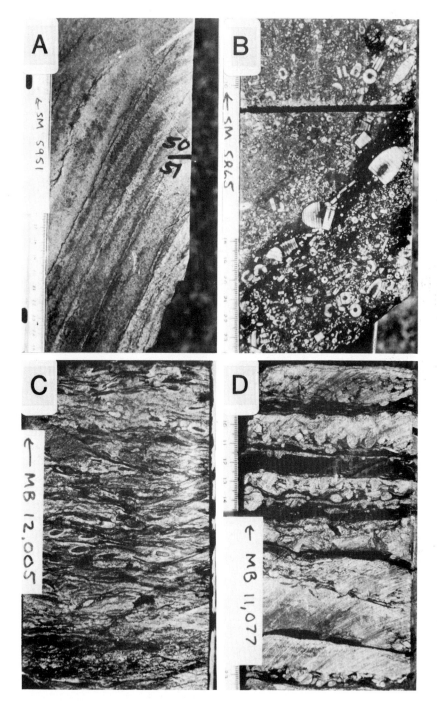

Fig. 8. Core photographs of facies in the Spiro sandstone. (A) Cross-bedded, quartzose sandstone of inferred fluvially dominated channel deposit (foreset dip is exaggerated by structural dip). (B) Cross-bedded, quartzose sandstone with abundant echinoderm fragments of inferred tidally dominated channel deposit (foreset dip is exaggerated by structural dip). (C) Thoroughly bioturbated interchannel facies. (D) Interbedded shale and bioturbated sandstone of interchannel facies. Ruler scales in centimetres; scale of (C) is the same as (D).

orogenic belt, which had already been uplifted and exposed to erosion along the south-western margin of the Black Warrior basin (Mack, Thomas & Horsey, 1983). Sand associated with that orogenic provenance contains a significant volume of lithic fragments and some feldspar, indicating derivation from metamorphic and volcanic source rocks (Mack *et al.*, 1983). Considerations of basin geometry (Figs 1, 4, 6 and 7) suggest that sand derived from the continental provenance to the north would have been transported more-or-less southwards into the basin and that sand derived from the other two areas would have entered the eastern portion of the basin and could only have reached other parts of the basin by westwards longitudinal transport.

Figure 9 summarizes framework grain composition of nearly 50 samples of Atoka (post-Spiro) sandstones collected in various parts of the basin. Those samples that plot within the field labelled Ozark petrofacies were collected at two locations indicated by triangles D and E in Fig. 5. Both the texture and composition of these samples are like the underlying Spiro and are consistent with derivation of sand from older platform strata north of the basin. The term 'Ozark petrofacies' is used to imply derivation from the north and to emphasize the widespread truncation of older Palaeozoic sandstones around the flanks of the Ozark dome as a likely source of this sand.

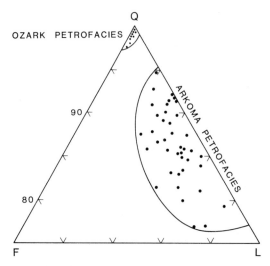

Fig. 9. QFL (quartz, feldspar, lithics) diagram showing framework grain composition of Ozark and Arkoma petrofacies. All lithic fragments are of metamorphic origin; feldspar is mostly albite. Composition is based on counting 300 points per sample.

Samples within the field labelled Arkoma petrofacies were collected from outcrops along all of the thrusted anticlines shown in Fig. 5 and from numerous wells in Oklahoma located south of the Milton and Backbone anticlines and north of the Ouachita frontal thrust belt. On a QFL diagram, these sandstones contain 75–90% quartz, and the quartz grains are very fine to fine grained, subelongate to subequant, and poorly to moderately rounded. Moreover, these sandstones contain 5–25% metamorphic lithic fragments (mostly slate and phyllite) and up to 8% feldspar (mostly albite). All of these characteristics significantly differ from those of the Ozark petrofacies and suggest derivation from an orogenic provenance, although the relatively quartzose samples within the Arkoma petrofacies may reflect a mixing of detritus from the orogenic source with either Ozark petrofacies or sand distributed through the Illinois basin. The framework grain composition of these sandstones is virtually identical to that of Atoka sandstones exposed in the Ouachita frontal thrust belt (Graham *et al.*, 1976; and unpublished data), suggesting a common provenance. The term 'Arkoma petrofacies' is used to emphasize the volumetric predominance of this sandstone composition within the basin.

In light of the foregoing discussion, it appears that most of the sand in the Atoka Formation was derived from the uplifted Ouachita orogenic belt to the east (SW of the Black Warrior basin), entered the eastern portion of the basin, and was transported longitudinally westwards to sites of deposition. Rapid subsidence associated with syndepositional normal faults undoubtedly promoted longitudinal dispersion of sediment. It is probably no coincidence that the Ozark petrofacies appears to be limited to that part of the basin north of the Mulberry fault and that the Arkoma petrofacies is pervasively distributed south of that fault (Fig. 5). Both its length and its position as the northernmost fault with significant displacement suggest that it may have exerted a major influence on the distribution of the two petrofacies.

Facies and environments of deposition

Synthesis of results from continuing research and previously published information allows reconstruction of the depositional system that prevailed during deposition of the Atoka Formation (for the sake of convenience, this discussion excludes the Spiro sandstone from the Atoka Formation). Figure 10 is a depositional model that illustrates relationships observed among Atoka facies and explains the influence

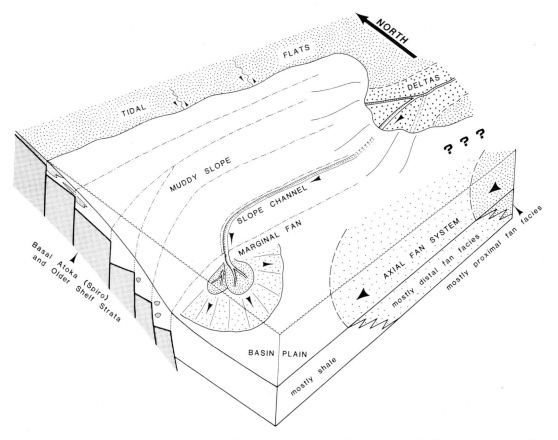

Fig. 10. Reconstruction of depositional system in which Atoka strata were deposited. Dash–dot lines represent the seafloor traces of syndepositional normal faults. The nature of the lateral relationship between deltaic facies and axial fan facies in the eastern portion of the basin is neither well documented nor well understood (as denoted by question marks), and should be considered schematic.

that syndepositional normal faulting exerted on sediment dispersal and the distribution of Atoka facies. The model is constrained by the fact that all facies illustrated are composed of Arkoma petrofacies and therefore must have been deposited within a system dominated by longitudinal sediment dispersal. It emphasizes sedimentation in that part of the basin where syndepositional faults were most active and where shallow to deep marine facies transitions occur over relatively short distances. It does not emphasize either fluvial-deltaic sedimentation along the northern margin of the basin (Zachry, 1983) or submarine fan sedimentation in the deepest part of the basin (Morris, 1974).

Shoreline facies

The northern limit of marine sedimentation in the Oklahoma portion of the basin is indicated by the

presence of tidal flat shoreline facies (Fig. 10), whose southernmost extent occurs in the subsurface just south of the Milton and Backbone anticlines (Fig. 5). Marine shales grade upward via increasing number and thickness of sandstone beds into sand-dominated tidal flat facies, and the entire sequence displays a rich and diverse, shallow marine trace fossil assemblage (Chamberlain, 1978). The tidal flat facies comprise fining upward sequences like the one shown in Fig. 11(A), in which sand-flat, mixed-flat, and mud-flat subfacies are recognized. Sand-flat subfacies are composed of nearly 100% sand and display ripple-scale cross-bedding, swash bedding, and locally abundant U-shaped burrows (Fig. 11). Within the sand-flat subfacies, broadly channel-shaped beds displaying epsilon cross-bedding and siderite pebble lags probably represent deposition in meandering tidal channels. These sand-flat subfacies grade upward

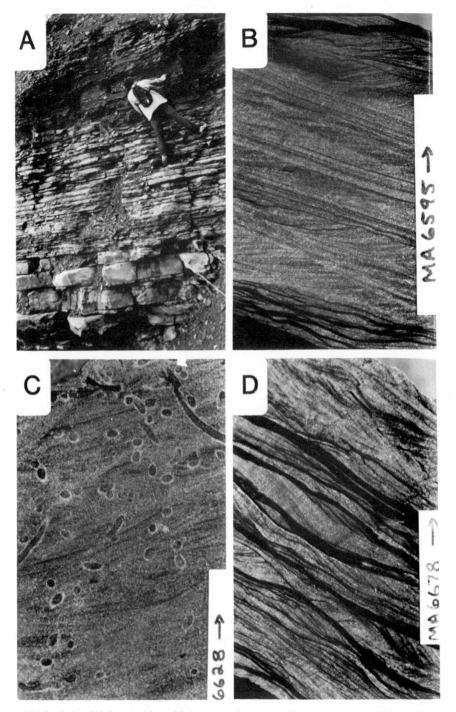

Fig. ....a tidal flat facies. (A) Outcrop view of fining upward sequence, showing upper part of sand-flat (bottom), mixed-flat (centre, at man's feet), and mud-flat (top, at man's head and hand) subfacies. (B) Core photograph of ripple-scale cross-bedding and swash bedding from sand-flat subfacies. (C) Ripple cross-bedded sand-flat subfacies displaying U-shaped burrows. (D) Ripple cross-bedded sandstone with shale interbeds from lower part of mixed-flat subfacies. Width of each core is 10 cm in (B), (C) and (D).

into mixed-flat subfacies that are composed of interbedded sandstone and shale (Fig. 11). The sandstone layers display ripple-scale cross-bedding, flaser bedding, and mud-draped, asymmetric ripples. With increasing mud content, the mixed-flat subfacies grade upward rather abruptly into mud-flat subfacies (Fig. 11A), which are composed mostly of silty mudstone. Individual, fining upward sequences are typically 5–8 m thick but vertical repetition of tidal flat sequences results in sand-rich, shoreline deposits that are locally more than 100 m thick.

Eastward in Arkansas, tidally dominated shoreline facies are exposed along all the anticlines shown in Fig. 5. These also display fining upward sequences that can be attributed to tidal flat sedimentation, but differ in that individual sequences range from 50 to 100 m in thickness and trace fossils are sparse. In every other detail, these facies are identical to those present in Oklahoma. Two factors are probably responsible for the presence of these overthickened sequences. First, they accumulated in that part of the basin most affected by syndepositional normal faulting and second, they were deposited closer to the location of major sediment input. Both the high rate of subsidence and the voluminous sediment influx would contribute to deposition of abnormally thick shoreline facies that contain sparse trace fauna.

Deltaic facies are also present in the Arkansas portion of the basin, although they are not as common as overthickened tidal flat facies. Marine shales grade upwards rather abruptly into sand-rich delta front subfacies, which display an abundance of ripple-scale cross-bedding and mud drapes. Locally such subfacies are interbedded with trough cross-bed sets that are typically about 50 cm thick. The latter probably represent deposition on distributary mouth bars by fluvial currents during high stage whereas the former represent tidal reworking of bar sediment during low stage. Delta front subfacies are typically overlain by distributary channel deposits, which display a predominance of trough cross-bedding and horizontal bedding. In those few outcrops where palaeocurrent data can be collected from distributrary channel subfacies, westwards palaeoflow is indicated. Atoka delta front and distributory channel deposits are virtually identical in both facies characteristics and indicated flow direction to Desmoinesian deltaic deposits of the basin described by Houseknecht *et al.* (1983). However, the limited distribution of Atoka deltaic facies relative to overthickened tidal flat facies suggests that tidal processes effectively reworked shoreline sediment and essentially destroyed many facies characteristics that are normally associated with deltaic sedimentation.

Slope facies

In the southern portion of the basin where the section thickens significantly across syndepositional faults, the Atoka comprises dark grey shale, which is unfossiliferous except for macerated and obviously transported plant debris, and numerous localized sandstones. Despite significant syndepositional displacement along many normal faults, numerous widespread key beds display continuity across the faults and there is absolutely no evidence of erosional truncation of strata upthrown to the faults (Figs 6 and 7). These relationships imply that the rate of mud deposition kept pace with the rate of subsidence and that the seafloor above the faults displayed little or no relief. However, it is equally apparent that these predominately pelitic facies grade southwards from shoreline facies discussed above into deep marine facies that are now exposed in the Ouachita frontal thrust belt. This combination of relationships suggests that deposition occurred on a gently dipping, muddy slope lacking a bathymetrically distinct shelf-slope-rise geometry (Fig. 10).

The localized sandstones that occur within these thick shales are difficult to analyse because they do not crop out anywhere around the basin and they are known primarily from wire-line log characteristics. Most of the sandstones that occur in this setting are lenticular (1–5 km wide; 10–50 m thick) and elongate (several kilometres to tens of kilometres long). Moreover, they are localized parallel and just downthrown to syndepositional normal faults. Fortunately, one such sandstone (the Red Oak) was extensively cored during its development as a major gas reservoir, and the preserved cores reveal much about deposition of sand on the muddy slope. The Red Oak, which is localized just below bed B and just downthrown to the San Bois fault in Fig. 7, is a multistoried sand body that has previously been interpreted as a submarine fan deposit (Vendros & Visher, 1978). In core, the base of each sandstone (story) is erosive, either into shale or a somewhat older sandstone bed. The sandstone is extremely uniform in grain size (very fine to fine sand) and sorting (moderately well to well sorted) and displays a limited range of sedimentary structures, with most cores appearing massive (Fig. 12). Other common structures include diffuse flat lamination and a diverse spectrum of dewatering structures, including dish structures, convolute lami-

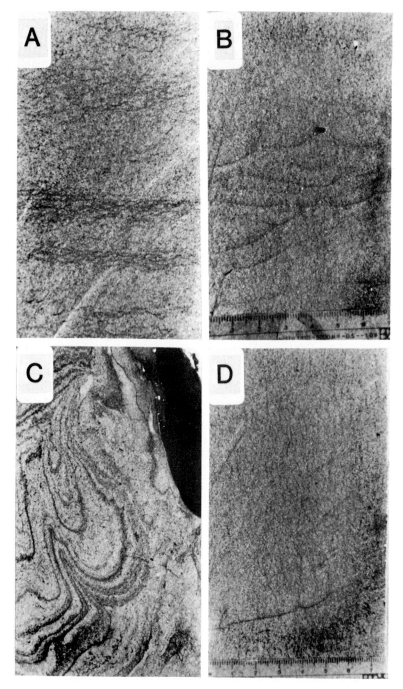

Fig. 12. Core photographs of sedimentary structures in the Red Oak sandstone. (A) Diffuse flat lamination overlain by massive sandstone. (B) Dish structures. (C) Convolute lamination. (D) Diffuse vertical lamination. Ruler scaled in centimetres in (B) and (D); (A) and (C) are the same scale.

nation, diffuse vertical lamination, swirled lamination, and dewatering pipes (Fig. 12) (Middleton & Hampton, 1976). Along the lateral margins of the lenticular sand bodies, silty sandstone displaying parallel lamination occurs, and identical facies are locally included as intraclasts within the sand body proper. There are no thin, widespread sandstone beds associated with these sand bodies, nor are there partial or complete Bouma sequences or any sort of graded beds present (over 350 m of core was described).

All of the characteristics enumerated above suggest deposition by channellized sediment gravity flows below storm wave base, with grain flow being the most likely specific process (Middleton & Hampton, 1976). It is proposed that deposition occurred in slope channels that were localized by rapid subsidence rates just downthrown to syndepositional normal faults (Fig. 10). Apparently the rotational sense of movement on certain faults (Fig. 6) resulted in at least a slight depression on the seafloor, which was sufficient to localize sediment gravity flows that entered the slope environment. The flows were erosive into unconsolidated slope muds, thereby channelizing the slope along the seafloor trace of syndepositional faults (Fig. 10). It appears that levee facies were locally deposited along the margins of the slope channels (parallel bedded silty sandstones), but there is no evidence of deposition by unconfined sediment gravity flows as would be expected on submarine fan lobes.

Atoka sandstones like the Red Oak have long been attributed to deposition in 'deep water' because of the predominance of sedimentary structures indicative of sediment gravity flow mechanisms. As a result, many previous workers have inferred that the syndepositional normal faults acted as fundamental bathymetric features within the depositional basin, with shelfal conditions prevailing upthrown and 'deep water' conditions prevailing downthrown to a major fault (e.g. Vedros & Visher, 1978). This inference does not seem compatible with the continuity of correlative beds across many syndepositional faults (Fig. 7) and needs to be re-evaluated in the light of voluminous recent work demonstrating that the only requisite for accumulation and preservation of facies deposited by sediment gravity flow mechanisms is water depths below storm wave base (e.g. Walker, 1979). Within a tidally dominated basin like the Arkoma, it is feasible for such deposits to accumulate in water depths that would not normally be considered 'deep'. In support of this interpretation, the tidal flat facies illustrated in Fig. 11 (B–D) occur in a sandstone that is only 200 m up section from the Red Oak sandstone illustrated in

Fig. 12, and both sandstones occur at the same location in the basin. Assuming that neither uplift of the basin nor a significant drop in sea-level occurred after deposition of the slope channel facies and before deposition of the tidal flat facies (neither of which is likely), 200 m is a reasonable estimate for the depth of water in which the Red Oak slope channel sand was deposited.

Submarine fan facies

As mentioned previously, numerous workers have published details of submarine fan facies that were deposited on the deep basin floor and are now exposed in the Ouachita frontal thrust belt. These are illustrated schematically and labelled 'axial fan system' in Fig. 10. However, there are also smaller submarine fan sequences that were apparently deposited at the toe of the slope as distal facies equivalents of slope channels discussed above, and these are labelled 'marginal fan' in Fig. 10. In outcrops along the frontal thrust belt, these facies can be distinguished from those associated with the axial fan system in three ways. First, the marginal fans are composed of 'cleaner' sandstone than the axial fans. Petrographic analysis reveals that the framework grain composition of sandstones in the two facies are identical, but samples of the axial fan system contain significantly more matrix and are more poorly sorted than samples of the marginal fan system. Second, deposits associated with the axial fan system invariably display westward directed palaeocurrent indicators whereas marginal fan facies display southwards to south-westwards indicators. Thirdly, the marginal fan deposits display internal facies architecture that suggests progradation from north to south. For example, channellized proximal lobe facies grade both westwards and eastwards along the strike of the frontal thrust belt into more distal lobe facies characterized by beds deposited by unconfined sediment gravity flows. The dimensions of such lateral facies sequences are typically 15–25 km, suggesting the original size of the marginal fans. As indicated in Fig. 10, there is abundant evidence for interfingering of marginal and axial fan facies, suggesting that the two systems evolved simultaneously but were fed by different sediment dispersal systems.

SUMMARY

The Arkoma foreland basin developed in response to convergent tectonism associated with the Ouachita orogenic event. During Atokan time, the tectonically stable shelf that had characterized the passive margin of southern North America was broken down by normal faults induced by obduction of the Ouachita accretionary prism on to the continental crust. The lower and middle portion of the Atoka Formation comprises a wedge of strata that displays thickness, petrofacies, and depositional facies patterns that were significantly influenced by syndepositional movement along these normal faults. By the end of Atokan time, fault movement had ceased and the basin was characterized by flexural subsidence induced by foreland-style thrust loading and shallow marine through non-marine sedimentation in a fully formed foreland basin.

ACKNOWLEDGMENTS

This work has benefited tremendously from interaction with, and funding from Tenneco Oil Company; I particularly appreciate the input of Mac McGilvery, Mike Albano, and W. D. Hollar. Significant contributions have been made by former students E. C. Williams and S. N. Williams. The project has also been supported with funding and/or data provided by ARCO, Marathon, Sampson Resources, Santa Fe Minerals, Sohio, and Texaco.

REFERENCES

BERRY, R.M. & TRUMBLY, W.D. (1968) Wilburton gas field, Arkoma basin, Oklahoma. In: *Geology of the Western Arkoma Basin and Ouachita Mountains* (Ed. by L. M. Cline), pp. 86–103. *Okla. City geol. Soc. Guidebook*.

BUCHANAN, R.S. & JOHNSON, F.K. (1968) Bonanza gas field—a model for Arkoma basin growth faulting. In: *Geology of the Western Arkoma Basin and Ouachita Mountains* (Ed. by L.M. Cline), pp. 75–85. *Okla. City geol. Soc. Guidebook*.

CHAMBERLAIN, C.K. (1978) *Trace fossils and paleoecology of the Ouachita geosyncline. Soc. econ. Paleont. Miner. Guidebook*, 68 pp.

DICKINSON, W.R. (1974) Plate tectonics and sedimentation. In: *Tectonics and Sedimentation* (Ed. by W. R. Dickinson). *Spec. Publ. Soc. econ. Paleont. Miner.*, Tulsa, **22**, 1–27.

DICKINSON, W.R. (1976) Plate tectonics and hydrocarbon accumulations. *Am. Ass. Petrol. Geol. Short Course 1*.

FLAWN, P.T., GOLDSTEIN, A. Jr, KING, P.B. & WEAVER, C.E. (1961) The Ouachita System. *Bur. econ. Geol., Univ. Texas Pub 6120*, 401 pp.

GRAHAM, S.A., DICKINSON, W.R. & INGERSOLL, R.V. (1975) Himalayan–Bengal model for flysch dispersal in the Appalachian–Ouachita system. *Bull. geol. Soc. Am.* **86**, 273–286.

GRAHAM, S.A., INGERSOLL, R.V. & DICKINSON, W.R. (1976) Common provenance for lithic grains in Carboniferous sandstones from Ouachita Mountains and Black Warrior basin. *J. sedim. Petrol.* **46**, 620–632.

HALEY, B.R. (1982) Geology and energy resources of the Arkoma basin, Oklahoma and Arkansas. *Univ. Mo.-Rolla J.* **3**, 43–53.

HOUSEKNECHT, D.W., ZAENGLE, J.F., STEYAERT, D.J., MATTEO, A.P. Jr & KUHN, M.A. (1983) Facies and depositional environments of the Desmoinesian Hartshorne Sandstone, Arkoma basin. In: *Tectonic–Sedimentary Evolution of the Arkoma Basin* (Ed. by D. W. Houseknecht). *Soc. econ. Paleont. Miner. Midcont. Sec.* **1**, 53–82.

HOUSEKNECHT, D.W. & KACENA, J.A. (1983) Tectonic and sedimentary evolution of the Arkoma foreland basin. In: *Tectonic–Sedimentary Evolution of the Arkoma Basin* (Ed. by D. W. Houseknecht). *Soc. econ. Paleont. Miner. Midcont. Sec.* **1**, 3–33.

HOUSEKNECHT, D.W. & MATTHEWS, S.M. (1985) Thermal maturity of Carboniferous strata, Ouachita Mountains. *Bull. Am. Ass. Petrol. Geol.* **69**, 335–345.

JORDAN, T.E. (1981) Thrust loads and foreland basin evolution, Cretaceous, western United States. *Bull. Am. Ass. Petrol. Geol.* **65**, 2506–2520.

KOINM, D.N. & DICKEY, P.A. (1967) Growth faulting in McAlester basin of Oklahoma. *Bull. Am. Ass. Petrol. Geol.* **51**, 710–718.

LILLIE, R.J., NELSON, K.D., DE VOOGD, B., BREWER, J.A., OLIVER, J.E., BROWN, L.D., KAUFMAN, S. & VIELE, G.W. (1983) Crustal structure of Ouachita Mountains, Arkansas: a model based on integration of COCORP reflection profiles and regional geophysical data. *Bull. Am. Ass. Petrol. Geol.* **67**, 907–931.

LUMSDEN, D.N., PITTMAN, E.D. & BUCHANAN, R.S. (1971) Sedimentation and petrology of Spiro and Foster sands (Pennsylvanian), McAlester basin, Oklahoma. *Bull. Am. Ass. Petrol. Geol.* **55**, 254–266.

MACK, G.H., THOMAS, W.A. & HORSEY, C.A. (1983) Composition of Carboniferous sandstones and tectonic framework of southern Appalachian–Ouachita orogen. *J. sedim. Petrol.* **53**, 931–946.

MIDDLETON, G. & HAMPTON, M. (1976) Subaqueous sediment transport and deposition by sediment gravity flows. In: *Marine Sediment Transport and Environmental Management* (Ed. by D. Stanley and D. Swift), pp. 197–218. Wiley, New York.

MOIOLA, R.J. & SHANMUGAM, G. (1984) Submarine fan sedimentation, Ouachita Mountains, Arkansas and Oklahoma. *Trans. Gulf-Cst Ass. geol. Soc.* **34**, 175–182.

MORRIS, R.C. (1974) Sedimentary and tectonic history of the Ouachita Mountains. In: *Tectonics and Sedimentation* (Ed. by W. R. Dickinson). *Spec. Publ. Soc. econ. Paleont. Miner.*, Tulsa., **22**, 120–142.

NICHOLAS, R.L. & WADDELL, D.E. (1982) New Paleozoic subsurface data from the north-central Gulf Coast (abstract). *Abstr. Prog. geol. Soc. Am.* **14**, 576.

POTTER, P.E. & GLASS, H.D. (1958) Petrology and sedimentation of the Pennsylvanian sediments in southern Illinois—a vertical profile. *Ill. geol. Surv. R.I.* **204**, 60 pp.

THOMAS, W.A. (1977) Evolution of Appalachian-Ouachita salients and recesses for reentrants and promontories in the continental margin. *Am. J. Sci.* **277**, 1233–1278.

THOMAS, W.A. (1985) The Appalachian–Ouachita connection: Paleozoic orogenic belt at the southern margin of North America. *Ann Rev. Earth planet. Sci.* **13**, 175–199.

VEDROS, S.G. & VISHER, G.S. (1978) The Red Oak sandstone: a hydrocarbon-producing submarine fan deposit. In: *Sedimentation in Submarine Canyons, Fans, and Trenches* (Ed. by D. J. Stanley & G. Kelling), pp. 292–308. Dowden, Hutchinson & Ross, Stroudsburg.

WALKER, R. G. (1979) Shallow marine sands. In: *Facies Models* (Ed. by R. G. Walker). *Geosc. Can. Reprint Ser.* **1**, 75–89.

ZACHRY, D.L. (1983) Sedimentologic framework of the Atoka Formation, Arkoma basin, Arkansas. In: *Tectonic–Sedimentary Evolution of the Arkoma Basin* (Ed. by D. W. Houseknecht). *Soc. econ. Paleont. Miner. Midcont. Sec.* **1**, 34–52.

Spec. Publs int. Ass. Sediment. (1986) **8**, 347–368

Sedimentary-tectonic development of the Marathon and Val Verde basins, West Texas, U.S.A.: a Permo–Carboniferous migrating foredeep

DIRCK E. WUELLNER*, LEE R. LEHTONEN *and* W. C. JAMES

Department of Geological Sciences, University of Texas at El Paso, El Paso, Texas 79968, U.S.A.

ABSTRACT

Upper Palaeozoic strata of West Texas record the convergent history of a portion of the southern North American continental margin and associated development of the Marathon and Val Verde basins.

The Tesnus, Dimple and Haymond Formations (Chesterian—Late Atokan; 3,400 m thick) of the Marathon area were formed within a probable remnant ocean basin setting in part developed adjacent to the more eastern Kerr peripheral basin. The Gaptank, Neal Ranch, Lenox Hills, 'Wolfcamp', and equivalent Permo–Pennsylvanian formations (Late Desmoinesian–Late Wolfcampian) were deposited in orogenic clastic wedge, starved basin, submarine fan, and shelf/platform settings (Val Verde Peripheral Basin Facies). A maximum preserved thickness of 6,200 m is present near the southern Val Verde basin margin.

We suggest that structural weaknesses developed in the Late Precambrian/Early Cambrian were reactivated in the late Palaeozoic as reverse faults in response to compressional stresses. These reactivated structural elements, in conjunction with Ouachita orogenesis, apparently influenced the distribution of intrabasin palaeohighs and concomitant patterns of sedimentation. Northward migration of the Val Verde basin axis was accompanied by continued fold-thrust belt development and reworking of earlier deposits of the Val Verde Peripheral Basin Facies.

INTRODUCTION

Rocks of the Marathon–Val Verde basin area were deposited in a migrant foredeep produced in part by collision and suturing during Ouachita orogenesis along the southern margin of North America (Fig. 1). However, to restrict observations to late Palaeozoic strata deposited within the foreland area is to ignore several of the underlying mechanisms that shaped basin evolution.

Most investigations of the Marathon–Val Verde basin region have focused on the structure and stratigraphy of the partly allochthonous late Palaeozoic sedimentary sequence in an effort to understand Ouachita tectonism. These late Palaeozoic rocks (Fig. 2) certainly offer invaluable information. However, the work of Jackson (1980), Cohen (1982) and Schmidt & Hendrix (1981) has shown that an appreciation of

*Present address: Tecton Energy, 277 Arthur Avenue, Shreveport, LA 71105, U.S.A.

pre-orogenic tectonism can be beneficial and in places essential when seeking to decipher subsequent synorogenic events.

Our principal objectives are to:

(1) document, as best as current data will allow, the influence of pre-orogenic structural and sedimentation patterns on subsequent late Palaeozoic foreland basin development in the Marathon–Val Verde basin region;

(2) summarize the sedimentologic character of the foreland basin fill (general lithologies and stratigraphy, dispersal patterns and sandstone composition) and compare it with strata of the closely associated remnant ocean basin setting; and

(3) delineate the northward migration of the foreland basin axis and document concomitant structural development along this part of the southern North American continental margin.

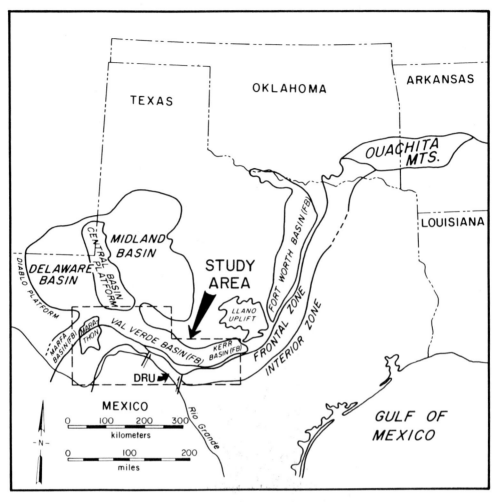

Fig. 1. Map of the Ouachita orogenic belt and associated foreland basins (FB) in Texas. Study area defined by dashed line. DRU = Devils River uplift.

HYPOTHESIS FOR INITIAL ESTABLISHMENT OF STRUCTURAL GRAIN

Information concerning the Precambrian tectonic history of West Texas is derived from a limited number of wells that have penetrated basement rock and sparse Precambrian exposures. These data suggest that two prominent trends of crustal weakness influenced subsequent Palaeozoic tectonic and depositional events. Hills (1984) speculated that NNW–NW trending zones of crustal weakness (i.e. Central Basin platform faults) were created by a left-lateral strike-slip fault zone and associated triple juncture

which developed during a Precambrian convergent event. Walper (1977), while recognizing the importance of an early convergent event between 900 and 1,000 Myr ago, ascribes much of the tectonic activity in the Delaware basin region to the development of a failed rift arm in the Late Precambrian, which he names the Delaware aulacogen (Fig. 3).

A second inherited trend of Precambrian crustal weakness has been obscured by Ouachita plate convergence. The existence of this trend is supported by the recognition of two structural elements with a similar and general east–west orientation. The first element is a group of faults with significant reverse motion. The better-known faults in this group include the Devils River uplift fault system and the Chalk Draw fault (Fig. 3; see Ewing, 1985). The second structural element is a series of crustal lineaments recognized by Bolden (1983) (Fig. 3). We postulate these lineaments have experienced movement at different times since the Precambrian, and are associated with zones of crustal weakness.

It is proposed that these two trends of weakness are linked to a single rifting event which shaped the southern margin of the lower and middle Palaeozoic North American continent. The exact nature and timing of this event is not known for certain. Briggs & Roeder (1978) proposed a rifting event around 530–507 Myr ago in the Ouachita Mountain–Oklahoma aulacogen region. Radiometric age dates of volcanic rocks from the Devils River uplift record possible rifting approximately 480–530 Myr ago (Nicholas, 1983) or 692–712 Myr ago (Denison *et al.*, 1977). The correspondence of the Briggs & Roeder (1978) and Nicholas (1983) dates, which are both from tectonic elements of the southern continental margin, suggests a Late Precambrian–Early Cambrian rifting episode.

The nature of this rifting event has been discussed by Thomas (1983), King (1975) and Cebull *et al.*, (1976). All of these investigators agree that the irregular shape of the Ouachita orogenic belt is directly linked to the pre-convergent or early Palaeozoic rift geometry of the North American continent. The geometry of the southern continental margin was characterized by an alternating series of continental embayments and promontories, which formed when segments of normally faulted rift zones were offset by large transform faults. Most notable to this study are the Texas Promontory and Marathon Embayment (Thomas, 1983) (Fig. 3). These two distinct bends in the Palaeozoic continental margin were apparently created by offset along a major transform fault near the position of the modern Llano uplift.

We postulate that prior to and during rifting, the Precambrian basement experienced extensive wrenching which led to the establishment of general east–west zones of basement weakness paralleling a transform margin. Crustal attenuation along this transform was less than that associated with a normally faulted, rifted margin which extended into Mexico. This is illustrated in Fig. 3 by the decrease in attenuated crustal area in the NW-trending segment. Structurally, the transform margin was characterized by a greater number of steeply dipping strike slip faults as opposed to normal faults. A significant bend in the rifted margin (the Marathon Embayment) formed at the position of an RRT (rift, rift, transform)

triple junction. One of the rifted margins projecting from this triple junction extended into Mexico and will not be discussed here. The other continental rift extended into the North American craton, but failed soon after developing (Delaware aulacogen). The exact history of this failed arm is not known and its existence is not agreed upon. Moreover, the Delaware aulacogen is not developed as well as its sister, the Oklahoma aulacogen. Yet, as proposed by Walper (1977), NNW–NW zones of crustal weakness resulted from the early extension associated with this failed rift arm. Therefore, in West Texas, two prominent trends of crustal weakness may be linked to a single Late Precambrian–Early Cambrian rifting event. We suggest this irregular plate boundary, hypothesized to have been associated with rifting, later influenced the location of initial suturing and subsequent foreland evolution.

PRE-OROGENIC PALAEOZOIC IMPRINT

Following continental rifting, a stable passive margin developed along the southern terminus of the North American landmass (Fig. 4). The position of sedimentary facies during the early to middle Palaeozoic was ultimately controlled by the position of the shelfbreak, which we believe was in turn influenced by the location of the zone of crustal attenuation probably established during Late Precambrian–Early Cambrian rifting. Shallow water carbonates and siliciclastics were deposited on largely stable cratonic continental crust (north of the shelfbreak). Rocks of this setting are assigned to the Passive Margin Shelf Facies (Fig. 2). South of the shelfbreak, rocks of a different character were deposited synchronously with the shallow water sequence of the passive margin shelf (Fig. 4). On the deep water slope to abyssal plain south of the shelfbreak, organic-rich mudrocks, cherts, carbonates, and minor amounts of sandstone were deposited on slowly subsiding transitional to ocean crust. These rocks, represented in the allochthonous Marathon sequence, are assigned to the Passive Margin Slope–Basin Facies (Fig. 2).

Close scrutiny of available data suggests that the influx of sediment on to the slope and basin was primarily influenced by sedimentary and tectonic conditions on the relatively stable craton. The greatest factor influencing sediment influx on to the continental slope and abyssal plain was a series of sea-level

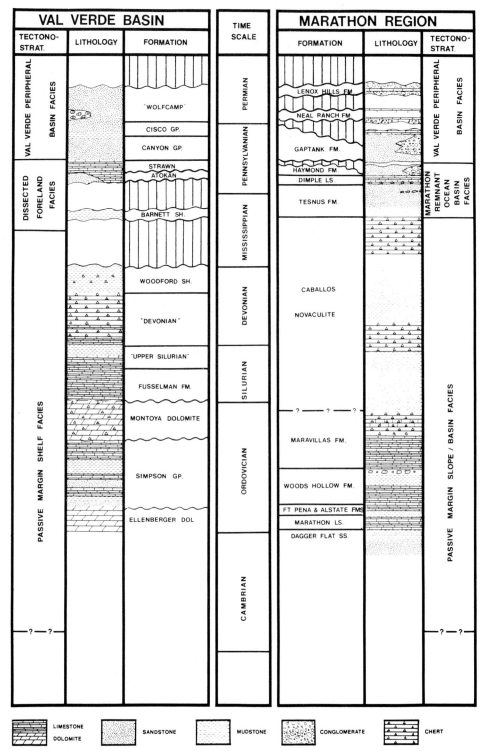

Fig. 2. Generalized stratigraphy, tectonostratigraphy and lithologic character for rocks of the Val Verde basin and Marathon region (adapted from McBride, 1970; Hills & Kottlowski, 1983; Nicholas & Rozendal, 1975).

fluctuations which led to alternating conditions of shelfal emergence and shallow water deposition. Periods of cratonal emergence correlate well with calcarenite deposition on the continental slope. This correlation of regional cratonal unconformities with calcareous slope sedimentation is probably due to the flushing of shelf carbonate debris on to the continental slope during low sea-level stand.

A second factor influencing sedimentation on the passive margin slope and basin was epeirogenic movement of the cratonal margin. The existence of 'boulder beds' or matrix-supported debris flow deposits containing clasts of a variety of compositions has been documented by Wilson (1954), Young (1970) and McBride (1969). King (1978) suggested that these deposits resulted from the slumping of an unstable cratonal margin into a deeper geosynclinal trough. However, what was the cause of this instability? One possibility is that the attenuated crustal margin was still undergoing substantial subsidence following Precambrian–Cambrian rifting. Instability resulting from this subsidence could have led to the slumping of crystalline basement rocks and overlying rocks of the Passive Margin Shelf Facies. A second hypothesis, however, considers the possible relation of conglomeratic horizons to epeirogenic movements during the developing stages of the Tobosa basin (Fig. 4). The temporal distribution of the boulder beds corresponds with periods of initial activation and subsequent subsidence of the Tobosa basin. The geometry of the Tobosa basin was such that the region undergoing the greatest amount of subsidence was to the south (Ammon, 1981), near the position of the shelfbreak. Associated with this region of relatively rapid subsidence, a series of high-angle faults with a down to the basin motion are postulated to have existed. Activation of these normal faults, which developed along zones of crustal weakness formed by the earlier failed rift arm, provided the instability necessary to initiate slumping. Unfortunately, the position of the shelfbreak and the coinciding location of these faults has been obscured by Ouachita thin-skinned tectonism, making it difficult to test the validity of this hypothesis.

We mention pre-orogenic Palaeozoic strata of the region for the following reasons. First, these rocks later serve as source terranes for clastic debris shed into orogenic basins. Secondly, we believe preorogenic, Palaeozoic facies patterns show evidence of partial control by structures initiated in the Late Precambrian–Early Cambrian, thus giving additional credence to the role of structural reactivation in the development of later Palaeozoic tectonic elements.

THE CARBONIFEROUS: A TIME OF TRANSITION

Dissected foreland development

The late Palaeozoic was characterized by a significant change in sedimentation on the formerly, primarily stable West Texas craton. Deposition of the Woodford Shale (Fig. 2) probably continued into the Early Mississippian. At about this same time the Tobosa basin began deepening on either side of a palaeohigh, the position of which is correlative with that of the modern Central Basin platform (Hills, 1972) (see Fig. 1). The development of this palaeohigh marked a period of foreland dissection, during which the previously stable craton experienced increasing amounts of tectonic disturbance. Sediments deposited in the foreland during this time are assigned to the Dissected Foreland Facies (Fig. 2).

By the Late Mississippian, the former Tobosa basin was divided into two basins, now known as the Delaware (western basin) and the Midland (eastern basin) basins (Fig. 1) (see Nicholas & Rozendal, 1975). Perhaps as early as the Late Mississippian, broad carbonate shelves developed around the margins of these basins, while organic-rich shales were deposited in deeper water and/or quieter water areas.

Throughout the Pennsylvanian, the Central Basin platform continued to rise, probably in response to increasing tectonism along the Ouachita front (Kluth & Coney, 1981). Reef and carbonate bank growth continued around the basin margins during the early Pennsylvanian, while Morrowan clastics were deposited in the deeper basin. Areas to the south (i.e. present position of the southern Delaware and Val Verde basins) were emergent during the Morrowan, and underlying Mississippian strata underwent minor erosion (Galley, 1958). In these southern regions Atokan clastics and carbonates were deposited directly atop the erosional surface developed on Mississippian rocks. Atokan clastics were subsequently overlain by carbonate platform rocks of the Strawn (Fig. 2).

Remnant ocean versus peripheral basins

By the Middle Ordovician (Walper, 1977) to Middle Silurian plate divergence had ceased, and older Palaeozoic and Proterozoic oceanic crust was actively subducted by a newly developed, probable southward-dipping subduction zone. The consumption of oceanic crust drew the North American plate closer to a poorly

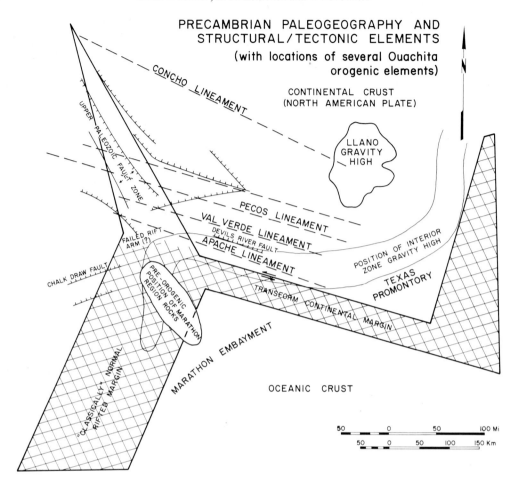

Fig. 3. Hypothesized Late Precambrian paleogeography (with key Ouachita orogenic elements superimposed) for the Marathon and Llano regions. Irregular continental margin (i.e. promontory/embayment geometry) resulted from the position of a triple junction and associated transform continental margin. Cross-hatched zone represents zone of crustal attenuation. Note lineaments which parallel transform margin (adapted from Bolden, 1983; Hills, 1984; Thomas, 1983).

understood landmass, Llanoria. Pindell (1985) speculated that in the Marathon region, this landmass was possibly the Yucatan Block, a block of continental crust which contains the present Yucatan Penninsula.

Previous tectonic summaries for the Marathon region have described the thick Carboniferous clastic sequence as a peripheral basin deposit (Ammon, 1981; Mauch, 1982). Undoubtedly, the great influx of clastic material during the Carboniferous heralded a change in the tectonic setting of the region. However, Carboniferous strata of the Marathon region are believed to have been deposited largely in a remnant ocean basin and not a peripheral basin. This is an

important distinction, because assigning these rocks to a peripheral basin setting implies that plate suturing had already occurred. In fact, the age of thin-skinned tectonism, the intensification of foreland dissection during the Pennsylvanian, and added sedimentologic evidence (i.e. shifting palaeocurrent directions and conglomerate influxes), all suggest a middle Pennsylvanian suturing for the area immediately south of the Marathon region. Therefore, the majority of the Carboniferous strata in the Marathon region was more probably deposited prior to suturing in a remnant ocean basin (Remnant Ocean Basin Facies) as depicted in Figs 5 and 6.

PASSIVE MARGIN DEVELOPMENT: LATE PRECAMBRIAN - MIDDLE MISSISSIPPIAN

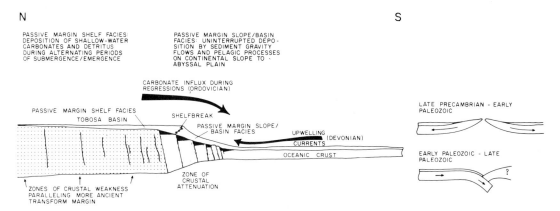

Fig. 4. Cross-sectional diagram showing attenuated margin and the relation between Passive Margin Shelf and Passive Margin Slope/Basin facies. Rocks of the Marathon region are part of the slope/basin facies.

First evidence of plate convergence: Tesnus Formation

In the Middle Mississippian, starved sedimentary conditions associated with Caballos Novaculite deposition (Fig. 2) were abruptly terminated with the influx of large amounts of clastic debris which comprise the lower Tesnus. Rocks of the Tesnus were deposited on a large submarine fan (McBride, 1969) by a variety of sedimentary gravity flow processes (Wuellner & James, in preparation). Initial deposition of these southerly derived (Johnson, 1962) clastics was predominantly in the form of a thick sequence of outer-fan mudstones, over which a series of interbedded sandstones and mudstones were deposited. Based upon a crude upward-coarsening (Cotera, 1969) and the predominance of mid-fan channel deposits in the upper Tesnus, it is probable that the Tesnus represents a south to north prograding fan system.

The interpreted tectonic setting during Tesnus deposition (Chesterian–Morrowan) is shown in Fig. 5. As stated, we consider the Tesnus to have been deposited in a remnant ocean basin that was undergoing active subduction. We believe the geological evolution of the approaching landmass, Llanoria, exerted a large influence on Tesnus deposition. During the early stages in the development of the Llanorian subduction zone, a volcanic arc undoubtedly existed which supplied arc-derived detritus to a newly developed forearc basin. As subduction continued into the Carboniferous, this volcanic arc is postulated to have migrated away from the trench and to the south. Older plutons and surrounding metamorphic terranes, which marked the position of the initial volcanic arc, were then exhumed. Detritus derived from these exhumed terranes flooded the forearc basin. By Early Mississippian the rate of sediment influx exceeded the rate of forearc basin subsidence. Overflowing clastic material bypassed the forearc basin and was transported by major channels across the subduction complex where it was deposited directly in the subduction trench and, more importantly, on adjacent remnant oceanic and transitional crust. An analogous situation was reported by Dickinson *et al.* (1982) for a portion of the Franciscan complex. In addition, detritus derived from the subduction complex fed into the Tesnus depositional system. In the remnant ocean basin, clastic debris was deposited on rapidly subsiding crust on a large submarine fan whose gradual progradation was enhanced by the continued influx of north-directed clastic debris, in addition to the southward consumption of oceanic crust upon which a portion of the fan was constructed.

The great amount of subsidence which the remnant ocean basin experienced during Tesnus deposition was likely due to the downward tilting of the oceanic crust immediately prior to subduction (Dickinson, 1981). In addition, the lithostatic load of the great

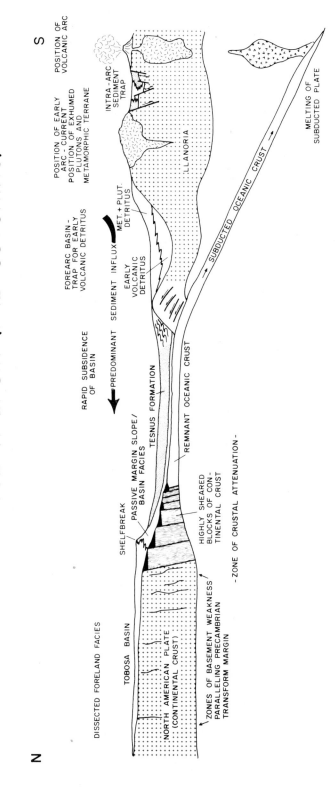

Fig. 5. Cross-sectional diagram showing tectonic conditions during Tesnus deposition. Tesnus Formation was deposited on remnant oceanic and transitional crust north of the approaching Llanoria subduction complex. Tesnus clastic debris was derived mainly from the approaching southern landmass and subduction complex.

DIMPLE DEPOSITION: MORROWAN - ATOKAN CRUSTAL "RETHICKENING" (LATE PRE-SUTURING)

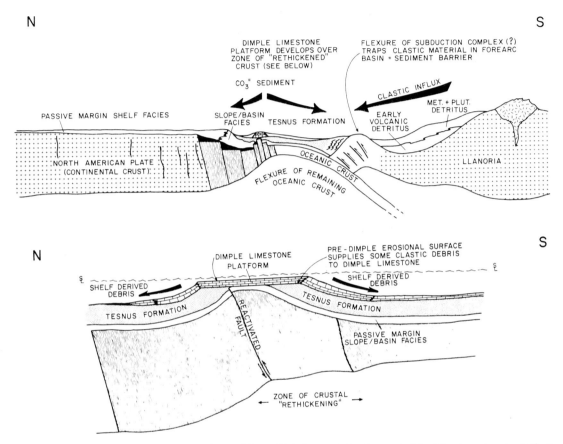

Fig. 6. Cross-sectional diagram shows tectonic conditions which favoured development of the Dimple carbonate platform. Highly sheared attenuated margin is 'rethickened' (see Cohen, 1982) as Llanoria was drawn closer to North America. Carbonate platform established on top of eroded palaeohighs produced by reverse faulting. Lower diagram is larger schematic conception of Dimple carbonate platform shown in upper diagram.

amounts of clastic debris further served to depress underlying oceanic and attenuated transitional crust.

Crustal rethickening and the sedimentologic response

The attenuated margin on which the Tesnus was deposited may have experienced minor tectonic disturbance as early as late Tesnus deposition (Morrowan?). McBride (1978) has documented the existence of 'olistostromes' in the upper Tesnus. These olistostromal horizons contain boulders of a variety of lithologies (McBride, 1978) including rocks of both the Passive Margin Shelf and Passive Margin Slope–Basin Facies. We suggest the nature of the disturbance which led to the deposition of these olistostromes was related to a process of crustal rethickening as generally outlined by Cohen (1982).

Crustal rethickening refers to a series of structural events that occur as a passive Atlantic-type continental margin experiences compressional forces associated with plate convergence. Central to the hypothesis of crustal rethickening is that previous zones of crustal weakness, related to an earlier extensional event, are reactivated as reverse faults. This reactivation of previous, possibly listric, normal faults migrates toward the craton as time proceeds. A net crustal shortening results from continued fault reactivation as previously thinned blocks of continental crust are rethickened. In response to this crustal shortening the sedimentary cover overlying the rethickened crust is folded and commonly undergoes significant detachment and thrusting.

The inclusions of boulders from the underlying Passive Margin Shelf and Slope–Basin Facies in the upper Tesnus is significant, because it signals uplift and erosion of previously stable regions. The disturbance of these regions is believed to be due to incipient crustal rethickening which manifested itself as minor block uplifts in the previously attenuated crustal zone. These disturbances, while probably minor, also affected sites of stable shelf deposition which developed over former regions of minor crustal attenuation.

Crustal rethickening: Dimple Limestone deposition

The carbonate deposits of the Dimple Limestone (Morrowan–Atokan) are enigmatic for two reasons. First, clastic deposition which characterizes most of the Carboniferous is interrupted by the Dimple carbonates, suggesting that clastic influx into the remnant ocean basin and adjacent areas was inhibited. Secondly, palaeocurrent data from the Dimple (Thomson & Thomasson, 1969) indicate that it was derived predominantly from a northern source as opposed to a southern or eastern source area that characterized Tesnus and Haymond provenance.

The Dimple Limestone has been investigated in detail by Thomson & Thomasson (1969), who found that this carbonate-dominated sequence was deposited in a shelf, slope and abyssal plain setting. These facies are arranged in a north–south direction further supporting sediment dispersal data which suggest a northern source. Compositionally, the Dimple varies from an oolitic or fossiliferous grainstone to a micrite. Mudstone, chert and conglomerate horizons are also present depending on the particular facies in question. Of particular importance, however, are the conglomeratic horizons which contain reworked novaculite and other chert varieties (Passive Margin Facies).

Any attempt to understand the tectonic conditions during Dimple deposition must consider several questions: (1) why was clastic deposition interrupted during Dimple deposition?; (2) how can the northern source area for the carbonate detritus be explained?, and (3) what is the origin of the reworked novaculite and other chert grains?

Contrary to the hypothesis that the Dimple Limestone signalled a period of tectonic quiescence (Mauch, 1982), we propose that the Late Morrowan and Early Atokan were ages of intense tectonic activity. During this time portions of the cratonal regions underlying the modern Val Verde and southern Delaware basins were subaerially exposed and underwent mild erosion. This period of emergence is believed related to the flexure of the southern cratonal margin of the North American plate immediately prior to continental collision and suturing. The attenuated crustal regions of the southern margin likewise experienced extensive crustal shortening. Regions which were sites of deeper water deposition during the lower Palaeozoic were uplifted and exposed to erosion during Dimple deposition, as suggested by the reworking of novaculite and other chert varieties probably derived in large part from the Caballos Novaculite (Fig. 2).

We believe crustal rethickening played an important role in the development of the Dimple carbonate platform. Two hypotheses are proposed to explain deposition of the northerly derived Dimple Limestone. The first hypothesis calls for the development of an intrabasinal carbonate platform (Fig. 6). In this case, crustal rethickening resulted in the development of an intrabasinal high upon which a carbonate platform developed. Note that in this situation carbonate detritus was shed both to the north and south, while predominantly northerly derived carbonate rocks are found in the Marathon region. However, subsequent thrusting, which would have propagated from the palaeohigh (Cohen, 1982), could have buried north-directed carbonate debris. The second hypothesis proposes that crustal rethickening immediately adjacent to the stable craton led to the establishment of a basin margin platform. In this latter case, however, it is difficult to place the pre-thrust position of the Marathon region in the attenuated crustal zone shown (Fig. 3). Dimple rocks in the Marathon region lie atop the Dagger Flat thrust and have, therefore, been transported a large distance (tens of kilometres) from a probable intrabasinal location. Note that in both of the two presented cases crustal rethickening established a fault-bounded high. This uplifted surface underwent erosion prior to Dimple deposition and contributed minor amounts of clastic debris to the Dimple.

The absence of large amounts of sand-sized terrigenous debris deposited during Dimple time is best explained by a flexure of the subduction complex. Underlying Tesnus clastic debris had to cross the Llanorian subduction complex before being deposited in the last vestiges of the remnant ocean basin. As subduction continued, increasing amounts of Tesnus and underlying rocks entered and were accreted to the subduction complex, thereby greatly increasing its size. The nature of the subduction complex was also altered by incipient continental collision at the Texas Promontory and associated flexure of the remnant ocean basin immediately prior to subduction. The combined effect of the aforementioned factors led to an increase in size of the subduction complex. As a result of this increase in size the subduction complex became an effective sediment barrier and no longer allowed Llanorian-derived clastics to enter the area.

Late Atokan tectonism: deposition of the Haymond Formation

By the Late Atokan, portions of the central Texas craton were sutured to the convergent landmass of Llanoria. We hypothesize the position of this suturing was principally controlled by the passive margin promontory/embayment geometry (Fig. 3). The Texas Promontory protruded from the North American plate, and so was the first geographical region to experience plate suturing during oblique collision. In conjunction with this suturing, the North American cratonal margin was drawn into the subduction zone and the Kerr peripheral basin established (see Fig. 7). This conclusion is based on the recognition of 2,500–2,750 m of Atokan clastics in the Kerr basin (Flawn *et al.*, 1961). Atokan clastics along the southern margin of the Val Verde peripheral basin (Flawn, 1959) suggest this area was also experiencing increasing amounts of subsidence. However, not all regions along the southern margin were experiencing rapid subsidence. The recognition of Atokan carbonate rocks on the Devils River uplift (Nicholas, 1983) suggests that this region was probably a site of shallow-water deposition whose origin was related to crustal rethickening (see Fig. 1).

Deposition of the Haymond Formation, some 300 km to the SW, was influenced by this Atokan suturing event. The Haymond is composed of a series of interbedded sandstones and mudstones which are capped by a *Chaetetes*-bearing limestone. Interpretations of depositional environment for the Haymond

vary from submarine fan (Thomson & McBride, 1969) to deltaic (Flores, 1974). Whatever the interpretation, however, the rocks of the Haymond exhibit a shoaling-upward sequence. The predominance of channel deposits in the upper Haymond in addition to the existence of a shallow-water patch reef limestone at the top of the formation support this interpretation.

Palaeocurrent data from the Haymond (McBride, 1966) show that the majority of the clastic debris was derived from an eastern source. This is a significant change from the underlying Dimple Limestone (northerly derived) and Tesnus Formation (southerly derived). Another important dissimilarity between the Haymond and the underlying Carboniferous rocks is conglomeratic debris flows which contain abundant crystalline basement and plutonic clasts (McBride, 1966) in addition to boulders of reworked passive margin shelf rocks (Palmer *et al.*, 1984).

The tectonic setting proposed for the Late Atokan is shown in Fig. 7. The majority of the clastic debris derived from the suture belt was fed into the Kerr and south-eastern Val Verde peripheral basins. However, large amounts of debris were also shed to the SW, along an axial trough that developed parallel to the subduction trench (Fig. 7). The Haymond trough, as it is here named, was separated from the Val Verde and Kerr peripheral basins by a palaeohigh (perhaps the incipient Devils River uplift). This palaeohigh is believed to have been a zone of rethickened crust which developed along the ancient transform margin.

The Haymond Formation is part of the clastic wedge that was deposited in this axial trough. The Haymond trough underwent a gradual evolution that was controlled by the scissor suturing of the Ouachita orogenic belt (see Graham, Dickinson & Ingersoll, 1975). As time proceeded, the Texas Promontory suture zone grew and was extended to the west as Llanoria continued its collision with the North American plate. During early Haymond deposition the central portion of the Haymond trough continued its rapid subsidence as the last portions of remnant oceanic crust were pulled into the subduction zone. Easterly derived clastic material fed into this portion of the trough was deposited predominantly by gravity flow processes. Subduction and associated subsidence slowed, as the remaining attenuated margin was partially consumed. Sedimentation rates, however, accelerated as sutured source terranes became more proximal. These changes in sedimentation and subsidence rates resulted in a shoaling-upward sequence and the influx of very coarse boulder-sized debris, both of which characterize the Haymond Formation.

LATE ATOKAN PALEOGEOGRAPHY AND TECTONIC SETTING: INITIAL PLATE SUTURING

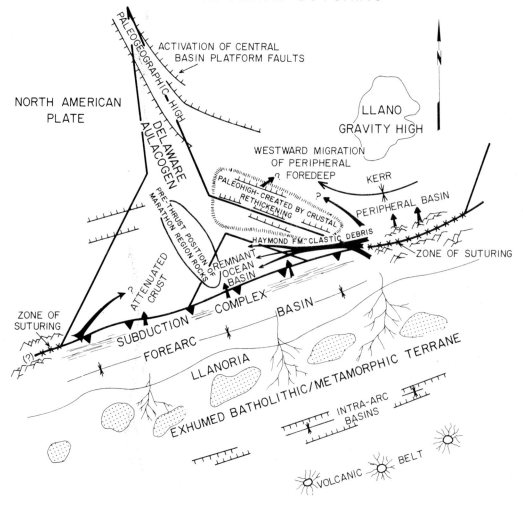

Fig. 7. Map view showing Late Atokan palaeogeography and tectonic setting. At Texas Promontory North American plate underwent initial suturing, with concurrent establishment of Kerr peripheral basin. Irregular continental margin geometry resulted in establishment of a scissor suture zone. Clastic debris of the Haymond Formation was derived largely from the eastern suture belt.

VAL VERDE PERIPHERAL BASIN FACIES

The progressive closure of the remnant ocean basin in a north-westerly direction resulted in the formation and filling of three peripheral basins along the western portion of the Ouachita front (Fig. 1): Kerr basin (Atokan; previously mentioned), Val Verde basin and Marfa basin (Late Desmoinesian to Leonard; Am-

mon, 1981). The Val Verde basin received orogenic detritus from the Late Desmoinesian through the Wolfcampian (Sanders, Boyce & Peterson, 1983).

We will focus on the tectonostratigraphic evolution of the Val Verde peripheral basin and peripheral basin facies of adjacent areas as they pertain to progressive deformation during Ouachita orogenesis.

There is a paucity of data (exclusive of industry files) published on rocks of the Val Verde basin.

Moreover, the strata of the Val Verde Peripheral Basin Facies are sometimes unfossiliferous, leading to problems of regional correlation. The phrase 'Wolfcamp clastics', as it is widely used in the literature, probably encompasses a stratigraphic interval from Late Desmoinesian to the Wolfcamp. Despite these shortcomings we have found it possible to piece together a reasonable history for the Val Verde peripheral basin area.

The major basin-filling stages of the Val Verde basin have been determined from detailed subsurface work on the Devils River uplift (Nicholas & Rozendal, 1975; Webster, 1980; Calhoun & Webster, 1983; Nicholas, 1983) and studies of the Gaptank, Neal Ranch, and Lenox Hills formations exposed in the Marathon uplift and Glass Mountains (King, 1930, 1937; Ross, 1962, 1963, 1967; Flores, McMillan & Walters, 1977) (Fig. 2).

Following deposition of the Haymond Formation (Marathon Remnant Ocean Basin Facies) and the shallow-water shelf carbonates of the Strawn (foreland shelf; Fig. 2), the axis of deposition migrated northward to a location along the present southern margin of the Val Verde basin (Fig. 8). We consider a first stage of basin fill to be approximately correlative with Gaptank and Neal Ranch deposition (Late Desmoinesian to Early Wolfcampian) (Fig. 2). A second stage of basin filling is assigned to the 'Wolfcamp' Formation (upper Wolfcamp) and approximately correlates with deposition of the Lenox Hills Formation. This latter stage of basin fill postdates final overthrusting on the Dugout Creek thrust of strata over the Late Desmoinesian to Early Wolfcampian basin axis (Moore, Mendenhall & Saultz, 1981; Ross, 1981).

Late Desmoinesian–Early Wolfcampian peripheral basin development

The initial filling of the extreme south-eastern part of the Val Verde peripheral basin probably took place in the Atokan coincident with development of the Kerr basin (Fig. 7). Atokan clastics in this area contain abundant metamorphic rock fragments derived from Ouachita sources (Flawn, 1959). However, the presence of Atokan limestones from wells on and adjacent to the Devils River uplift (Nicholas, 1983) precludes development of a terrigenous clastic-dominated basin throughout the area.

The first major (Late Desmoinesian–Early Wolfcampian) basin-filling episode followed a relatively stable period of Strawn carbonate deposition over the foreland (Early Desmoinesian; Crosby & Mapel, 1975) (Fig. 2). To the south, localized carbonates were deposited over a shoaling and semi-stable, deformed Marathon sequence (*Chaetetes*-bearing member, Haymond Formation; Ross, 1967). In the Marathon region, an unconformity separates these Haymond carbonates from the Gaptank Formation of the Val Verde Peripheral Basin Facies. Ross (1967, 1981) reports that the *Chaetetes*-bearing limestone contains early Desmoinesian fossils, whereas fossils in the overlying Gaptank indicate an early Missourian age.

We believe the existence of this unconformity and associated structural activity signalled the suturing event which welded the North American plate to Llanoria in the area immediately south of the present Marathon region. In response to this suturing and crustal shortening, rocks of the Marathon region were thrust to the north, most likely by a gravity slide mechanism (see Cohen, 1982; Figs 1 and 3).

Foreland crustal response to intensified middle Pennsylvanian tectonic activity likewise increased in magnitude. The earlier Tobosa basin was further subdivided into a number of foreland basins (Val Verde, Marfa, Delaware, Midland) bordered by foreland uplifts (Diablo, Central Basin, Devils River) and shelves (Eastern and Northwestern) (Figs 1 and 8). However, most importantly, the Val Verde basin became a distinct tectonic feature which received detritus from Ouachita orogenic highlands to the south (Oriel, Myers & Crosby, 1967; Crosby & Mapel, 1975; Ammon, 1981; Sanders *et al.*, 1983).

The rapid increase in subsidence of the Val Verde basin is believed to be due to two factors. First, during and immediately after plate suturing the southern cratonal margin was probably drawn into the subduction zone, thereby increasing the overall depth of the Val Verde basin. In addition, however, thrust faults were transporting significant volumes of rock to the north. The lithostatic load exerted by these allochthonous strata undoubtedly served to depress the underlying crust further (Jordon, 1981).

Four general settings were associated with sediment filling in the Val Verde peripheral basin: orogenic clastic wedge, starved basin, submarine fan, and carbonate shelf/platform (Fig. 8). In Late Desmoinesian, perhaps in response to early crustal downwarping, water depth increased across the basin. Thereafter, carbonate sedimentation was restricted to positive elements rimming the Val Verde basin (i.e. Central Basin platform and Eastern shelf). These shelf-edge buildups entrapped much of the northerly and easterly derived detritus (Adams *et al.*, 1951);

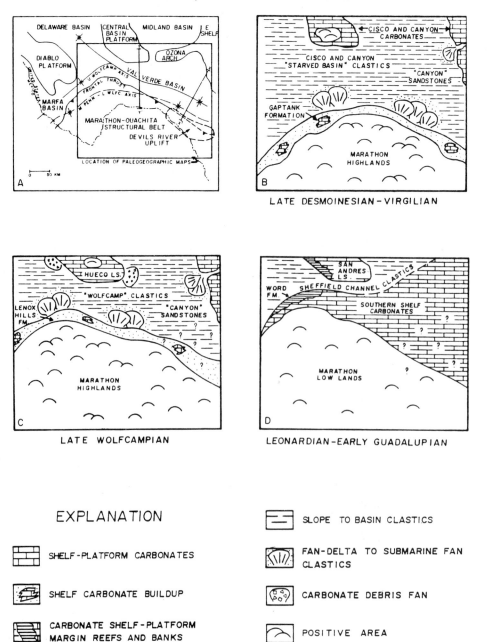

Fig. 8. (A) Map of tectonic elements in the Val Verde peripheral basin area during middle Pennsylvanian–upper Wolfcampian (adapted from Ewing, 1985; Nicholas, 1983; Ammon, 1981; Crosby & Mapel, 1975; Flawn *et al.*, 1961; and Rall & Rall, 1958). Note positions of cross-sections X–X' and Y–Y' (Figs 9 and 10) and location of included palaeogeographic maps (B–D). (B–D) Generalized palaeogeographic maps showing progressive changes in sedimentation patterns for the Val Verde peripheral basin area during late Desmoinesian–early Guadalupian (adapted from Ross, 1981; Crosby & Mapel, 1975; Oriel *et al.*, 1967; King, 1934, 1942).

however, some terrigenous clastics (upper Pennsylvanian Canyon sandstones) by-passed the Eastern shelf and entered the northern Val Verde basin, building turbidite fan complexes (Brown *et al.*, 1973). While carbonate and coarser clastic sedimentation predominated over positive regions, unfossiliferous mudrocks (≤ 100 m), which characterized the starved basin facies, were deposited over much of the northern and central Val Verde basin. The large areal extent of this starved basin facies suggests that most of the thick orogenic detritus (up to 4,600 m near Devils River uplift) was deposited in a narrow trough proximal to the early fold and thrust belt and Devils River uplift (Fig. 8).

There is very little detailed data concerning depositional environments associated with these thick orogenic clastics. The cross-section (Fig. 9, X–X') from the Devils River uplift to the Eastern shelf and palaeogeographic map (Fig. 8) illustrate our interpretation of depositional conditions. We speculate that alluvial fan-fan delta-submarine fan complexes existed adjacent to the Devils River uplift which is reported to have had 4,600–6,200 m of structural relief relative to the basin (Nicholas, 1983). Deep drilling in the area has determined that much of this thick clastic fill in the south-eastern Val Verde basin is Middle–Late Pennsylvanian. Cross-section Y–Y' across the early Marathon fold and thrust belt (Fig. 9) along with the interpretative palaeogeographic map (Fig. 8) show the distribution of depositional environments based on studies of the Gaptank Formation in the Glass Mountains and western fold belt of the Marathon area (Ross, 1967, 1981).

The Gaptank consists of about 525 m of sandstones, shales, conglomerates, and limestones deposited in a generally prograding nearshore-shelf-slope environment (Fig. 2). Clasts within the Gaptank are derived from the Maravillas Chert, Caballos Novaculite and Dimple Limestone (King, 1930) as well as other rocks of the Marathon Remnant Ocean and Passive Margin Slope–Basin Facies. The lack of major unconformities within the Gaptank suggests that a shallow shelf complex was established over a semi-stabilized fold and thrust belt, which at the time of deposition was subject to only local structural modifications and minor sea-level changes.

Approximate Gaptank equivalents in the western fold belt (King, 1930, 1937) are interpreted in part to be slope to deep basin clastics deposited by turbidity currents. These rocks are mainly interbedded, thin units of dark shale, siltstone, sandstone, and limestone (Ross, 1981). Thicknesses of the Gaptank equivalents

in this area cannot be accurately determined due to structural complexities. Ross (1981) estimates exposures of Late Pennsylvanian age are at least 60–100 m thick. Wells drilled in the area report thicknesses up to 2,450 m for a combined interval of Wolfcampian and Pennsylvanian clastics beneath the Dugout Creek thrust (King, 1978; Moore *et al.*, 1981; Fig. 10).

Meanwhile, clastic deposition continued along the southern margin of the Val Verde basin into the Early Wolfcampian, but under increasingly unstable conditions. Rocks of this latter age were probably deposited over much of the Devils River uplift area (Fig. 9). However, Mesozoic erosion has removed much of these strata. Vitrinite reflectance data (Sanders *et al.*, 1983) indicate that a minimum of 2,500–3,000 m of Permian rocks may have been removed at this Mesozoic unconformity. To the west in the Glass Mountains, the Neal Ranch Formation (Lower Wolfcampian; Fig. 2) is a thin, highly fossiliferous unit consisting of biohermal limestones, shales and sandstones (Ross, 1981). The Neal Ranch rests uncomformably on a deformed Gaptank sequence, and is uncomformably overlain by conglomerates of the Lenox Hills Formation (Upper Wolfcamp). The Neal Ranch is interpreted to have been deposited on a shallow shelf fringing the Marathon highlands. Neal Ranch equivalents in the western fold belt, however, conformably overlie Gaptank equivalents with no apparent change in texture or lithology. This relation suggests continuous deposition persisted locally on the Gaptank–Neal Ranch shelf while unconformities developed over structural highs created by movements within the Marathon fold and thrust belt.

Late Wolfcampian peripheral basin development

The final episode of filling ($> 3,700$ m) in the Val Verde peripheral basin followed the climax of Ouachita orogenesis in West Texas. The evidence for migration of the Middle Pennsylvanian–Early Wolfcampian basin axis is represented in the Glass Mountains by conglomerates and shallow water strata of the Lenox Hills Formation deposited unconformably over a thrust plate containing earlier Val Verde Peripheral Basin Facies (Gaptank and Neal Ranch formations) (Oriel *et al.*, 1967; Ross, 1981). Numerous wells drilled through the overthrust have encountered a similar stratigraphic sequence in addition to penetrating lower Wolfcampian and Pennsylvanian (?) clastics beneath the thrust plate (Decker, 1981; Moore *et al.*, 1981; Fig. 10, cross-section Y–Y'). Northward

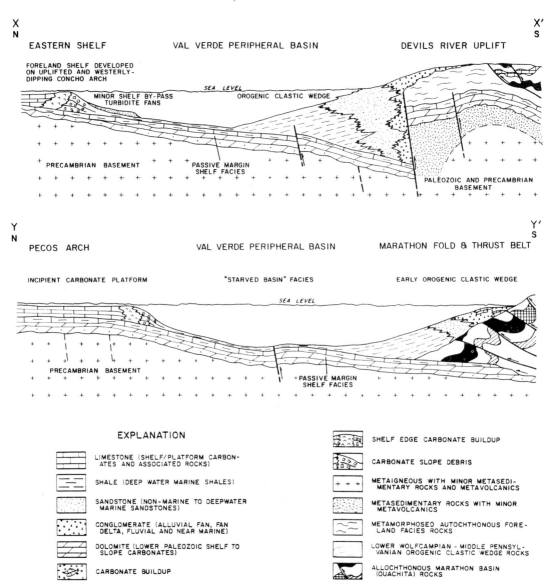

Fig. 9. Schematic cross-sections (X–X′ and Y–Y′) showing interpretive structural development and sedimentation in the middle Pennsylvanian–lower Wolfcampian Val Verde peripheral basin (adapted from Hills, 1970, 1984; Bass, 1983; Calhoun & Webster, 1983; Nicholas, 1983; Ammon, 1981; Decker, 1981; Moore *et al.*, 1981; Ross, 1981; Nicholas & Rozendal, 1975; Feldman, 1962; Rall & Rall, 1958). These cross-sections illustrate initial basin filling as influenced by foreland uplifts (especially the Devils River uplift) and early development of a fold and thrust belt. Reverse faults depicted are considered to have formed along zones of weakness set up during the late Precambrian rifting event. Location of sections shown on Fig. 8(A).

migration of the basin axis (Fig. 8) on the order of tens of kilometres seems reasonable based on limited published data (Ammon, 1981; Decker, 1981; Moore *et al.*, 1981).

Continued foreland uplift (Diablo and Central

Basin platforms) accompanied this Permian tectonism with increased movements and probable reactivation along Precambrian–Cambrian zones of weakness (Fig. 10, cross-section Y–Y′; Galley, 1958; Oriel *et al.*, 1967; Hills, 1972; Orr, 1984). Subaerial exposure of

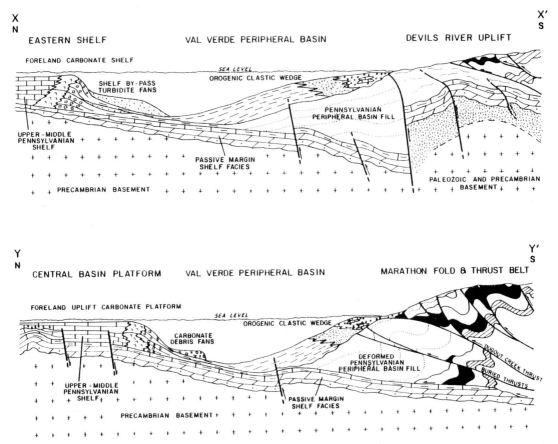

Fig. 10. Schematic cross-sections (X–X′ and Y–Y′) showing interpretive structural development and sedimentation in the upper Wolfcampian Val Verde peripheral basin (adapted from references cited in Fig. 9). These cross-sections illustrate progressive deformation along the Ouachita–Marathon orogenic front and corresponding northward migration of the basin's axis (see Fig. 8). Further uplift on the Central Basin platform resulted in an unstable shelf-break supplying carbonate detritus to the basin from the north. See Fig. 9 for explanation of symbols.

folded and faulted strata on the Central Basin platform resulted in local deposition of coarser clastics in the adjacent basins. Additional movement along faults of the Devils River uplift is hypothesized in response to late orogenic thrusting (Fig. 10). Permian isotopic dates of metamorphic rocks on the Devils River uplift (Nicholas & Rozendal, 1975; Nicholas, 1983) are partially supportive of this interpretation.

The major depositional settings established in the earlier stage of basin fill (Late Desmoinesian–Early Wolfcampian) continued into the Late Wolfcampian with the exclusion of the starved basin facies. Substantial narrowing of the basin (to 50 km) resulted in rapid filling and establishment of the Southern shelf (Fig. 8) by early Leonard time.

Depositional environments are largely interpretive for the thick orogenic fill (>3,700 m; Oriel *et al.*, 1967) of the upper Wolfcampian. Wolfcamp sandstones of the Brown–Bassett Field (south-central Val Verde Basin) are described as poorly sorted, conglomeratic and lenticular (Vinson, 1959). These sandstones were probably deposited in turbidite fan networks, and may be representative of orogenic basinal sandstones throughout the area. Nearshore, shallow-marine environments in the Marathon area are better documented based on exposure of the Lenox Hills Formation in the Glass Mountains (King, 1930; Ross, 1963, 1981; Flores *et al.*, 1977; Lehtonen, in progress). A variety of depositional environments are interpreted for the Lenox Hills including deltaic and lagoonal terrigenous clastics and biohermal carbonates. Rapid lateral facies changes within the Lenox Hills Formation suggest several environments were juxtaposed depending on structural topography and localized clastic influx from orogenic highs.

Carbonate sedimentation continued to dominate the Eastern shelf and Central Basin platform (Fig. 8); however, increasing amounts of carbonate debris and terrigenous clastics contributed to the fill of the Val Verde basin. Allochthonous carbonate deposits flanking the Wolfcampian Central Basin platform have been recently documented (Yusas, 1984; Hobson *et al.*, 1985a, b). Unstable shelf breaks surrounding foreland uplifts provided allochthonous material, some of which may have been derived from exposed Precambrian to Pennsylvanian rocks. This material cascaded into the basin mainly as debris flows (Fig. 10). Adjacent to the Eastern shelf, by-pass of clastics persisted into the Late Wolfcampian resulting in submarine fan deposition (Mitchell, 1975; Berg & Mitchell, 1976).

Limestones in the upper part of the 'Wolfcamp' Formation indicate that shallow water shelf conditions developed after rapid Wolfcamp basin filling (Oriel *et al.*, 1967). During the Leonard, carbonates and clastics were deposited over the Southern shelf (Fig. 8) that had prograded over the Late Wolfcampian basin axis. Extensive carbonate reefs and banks, which continued to develop on the margins of the Central Basin platform and other shelf areas, severely restricted marine circulation (Hills, 1972). The Sheffield Channel, Late Permian successor of the Val Verde peripheral basin, served as a passageway for marine circulation between the Delaware and Midland basins, but by the late Guadalupian was closed or severely restricted. Increasing marine restriction throughout the remainder of the Permian resulted in carbonates and evaporites filling the remaining basins (Delaware and Midland basins) of the foreland.

SANDSTONE COMPOSITION: REMNANT OCEAN BASIN VERSUS PERIPHERAL BASIN DETRITAL MODES

A comparison is in order regarding the framework composition of sandstones deposited during remnant ocean basin closure and detrital rocks of the Val Verde peripheral basin proper. Such questions arise as: Are sandstone composition signatures recognizable, explainable, and thus of utility in differentiating these

two closely related depositional settings? We submit they are at least partially successful in this case and offer as evidence data from the Tesnus Formation (remnant ocean basin deposit) and Val Verde basin fill (peripheral basin).

Data from the Tesnus and Val Verde basin sandstones (Late? Wolfcampian) were plotted on the provenance diagrams of Dickinson *et al.* (1983) and Ingersoll & Suczek (1979) (Fig. 11). For Q-F-L and Qm-F-Lt triangular diagram plots, both sets of data fall mainly within the recycled orogen field. However, the composition of the Tesnus and Val Verde sandstones tend to be segregated, with samples from the peripheral basin having less of a feldspar content. Lm-Lv-Ls plots fall within the suture belt and/or remnant ocean basin field dominated by samples very low in volcanic clasts (Fig. 11). Again there is a distinct compositional difference between these two data sets with Val Verde basin sandstones richer in sedimentary rock fragments.

A clearer appreciation of provenance differences is discernible when specific rock fragment types (including carbonate clasts) are used as end members (Fig. 11). The Tesnus sandstones contain a chert-shale-metamorphic-rock fragment-dominated suite. This combination of grain types is consistent with derivation from a subduction complex. However, such a clast suite is not unique to subduction complexes, as fold-thrust belts can yield similar grain types. Other arguments listed earlier (shifting palaeocurrent patterns, conglomerate influxes, intensification of foreland dissection, etc.), independent of sandstone composition, support derivation from the former. Microprobe data from detrital feldspars (following Trevena & Nash, 1981) in part support derivation of some clasts from areas outside the subduction complex, such as other metamorphic or plutonic terranes. There is a paucity of detrital feldspars (based on composition) derived from volcanic sources. However, it is difficult to evaluate accurately the likelihood of the presence of feldspars reworked from sedimentary sources.

Rock fragment compositions within the Val Verde peripheral basin sandstones not only include shale and chert but sandstone, coarse siltstone and carbonate clasts. Some of the more conglomeratic samples (Lenox Hills Formation) yield clasts identifiable as having been reworked from earlier peripheral basin, remnant ocean basin, or older Palaeozoic strata (Tesnus, Caballos Novaculite, etc.). Such a grain population is consistent with derivation from an active fold-thrust belt terrane.

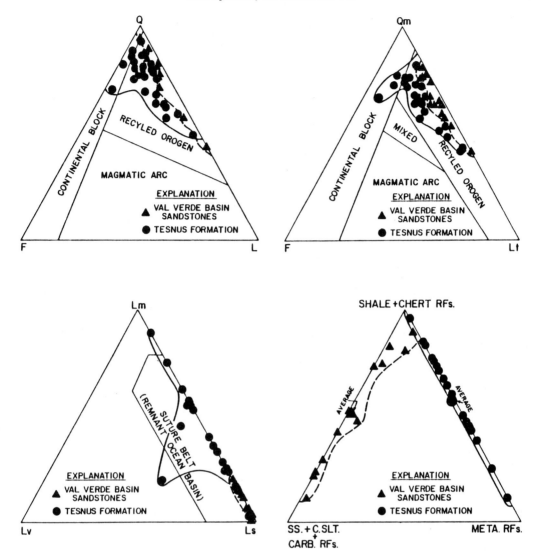

Fig. 11. Sandstone composition diagrams for Tesnus (Remnant Ocean Basin Facies; solid boundary) and Val Verde peripheral basin sandstones (dashed boundary). Q-F-L, Qm-F-Lt and Lm-Lv-Ls poles defined in Dickinson *et al.* (1983) and Ingersoll & Suczek (1979).

CONCLUSIONS

Depositional patterns within the Marathon and Val Verde basins were likely influenced by reactivated structures (initiated in Late Precambrian–Early Cambrian) in concert with plate convergence during Ouachita orogenesis. In conjunction with convergence and plate suturing, Permo–Carboniferous rocks were probably deposited in remnant ocean basin and peripheral basin settings.

The Val Verde Peripheral Basin Facies (Late Desmoinesian–Late Wolfcampian) is up to 6,200 m thick and accumulated in orogenic clastic wedges, starved basin, submarine fan, and shelf/platform settings. With continued continental suturing the Val Verde basin axis shifted northward during plate margin evolution. With axis migration, earlier peripheral basin rocks were exposed in a fold and thrust belt terrane, partially reworked, and redeposited in this more northward foreland area.

ACKNOWLEDGMENTS

We are grateful to John Hills in addition to numerous others for discussions concerning the geology of West Texas. R. L. Nicholas read an early version of this manuscript and made several constructive comments. R. J. Moiola and R. E. Denison are acknowledged as reviewers. The Society of the Sigma Xi and the American Association of Petroleum Geologists provided financial support for one of us (D.W.). We are grateful to Linda Marston and Doug Madden for drafting of figures. We acknowledge Jean Hocking for typing this manuscript. The University of Texas at El Paso financed a portion of the travel.

Despite the numerous constructive additions by those individuals mentioned above, we take full responsibility for the conclusions of this study.

REFERENCES

ADAMS, J.E., FRENZEL, H.N., RHODES, M.L. & JOHNSON, D.F. (1951) Starved Pennsylvanian Midland Basin. *Bull. Am. Ass. Petrol. Geol.* **35**, 2600–2607.

AMMON, W.L. (1981) Geology and plate tectonic history of the Marfa Basin, Presidio County, Texas. In: *Marathon–Marfa Region of West Texas, Symposium and Guidebook. Publ. Permian Basin Sect. Soc. econ. Paleont. Mineral. 81–20*, 75–102.

BASS, R.O. (1983) Significant wells—Val Verde Basin—1959 to 1982. In: *Stratigraphy and Structure of the Devil's River Uplift, Texas. Publ. West Texas geol. Soc. 83–77*, 147–149.

BERG, R.R. & MITCHELL, M.M. (1976) Turbidite reservoir in Canyon sandstone, Val Verde Basin, Texas (abstract). *Bull. Am. Ass. Petrol. Geol.* **60**, 324.

BOLDEN, G.P. (1983) Val Verde basin structural geology related to the rest of the Permian Basin. In: *Stratigraphy and Structure of the Devil's River Uplift. Publ. West Texas geol. Soc. 83–77*, 125–137.

BRIGGS, G. & ROEDER, D. (1978) Sedimentation and plate tectonics, Ouachita Mountains and Arkoma Basin. In: *Trace Fossils and Paleoecology of the Ouachita Geosyncline* (Ed. by C. K. Chamberlain), pp. 1–22.

BROWN, L.F., Jr, CLEAVES, A.W., II & ERXLEBEN, A.W. (1973) Pennsylvanian depositional systems in north-central Texas. *Guidebook. No. 14. Bur econ. Geol. Univ. Texas Austin*, pp. 57–73.

CALHOUN, G.G. & WEBSTER, R.E. (1983) Surface and subsurface expression of Devil's River Uplift, Kinney and Val Verde counties, Texas. In: *Structure and Stratigraphy of the Devil's River Uplift, Texas. Publ. West Texas geol. Soc. 83–77*, 101–118.

CEBULL, S.E., SHURBET, D.H., KELLER, G.R. & RUSSELL, L.R. (1976) Possible role of transform faults in the development of apparent offsets in the Ouachita–Southern Appalachian tectonic belt. *J. Geol.* **84**, 107–114.

COHEN, C.R. (1982) Model for a passive to active continental margin transition: implications for hydrocarbon exploration. *Bull. Am. Ass. Petrol. Geol.* **66**, 708–718.

COTERA, A.S., Jr (1969) Petrology and petrography of the Tesnus Formation. In: *Stratigraphy, Sedimentary Structures and Origin of Flysch and Pre-flysch Rocks* (Ed. by E. F. McBride), pp. 66–71. Dallas Geological Society.

CROSBY, E.J. & MAPEL, W.J. (1975) Paleotectonic investigations of the Pennsylvanian System in the United States—central and west Texas. In: *Paleotectonic Investigations of the Pennsylvanian System in the United States. Prof. Pap. U.S. geol. Surv. 853*, 197–232.

DECKER, G.M. (1981) *A surface and subsurface study along the northwestern margin of the Val Verde Basin, Pecos, Terrell and Brewster counties, Texas.* Unpublished M.S. Thesis, University of Texas at El Paso.

DENISON, R.E., BURKE, W.H., OTTO, J.B. & HETHERINGTON, E.A. (1977) Age of igeous and metamorphic activity affecting the Ouachita foldbelt. In: *Symposium on the Geology of the Ouachita Mountains, Vol. 1* (Ed. by C. G. Stone), pp. 25–40.

DICKINSON, W.R. (1981) Plate tectonic evolution of sedimentary basins. In: *Plate Tectonics and Hydrocarbon Accumulations* (Ed. by W. R. Dickinson & H. Yarborough), pp. 1–62.

DICKINSON, W.R., BEARD, L.S., BRAKENRIDGE, G.R., ERJAVEC, J.L., FERGUSON, R.C., INMAN, K.F., KNEPP, R.A., LINDBERG, R.A. & RYBERG, P.T. (1983) Provenance of North American Phanerozoic sandstones in relation to tectonic setting. *Bull. geol. Soc. Am.* **94**, 222–235.

DICKINSON, W.R., INGERSOLL, R.V., COWAN, D.S., HELMOND, K.P. & SUCZEK, C.A. (1982) Provenance of Franciscan graywackes in coastal California. *Bull geol. Soc. Am.* **93**, 95–107.

EWING, T.E. (1985) Westward extension of the Devils River uplift—implications for the Paleozoic evolution of the southern margin of North America. *Geology*, **13**, 433–436.

FELDMAN, M.L., Jr (1962) Southwest–northeast cross-section, Marathon region to Midland Basin. *Publ. W. Texas geol. Soc. 62–47*.

FLAWN, P.T. (1959) Devil's River Uplift. In: Geology of the Val Verde Basin. *Publ. W. Texas geol. Soc. 59–43*, 74–78.

FLAWN, P.T., GOLDSTEIN, A., Jr, KING, P.B. & WEAVER, C.E. (1961) The Ouachita System. The Univ. of Texas at Austin, *Publ. Bur econ. Geol. 6120*, 401 pp.

FLORES, R.M. (1974) Characteristics of Pennsylvanian lower-middle Haymond delta-front sandstones, Marathon Basin, West Texas. *Bull. geol. Soc. Am.* **85**, 709–716.

FLORES, R.M., MCMILLAN, T.L. & WALTERS, G.E. (1977) Lithofacies and sedimentation of Lower Permian carbonates of the Leonard Mountains area, Glass Mountains, western Texas. *J. sedim. Petrol.* **47**, 1610–1622.

GALLEY, J.E. (1958) Oil and geology in the Permian Basin of Texas and New Mexico. In: *Habitat of Oil* (Ed. by L. G. Weeks). *Am. Ass. Petrol. Geol., Tulsa*, 395–446.

GRAHAM, S.A., DICKINSON, W.R. & INGERSOLL, R.V. (1975) Himalayan-Bengal model for flysch dispersal in the Appalachian-Ouachita system. *Bull. geol. Soc. Am.* **86**, 273–286.

HILLS, J.M. (1970) Gas in Delaware and Val Verde basins, west Texas and southeastern New Mexico. In: *Natural Gases of North America. Mem. Am. Ass. Petrol. Geol., Tulsa*, **9**, 1394–1432.

HILLS, J.M. (1972) Late Paleozoic sedimentation in West Texas Permian Basin. *Bull. Am. Ass. Petrol. Geol.* **56**, 2303–2322.

HILLS, J.M. (1984) Sedimentation, tectonism, and hydrocarbon generation in Delaware Basin, West Texas and southeastern New Mexico. *Bull. Am. Ass. Petrol. Geol.* **68**, 250–267.

HILLS, J.M. & KOTTLOWSKI, F.E. (1983) *Correlation of stratigraphic units of North America (COSUNA) Project, Southwest/Southwest Mid-Continent Region.* Am. Ass. Petrol. Geol., Tulsa.

HOBSON, J.P., Jr, CALDWELL, C.D. & TOOMEY, D.F. (1985a) Sedimentary facies and biota of early Permian deep-water allochthonous limestone, southwest Reagan County, Texas. In: *Deep-Water Carbonates—a Core Workshop* (Ed. by P. D. Crevello & P. M. Harris). *Soc. econ. Paleont. Mineral., SEPM Core Workshop No. 6*, 93–139.

HOBSON, J.P., Jr, CALDWELL, C.D. & TOOMEY, D.F. (1985b) Early Permian deep-water allochthonous limestone facies and reservoir, Reagan and Crockett counties, West Texas. *Bull. Am. Ass. Petrol. Geol.* **69**, 2130–2147.

INGERSOLL, R. V. & SUCZEK, C.A. (1979) Petrology and provenance of Neogene sand from Nicobar and Bengal fans, DSDP Sites 211 and 218. *J. sedim. Petrol.* **49**, 1217–1228.

JACKSON, J.A. (1980) Reactivation of basement faults and crustal shortening in orogenic belts. *Nature*, **283**, 343–346.

JOHNSON, K.E. (1962) Paleocurrent study of the Tesnus Formation, Marathon Basin, Texas. *J. sedim. Petrol.* **32**, 781–792.

JORDAN, T.E. (1981) Thrust loads and foreland basin evolution, Cretaceous, Western United States. *Bull. Am. Ass. Petrol. Geol.* **65**, 2506–2520.

KING, P.B. (1930) The Geology of the Glass Mountains, Texas, Pt. 1. *Bull. Univ. Texas*, *3038*, 167 pp.

KING, P.B. (1934) Permian stratigraphy of Trans-Pecos, Texas. *Bull. geol. Soc. Am.* **45**, 697–798.

KING, P.B. (1937) Geology of the Marathon Region, Texas. *Prof. Pap. U.S. geol. Surv. 187*, 148 pp.

KING P.B. (1942) Permian of West Texas and southeastern New Mexico: West Texas–New Mexico Symposium, Part II. *Bull. Am. Ass. Petrol. Geol.* **26**, 535–563.

KING, P.B. (1975) Ancient southern margin of North America. *Geology*, **3**, 132–134.

KING, P.B. (1978) Tectonics and sedimentation of the Paleozoic rocks in the Marathon Region, West Texas. In: *Tectonics and Paleozoic Facies of the Marathon Geosyncline, West Texas* (Ed. by S. J. Mazzullo). *Permian Basin Sect Soc. econ. Paleont. Mineral. 78–17*, 5–37.

KLUTH, C.F. & CONEY, P.J. (1981) Plate tectonics of the Ancestral Rocky Mountains. *Geology*, **9**, 10–15.

MAUCH, J.M. (1982) *The Late Paleozoic tectono-sedimentary history of the Marfa Basin, West Texas.* Unpublished M.S. Thesis, Texas Christian University.

MCBRIDE, E.F. (1966) Sedimentary petrology and history of the Haymond Formation (Pennsylvanian), Marathon Basin, Texas. *Publ. Bur. econ. Geol. Univ. Texas Austin*, **57**, 101 pp.

MCBRIDE, E.F. (1969) Stratigraphy and sedimentology of the Fort Pena Formation. In: *Stratigraphy, Sedimentary Structures and Origin of Flysch and Pre-flysch rocks, Marathon Basin, Texas* (Ed. by E. F. McBride), pp. 43–46. Dallas Geological Society.

MCBRIDE, E.F. (1970) Flysch sedimentation in the Marathon Region, Texas. In: *Spec. Pap. geol. Soc. Can.* **7**, 67–83.

MCBRIDE, E.F. (1978) Olistostromes in the Tesnus Formation (Mississippian–Pennsylvanian), Payne Hills, Marathon Region, Texas. *Bull. geol. Soc. Am.* **89**, 1550–1558.

MITCHELL, M.M. (1975) *Depositional environment and facies relationships of the Canyon sandstone, Val Verde Basin, Texas.* Unpublished M.S. thesis, Texas A & M University.

MOORE, G.E., MENDENHALL, G.V. & SAULTZ, W.L. (1981) Northern extent of Marathon thrust Elsinore area, Pecos County, Texas. In: *Marathon-Marfa Region of West Texas, Symposium and Guidebook. Publ. Permian Basin Sect Soc. Econ. Paleont. Mineral. 81–20*, 111–128.

NICHOLAS, R.L. (1983) Devil's River Uplift. In: *Stratigraphy and structure of the Devil's River Uplift. Publ. W. Texas geol. Soc. 83–77*, 125–137.

NICHOLAS, R.L. & ROZENDAL, R.A. (1975) Subsurface positive elements within the Ouachita foldbelt in Texas and their relation to Paleozoic cratonic margin. *Bull. Am. Ass. Petrol. Geol.* **59**, 193–216.

ORIEL, S.S., MYERS, D.A. & CROSBY, E.J. (1967) Paleotectonic investigations of the Permian System in the United States—West Texas Permian Basin region. In: *Paleotectonic Investigations of the Permian System. Prof. Pap. U.S. geol. Surv. 515*, 21–60.

ORR, C.D. (1984) *A seismotectonic study and stress analysis of the Kermit seismic zone, Texas.* Unpublished Ph.D. Dissertation, The University of Texas at El Paso.

PALMER, A.R., DEMIS, W.D., MUEHLBERGER, W.R. & ROBINSON, R.A. (1984) Geological implications of Middle Cambrian boulders from the Haymond Formation (Pennsylvanian) in the Marathon Basin, West Texas. *Geology*, **12**, 91–94.

PINDELL, J.L. (1985) Alleghenian reconstruction and subsequent evolution of the Gulf of Mexico. *Tectonics*, **4**, 1–40.

RALL, R.W. & RALL, E.P. (1958) Pennsylvanian subsurface geology of Sutton and Schleicher Counties, Texas. *Bull. Am. Ass. Petrol. Geol.* **42**, 839–870.

ROSS, C.A. (1962) Permian tectonic history in Glass Mountains, Texas. *Bull. Am. Ass. Petrol. Geol.* **46**, 1728–1733.

ROSS, C.A. (1963) Standard Wolfcampian Series (Permian), Glass Mountains, Texas. *Mem. geol. Soc. Am.* **88**, 205 pp.

ROSS, C.A. (1967) Stratigraphy and depositional history of the Gaptank Formation (Pennsylvanian), West Texas. *Bull. geol. Soc. Am.* **78**, 369–384.

ROSS, C.A. (1981) Pennsylvanian and Early Permian history of the Marathon Basin, West Texas. In: *Marathon-Marfa Region of West Texas, Symposium and Guidebook. Publ. Permian Basin Sect. Soc. econ. Paleont. Mineral. 81–20*, 135–144.

SANDERS, D.E., BOYCE, R.G. & PETERSON, N. (1983) The structural evolution of the Val Verde Basin, West Texas. In: *Stratigraphy and Structure of Devil's River Uplift, Del Rio, Texas. Publ. W. Texas geol. Soc. 83–77*, 123.

SCHMIDT, C.J. & HENDRIX, T.E. (1981) Tectonic controls for thrust belt and Rocky Mountain foreland structures in the northern Tobacco Root Mountains–Jefferson Canyon area, Southwestern Montana. In: *Montana Geol. Soc. Field Conf. and Symposium Guidebook to Southwest Montana* (Ed. by T. E. Tucker), pp. 167–179.

THOMAS, W.A. (1983) Continental margins, orogenic belts and intracratonic structures. *Geology*, **11**, 270–272.

THOMSON, A. & MCBRIDE, E.F. (1964) Summary of the geologic history of the Marathon geosyncline. In: *The*

Filling of the Marathon Geosyncline, Symposium and Guidebook. Publ. Permian Basin Sect. Soc. econ. Mineral. Paleont. 64–9, 52–60.

THOMSON, A.F. & THOMASSON, M.R. (1969) Shallow- to deep-water facies development in the Dimple Limestone (Lower Pennsylvanian), Marathon Region, Texas. In: *Depositional Environments in Carbonate Rocks. Spec. Publ. Soc. econ. Paleontol. Mineral.* **14**, 57–78.

TREVENA, A.S. & NASH, W.P. (1981) An electron microscope study of detrital feldspar. *J. sedim. Petrol.* **51**, 137–150.

VINSON, M.C. (1959) Brown-Bassett field, Terrell County. In: *Geology of the Val Verde Basin, Field Trip Guidebook. Publ. W. Texas geol. Soc. 59–43*, 85–86.

WALPER, J.L. (1977) Paleozoic tectonics of the southern margin of North America. *Trans. Gulf-Cst Ass. geol. Soc.* **27**, 230–241.

WILSON, J.L. (1954) Late Cambrian and Early Ordovician trilobites from the Marathon uplift, Texas. *J. Paleontol.* **28**, 349–385.

WEBSTER, R.E. (1980) Structural analysis of the Devil's River Uplift—southern Val Verde Basin, southwest Texas. *Bull. Am. Ass. Petrol. Geol.* **64**, 230–241.

YOUNG, L.M. (1970) Early Ordovician sedimentary history of the Marathon geosyncline, Trans-Pecos, Texas. *Bull Am. Ass. Petrol. Geol.* **54**, 2303–2316.

YUSAS, M.R. (1984) Mineralogy, petrology, and stratigraphy of Wolfcampian strata in the Getty University 7–21 #1, Winkler Co., Texas. *Bull. W. Texas geol. Soc.* **24** (2) 4–12.

Spec. Publs int. Ass. Sediment. (1986) **8**, 369–392

On the depositional response to thrusting and lithospheric flexure: examples from the Appalachian and Rocky Mountain basins

ANTHONY J. TANKARD

Petro-Canada Inc., Box 2844, Calgary, Alberta T2P 3E3, Canada

ABSTRACT

The Appalachian and Cordilleran foreland basins of North America resulted from regional isostatic adjustments of the lithosphere to thrust-belt loading. Flexural subsidence of these basins was accompanied by upwarping along their cratonward margins; subtle irregularities of the foreland lithosphere reflect variations in composition and inherited structural fabric. It is suggested that viscoelastic relaxation of plate bending stresses during tectonically quiescent interludes resulted in overdeepening of the shale-prone basins and upwarping along the basin margin arch systems.

On a large scale these basins evolved through three distinct phases as overthrust terranes migrated across older passive margins. Initially, exotic terranes were accreted on to the attenuated passive margin ramps, loading relatively thin lithosphere, and creating the deep Taconic (Appalachian) and Columbian (Cordilleran) foreland basins. The intermediate Acadian and mid-Cretaceous episodes were more passive in style, probably because the continental margin ramps impeded overthrusting for tens of millions of years. This was a period characterized by subtle adjustments and tectonic thickening of the overthrust loads, viscoelastic relaxation of the lithosphere, overdeepening of the shale-dominated basins, and upwarping of basin-margin arch systems. Marine environments were widespread. Eventually, tectonic overthickening resulted in topographic head sufficient to drive thin thrust sheets across the hingeline, loading thick, thermally mature, and rigid continental crust. The Alleghenian and Laramide basins oscillated between underfilled and overfilled conditions. Periodic abandonment of the depositional landscape is reflected in accumulation of thick and regionally persistent coal seams.

Many of the stratigraphic patterns predicted by viscoelastic models of the lithosphere are supported by field studies. Marine inundation (overdeepening) and dark shale deposition were most common during periods of relative orogenic quiescence. 'Shelf' sand-ridge and carbonate shoal deposition were controlled by actively rising basin margin arches during the quiescent episodes. Renewed overthrusting (Alleghenian and Laramide) introduced a flood of river-borne sediments, but the basement arches were no longer prominent.

INTRODUCTION

The foreland thrust and fold belts of the North American Appalachians and Cordillera are hundreds of kilometres wide, and record long histories of intermittent shortening spanning 100–200 Myr. Detailed field studies and deep seismic reflection profiling have shown that older attenuated passive margins underlie these thrust systems, and that this lithospheric configuration has profoundly influenced structural style and level (Price, 1981; Ando *et al.*, 1984; Cook, 1984). Thrust-sheets impose immense loads on the margins of the craton. Foreland basins are a conse- quence of lithospheric flexure beneath these loads. The pattern of basin evolution should reflect subtle adjustments to the size of the overthrust load, thickness and rheology of the lithospheric plate beneath the load, rate of thrusting, as well as the time available for isostatic adjustment between active thrusting epi- sodes.

Attempts to model these responses quantitatively are adding a vital new dimension to foreland basin studies (Beaumont, 1981; Jordan, 1981; Karner & Watts, 1983; Quinlan & Beaumont, 1984; Royden &

Karner, 1984; Schedl & Wiltschko, 1984; Stockmal, Beaumont & Boutilier, 1986). These models address the inferred rheological characteristics of the deformable lithospheric plates in terms of continuous elastic layers, continuous viscoelastic layers, and viscoelastic layers with vertical variations in temperature-dependent viscosity. To a considerable extent these rheology-dependent models remain mathematical abstractions requiring deliberate stratigraphic testing.

These numerical models conveniently bridge the realms of the structural geologist and stratigrapher. Tectono-stratigraphy integrates concepts of thrust-sheet behaviour, basin deformation, and sedimentology. On a large scale the unconformity-bounded sequences of Sloss (1963, 1982) record major episodes in basin evolution (events lasting tens of millions of years). At the other extreme, facies analysis has proved invaluable in reconstructing depositional events on the scale of the outcrop, mine property, or hydrocarbon reservoir (events generally shorter than a million years). The depositional system is intermediate in scale, and is capable of recording basin responses to more subtle changes in thrust-sheet behaviour (million-year scale).

Because sedimentological interpretation relies to a considerable extent on modern analogues in reconstructing palaeogeographies, it frequently fails to recognize unique landscapes that have no Recent counterparts. An example is the enigmatic 'shelf' sand-ridge province of the Rocky Mountain foreland basin (Campbell, 1973; Brenner, 1980; Walker, 1983; Tillman & Martinsen, 1984). In the Appalachian and Rocky Mountain foreland basins lithospheric dynamics may have been the primary control on the distribution of depositional systems. It is surely short-sighted to attempt palaeogeographic reconstructions without taking cognizance of the effect of thrust-sheet loading on basin dynamics.

The intention of this paper is to review selected results of quantitative geodynamic modelling, and to illustrate their stratigraphic expression. In the Palaeozoic, overthrusts of the Taconic, Acadian and Alleghenian orogenies caused subsidence of the Appalachian basin. Each orogenic episode drove the thrust-sheets further across an attenuated continental margin, progressively loading thicker and stronger lithosphere, and resulting in shallowing of successive foreland basins and greater relief of the orogens. Shoaling across the forebulge produced carbonate environments in the Ordovician, and reworked quartz arenites in the Silurian to Carboniferous basins.

A suite of Carboniferous depositional systems in eastern Kentucky is described in detail to show the more subtle lithospheric and stratigraphic responses to episodic thrust-sheet behaviour and sediment supply. Finally, intermittent upwarping of basin-margin arches is suggested as a cause of shoaling and deposition of the marine sand-ridge province in the mid-Cretaceous Rocky Mountain basin.

TECTONIC FRAMEWORK

Quantitative models

Downward flexure in foreland basins is primarily a regional isostatic adjustment to thrust-belt loading, with a subordinate increment of subsidence resulting from sediment loading. The dynamics of basin subsidence and upwarping of basin margin arch systems reflects the flexural properties of the lithosphere. Quantitative models of lithospheric flexure depend upon these assumed rheological properties. In the Appalachian basin Quinlan & Beaumont (1984) have used the known stratigraphic record to constrain lithospheric behaviour, as well as timing, distribution, and thicknesses of thrust sheets. COCORP profiles (Cook, 1984) independently predict overthrust thicknesses very similar to those derived from quantitative models (Quinlan & Beaumont, 1984; Stockmal *et al.*, 1986), thus providing confidence in the modelling techniques.

The ability of a flexural model to account for stratigraphic characteristics depends upon the specified rheological properties. Quinlan & Beaumont (1984) examined three lithospheric types: continuous elastic layer, continuous uniform viscoelastic layer, and a continuous layer with temperature-dependent viscosity. (Modelling procedures are described by Beaumont, 1981; Wu & Peltier, 1982; Quinlan & Beaumont, 1984.)

Elastic and uniform viscoelastic representations of the lithosphere were unable to predict the stratigraphic record, the first because it allowed no stress relaxation, the second because it responded too generously over long time periods. The temperature-dependent viscosity model is a compromise that depends primarily on geothermal gradient and activation energy for creep (Quinlan & Beaumont, 1984). These parameters affect the effective elastic thickness, which in turn controls the position of the forebulge relative to the overthrust load, the distance increasing as the 3/4 power of lithospheric thickness. Thrust-sheet loading of strong (thick) lithosphere would produce a wider and shallower basin than it would on weak (thin) lithosphere.

The initial elastic response of the lithosphere to loading is independent of any assumed time-dependent rheological properties, and results in a downwarped flexural basin adjacent to the load, and peripheral upwarping along the cratonward edge of the basin (Fig. 1). If the lithosphere behaved as an elastic layer only, there would be no further lithospheric deformation if the load remained unchanged. In contrast, a viscoelastic rheology would enable relaxation of the plate bending stress by lithospheric flow (Quinlan & Beaumont, 1984). Relaxation thus

results in deepening of the basin with the forebulge uplifting and migrating towards the load (Fig. 2). This process is repeated as each new thrust-sheet advance causes an initial elastic flexural response which is superimposed over the previous relaxation geometry. The temperature-dependent viscoelastic model thus emphasizes stress relaxation of a lithosphere that initially responds elastically to loading, with the forebulge also reflecting flexural and relaxation modes as a consequence.

Relaxation of the load-induced stress is a necessary consequence of viscoelasticity. What is contentious is whether the lithosphere behaves elastically or viscoelastically. A potentially useful constraint on the various flexural models is the stratigraphic relationship to basin margin arches (Tankard, 1986). These arches are sensitive indicators of changing flexural properties. Although their scope for vertical movement is small compared with the distances travelled by thrust sheets, just a few metres of uplift could be the difference between blanket deposition of fines and reworking within wave-base.

Recent studies have shown that the Appalachian and Cordilleran orogenic belts were obducted across earlier attenuated passive margins (Price, 1981; Price & Hatcher, 1983; Ando *et al.*, 1984; Cook, 1984). Stockmal *et al.* (1986) have extended previous geodynamic models to include inherited passive margin configuration and age, topographic expression of the overthrust load, and foreland depth. They have treated the lithosphere as a non-uniform elastic plate whose strength is controlled thermally, and whose equilibrium time-averaged lithospheric thickness (70 km, or

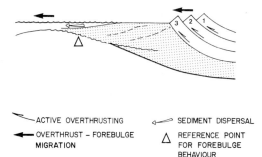

1. OVERTHRUST LOADING - Flexural deformation

2. RELAXATION PHASE - Viscoelastic response

3. RENEWED OVERTHRUST LOADING - Flexural deformation

ACTIVE OVERTHRUSTING

OVERTHRUST – FOREBULGE MIGRATION

SEDIMENT DISPERSAL

△ REFERENCE POINT FOR FOREBULGE BEHAVIOUR

Fig. 1. Elements of the foreland deformational model in which basin subsidence and peripheral upwarping are a response to thrust-belt loading. (1) The lithosphere responds elastically to initial loading. (2) As the thrust-sheet load remains in place for a long period of time the lithosphere adjusts viscoelastically, resulting in basin deepening, and accentuated upwarping of the forebulge and its contraction towards the load. Shoaling results in reworking about the crest of the forebulge and development of unconformities. (3) With renewed thrust-sheet encroachment the lithosphere responds elastically, suppressing the forebulge, and forcing it to migrate ahead of the advancing load. (After Quinlan & Beaumont, 1984.)

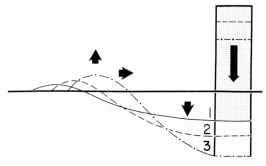

Fig. 2. Deformation of a continuous viscoelastic plate under an applied load. The initial response (profile 1) is essentially elastic. As the load remains stationary for a long period of time, lithospheric flow at depth relaxes the plate bending stress, creating a deeper and narrower moat. Viscoelastic relaxation thus results in progressive uplift of the forebulge and its migration towards the load (profiles 2 and 3). (After Quinlan & Beaumont, 1984.)

a relaxation isotherm of 750°C) is that derived for the Appalachians by Quinlan & Beaumont (1984). However, the instantaneous elastic thickness of the plate will reflect the thermal state of the earlier passive margin (Courtney & Beaumont, 1983). The specified passive margin bathymetry and configuration is based loosely on the modern, 'typical' margin of Nova Scotia. Stockmal *et al.* (1986) constructed four archetypal models involving overthrusting of thermally young (~10 Ma) or thermally old (~120 Ma) passive margins by topographically high or topographically low overthrust wedges. Their calculations show that for continental margins older than about 30 Ma, the size of the overthrust wedge has a greater influence than age on foreland basin geometry. Because Appalachian and Cordilleran deformation was dominated by thick overthrusts on thermally mature continental margins (Price, 1981; Price & Hatcher, 1983; Ando *et al.*, 1984) only the 'high/old' archetype of Stockmal *et al.* is emphasized here (Fig. 3).

The seaward flank of the continental margin on to which the first overthrust loads were emplaced was significantly below sea-level. A 14 km thick overthrust wedge (e.g. Ando *et al.*, 1984) obducted across an ocean floor initially 4–5 km deep would create very little relief above sea-level simply because of Airy regional compensation (Quinlan & Beaumont, 1984; Stockmal *et al.*, 1986). In the earliest stages of ocean closure very thick overthrust wedges can be accommodated with little topographic expression.

Figure 3 illustrates schematically just four of the many timesteps in the general model of Stockmal *et al.* (1986). At time 0, a passive margin sequence has been deposited on a stretched and rifted continental margin. At timestep 1, a thick accretionary terrane has overridden the oceanic crust and lithosphere, structurally deformed and incorporated slope and rise sediments of the miogeoclinal wedge, and abutted against the large crustal ramp immediately outboard of the hingeline. The topographic expression of the accretionary terrane may be very low at this stage. At timestep 2, the accretionary terrane has thickened considerably, but has not been pushed much further over the hingeline. This thickening stage may be due to the mechanical necessity of building sufficient topography (cf. Chapple, 1978) to drive the thin-skinned thrusts of timestep 3 over the hingeline. This increase in topographic expression is suggested in quantitative models of the Canadian Cordillera and the Alps by Stockmal & Beaumont (in preparation). Tectonic thickening may also initiate granitic plutonism deep in the structural pile (cf. Fyffe, Pajari &

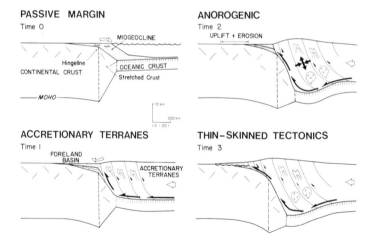

Fig. 3. Progressive evolution of a foreland basin as allochthonous terranes migrated across an inherited passive margin. These illustrations are schematic, but based on the predictions of thermo-mechanical models which have been determined quantitatively by Stockmal *et al.* (1986). The pre-convergence passive margin sequence was deposited on a stretched and rifted continental margin (timestep 0). By timestep 1 the earliest accretionary terrane had overridden oceanic crust and lithosphere, and abutted against the crustal ramp outboard of the hingeline. Topographic relief was negligible and the foreland basin very deep. At timestep 2 the ramp impeded further advance, but the accretionary terrane had thickened considerably. Lithospheric relaxation created shale-prone basins. By timestep 3 tectonic thickening had established sufficient topography to drive thin-skinned thrusts over the buried hingeline (G. S. Stockmal, personal communication, 1985). The thicker and stronger lithosphere resulted in shallower foreland basin environments.

Cherry, 1981). The last timestep shows the culmination of structural development, where a thin-skinned thrust and fold belt is well-developed inboard of the buried hingeline.

Geological setting

Quantitative models have attempted to account for the broadscale responses of the lithosphere to thrust-belt loading. Of necessity they have assumed idealized crustal properties, varying only in configuration and rheological characteristics. In reality, the lithosphere is inhomogeneous and has inherited structural fabrics, lithologic variations, and different ages. Subsequent geodynamic behaviour as well as structural styles and levels reflect the inherited geology. The Appalachians and Cordillera have both evolved through the accretion of exotic terranes over attenuated Atlantic-type continental margins, with a final emphasis on extreme structural telescoping as thin thrust sheets were pushed across the hingeline. It is hardly surprising that history has repeated itself in these two mountain belts.

Regional monoclinal flexures expressed in the Appalachian Kings Mountain Belt and Cordilleran Kootenay Arc occur where the allochthons are draped over crustal ramps (Price & Hatcher, 1983). The crests of these ramps are believed to mark the hingeline or kinematic boundary between ancient stretched and rifted continental margins and the North American craton (cf. Watts, 1981). Field studies, Bouguer gravity trends, and seismic data confirm the attenuated geometry of these pre-convergence passive margins (Price, 1981; Cook *et al.*, 1979, 1981; Ando *et al.*, 1984). In the southern Appalachians COCORP profiles display shallow easterly dipping reflections beneath the allochthon suggesting that detachment occurred within the miogeoclinal wedge (Cook *et al.*, 1979, 1981; Ando *et al.*, 1984). The thickest and most extensive parts of the Appalachian and Rocky Mountain foreland basins occur where overthrusting of the continental margins is greatest (Price & Hatcher, 1983).

The ramp-miogeoclinal wedge couplets on either side of the North American craton have influenced the patterns of tectonic evolution and controlled structural levels. Three distinct tectonic styles reflect initial convergence and accretion of exotic terranes, isostatic adjustment, and finally the movement of the overthrust wedge across the hingeline (Table 1).

Passive margin deposition was terminated by the mid-Ordovician Taconic orogeny in the southern Appalachians and by the Late Jurassic Columbian orogeny in the Cordillera (Table 1). In both regions early collision was characterized by magmatic arc convergence and accretion of a mosaic of exotic terranes, each docking outboard of the previous (Coney, Jones & Monger, 1980; Williams & Hatcher, 1983). This tectonic style persisted until the exotic terranes encountered the crustal ramps. The evolution of trench-fill basin to foreland basin reflects the

Table 1. Tectonic evolution of the Appalachian and Cordilleran foreland basins (modified after Price & Hatcher, 1983)

Tectonic evolution	Appalachians	Cordillera
Passive margin Attenuated lithosphere, rift clastics and post-rift continental terrace wedge of carbonates and siliciclastics	Grenville basement, late Precambrian to Early Ordovician passive margin	Hudsonian basement, late Precambrian to Middle Jurassic passive margin
Early convergence Accretion of exotic terranes on to passive margin ramp. Flexural deformation of stretched lithosphere. Deep foreland basin, prominent unconformity at base	Taconic orogeny, mid-Ordovician to Early Silurian	Columbian orogeny, Middle Jurassic to Early Cretaceous
Intermediate stage Relaxation of load-induced stress, tectonic overthickening, granitic magmatism. Shale-prone foreland basin, reworking across forebulge, prominent unconformity at top	Acadian, Middle Devonian to Early Mississippian	Mid-Cretaceous, Aptian to Campanian
Terminal convergence Thin-skinned tectonics and transport of overthrusts across hingeline of buried passive margin. Very shallow foreland basin, molasse, coal	Alleghenian orogeny, Early Pennsylvanian to Permian	Laramide orogeny, Maastrichtian to Palaeocene

tectonic loading of progressively thicker continental crust. COCORP data show that the Taconic-age overthrust terrane is up to 14 km thick (Cook, 1984), but was insufficient to establish much relief above sea-level when emplaced against a 4–5 km deep continental margin (Quinlan & Beaumont, 1984; Stockmal *et al.*, 1986).

The Devonian Acadian event of the southern Appalachians and mid-Cretaceous event of the Cordillera share similarities which characterize the second tectonic style. Unlike the preceding orogenic phases they were not markedly compressive, but represented subtle changes and regional isostatic adjustments to the accretionary terranes. (In other areas, such as the New England states, orogenesis may have persisted.) These intermediate stages were characterized by voluminous granitic plutonism in the orogens, major subsidence of the foreland basins, deposition of argillaceous sediments, and incision of prominent unconformities along the unwarped margins of the basins (Price & Hatcher, 1983). This intermediate stage is attributed to viscoelastic relaxation of the regional stresses exerted by loading (Quinlan & Beaumont, 1984). The plutonism probably resulted from melting of compressed and tectonically over-thickened crust (cf. Fyffe *et al.*, 1981).

The third style, expressed in the Carboniferous Alleghenian orogeny of the southern Appalachians and late Cretaceous Laramide orogeny of the Cordillera, is characterized by large structural telescoping and transport distances of thin thrust sheets across the hingeline (Cook *et al.*, 1979, 1981; Price & Hatcher, 1983). The foreland thrust and fold belts developed mainly during these terminal stages (Price & Hatcher, 1983). This episode of loading was characterized by displacement of relatively thin thrust sheets above thick, unstretched lithosphere. As a consequence, the foreland basins were at their shallowest and generally filled to depositional baseline by coarser terrigenous clastics. These depositional landscapes were also ideal for coal accumulation.

THE APPALACHIAN FORELAND BASIN

In the eastern interior of the United States three major depocentres are separated by a system of arches and domes. The Appalachian, Illinois, and Michigan basins (Fig. 4) share some similarities as periodic upwarping and subsidence of the arch or forebulge

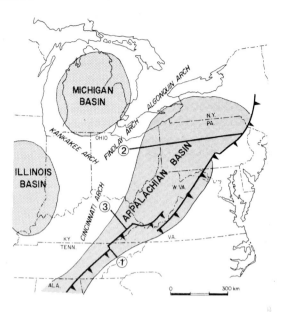

Fig. 4. The Appalachian foreland basin resulted from Palaeozoic thrusting and lithospheric flexure. A system of arches and domes separates the Appalachian basin from the Illinois and Michigan basins. Locations of Fig. 5 cross-sections are indicated.

system alternately decoupled or yoked them together (Quinlan & Beaumont, 1984). Appalachian stratigraphy comprises four major unconformity-bounded sequences which are attributed to pre-convergence deposition, and successive though diachronous episodes of Taconic, Acadian and Alleghenian deformation (Table 1). The basin fluctuated between overfilled and underfilled conditions as the lithosphere responded to overthrust encroachment and subsequent viscoelastic relaxation of the plate bending stresses (Quinlan & Beaumont, 1984; Tankard, 1986). Thick wedges of non-marine sediments prograded across the basin during Late Ordovician–Early Silurian, Late Devonian, and Pennsylvanian episodes of orogenesis. At other times carbonates and argillaceous sediments characterized the underfilled basin (Meckel, 1970; Laporte, 1971). Buildups were commonly associated with the basin-margin arch system.

The following discussion will review some highlights of Taconic and Acadian history, and then focus in more detail on Mississippian and Pennsylvanian stratigraphy which records the transition from relatively passive Acadian events to renewed mountain building in Alleghenian time.

Taconic stratigraphy

Middle Ordovician transition from passive margin to convergent tectonics

A continental terrace wedge, comprising shelf carbonates and argillaceous slope and rise deposits, built seawards (eastwards) in Cambro–Ordovician time. COCORP reflection profiles show that the structural detachment beneath the Taconic terranes has followed the shallowly dipping continental margin (Cook *et al.*, 1979; Ando *et al.*, 1984). The Taconic orogeny was most intense in the southern and central Appalachians where it resulted in the thickest and most extensive foreland basin sediment accumulation.

The Middle Ordovician Sevier basin of Tennessee and Virginia records the transition from passive margin through trench and foreland basin as the earliest Taconic terranes first encountered the distal edge of the continental margin, and continued to load progressively thicker and more rigid parts of the cratonic plate (Fig. 5). Subduction and flexure of the lithosphere also resulted in development of a forebulge system with widespread erosion of the Knox–Beek-mantown shelf carbonates (Jacobi, 1981; Shanmugam & Lash, 1982). This unconformity essentially marks the change from passive to convergent tectonics. Jacobi estimates that forebulge migration in the northern Appalachians could have induced erosion and removal of at least 180 m of section. Theoretical considerations predict as much as 800 m of uplift in certain circumstances (Stockmal *et al.*, 1986).

Although the continental margin subsided beneath encroaching overthrust loads, carbonate accumulation and buildup continued along a rising forebulge (Read, 1980; Shanmugam & Lash, 1982; Walker, Shanmu-gam & Ruppel, 1983). The skeletal sand banks and cross-bed patterns show that wave and current agitation were endemic along the crest of the forebulge. These onshelf carbonates and marginal ramp facies intertongued basinward with anoxic black shales and shaly limestones (Fig. 5). Starved basin conditions persisted in the early foreland basin (Blockhouse and Sevier formations) with pelagites and fine-grained turbidites accumulating in water 700–2000 m deep (Shanmugam & Lash, 1982; Walker *et al.*, 1983).

Younger stratigraphic intervals suggest that over-thrusting of the crustal ramp persisted. The overthrust load gradually emerged, shedding vast amounts of siliciclastics which built the extensive submarine fan complex of the Tellico Formation (Bowlin & Keller, 1980). Along the western margin of the basin carbonate depocentres migrated cratonward episodi-

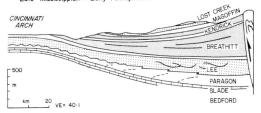

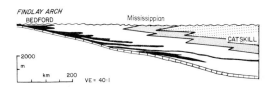

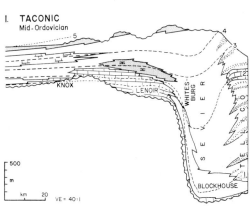

Fig. 5. Schematic cross-sections of the Appalachian basin illustrating three tectono-stratigraphic styles. (1) The Taconic basin records the earliest approach of accretionary terranes which loaded weak (thin) lithosphere (Fig. 3), resulting in a deep foreland basin. Peripheral upwarping and erosion of the pre-convergence shelf created the Knox–Beekmantown unconformity. Carbonate buildups continued along the forebulge. (After Walker *et al.*, 1983.) (2) Acadian events were relatively passive and given to thickening of the accretionary terranes against the large crustal ramp, and relaxation of the foreland lithosphere. Basinal environments were generally shale-dominated. Carbonate and sandstone deposition were common along the basin-margin arches. (After Broadhead *et al.*, 1982.) (3) At the end of Acadian time forebulge uplift was pronounced and carbonate deposition widespread. The regional mid-Carboniferous uncon-formity terminated this event. Renewed terrigenous influx is correlated with Alleghenian tectonism. Thin thrust sheets were pushed across the hingeline over strong (thick) lithosphere. The foreland basin was generally shallow. The influence of the forebulge decreased progressively. (After Tankard, 1986.) Locations of these three cross-sections are shown in Fig. 4.

cally. Water depths gradually decreased in the Sevier basin, and upper Tellico facies were deposited in tidal flat and lagoonal environments. Westward progradation of deltaic clastics eventually smothered carbonate production and filled the basin to depositional baseline (Read, 1980; Walker *et al.*, 1983).

The Sevier basin of Virginia–Tennessee and the Martinsburg basin of Pennsylvania evolved in a very similar manner, although there were some differences in dispersal patterns. In Pennsylvania turbidite fan systems prograded in a south-westerly direction down the axis of the trough. In contrast, submarine fans in the deeper Sevier basin developed transverse to strike (Shanmugam & Lash, 1982).

The rock record agrees well with theoretical predictions of overthrusting and basin evolution. By timestep 1 of Fig. 3 the overthrust load had encountered the distal edge of attenuated continental crust, causing an inboard shift and emergence of the forebulge, and widespread erosion. Facies similar to those onlapping the Knox unconformity (Read, 1980) are consistent with the configuration predicted in timestep 1. Seismic reflection data show that the overthrust terranes were as much as 14 km thick (Ando *et al.*, 1984). If obducted across a stretched (weak) continental margin these overthrusts could have resulted in substantial subsidence without creating much relief above sea-level (Fig. 3) (Quinlan & Beaumont, 1984; Stockmal *et al.*, 1986). The corresponding Sevier trench was 700–2000 m deep and dominated by pelagic sedimentation. The accretionary terranes gradually emerged as they encountered progressively thicker parts of the crustal ramp, and the foreland basin shallowed; this is attributed to greater flexural rigidity of the thicker lithosphere. In the Sevier basin these characteristics are reflected in the Tellico submarine fans and deltas and westward migration of the forebulge-carbonate tract. This tectono-sedimentary cycle terminated with filling of the Sevier basin (Shanmugam & Walker, 1980).

Taconic successor basins

Renewed transgression in Late Ordovician time established a new blanket of shelf carbonates which were unaffected by uplift along the Cincinnati Arch (Weir, Peterson & Swadley, 1984); eustatic processes are inferred. This respite was short-lived as overthrusting again resulted in massive subsidence of the southern Appalachian basin and deposition of vast wedges of deltaic and alluvial sandstones. The

Tuscarora sandstones and conglomerates were deposited by high-gradient braided streams, but were also subjected to periodic wave-reworking in the shore zone (Folk, 1960; Cotter, 1983). However, there is little evidence of reworking along the cratonward margin of the basin, suggesting that the orogenic climax coincided with a relatively deflated Cincinnati–Findlay arch system.

Tectonic activity waned from Middle Silurian to Early Devonian time, resulting in erosional decay of the orogen and shale-prone depositional environments. Marine rock sequences were best developed in the northern Appalachians. Rochester shales at the apex of the Appalachian basin in New York and Ontario accumulated between the Taconic source terrane and the Algonquin Arch. High energy environments dominated the crest of the Algonquin Arch, including carbonate platform, crinoid shoals, and cross-bedded bioclastics reworked above wave base. The axis of the arch was an area of relative uplift, and erosion surfaces punctuate the carbonate sequence (Crowley, 1973; Brett, 1983).

In summary, during early Silurian overthrusting the forebulge was relatively suppressed, suggesting an initial elastic response to active loading (cf. Quinlan & Beaumont, 1984). In contrast, the Middle Silurian to Early Devonian Appalachian basin was characterized by more or less stationary allochthons, mud-prone foredeep environments, and uplift along the basin-margin arch system which effectively dampened the wave climate. Uplift along the forebulge system during episodes of tectonic quiescence is characteristic of viscoelastic models of the lithosphere.

Acadian stratigraphy

The Acadian interlude was characterized by long periods of relative quiescence during which vast quantities of shale and mudstone accumulated in an overdeepened basin. The Devonian depocentre was in the northern Appalachians where over 3000 m of sediment is preserved (Johnson & Friedman, 1969). Regional studies show that the Cincinnati–Findlay arch subsided and uplifted episodically, alternately yoking together and decoupling the Appalachian and Illinois basins (Potter, Maynard & Pryor, 1981; Kepferle & Pollock, 1983). Dillman & Ettensohn (1980) found that during periods of black shale deposition the Cincinnati Arch was a positive feature. A substantial source terrain formed the eastern margin of the basin for some of Late Devonian time, shedding voluminous clastics across the Catskill deltas (Fig. 3, timestep 2).

The Hamilton siliciclastics bridge the interval between Lower Devonian carbonate platform environments (Helderberg) and large-scale progradation of Late Devonian Catskill deltas (Brett & Baird, 1985). Hamilton shales are mainly associated with the axis of the foreland basin; carbonates accumulated continuously in the shallower water along the north-western margin.

By the Late Devonian the northern Appalachian basin was dominated by Catskill deltaic facies (Fig. 5). These deltaic facies prograded basinwards over marine deposits, which are comprised of turbidites and black shales at the toe of the delta (Johnson & Friedman, 1969; Allen & Friend, 1968; Potter *et al.*, 1981). Water depths at this time are believed to have exceeded 200 m (Broadhead, Kepferle & Potter, 1982). Parts of the Catskill sequence have been attributed to rapid progradation of muddy shorelines, suggesting the conspicuous absence of winnowing agents such as waves and tides (Walker & Harms, 1971).

In summary, the Devonian Appalachian basin was shale-prone, indicating long periods of quiescence. Uplift of the basin margin arch system has been correlated with periods of black shale deposition (Dillman & Ettensohn, 1980), and locally induced carbonate accumulation and buildup (Brett & Baird, 1985). During the orogenic climax (e.g. Catskill) there was little evidence for relative uplift of the forebulge (Potter *et al.*, 1981).

Transition from Acadian relaxation to Alleghenian overthrusting

The Carboniferous Appalachian basin of eastern Kentucky evolved into its present form as thin-skinned overthrusts were pushed across the hingeline and loaded thick and thermally mature continental crust (Fig. 3, timestep 3). This activity resulted in over 200 km of structural shortening (Roeder, Gilbert & Witherspoon, 1978). Isostatic flexure of the foreland lithosphere maintained the Cincinnati Arch along the landward margin of the basin (Fig. 6). Forebulge behaviour may have been modified to a considerable extent by crustal and lithospheric inhomogeneities and may not have behaved as predicted for a simple, homogeneous plate. These same inhomogeneities are also expressed in the Kentucky River fault system and Waverly Arch (Woodward, 1961; Dever *et al.*, 1977). Transbasin faults originated in Cambrian–Ordovician rifts, but were periodically rejuvenated, sometimes

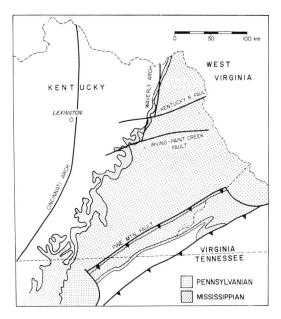

Fig. 6. The Appalachian basin in eastern Kentucky showing major structural elements.

with a strike-slip component (Ammerman & Keller, 1979; Wehr & Glover, 1985). The relatively small Waverley Arch is a local aberration of the Cincinnati Arch; it does, however, preserve a unique stratigraphic record which reflects the response of the forebulge to loading (Tankard, 1986).

Late Acadian stratigraphy and tectonic quiescence

An inactive orogen and decaying Catskill deltas starved the Appalachian basin of coarse material. Throughout this period the basin was underfilled. Until the Late Mississippian the Appalachian and Illinois basins were generally yoked together and most stratigraphic units were regionally persistent (Fig. 7). However, at times latest Devonian and Mississippian palaeogeographies were significantly affected by shoaling and emergence of the Cincinnati Arch in the south-central Appalachians (Pepper, De Witt & Demarest, 1954; Pryor & Sable, 1974). Black shales such as the Ohio and Chattanooga sequences were deposited under anaerobic conditions. Shale deposition persisted into the Mississippian. On one side of the basin the Bedford shales were fed by small deltas, but along the cratonward margin periodic shoaling of the Cincinnati Arch resulted in wave and current reworking, and deposition of elongate sand bars or barrier islands up to 130 km long and 5–24 km wide.

EARLY ALLEGHENIAN

UPPER BREATHITT — FLEXURAL DEFORMATION
Overfilled basin

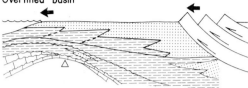

LOWER BREATHITT — FLEXURAL DEFORMATION
Underfilled basin

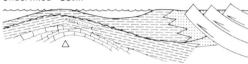

LATE ACADIAN

CARTER CAVES — RELAXATION PHASE 2
Uplift, reworking of mudsheet, barrier quartz arenites,
underfilled basin, major unconformity

PARAGON — FLEXURAL DEFORMATION
Progradation of sandy mudsheet

SLADE — RELAXATION PHASE I
Uplift, shoaling, unconformities

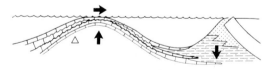

SLADE — STABLE PLATFORM
Blanket limestone sedimentation

The so-called Ohio Bay, between the Cincinnati shoals and the delta fringe along the orogen, was up to 250 km wide and 70 m deep (Pepper *et al.*, 1954).

By the end of the Mississippian uplift along the arch complex resulted in shoaling, erosion, and irregular thinning of the Slade and Paragon* successions (De Witt & McGraw, 1979). Limestones in the basal Slade were deposited as a blanket in relatively deep water. Younger bioclastic and oolitic limestones, deposited in shoal water, onlap the flank of the Waverly Arch (Figs 7 and 8). Periodic emergence was most pronounced along the crest of the arch, as evidenced by breccia, karst, and caliche development (Dever *et al.*, 1977; Ettensohn, 1981).

On either side of the Waverly Arch the Slade is punctuated by a wedge of unconformities (Tankard, 1986). Merging of successive unconformities resulted from progressive uplift. The wedge of unconformities along the western flank of the arch is also distinctive in that the intersection points of merging unconformities overstep each other towards the arch axis (Fig. 8). This implies progressive uplift and eastward migration of the arch during this phase.

The Slade limestones were eventually smothered by the Paragon succession, a westerly prograding wedge of sandy mudstones (Fig. 7) which De Witt & McGraw (1979) related to rejuvenation of the orogenic source terrane. The lateral persistence and small thickness variations are attributed to arch deflation and yoking together of the Appalachian and Illinois basins (Sable, 1979). However, there was a brief repeat of Bedford-Berea history when local uplift of the Cincinnati-Waverly arch complex caused shoaling, reworking, and deposition of Carter Caves quartz arenites (Fig.

*The terms Slade Formation and Paragon Formation have recently replaced the Newman and Pennington of eastern Kentucky (Ettensohn *et al.*, 1984).

Fig. 7. Cartoon summarizing some elements of Carboniferous stratigraphy in terms of episodic foreland basin subsidence and peripheral upwarping. The sequence begins in late Acadian (Mississippian) time when the Appalachian and Illinois basins were yoked together and blanket shale and limestone deposition predominated. The lithosphere responded viscoelastically to the stationary thrust-load; uplift and contraction of the forebulge caused shoaling, reworking, and developed wedges of unconformities. The Paragon sandy mud-sheet prograded across a flexurally subdued lithosphere. Brief periods of arch uplift resulted in reworking and deposition of sand bodies such as the Carter Caves sandstone. Alleghenian (Pennsylvanian) deposition was a response to active thrust loading as the basin fluctuated between underfilled and overfilled conditions.

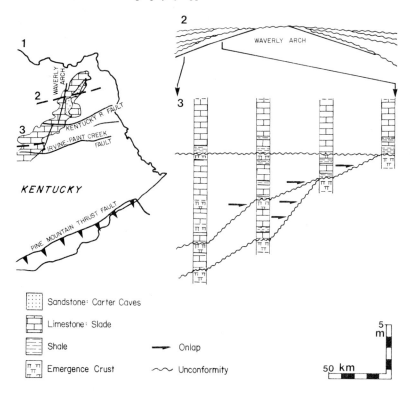

Fig. 8. Late Acadian stratigraphic relations on the Waverly Arch. (1) Distribution of Slade and Carter Caves sequences (after Dever *et al.*, 1977). (2) Wedges of unconformities punctuate the limestones on both sides of the arch (modified after Ettensohn, 1981). (3) The wedge of unconformities on the western flank shows overstepping intersection points of merging unconformities (Dever *et al.*, 1977); together with evidence for shoaling, this implies arch uplift and migration towards the orogen. The locations of these two cross-sections are shown on the map.

8) as channelized shoals and barrier bars (Dever *et al.*, 1977; Horne *et al.*, 1978; Tankard, 1986). Similar sand-shoal facies extended southwards into Tennessee (Milici *et al.*, 1979).

The stratigraphic record of the arch complex appears to be a sensitive indicator of lithospheric behaviour. In the southern Appalachians the Cincinnati Arch maintained a low relief throughout much of Late Devonian–Early Mississippian time, episodically uplifting and subsiding as the lithosphere responded to overthrust loads. Basin starvation and black shale deposition coincided with increased subsidence and contraction of the depocentre towards the orogen (Pepper *et al.*, 1954). And along the margin of the basin occasional uplift of the Cincinnati–Waverly arch complex induced shoal-water reworking and deposition of linear tracts of mature sandstones. Slade limestones along the Waverly Arch also preserve a subtle record of arch activity. Stratigraphic relations and the internal geometry of the wedge of unconform-

ities indicate progressive uplift of the relaxation arch and basinward (eastward) migration of its axis; this was a period of orogenic quiescence which starved the basin of terrigenous clastics (De Witt & McGraw, 1979).

In summary, the rock record suggests that periods of orogenic quiescence and terrigenous sediment starvation coincided with basin overdeepening, uplift of the forebulge system, and migration of the forebulge and depocentre *towards* the stationary overthrust load. These characteristics support a viscoelastic model of the lithosphere (Quinlan & Beaumont, 1984; Schedl & Wiltschko, 1984). Late Devonian–Mississippian palaeoenvironments were almost entirely marine.

Alleghenian stratigraphy and response to overthrust loading

At the climax of Acadian relaxation a regional unconformity was carved along the margin of the

Appalachian basin, and in places may separate Mississippian and Pennsylvanian strata across the width of the basin (Ferm, 1974; Ettensohn, 1980, 1981). Alleghenian overthrust loading imposed a new successor basin across this denuded landscape (Fig. 5C). Arrival of these overthrusts, substantial relief, and higher gradients than during Acadian time are reflected in the calibre and composition of the molasse wedge (Davis & Ehrlich, 1974; Miller, Bargar & Jackson, 1984). Two distinct styles of sedimentation characterize the succession (Fig. 7). During episodes of basin overfilling fluvial sands filled the basin beyond the depositional base level. The underfilled basin was dominated by shallow marine bayfill deposition in water which seldom exceeded 20 m in depth (Tankard, 1986). Commercial coal seams are common to both styles.

The onset of Alleghenian overthrusting coincided with a flood of fluvial sandstones. These Lee sandstones (Fig. 5C) were deposited principally by braided rivers which flowed in a south-westerly direction, down the axis of the foreland basin, terminating in shallow marine embayments. These river systems scoured 60 m deep palaeovalleys into a Mississippian platform (C. Rice, 1984).

Deposition in an underfilled basin

The Breathitt coastal plain depositional system (Fig. 7) is characterized by relatively thin bayfill sequences, sandy bayhead deltas, and coal seams which are numerous and persistent. Three transgressive shales are attributed to basin overdeepening: Kendrick, Magoffin, and Lost Creek (Fig. 9; Tankard, 1986). The major shoreline probably coincided with the Cincinnati Arch. Along this margin of the basin lagoonal shales intertongue with quartz arenites of tidal channel, flood tidal delta, and barrier washover origin (Smith *et al.*, 1971; Barwis & Horne, 1979).

Upward-coarsening bayfill sequences of mudstone, siltstone, and fine-grained sandstone are 2–20 m thick, and are believed to have resulted from encroachment of bay margins and local development of bayhead deltas and crevasse subdeltas. Bioturbation is common. Seat-earths and thin coal seams generally cap each sequence. Fluvial sands form thin, isolated bodies interspersed in bayfill lithologies.

Bayhead deltas form an important association with transgressive intervals (Fig. 9). These bayhead delta facies coalesce along strike to form broad platforms. Rapid subsidence and very rapid deposition is inferred for this sandstone facies which buries peat swamp

(coal) surfaces (Fig. 10): sedimentary structures include turbidite-like graded beds, large flow rolls, ripple drift, convolute bedding, and liquefaction structures. These episodes of rapid sedimentation seldom terminated with coal accumulation, normally a consequence of lobe abandonment, but were inundated by transgressive shales, such as the Kendrick and Magoffin intervals (Tankard, 1986).

The Kendrick (late Morrowan) and Magoffin (early Atokan) transgressive systems record marine inundation of the foreland basin as it subsided beneath the load of an advancing thrust-sheet complex (cf. Graham, Dickinson & Ingersoll, 1975). Preceding bayhead delta facies have the attributes of rapid subsidence rather than sea-level rise. The Kendrick and Magoffin paralic drapes are regionally persistent, their scale, geometry, and facies associations precluding an origin by abandonment and decay of numerous small delta lobes. Rich invertebrate faunas are typically normal marine, and the underfilled basin was possibly as wide as 300 km (cf. Roeder *et al.*, 1978).

The combined effect of bayfill and baymargin sedimentation was to establish a broad platform of low relief. Each platform succession is capped by a seat-earth and thin coal. The great lateral extent of these blanket coals implies very low overall rates of detrital influx and abandonment of the depositional landscape (Ferm, 1970; Tankard, 1986). Episodes of peat swamp accumulation were generally terminated by marine inundation which deposited transgressive siltstone veneers and black carbonaceous shales.

In summary, the persistent bayfill deposits (Fig. 9) show that the basin was generally underfilled and dominated by shallow bays whose widths were limited only by the orogen and the confining forebulge system. The sediments were predominantly argillaceous; periodic abandonment of the depositional landscape resulted in blanket coal accumulation. The major shoreline apparently coincided with the Cincinnati or Waverly Arch (Smith *et al.*, 1971; Barwis & Horne, 1979). Behind it the landscape of broad bays, continuous bay margins, and bayhead deltas has no obvious modern counterpart. Basin overdeepening resulted in marine transgression (Kendrick and Magoffin; Fig. 9). Inundation is attributed to increased subsidence rates rather than eustasy.

Deposition in an overfilled basin

The transition from paralic sedimentation in the Magoffin trough to sand-dominated alluvial plain

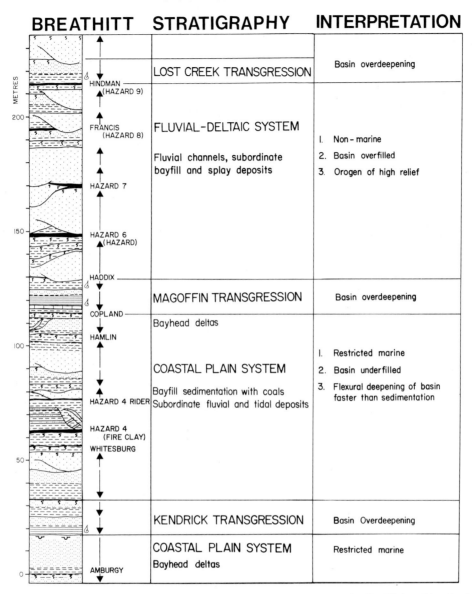

Fig. 9. Composite stratigraphic column for the Breathitt Formation in eastern Kentucky. Bayfill deposition dominated the coastal plain system. In contrast, the succeeding fluvial-deltaic system is dominated by fluvial sandstones. Arrows point in direction of decreasing grain size.

deposition was sudden. Coarse clastics were shed from the rising orogen faster than the subsiding foreland basin could accommodate them, resulting in an overfilled basin (Figs 7 and 9) (De Witt & McGraw, 1979; Tankard, 1986). However, gradients were sufficiently low to accommodate meandering rivers. These rivers drained an orogen no more than 160 km

away (Wanless, 1975; Gardner, 1983), and flowed across the axis of the basin. Subordinate facies include flood-basin mudstones and coals, as well as crevasse splays. The shallowness of the basin precluded well-developed, progradational delta front deposits.

Deposition of this alluvial system essentially terminated with transgression by the Lost Creek limestone

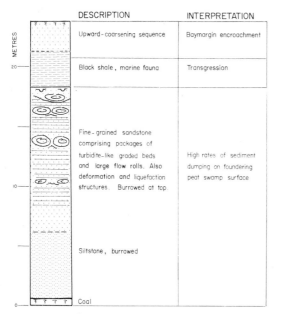

Fig. 10. Rapid subsidence and marine inundation are reflected in the two-part sequence comprising a platform of bayhead delta deposits and transgressive shales. Gravity sedimentation, deformation and liquefaction processes dominated bayhead delta deposition as they buried foundering peat swamp surfaces.

and shale in yet another period of basin overdeepening (Fig. 9).

Acadian–Alleghenian transition and tectonic implications

Compared with the Alleghenian orogeny, the Acadian was generally quiescent. Long periods dominated by plate relaxation and overdeepening ensured that the basin was generally underfilled, and that the predominantly marine trough was usually deeper than 70 m. Episodic uplift and shoaling of the forebulge resulted in distal reworking of the argillaceous wedges, the residues accumulating as mature sand shoals, offshore or 'shelf' sand bars, and even barrier island complexes if emergent; the shoreline at the toe of the orogen was hundreds of kilometres away. At other times the foreland basin was so starved of terrigenous detritus that shoal-water carbonate environments dominated the Cincinnati Arch (Fig. 11). The geometry of wedges of unconformities suggests that while the forebulge was rising it also migrated towards the stationary overthrust loads. A viscoelastic model for the lithosphere (Quinlan & Beaumont, 1984) is

inferred to explain basin overdeepening, and withdrawal of the depocentre and forebulge towards the orogen at times of relative quiescence. At the climax of relaxation and erosional unloading of the orogen a regional unconformity truncated the Mississippian landscape.

In the Pennsylvanian, Alleghenian thrust sheets were transported considerable distances beyond the hingeline, loading thick and thermally mature continental crust. Because of the greater flexural rigidity of the crust broad marine embayments were usually shallower than 20 m. Palaeoenvironments varied between coastal plain with broad embayments and alluvial plain (Fig. 11). The initial response of the lithosphere to Alleghenian loading was sluggish, and braided rivers incised deep palaeovalleys into the Mississippian landsurface. But respond it did, and the subsequent basin fill never deviated far from depositional base level.

The great lateral extent of thin bayfill deposits indicates that the bays were very broad and shallow (Fig. 11). Reworking processes were relatively unimportant because shoaling over the Cincinnati–Waverly Arch dampened tide and wave energy. A coastal plain interpretation is intended to distinguish this Breathitt depositional system, deposited in a foreland basin, from the better known modern deltas associated with passive continental margins. The coals are also more abundant and thicker than the peats of modern deltas (Ferm, 1970). Their great lateral extent and the characteristics of the encapsulating facies argue against an interdistributary bay origin or accumulation by delta lobe switching and abandonment. Although the individual sedimentary facies can be interpreted by comparison with Recent counterparts, the succession and its tectonic setting have no known modern analogues.

The characteristics and associations of several of the facies have tectonic implications. Vigorous bayhead delta deposition across rapidly subsiding peat swamp surfaces preceded major episodes of transgression. Paralic Magoffin sediments record an Atokan episode of basin deepening. Transition to alluvial plain deposition was sudden due to the rejuvenation of the source terrane. Bypassing of sediment from this overfilled basin fed flysch and turbidites longitudinally into the deep-marine Ouachita trough at the southern terminus of the Appalachians (Graham *et al.*, 1975; Graham, Ingersoll & Dickinson, 1976). Atokan overthrusting in the Ouachita orogenic belt suggests that basin subsidence and transgression were a response to thrust sheet loading.

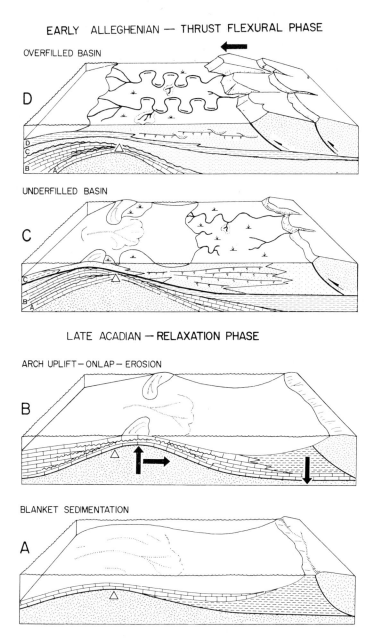

EARLY ALLEGHENIAN — THRUST FLEXURAL PHASE

OVERFILLED BASIN

D

UNDERFILLED BASIN

C

LATE ACADIAN — RELAXATION PHASE

ARCH UPLIFT — ONLAP — EROSION

B

BLANKET SEDIMENTATION

A

Fig. 11. The evolution of Appalachian landscapes during transition from Acadian relaxation to Alleghenian tectonism. Late Acadian palaeogeographies were characterized by quiescence of the orogenic belt which starved the basin of terrigenous sediments. Deposition on the Waverly Arch was for a while dominated by limestones. Relaxation caused uplift, shoaling, and resulted in multiple unconformities. In the Pennsylvanian Alleghenian overthrust loading and arch depression are correlated with deposition of immature terrigenous clastics. The basin fluctuated between underfilled and overfilled conditions. The Waverly Arch initially localized the major shoreline, but uplift of the flexural arch was insufficient to form major unconformities or large tracts of reworked sandstones. Bold arrows indicate overthrust and forebulge migration; half arrows show active overthrusting; open triangles are reference points for forebulge behaviour. (After Tankard, 1986.)

THE CORDILLERA AND THE MID-CRETACEOUS FORELAND BASIN

Cordilleran landscapes evolved in a predictable pattern as the original passive margin succumbed to intermittent loading of Columbian–Nevadan exotic terranes. The development of this orogeny apparently slowed as the thrust terranes overrode the hingeline, and there followed a long period from Albian to Maastrichtian time when the lithosphere adjusted isostatically to the overthrust loads. Relaxation of the lithosphere resulted in increased basin subsidence and marine inundation. Thin-skinned tectonics of the Laramide orogeny terminated this underfilled shale-basin episode, and initiated a flood of coarser non-marine clastics across an eroded landsurface (Fig. 12). This history of accretion, relaxation, and thin-skinned tectonics parallels Appalachian history (Table 1). One aspect that is better defined in the Cordillera is the relationship between magmatism and orogenesis (Fig. 12). The first phase of plutonism ended near 147 Ma during early accretion. A new suite of batholiths was emplaced in the Sierra Nevada from 120 to 80 Ma, an anorogenic episode characterized by tectonic thickening. The final phase of magmatism post-dates the Laramide (Engebretson, Cox & Thompson, 1984; Page & Engebretson, 1984). These events have been correlated with Pacific plate motions.

The cratonic platform inboard of the passive margin hingeline was dissected by a network of epeirogenic arches and basins, reflecting inhomogeneities in crustal composition and fabric of the Hudsonian basement (Porter, Price & McCrossan, 1982). During Mesozoic tectonism many of these arches interfered constructively with flexible foreland lithosphere, forming subtle irregularities. They were an important influence on sedimentation along the shallow margins of the basins. Some, such as the Moxa Arch of Wyoming (Jordan, 1981) and the Sweetgrass Arch of southern Alberta and NW Montana (Lorenz, 1982), have been interpreted as forebulges. These arches were inherited structures, but they did respond isostatistically to loading by rising and subsiding in a non-uniform pattern.

Stratigraphic framework

The earliest record of substantial subsidence of the foreland basin (Fig. 13) is dated to the Late Jurassic when Columbian terranes progressively loaded the craton margin (Bally, Gordy & Stewart, 1966; Porter *et al.*, 1982). Since then the basin has migrated

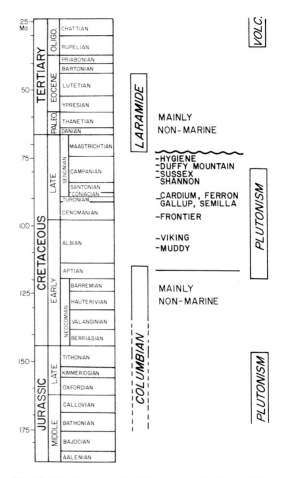

Fig. 12. The Mesozoic Cordillera evolved in three distinct stages. (1) Columbian exotic terranes were accreted on to the inherited passive margin ramp in Late Jurassic to Early Cretaceous time. (2) The succeeding mid-Cretaceous episode, spanning the Albian–Campanian interval, was largely anorogenic, presumably because the crustal ramp impeded further transport. A suite of batholiths was emplaced during this period. The relaxation foreland basin was characterized by marine inundation and shale deposition. Uplift of basin margin arches resulted in shoal-water reworking and deposition of elongate sand ridges, the most important of which are illustrated. (3) In the Maastrichtian thin Laramide thrust sheets were pushed across the buried hingeline, essentially terminating the widespread mid-Cretaceous seaways. (Based on Bally *et al.*, 1966; Brenner, 1980; Price, 1981; Porter *et al.*, 1982; Price & Hatcher, 1983; Weimer, 1983; Engebretson *et al.*, 1984; Page & Engebretson, 1984.)

eastward. Until the Aptian the foreland basin was filled to depositional baselevel with arenaceous siliciclastic sediments.

By late Neocomian time the Columbian orogenic cycle was gradually ending, and mixed marine and

Fig. 13. Palaeogeographic reconstruction of the Cretaceous seaway of North America at the climax of lithospheric relaxation. (Modified after Brenner, 1978.)

non-marine conditions prevailed. A long period of relative quiescence lasted until the Maastrichtian (Fig. 12). Downwarping of the foreland basin persisted and muddy depositional environments predominated. Uplift along the cratonward margin of the basin accentuated arches such as the Sweetgrass Arch of Montana and Alberta (Williams & Burk, 1966), and resulted in numerous unconformities (Weimer, 1983). Because this was a magmatic period bentonitic and porcelaneous sediments are common.

The first major marine invasion occurred in the mid-Aptian, and by the early Albian the narrow Skull Creek seaway connected the Gulf of Mexico with the Arctic Ocean, depositing thick sequences of dark shales (McGookey *et al.*, 1972). Until the Maastrichtian, alternately transgressive and regressive conditions prevailed. Also characterizing the relaxation foreland basin is a suite of 'shelf' sand ridges (Fig. 12) which were deposited often several hundred kilometres from the shoreline with a muddy 'shelf' intervening. Like the forebulge sandstones of the Acadian Appalachian basin, they resulted from shoal-water reworking in a unique tectonic setting.

In the final episode of mountain building Laramide thrust sheets crossed the hingeline, loading the foreland lithosphere. The Absaroka thrust sheet is known to have moved many times faster in the Maastrichtian than in the Santonian (Jordan, 1981). Sheets of coarse detritus deposited by high-gradient braided rivers overfilled the Rocky Mountain basin, building wedges such as Belly River and Beaverhead. Important coal seams, some more than 30 m thick, accumulated at this time. Basin margin arches were no longer important (McGookey *et al.*, 1972).

The tectonics of 'shelf' sand ridge accumulation

From Aptian to Maastrichtian time the Cordillera was generally quiescent, and the foreland basins invariably inundated by epeiric seas of varying depths. Many thousands of metres of organic-rich shales were deposited. Associated with these mud-prone basins are numerous elongate sand bodies which have commanded attention because they form important reservoirs. These sand bodies are generally NW–SE trending, 4–20 m thick, and up to 115 km long (Brenner, 1980; Slatt, 1984; Swift & Rice, 1984). Perhaps the best studied of these are the Muddy, Gallup, Shannon, Sussex, Cardium, and Viking sandstones (Fig. 12).

These sandstone bodies all resemble the storm, wave, and tide built sand ridges of modern shelf seas. But they differ in that they were built on muddy shelf surfaces and are completely enclosed in shale, usually hundreds of kilometres from their palaeoshorelines. In most cases these sandstones were deposited below daily effective wave base in inner shelf depths (Berg, 1975; Spearing, 1976; Martinsen & Tillman, 1979; Walker, 1983; Tillman & Martinsen, 1984). Swift & Rice (1984) were surely correct in asserting that these sandstone bodies have no known modern analogues.

The bar-shaped marine sandstones are frequently cross-stratified and exhibit offlapping relationships, suggesting that they were built by migrating bedforms. Without modern analogues sedimentologists have appealed to most shallow marine processes: tidal currents, longshore currents, wave reworking, storm-intensified currents, fairweather processes, geostrophic currents, etc. (Spearing, 1976; Brenner, 1980; Boyles & Scott, 1982; D. Rice, 1984). What most of these sandstone ridges have in common is that they were preceded by progradation of a mudsheet across the foreland basin, and that shallow marine reworking and winnowing during stillstands piled sand residues into bars (Brenner, 1980; Beaumont, 1984).

These sandstone ridges share another important characteristic, an association with basement arch crests (Stelck, 1975; Slack, 1981; Slatt, 1984; Tillman & Martinsen, 1984). Active uplift along the Salt Creek anticline of Wyoming localized Shannon sand ridge deposition and resulted in vertical stacking of sandstones (Brenner, 1978, 1980; Tillman & Martinsen, 1984). And on the Casper Arch of Wyoming Shannon and Sussex deposition succeeded the Niobrara limestones (McGookey *et al.*, 1972). Regional structural highs extend into the Alberta basin where arch crests have localized erosion and winnowing (Stelck, 1975). Unconformities are also numerous along these basin margin arches (Weimer, 1983).

In summary, sand ridge accumulation in a foreland basin setting was most pronounced along actively rising basin-margin arch systems at a time when basinal environments were shale-prone. Periodic progradation of a sandy mud-sheet was the immediate source of sediment (Brenner, 1980). Figure 14 summarizes this tectono-stratigraphic setting where progradational muddy sediments are reworked along the crest of a relaxation arch. The composition of the sand ridges may reflect only the quality and quantity of arenaceous material available in the eroded sandy mud-sheet. This interpretation may not require any extraordinary processes to account for gravels and conglomerates that cap some Cardium sequences (cf. Walker, 1983). The important point emphasized here is that shoaling and reworking were strongly influenced by persistent uplift of basin-margin arches. Separating these sand ridges from their palaeoshorelines are underfilled, shaly foreland basins. Although the processes may be interpreted from Recent environments, there is no modern analogue for the tectono-sedimentary setting.

TECTONO-STRATIGRAPHIC OVERVIEW

The Appalachian and Cordilleran foreland basins are separated by the North American continent and evolved in different geological eras. Nevertheless, they are conspicuously similar and architecturally very simple. These similarities include structural relationships, magmatism, sedimentation, erosion, and resource distribution. On a large scale each evolved through three distinct phases (Table 1):

REWORKING ON UPWARPED FOREBULGE ARCHES

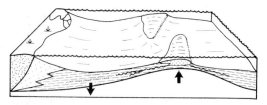

PROGRADING SANDY MUDSHEET

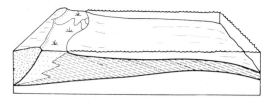

Fig. 14. Depositional model for 'shelf' sand ridges of the Albian–Campanian seaway of the Rocky Mountain foreland basin. In the first stage, a prograding sandy mudsheet blanketed the basin and spilled across the forebulge. The next stage was characterized by plate relaxation, basin deepening, and increased upwarping of the forebulge into the surf zone. Shoaling and reworking of the mudsheet along the basin margin arches removed considerable thicknesses of mudstone, the arenaceous residues accumulating as sand ridges (or barrier islands if emergent). These sand ridges were separated from their palaeoshorelines by underfilled, shaly foreland basin (not shelves).

(1) Early accretion of exotic terranes, abutting against an inherited passive margin ramp (Taconic and Columbian orogenies). Water depths in the Appalachian foredeep were as much as 2000 m.

(2) The continental margin ramp apparently impeded overthrusting for several tens of millions of years. This was a period given to subtle adjustments and tectonic thickening of the overthrust loads, erosion of the orogens, and relaxation and overdeepening of the shale-prone foreland basins (Acadian and mid-Cretaceous). Shallow-water sand ridges, carbonates, and wedges of unconformities characterized the basin margin arches. Marine environments predominated. Water depths in the relaxation basin varied up to 200 m. This passive style of basin development is also correlated with magmatism; for example, the Sierra Nevada batholiths (Fig. 12).

(3) In Alleghenian and Laramide times thin thrust sheets overrode the hingelines, travelling great distances across thick, thermally mature, and rigid

continental crust. The foreland basins oscillated between underfilled and overfilled conditions in response to the vagaries of thrust sheet behaviour. Shallow marine bays and alluvial plains were common. Water depths seldom exceeded 20 m.

Quantitative models

This pattern of basin evolution was also predicted on the basis of quantitative models. In the first type of model the response of an attenuated continental margin to overthrusting was examined. On a broad scale the agreement between these predictive lithospheric models and the rock record is very satisfying, accounting for relief, bathymetry, calibre of sediments, and unconformities. In the second type of model the rheological properties of lithospheric plates was examined. Both the elastic (Jordan, 1981) and viscoelastic (Quinlan & Beaumont, 1984) end-members have satisfactorily accounted for basin morphology, and even stratigraphy at the scale of the unconformity-bounded sequence. However, on smaller scales, such as the depositional system, viscoelastic lithospheric models appear to be more predictive. Resolution of the elastic versus viscoelastic argument depends more on the tectono-stratigraphic record than on mathematical contortions.

If the lithospheric response to loading were entirely *elastic*, forebulge elevation would be proportional to the size of the load. It would be most elevated when loading was greatest and coarse clastic wedges prograded across the basin, filling it to the depositional baseline. As the orogen load is reduced by erosion and sediment removed from the area, forebulge deflation would be expected.

A *viscoelastic* foreland basin would evolve through two steps (Figs 1 and 2). The immediate response of the lithosphere to thrust-sheet loading would involve elastic flexure. The difference between the two models is apparent as the load remains in place for an extended period of time and relaxation of the lithospheric bending stress occurs, resulting in basin deepening and forebulge upwarping. The relaxation basin should be characterized by finer grained deposition, and even basin starvation, and reworking along the forebulge.

The stratigraphic record clearly supports a viscoelastic model for the lithosphere. In the Appalachian and Cordilleran foreland basins marine inundation (overdeepening) and dark shale deposition were most

common during periods of relative orogenic quiescence. Marine sand-ridge and carbonate shoal deposition was most influenced by actively rising basin margin arches. In eastern Kentucky wedges of unconformities with overstepping intersection points have been cited as evidence that the Waverly Arch may have migrated back towards the stationary load during an inactive interlude of the Acadian event (Tankard, 1986). Renewed overthrusting (Alleghenian and Laramide) introduced a flood of river-borne sediments, but the basement arches were no longer prominent. Many of the stratigraphic patterns predicted by viscoelastic models of the lithosphere are thus supported by smaller-scale field studies.

Quantitative studies assign homogeneous properties to their modelled lithospheric plates. The inherited cratonic platforms on which the overthrusts were emplaced were, in fact, inhomogeneous, reflecting variations in crustal composition and fabric dating from Proterozoic events. Many of the arches were pre-convergence in origin, and formed subtle irregularities on a flexible foreland lithosphere when thrust-sheet loading did occur. Perhaps they rode piggyback on broader forebulges, uplifting and downwarping as crustal dynamics required. They may even have been able to migrate very short distances (Dever *et al.*, 1977). Old listric normal faults, such as the Kentucky River fault system, have been reactivated as transfer faults during convergence (Ammerman & Keller, 1979) with shoal-water reworking along the crest. Numerical models do not at present account for these irregularities and aberrations.

Unconformities

On a large scale, episodic tectonism is implied in Sloss's (1963) unconformity-bounded sequences, and Quinlan & Beaumont (1984) have correlated these unconformities with periods of viscoelastic relaxation at the termination of overthrusting. In the Appalachians two major unconformities appear to be regional in extent. The first (Knox–Beekmantown) marks the transition to convergent tectonics when peripheral upwarping of the passive margin shelf resulted in widespread erosion. The second separates Mississippian and Pennsylvanian strata and marks the end of a long period of Acadian relaxation, erosional unloading of the orogen and isostatic rebound. Other unconformities abound, but not on the same scale because relaxation periods were shorter. In the Rocky

Mountain foreland basin a prominent unconformity preceded the Laramide orogeny.

Then on a smaller scale the basin margins are frequently dissected by wedges of unconformities with complex cross-cutting relationships (Bally *et al.*, 1966; Tankard, 1986). Most of these unconformities have been attributed to arch uplift, tilting, and migration. Were a forebulge to migrate and leave these unconformity bundles stranded and suspended in a foreland basin succession, what configuration would they assume? Depending on the manner in which stratigraphic sections are hung and presented, it would appear that a lot of insight into unconformities is lost!

Eustasy

Carboniferous 'cyclothems' are sometimes attributed to eustatic sea-level fluctuations. Three examples frequently attributed to eustasy are the Kendrick, Magoffin, and Lost Creek marine zones in the Pennsylvanian of eastern Kentucky. The association with catastrophic bayhead delta accumulations suggests that these marine zones resulted principally from basin overdeepening (Tankard, 1986). Atokan overthrusting in the Ouachita orogenic belt (Graham *et al.*, 1975) was contemporaneous with Magoffin deposition, suggesting that subsidence and transgression were a response to overthrust loading. This interpretation obviously does not exclude a eustatic component. In the Cordillera overthrust episodes have been correlated with Pacific plate motions (Engebretson *et al.*, 1984; Page & Engebretson, 1984). So it is not inconceivable that transgressive shales in foreland basins include a eustatic component. However, the amount of foreland basin subsidence resulting from thrust-sheet loading could overwhelm the eustatic contribution. In those parts of the basin remote from the orogen, the effect of sea-level changes should be most pronounced (Beaumont, 1981). But even on the interbasin arches of eastern Kentucky the geometries of most stratigraphic units reflect a strong tectonic overprint. It is thus difficult to separate eustatic from tectonic processes.

Stratigraphic concepts

The foreland basins fluctuated between underfilled and overfilled conditions as the lithosphere responded to the vagaries of thrust-sheet behaviour and sediment supply. At times high rates of sediment supply from a rugged orogen overwhelmed the basin, building small deltas and alluvial plains. These deltas were fed by short-headed streams, unlike their counterparts on modern passive margins.

At other times the foreland basins were clearly underfilled and accumulated great thicknesses of organic-rich shales. These periods were accompanied by marked uplift of basin margin arch systems where wave and current reworking deposited elongate sand ridges hundreds of kilometres from their palaeoshorelines. The intervening area was not a muddy shelf, but an underfilled mud-prone foreland basin (Fig. 14). These landscapes have no known modern counterparts. The Mississippian basin edge and major shoreline in eastern Kentucky coincided with the emergent Waverly Arch where wave and tidal energy was dampened. This tectono-sedimentary landscape in which the major shoreline was fixed by a basin-margin arch (Fig. 11), behind which the embayments had foreland basin dimensions, again has no modern analogue. Modern barrier-lagoon coasts are characteristic of passive margins.

Depositional environments in these ancient foreland basins are repeatedly without modern analogue: muddy Catskill deltas, muddy Catskill shorelines, forebulge sand ridges, forebulge barrier islands forming the major shorelines of foreland basins, the mud-rich foreland basins themselves, and basin wide coal seams which reflect periodic abandonment of the depositional landscapes. Some insight into depositional processes can be gained from the application of facies models, but interpretation of the tectono-sedimentary systems requires a larger scale of interpretation and imagination.

Geochemistry

The intermediate stages in foreland basin evolution resulted in thick accumulations of organic-rich source rocks (Deroo *et al.*, 1977; Potter *et al.*, 1981; Broadhead *et al.*, 1982; Tainter, 1984; Moshier & Waples, 1985). Basin margin sandstone ridges have formed prolific hydrocarbon reservoirs in the Appalachians and Cordillera.

Cretaceous forebulges may play an important role in the biodegradation of oils, principally because they are regions susceptible to uplift. Episodic uplift and erosion probably provided the conduits by which meteoric recharge waters were carried into the subsurface (Bockmeulen, Barker & Dickey, 1983). These meteoric waters are the source of free oxygen, bacteria, and nutrients. The first evidence of biodegradation was, appropriately, found in the Bell Creek field of the Muddy sandstone (Fig. 12) (Winters & Williams, 1969). Biodegradation along basin margin arch systems can be devastating, and in Alberta has generated the tar sands of the Athabasca anticline.

OBSERVATION

The history of most continental margins and large sedimentary basins is a monotonous repetition of only two principle themes, the extensional regimes of passive margins, and convergent tectonics that mark ocean closing (strike-slip movements not withstanding). Every foreland thrust and fold belt is probably built on an earlier Atlantic-type passive margin. The tectonic processes of extension and convergence are fairly well understood. Quantitative modelling of the lithosphere is adding a vital new dimension. In contrast, stratigraphers have been less successful in interpreting foreland basins, probably because their science relies more on modern analogue than on principle. Modern examples of deltas, coasts and shelves are concentrated on passive margins. Tectono-stratigraphy, combining tectonics, theoretical models of the lithosphere, and meso-scale stratigraphy, may be a potent approach to basin analysis.

ACKNOWLEDGMENTS

This study has benefitted tremendously from the enthusiastic support of Glen Stockmal. Discussions with Hugh Balkwill, Bert Bally, John Barwis, Chris Beaumont, Dave Lawrence, Peter McCabe, and Herman Welsink are also gratefully acknowledged. Drafting and typing were done by Terri Haber and Dawn Holmes. I thank Petro-Canada for permission to publish.

REFERENCES

ALLEN, J.R.L. & FRIEND, P.F. (1968) Deposition of the Catskill facies, Appalachian region: with notes on some other Old Red Sandstone basins. In: *Late Paleozoic and Mesozoic Continental Sedimentation, Northeastern North America* (Ed. by G. deV. Klein). *Spec. Pap. geol. Soc. Am.* **106**, 21–74.

AMMERMAN, M.L. & KELLER, G.R. (1979) Delineation of Rome Trough in eastern Kentucky by gravity and deep drilling data. *Bull. Am. Ass. Petrol. Geol.* **63**, 341–353.

ANDO, C.J., CZUCHRA, B.L., KLEMPERER, S.L., BROWN, L.D., CHEADLE, M.J., COOK, F.A., OLIVER, J.E., KAUF-

MAN, S., WALSH, T., THOMPSON, J.B., LYONS, J.B. & ROSENFELD, J.L. (1984) Crustal profile of mountain belt: COCORP deep seismic reflection profiling in New England Appalachians and implications for architecture of convergent mountain chains. *Bull. Am. Ass. Petrol. Geol.* **68**, 819–837.

BALLY, A.W., GORDY, P.L. & STEWART, G.A. (1966) Structure, seismic data, and orogenic evolution of the southern Canadian Rocky Mountains. *Bull. Can. Soc. Petrol. Geol.* **14**, 337–381.

BARWIS, J.H. & HORNE, J.C. (1979) Paleotidal range indicators in the Carboniferous barriers of northeastern Kentucky. In: *Carboniferous Depositional Environments in the Appalachian Region* (Ed. by J. C. Ferm & J. C. Horne), pp. 460–471. University of South Carolina, Columbia.

BEAUMONT, C. (1981) Foreland basins. *Geophys. J. R. astr. Soc.* **65**, 291–329.

BEAUMONT, E.A. (1984) Retrogradational shelf sedimentation: Lower Cretaceous Viking Formation, central Alberta. *Spec. Publ. Soc. econ. Paleont. Mineral., Tulsa*, **34**, 163–177.

BERG, R.R. (1975) Depositional environments of the Upper Cretaceous Sussex Sandstone, House Creek Field, Wyoming. *Bull. Am. Ass. Petrol. Geol.* **59**, 2099–2110.

BOCKMEULEN, H., BARKER, C. & DICKEY, P.A. (1983) Geology and geochemistry of crude oils, Bolivar Coastal fields, Venezuela. *Bull. Am. Ass. Petrol. Geol.* **67**, 242–270.

BOWLIN, B.K. & KELLER, F.B. (1980) Incised submarine channel-fan deposits in the Tellico Formation, South Holston Dam, Tennessee. In: *Middle Ordovician Carbonate Shelf to Deep Water Basin Deposition in the Southern Appalachians* (Ed. by K. R. Walker, T. W. Broadhead & F. B. Keller), pp. 91–107. Field Trip Guide geol. Soc. Am.

BOYLES, J.M. & SCOTT, A.J. (1982) A model for migrating shelf-bar sandstones in Upper Mancos Shale (Campanian), northwestern Colorado. *Bull. Am. Ass. Petrol. Geol.* **66**, 491–508.

BRENNER, R.L. (1978) Sussex Sandstone of Wyoming-example of Cretaceous offshore sedimentation. *Bull. Am. Ass. Petrol. Geol.* **62**, 181–200.

BRENNER, R.L. (1980) Construction of process-response models for ancient epicontinental seaway depositional systems using partial analogs. *Bull. Am. Ass. Petrol. Geol.* **64**, 1223–1244.

BRETT, C.E. (1983) Sedimentology, facies relations, and depositional environments of the Rochester Shale (Silurian; Wenlockian) in western New York and Ontario. *J. sedim. Petrol.* **53**, 947–971.

BRETT, C.E. & BAIRD, G.C. (1985) Carbonate-shale cycles in the Middle Devonian of New York: an evaluation of models for the origin of limestones in terrigenous shelf sequences. *Geology*, **13**, 324–327.

BROADHEAD, R.F., KEPFERLE, R.C. & POTTER, P.E. (1982) Stratigraphic and sedimentologic controls of gas in shale-example from Upper Devonian of Northern Ohio. *Bull. Am. Ass. Petrol. Geol.* **66**, 10–27.

CAMPBELL, C.V. (1973) Offshore equivalents of Upper Cretaceous Gallup beach sandstone, northwestern New Mexico. In: *Cretaceous and Tertiary Rocks of the Southern Colorado Plateau* (Ed. by J. E. Fasset). *Mem. Four Corners geol. Soc.* 78–84.

CHAPPLE, W.M. (1978) Mechanics of thin-skinned fold-and-thrust belts. *Bull. geol. Soc. Am.* **89**, 1189–1198.

CONEY, P.J., JONES, D.L. & MONGER, J.W.H. (1980) Cordilleran suspect terranes. *Nature*, **288**, 329–333.

COOK, F.A. (1984) Geophysical anomalies along strike of the southern Appalachian piedmont. *Tectonics*, **3**, 45–61.

COOK, F.A., ALBOUGH, D.S., BROWN, L.D., KAUFMAN, S., OLIVER, J.E. & HATCHER, R.D. (1979) Thin-skinned tectonics in the crystalline southern Appalachians; CO-CORP seismic reflection profiling of the Blue Ridge and Piedmont. *Geology*, 7, 563–567.

COOK, F.A., BROWN, L.D., KAUFMAN, S., OLIVER, J.E. & PETERSEN, T.A. (1981) COCORP seismic profiling of the Appalachian orogen beneath the coastal plain of Georgia. *Bull. geol. Soc. Am.* **92**, 738–748.

COTTER, E. (1983) Shelf, paralic, and fluvial environments and eustatic sea-level fluctuations in the origin of the Tuscarora Formation (Lower Silurian) of central Pennsylvania. *J. sedim. Petrol.* **53**, 25–49.

COURTNEY, R.C. & BEAUMONT, C. (1983) Thermally-activated creep and flexure of the oceanic lithosphere. *Nature*, **305**, 201–204.

CROWLEY, D.J. (1973) Middle Silurian patch reefs in the Gasport member (Lockport Formation), New York. *Bull. Am. Ass. Petrol. Geol.* **52**, 283–300.

DAVIS, M.W. & EHRLICH, R. (1974) Late Paleozoic crustal composition and dynamics in the southeastern United States. *Spec. Pap. geol. Soc. Am.* **148**, 171–185.

DEROO, G., POWELL, T.G., TISSOT, B. & McCROSSAN, R.G. (1977) The origin and migration of petroleum in the western Canadian sedimentary basin, Alberta. *Bull. Can. geol. Surv.* **262**, 136 pp.

DEVER, G.R., HOGE, H.P., HESTER, N.C. & ETTENSOHN, F.R. (1977) *Stratigraphic evidence for late Paleozoic tectonism in northeastern Kentucky.* Field trip guide: Lexington, Kentucky Geological Survey.

DE WITT, W. & McGRAW, L.W. (1979) Appalachian basin region. In: *Paleotectonic Investigations of the Mississippian System in the United States (Pt 1)* (Ed. by L. C. Craig and C. W. Connor). *Prof. Pap. U.S. geol. Surv. 1010*, 13–48.

DILLMAN, S.B. & ETTENSOHN, F.R. (1980) Structure contour map on the base of the Three Lick Bed (Unit 2) of the Ohio Shale in eastern Kentucky. *U.S. Dept Energy METC/EGSP Series*, **513**.

ENGEBRETSON, D.C., COX, A. & THOMPSON, G.A. (1984) Correlation of plate motions with continental tectonics: Laramide to Basin-Range. *Tectonics*, **3**, 115–119.

ETTENSOHN, F.R. (1980) An alternative to the barrier-shoreline model for deposition of Mississippian and Pennsylvanian rocks in northeastern Kentucky: Summary. *Bull. geol. Soc. Am.* **91**, 130–135.

ETTENSOHN, F.R. (1981) Mississippian-Pennsylvanian boundary in northeastern Kentucky. *Geol. Soc. Am. Cincinnati '81 Field Trip Guidebook*, pp. 195–257.

ETTENSOHN, F.R., RICE, C.L., DEVER, G.R. & CHESNUT, D.R. (1984) Slade and Paragon formations—new stratigraphic nomenclature for Mississippian rocks along the Cumberland escarpment in Kentucky. *Bull. U.S. geol. Surv. 1605-B*, 41 pp.

FERM, J.C. (1970) Allegheny deltaic deposits. *Spec. Publ. Soc. econ. Paleont. Mineral., Tulsa*, **15**, 137–146.

FERM, J.C. (1974) Carboniferous environmental models in eastern United States and their significance. *Spec. Pap. geol. Soc. Am.* **148**, 79–95.

FOLK, R.L. (1960) Petrography and origin of the Tuscarora,

Rose Hill, and Keefer Formations, Lower and Middle Silurian of eastern West Virginia. *J. sedim. Petrol.* 30, 1–58.

FYFFE, L.R., PAJARI, G.E. & CHERRY, M.E. (1981) The Acadian plutonic rocks of New Brunswick. *Marit. Sedim. Atlantic Geol.* **17**, 23–36.

GARDNER, T.W. (1983) Paleohydrology and paleomorphology of a Carboniferous, meandering, fluvial sandstone. *J. sedim. Petrol.* **53**, 991–1005.

GRAHAM, S.A., DICKINSON, W.R. & INGERSOLL, R.V. (1975) Himalayan–Bengal model for flysch dispersal in the Appalachian–Ouachita system. *Bull. geol. Soc. Am.* **86**, 273–286.

GRAHAM, S.A., INGERSOLL, R.V. & DICKINSON, W.R. (1976) Common provenance for lithic grains in Carboniferous sandstones from Ouachita Mountains and Black Warrior basin. *J. sedim. Petrol.* **46**, 620–632.

HORNE, J.C., FERM, J.C., CARUCCIO, F.T. & BAGANZ, B.P. (1978) Depositional models in coal exploration and mine planning in Appalachian region. *Bull. Am. Ass. Petrol. Geol.* **62**, 2379–2411.

JACOBI, R.D. (1981) Peripheral bulge—a causal mechanism for the lower/middle Ordovician unconformity along the western margin of the northern Appalachians. *Earth planet. Sci. Lett.* **56**, 245–251.

JOHNSON, K.G. & FRIEDMAN, G.M. (1969) The Tully clastic correlatives (Upper Devonian) of New York State: A model for recognition of alluvial, dune (?), tidal, nearshore (bar and lagoon), and offshore sedimentary environments in a tectonic delta complex. *J. sedim. Petrol.* **39**, 451–485.

JORDAN, T.E. (1981) Thrust loads and foreland basin evolution, Cretaceous, western United States. *Bull. Am. Ass. Petrol. Geol.* **65**, 2506–2520.

KARNER, G.D. & WATTS, A.B. (1983) Gravity anomalies and flexure of the lithosphere at mountain ranges. *J. geophys. Res.* **88**, 10 449–10 477.

KEPFERLE, R.C. & POLLOCK, D. (1983) Correlations of the lower part of the Devonian oil shale in Kentucky. In: *Proc. Eastern Oil Shale Symp*, pp. 35–40. Univ. Kentucky Inst. Mining & Minerals Res., Lexington.

LAPORTE, L.F. (1971) Paleozoic carbonate facies of the central Appalachian shelf. *J. sedim. Petrol.* **41**, 724–740.

LORENZ, J.C. (1982) Lithospheric flexure and the history of the Sweetgrass Arch, northwestern Montana. In: *Geologic Studies of the Cordilleran Thrust Belt* (Ed. by R. B. Powers), pp. 77–89. Rocky Mtn. Assoc. Geol., Denver.

MARTINSEN, R.S. & TILLMAN, R.W. (1979) Facies and reservoir characteristics of shelf sandstones. Hartzog Draw Field, Powder River basin (abstract). *Bull. Am. Ass. Petrol. Geol.* **63**, 491.

McGOOKEY, D.P., HAUN, J.D., HALE, L.A., GOODELL, H.G., McGUBBIN, D.G., WEIMER, R.J. & WULF, G.R. (1972) Cretaceous System. In: *Geologic Atlas of the Rocky Mountain Region, United States of America*, pp. 190–232. Rocky Mtn. Assoc. Geol., Denver.

MECKEL, L.D. (1970) Paleozoic alluvial deposition in the central Appalachians: a summary. In: *Studies of Appalachian Geology* (Ed. by G. W. Fisher, F. J. Pettijohn, J. C. Reed & K. N. Weaver), pp. 49–67. Wiley (Interscience), New York.

MILICI, R.C., BRIGGS, G., KNOX, L.M., SITTERLY, P.D. & STATLER, A.T. (1979) The Mississippian and Pennsylvanian (Carboniferous) Systems in the United States–Tennessee. *Prof. Pap. U.S. geol. Surv. 110–G*, 38 pp.

MILLER, M.F., BARGAR, E.A. & JACKSON, S.R. (1984) Early Pennsylvanian thrust faulting inferred from sandstone composition and sequence of depositional environments (northern Cumberland plateau, Tennessee) (abstract). *Ann. mtg geol. Soc. Am.* p. 597.

MOSHIER, S.O. & WAPLES, D.W. (1985) Quantitative evaluation of Lower Cretaceous Mannville Group as source rock for Alberta's oil sands. *Bull. Am. Ass. Petrol. Geol.* **69**, 161–172.

PAGE, B.M. & ENGEBRETSON, D.C. (1984) Correlation between the geologic record and computed plate motions for central California. *Tectonics*, **3**, 133–155.

PEPPER, J.F., DE WITT, W. & DEMAREST, D.F. (1954) Geology of the Bedford Shale and Berea Sandstone in the Appalachian Basin. *Prof. Pap. U.S. geol. Surv. 259*, 111 pp.

PORTER, J.W., PRICE, R.A. & McCROSSAN, R.G. (1982) The Western Canada sedimentary basin. *Phil. Trans. R. Soc. A*, **305**, 169–192.

POTTER, P., MAYNARD, J. & PRYOR, A. (1981) Sedimentology of gas-bearing Devonian shales of the Appalachian basin. *U.S. Dept Energy DOE/METC*, **114**, 43 pp.

PRICE, R.A. (1981) The Cordilleran foreland thrust and fold belt in the southern Canadian Rocky Mountains. In: *Thrust and Nappe Tectonics* (Ed. by K. R. McClay & N. J. Price). *Spec. Publ. geol. Soc. London*, **9**, 427–448. Blackwell Scientific Publications, Oxford.

PRICE, R.A. & HATCHER, R.D. (1983) Tectonic significance of similarities in the evolution of the Alabama-Pennsylvania Appalachians and the Alberta-British Columbia Canadian Cordillera. In: *Contributions to the Tectonics and Geophysics of Mountain Chains* (Ed. by R. D. Hatcher, H. Williams & I. Zietz). *Mem. geol. Soc. Am.* **158**, 149–160.

PRYOR, W.A. & SABLE, E.G. (1974) Carboniferous of the eastern interior basin. In: *Carboniferous of the Southeastern United States* (Ed. by G. Briggs). *Spec. Pap. geol. Soc. Am.* **148**, 281–313.

QUINLAN, G.M. & BEAUMONT, C. (1984) Appalachian thrusting, lithospheric flexure, and the Paleozoic stratigraphy of the eastern interior of North America. *Can. J. Earth Sci.* **21**, 973–996.

READ, J.F. (1980) Depocenters, carbonate facies and foreland basin evolution, middle Ordovician, Virginia. In: *Proceedings 'The Caledonides in the USA'* (Ed. by D. R. Wones). *Mem. Virginia Polytech. Inst. State Univ.* **2**, 19–26.

RICE, C.L. (1984) Sandstone units of the Lee Formation and related strata in eastern Kentucky. *Prof. Pap. U.S. geol. Surv. 1151–G*, 53 pp.

RICE, D.D. (1984) Widespread, shallow-marine, storm-generated sandstone units in the Upper Cretaceous Mosby Sandstone, central Montana. *Spec. Publ. Soc. econ. Paleont. Mineral., Tulsa*, **34**, 143–161.

ROEDER, D., GILBERT, O.E. & WITHERSPOON, W.D. (1978) Evolution and macroscopic structure of valley and ridge thrust belt, Tennessee and Virginia. *Univ. Tenn. Dept geol. Sci. Stud. Geol.* **2**, Knoxville.

ROYDEN, L. & KARNER, G.D. (1984) Flexure of lithosphere beneath Appenine and Carpathian foredeep basins: evidence for an insufficient topographic load. *Bull. Am. Ass. Petrol. Geol.* **68**, 704–712.

SABLE, E.G. (1979) Eastern interior basin region. In:

Paleotectonic Investigations of the Mississippian System in the United States (Pt 1) (Ed. by L. C. Craig & C. W. Connor). *Prof. Pap. U.S. geol. Surv. 1010*, 59–105.

SCHEDL, A. & WILTSCHKO, D.W. (1984) Sedimentological effects of a moving terrain. *J. Geol.* **92**, 273–287.

SHANMUGAM, G. & LASH, G. (1982) Analogous tectonic evolution of the Ordovician foredeeps, southern and central Appalachians. *Geology*, **10**, 562–566.

SHANMUGAM, G. & WALKER, K.R. (1980) Sedimentation subsidence, and evolution of a foredeep basin in the middle Ordovician, southern Appalachians. *Am. J. Sci.* **280**, 479–496.

SLACK, P.B. (1981) Paleotectonics and hydrocarbon accumulation, Powder River basin, Wyoming. *Bull. Am. Ass. Petrol. Geol.* **65**, 730–743.

SLATT, R.M. (1984) Continental shelf topography: key to understanding distribution of shelf sand-ridge deposits from Cretaceous western interior seaway. *Bull. Am. Ass. Petrol. Geol.* **68**, 1107–1120.

SLOSS, L.L. (1963) Sequences in the cratonic interior of North America. *Bull. geol. Soc. Am.* **74**, 93–113.

SLOSS, L.L. (1982) The midcontinent province: United States. In: *Perspectives in Regional Geological Synthesis, planning for the Geology of North America* (Ed. by A. R. Palmer). *Spec. Publ. geol. Soc. Am. D-NAG* **1**, 27–39.

SMITH, G.E., DEVER, G.R., HORNE, J.C., FERM, J.C. & WHALEY, P.W. (1971) *Depositional environments of eastern Kentucky coals.* Field guide: Lexington, Kentucky Geological Survey.

SPEARING, D.R. (1976) *Upper Cretaceous Shannon Sandstone*, pp. 65–72. Guidebook Wyoming Geol. Assoc.

STELCK, C.R. (1975) Basement control of Cretaceous sand sequences in Western Canada. *Spec. Pap. geol. Ass. Can.* **13**, 427–440.

STOCKMAL, G.S., BEAUMONT, C. & BOUTILIER, R. (1986) Geodynamic models of convergent margin tectonics: the transition from rifted margin to overthrust belt and the consequences for foreland basin development. *Bull. Am. Ass. Petrol. Geol.* **70**, 181–190.

SWIFT, D.J.P. & RICE, D.D. (1984) Sand bodies on muddy shelves: a model for sedimentation in the western interior Cretaceous seaway, North America. *Spec. Publ. Soc. econ. Paleont. Mineral., Tulsa*, **34**, 43–62.

TAINTER, P.A. (1984) Stratigraphic and paleostructural controls on hydrocarbon migration in Cretaceous D and J sandstones of the Denver basin. In: *Hydrocarbon Source Rocks of the Greater Rocky Mountain Region* (Ed. by J. Woodward, F. F. Meissner & J. L. Clayton), pp. 339–354. Rocky Mtn. Assoc. Geol., Denver.

TANKARD, A.J. (1986) Depositional response to foreland deformation in the Carboniferous of eastern Kentucky. *Bull. Am. Ass. Petrol. Geol.* **70**, 853–868.

TILLMAN, R.W. & MARTINSEN, R.S. (1984) The Shannon shelf-ridge sandstone complex, Salt Creek anticline area, Powder River basin, Wyoming. In: *Siliclastic Shelf Sediments* (Ed. by R. W. Tillman & C. T. Siemers). *Spec. Publ. Soc. econ. Paleont. Mineral., Tulsa*, **34**, 85–142.

WALKER, K. R., SHANMUGAM, G. & RUPPEL, S.C. (1983) A model for carbonate to terrigenous clastic sequences. *Bull. geol. Soc. Am.* **94**, 700–712.

WALKER, R.G. (1983) Cardium Formation 3. Sedimentology and stratigraphy in the Garrington-Caroline area, Alberta. *Bull. Can. Petrol. Geol.* **31**, 213–230.

WALKER, R.G. & HARMS, J.C. (1971) The 'Catskill Delta': a prograding muddy shoreline in central Pennsylvania. *J. Geol.* **79**, 381–399.

WANLESS, H.R. (1975) Paleotectonic investigations of the Pennsylvanian System in the United States. Part 1. Appalachian Region. *Prof. Pap. U.S. geol. Surv.* **853**, 17–62.

WATTS, A.B. (1981) The U.S. Atlantic continental margin. In: *Geology of Passive Continental Margins* (Ed. by A. W. Bally *et al.*). *Am. Ass. Petrol. Geol. Course Notes Series*, **19**, 2(1–75).

WEHR, F. & GLOVER, L. (1985) Stratigraphy and tectonics of the Virginia–North Carolina Blue Ridge: evolution of a late Proterozoic–early Paleozoic hinge zone. *Bull. geol. Soc. Am.* **96**, 285–295.

WEIMER, R.J. (1983) Relation of unconformities, tectonics, and sea level changes, Cretaceous of the Denver basin and adjacent areas. In: *Mesozoic Paleogeography of the West-Central United States* (Ed. by M. W. Reynolds & E. D. Dolly), pp. 359–376. Rocky Mtn. Sectn. Soc. econ. Paleont. Mineral.

WEIR, G.W., PETERSON, W.L. & SWADLEY, W.C. (1984) Lithostratigraphy of Upper Ordovician strata exposed in Kentucky. *Prof. Pap. U.S. geol. Surv. 1151-E*, 121 pp.

WILLIAMS, G.D. & BURK, C.F. (1966) Upper Cretaceous. In: *Geological History of Western Canada* (Ed. by R. G. McCrossan & R. P. Glaister), pp. 169–189. Alberta Soc. Petrol. Geol., Calgary.

WILLIAMS, H. & HATCHER, R.D. (1983) Appalachian suspect terranes. In: *Contributions to the Tectonics and Geophysics of Mountain Chains* (Ed. by R. D. Hatcher, H. Williams & I. Zietz). *Mem. geol. Soc. Am.* **158**, 33–53.

WINTERS, J.D. & WILLIAMS, J.A. (1969) Microbiological alteration of crude oil in the reservoir. In: *Symposium on Petroleum Transformation in Geologic Environments. Am. Chem. Soc. (Div. Petrol. Chem.)*, New York, *Pap. PETR* **86**, E22–E31.

WOODWARD, H.P. (1961) Preliminary subsurface study of southeastern Appalachian interior plateau. *Bull. Am. Ass. Petrol. Geol.* **45**, 1634–1655.

WU, P. & PELTIER, W.R. (1982) Viscous gravitational relaxation. *Geophys. J. R. astr. Soc.* **70**, 435–486.

Petrography and stratigraphic techniques

Spec. Publs int. Ass. Sediment. (1986) **8**, 395–410

Sedimentary 'signatures' of foreland basin assemblages: real or counterfeit?

FREDERIC L. SCHWAB

Department of Geology, Washington and Lee University, Lexington, VA 24450, U.S.A.

ABSTRACT

Can various aspects of the sedimentary fill of foreland basins be used to characterize such basins and distinguish them from other basin varieties? Data selected from ancient assemblages of the Ouachita, Appalachian, Cordilleran, and Alpine belts are compared with one another and with modern deep-sea and continental margin deposits.

(1) The mineralogical composition of foreland basin sandstones (modes compiled using QFL, QmFLt, QpLvLs and QmPK plots) show a wide range of provenance types. Quartz-rich, feldspar-poor detritus derived from continental blocks (intensely weathered cratons as well as locally uplifted basement) comprises large portions of the *early* fill of foreland basins, but the bulk of the detritus is less quartz-rich. Subsequent material is richer in rock fragments and is derived from subjacent orogenic source source areas: *either much older* continental margin deposits exposed in sedimentary and metasedimentary nappes and thrusts *or from slightly older* deformed sediments exposed within the fold-thrust belt itself. Only small amounts of material are derived from tectonically uplifted subduction complexes or from magmatic arcs, either because the fold-thrust uplands topographically shield such sources, or thanks to mainly longitudinal dispersal along the strike of the suturing orogenic belt.

(2) Systematic, through time *secular* variations in provenance related to unroofing and to secular dominance of distinct source rocks are not obvious based on bulk framework mineralogy, even though some orderly evolution appears to occur when more precise mineral suites can be defined.

(3) The estimated bulk chemical composition of the sediment fill of foreland basins most resembles the fill of ancient geosynclines in general, and eugeoclines (eugeosynclines) specifically, and markedly contrasts with the fill of aulacogens, forearc basins, ancient miogeoclines, and modern trench-abyssal plain sediments.

(4) The rate at which sediment accumulates in foreland basins (0·044 to 0·927 m/1000 yr; mean is 0·186 m/1000 yr) exceeds sediment accumulation rates for cratonic basins and the abyssal ocean floor by a factor of 5–30. Comparably high rates of sediment accumulation characterize ancient aulacogens and modern rift valleys, successor basins, and (surprisingly) some modern trenches.

INTRODUCTION

This paper emphasizes the sedimentary-stratigraphic characteristics of foreland basin assemblages in general. Four questions are specifically addressed:

(1) Is the sand-sized detritus within foreland basins compositionally distinctive (modes compiled using QFL, QmFLt, QpLvLs and QmPK plots) and does it indicate a particular suite of provenance types?

(2) Can systematic, secular (through time) variations in provenance detected by studying distinctive source mineral suites be recognized from more generalized analyses of bulk sandstone mineralogy?

(3) What similarities and differences exist between the estimated bulk chemical composition of foreland basin fill and the fill of other basin types?

(4) Is the rate at which sediment accumulates in foreland basins (in m/1000 yr) sufficiently distinctive from sediment accumulation rates in other basin types?

DETRITAL MODES OF FORELAND BASIN SANDSTONES: IMPLICATIONS FOR PROVENANCE

Sandstone framework mineralogy primarily reflects tectonic setting and provenance terranes. Dickinson & Suczek (1979) have clearly demonstrated that quantitative sandstone detrital modes based on point counts of thin sections can be used to index provenance. The specific provenance setting is controlled principally by plate tectonic setting (Dickinson & Valloni, 1980). Throughout this paper, sandstone framework mineralogy modes expressed as QFL, QmFLt, QpLvLs and QmPK are calculated using: Q = total quartzose grains (monocrystalline and polycrystalline quartz); F = total feldspar; L = unstable lithic fragments; Qm = monocrystalline quartz; Lt = total lithic fragments (stable and unstable); Qp = polycrystalline quartz; Lv = volcanic rock fragments; Ls = sedimentary rock fragments; P = plagioclase feldspar; and K = K feldspar.

Plate tectonic setting

Table 1 duplicates the sandstone composition data of Dickinson & Valloni (1980) relating changes in sandstone mineralogy to variations in plate tectonic setting. Dickinson & Valloni defined eight plate tectonic margin types: (a) rifted margins of continental blocks within which there were no bordering or interior plate boundaries, essentially consisting only of cratonic source rocks (for example, Antarctica); (b)

rifted margins of continents which contain incipient rift belts or transform ruptures (cratonic source rocks plus detritus from upthrown fault-bounded basement blocks, i.e. Africa); (c) rifted margins of continents with orogenic highlands developed by subduction along one margin (North America); (d) rifted continental margins of composite continents traversed with collision belts (Eurasia), essentially cratonic source rocks plus detritus from orogenic belts); (e) active continental margins bordered with transform fault systems (like the California coast), with complex highland source rocks; (f) active margins with arc-trench systems shielding the adjacent ocean basin from cratonic sources, essentially entirely magmatic arc source rocks (western South America); (g) island arc margins of subduction zones adjacent to marginal seas (western Pacific arc complexes); and (h) intraplate oceanic islands that combine volcanic and coralline source rocks. The detrital modes of sands derived from each of these distinctive plate tectonic settings differ markedly from one another and, in many cases, from the mean sandstone modes representing various foreland basin deposits. Table 1 should be referred to throughout this paper.

Provenance

Table 2 and Fig. 1 duplicate the sandstone compositional data published by Dickinson & Suczek (1979) relating changes in sandstone mineralogy to variations in provenance. Dickinson & Suczek recognized four major source rock varieties: (i) *stable cratons* generat-

Table 1. Mean detrital modes of modern seafloor sands from different plate tectonic settings (Dickinson & Valloni, 1980)

Type	Ocean basin adjacent to	Q %	F %	L %	Qm %	F %	Lt %	Qp %	Lv %	Ls %	Qm %	P %	K %
	1. Rifted continental margin												
A.	Cratonic source rocks	78	18	6	71	18	11	—	—	—	80	5	15
B.	Cratonic source rocks + uplifted basement blocks	69	26	5	62	26	12	—	—	—	70	13	17
C, D.	Cratonic sources + orogenic belt sources	63	26	11	51	26	23	52	3	45	66	16	18
	2. Orogenic continental margin												
E.	Active margin bordered with transform faults (California)	31	45	24	22	45	33	27	26	47	33	42	25
F.	Active margin with arc-trench systems shielding adjacent ocean basin (mainly magmatic arc sources)	20	41	39	15	41	44	11	79	10	27	55	18
	3. Oceanic island chain												
G.	Island arc adjacent to subduction zone	11	34	55	8	34	58	5	90	5	19	74	7
H.	Intraplate ocean islands	0	5	95	0	5	95	0	100	0	0	100	0

Q = total quartzose grains (monocrystalline and polycrystalline); F = total feldspar; L = unstable lithic fragments; Qm = monocrystalline quartz; Lt = total lithic fragments (stable and unstable); Qp = polycrystalline quartz; Lv = volcanic rock fragments; Ls = sedimentary rock fragments; P = plagioclase feldspar; K = K-feldspar.

Table 2. Mean sandstone detrital modes and provenance (Dickinson & Suczek, 1979)*

Provenance type	Q %	F %	L %	Qm %	F %	Lt %	Qp %	Lv %	Ls %	Qm %	P %	K %
I. Continental blocks												
1. Mainly stable shield and sediment cover sources	93·5	5·5	1	89	6	5	82·5	10	7·5	94	1	5
2. Transitional between 1 and 2	74	23	3	71	22	7	66	21	13	77	8	16
3. Mainly uplifted blocks of basement rocks	51	44	5	44	44	12	42	32	26	51	29	20
II. Magmatic arcs												
4. Undissected arc	5	29	66	4	29	67	1	91	8	15	82	3
5. Transitional between 4 and 6	19	28	53	15	28	57	6	75	19	35	55	10
6. Dissected arc	33	37	30	31	36	33	8	57	35	46	39	15
III. Recycled orogens												
7. Mainly subduction complexes	45	13·5	41·5	7·5	13·5	79	48	25	27	51·5	46	2·5
8. Mainly collision orogens	71	12	17	63	12	25	35	5	60	82	12	6
9. Mainly foreland uplifts	67	7	26	50·5	7·5	42	45	12	43	90	5	5

* Listed values represent means calculated from available data in Dickinson & Suczek (1979, table 1).

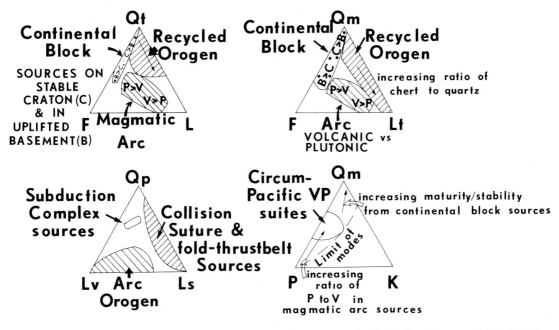

PROVENANCE AND DETRITAL SANDSTONE MODES

Fig. 1. Actual reported distribution of mean detrital modes for sandstone suites derived from different types of provenances plotted on standard triangular diagrams (after Dickinson, 1984): Qt = total quartzose grains; Qm = monocrystalline quartz (>0·625 mm); Qp = polycrystalline quartz or chalcedony; F = total feldspar grains; P = plagioclase grains; K = Fspar grains; L = total unstable lithic fragments; Lv = volcanic/metavolcanic lithic fragments; Ls = sedimentary/metasedimentary lithic fragments; Lt = total lithic fragments (l + Qp).

Table 3. Detrital sandstone modes, various kinds of recycled orogenic source types (see also Fig. 2)

Type of recycled orogenic source	Q %	F %	L %	Qm %	F %	Lt %
1. Backarc thrustbelt sources (classical foreland basin source areas)						
I. Antler foreland basin						
FA-1 Mississippian Guyet Formation, British Columbia	57	14	29			
FA-2 Mississippian Chainman Shale and Diamond Peak Formation, deltaic and turbidite clastics from sand-rich chert-argillite subduction complex	73	2	25	47	2	51
II. Wyoming foreland basin (FW) Upper Cretaceous deltaic and contourite clastics (Cody Shale and Parkman Sandstone) derived from sedimentary-metasedimentary thrustbelt	72	7	21	48	2	45
III. Mexican foreland basin (FM) Cretaceous–Palaeocene deltaic clastics derived mainly from arc volcanics	36	28	36	32	28	40
2. Uplifted supracrustal source rocks						
s-1 Upper Triassic Vester Formation, Oregon, subduction complex source	50	13	37	5	13	82
s-2 Lower Siwalik Group, India	66	3	31	59	3	38
s-3 Middle Siwalik Group, India	48	10	42	43	10	47
s-2 and s-3 derived from fold-thrust system of the Himalayas						
3. Collision orogen sources						
Col. H.1 (Himalayas 1), Indus cone- Cenozoic sand, Arabian Sea	45	32	23	44	32	24
Col. H.2 Neogene sand, Bengal–Nicobar Fan, Indian Ocean	57	28	21	56	28	15
Col. 0. Ouachita orogen, Haymond Formation	71	17	12	68	17	15
4. Ouachita sandstone sequence (from the top down) (g–a)						
g. Morrowan sandstone, Oklahoma	69	10	21			
f. Carboniferous sandstone, Ouachitas	79	3	18			
e. Pennsylvanian Atoka Formation	77	7	16			
d. Pennsylvanian Jackfort Formation	96	1	3			
c. Middle Devonian–Middle Carboniferous Stanley Formation	84	14	2			
b. Silurian Missouri Mountain Formation	100	0	0			
a. Ordovician Crystal Mtn. and Blackely Sandstone	97	2	1			

All data from Dickinson (1984) except Ouachita data from Dickinson *et al.* (1983).

ing quartz-rich sands derived from low-lying granitic and gneissic source rocks and the platform sediments that overlie them; (ii) *basement uplifts* that commonly border rift valleys and transform faults and which generate arkosic sands; (iii) *magmatic arcs* generating volcanoclastic detritus and/or (with deep dissection) quartzofeldspathic sands eroded from their batholithic roots; and (iv) *recycled orogens* (deformed, uplifted sedimentary rocks, for example, subduction complexes, with oceanic and trench-slope deposits and melange; backarc thrustbelts with folded, continentally-derived sedimentary and metasedimentary rocks; suture belts with oceanic and continental sediments) that generate sands of widely variable composition (depending on the particular source).

Dickinson & Suczek further grouped the source terrane types into three fundamentally different provenance types:

(I) *Continental block provenance*, with sediment sources either stable shields and platforms (1), and/or uplifted blocks of basement rock (3), or transitional between these two extremes (2).

(II) *Magmatic arc provenance*, with sediment sources either the volcanic cover of the arc itself (4), and/or the dissected plutonic roots to the arc (6), or transitional between these two extremes (5).

(III) *Recycled orogen provenance*, with sediment sources mainly uplifted and deformed sedimentary rocks, with several specific possibilities, including (7) subduction complexes of arc orogens, (8) folded and thrusted sediments of sutured collision orogens, and (9) thin-skinned foreland thrust-fold belts of arc and collision orogens.

The detrital modes derived from these three contrasting provenance types permit three, somewhat distinctive fields to be defined as shown in Fig. 1. Miniaturized versions of these distinctive provenance fields are reproduced in later figures for reference.

Recycled orogen provenance—general characteristics

Because foreland basins almost invariably involve the deformation, uplift, and erosion of subjacent sedimentary rock terranes, quite obviously most sand detritus should fall within the 'recycled orogen' field of Dickinson & Suczek. Dickinson (1984) has attempted to document the wide compositional variation in sandstones derived from various specific varieties of recycled orogens. These variations include (Table 3 and Fig. 2):

(I) *Backarc thrustbelt sources* (the classical provenance for foreland basins) composed mainly of continentally-derived sedimentary and metasedimentary strata located immediately adjacent to overridden foreland basins. These sands are

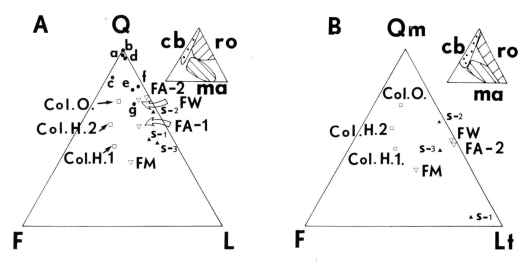

Fig. 2. (A) QFL plots for sandstones reflecting various recycled orogen source types (Table 3, this paper): foreland basin assemblages—FA-1 and FA-2 (Antler foreland basin), FW (Wyoming foreland basin, FM (Mexican foreland basin); sandstones derived from various supracrustal source rocks—s-1, s-2, and s-3; sandstones derived from various collision orogen sources—(Col. H.1 and Col. H.2-Himalaya, Col. 0.-Ouachitas); and a sequence of sandstones from the Ouachita foreland basin (a–g). Smaller adjacent triangle shows the three main provenance area fields shown in Fig. 1: cb = continental block; ma = magmatic arc; ro = recycled orogen. (B) QmFLt plots for sandstones derived from various recycled orogen sources (Table 3, this paper): various foreland basin assemblages (FA-2 Antler foreland basin; FW Wyoming foreland basin; FM Mexican foreland basin); sandstones derived from various supracrustal sources (s-1, s-2, s-3); and sandstones derived from various collision orogen sources (Col. H.1 and Col. H.2-Himalaya; Col. 0.-Ouachitas). Smaller adjacent triangle shows the three main provenance area fields shown in Fig. 1: cb = continental block; ma = magmatic arc; ro = recycled orogen.

typically low in feldspar and volcanic rock fragments and rich in quartz and sedimentary rock fragments (FA-1, FA-2, FW, FM).

(II) *Various uplifted supracrustal sources* (S-1, S-2, S-3) generate sand-sized detritus that varies appreciably in composition reflecting sources that include a chert-rich subduction complex floored by ocean crust (S-1) and the Himalayan fold-thrust system (S-2 and S-3).

(III) *Intercontinental collision orogens* (like S-2 and S-3 above) or 'suture belts' (Col. H.1, Col. H.2, Col. 0.) typically are composed of structurally uplifted and juxtaposed oceanic and continental sedimentary sequences that are commonly drained by longitudinal dispersal systems feeding remnant ocean basins (not unlike the transverse dispersal across foreland basins).

Figure 2 and Table 3 also show data from a sequence of sandstones of the Ouachita belt (Dickinson *et al.*, 1983, appendix). As a first approximation, the Ouachita succession (a–g) appears to reflect the initial importance of continental block provenance (a–d, with stable cratonic platform cover rocks or low-lying shield rocks far more important than uplifted basement), later replacement with collision orogen sources (e, f), and a final similarity to classical foreland basin sources (g). Subduction complexes and magmatic arcs (dissected or undissected) are never significant sources for the Ouachita detritus.

Appalachian foreland basin

Table 4 and Fig. 3 summarize the mineralogy of various sands in the Appalachian foreland basin (data from Dickinson *et al.*, 1983, appendix). These sandstones fall into two different provenance fields: (i) older, quartz-rich, feldspar- and lithic-poor sands of the Chilhowee Group (a, b, c) were derived from low-lying cratonic source areas located west and NW of the future site of the foreland basin; (ii) younger sandstones (1–10) were derived from uplifted sedimentary and metasedimentary source rocks with perhaps some very late, reactivated cratonic source rocks (9 and 10?). There is little evidence that subduction complexes and/or magmatic arcs provided much detritus to the Appalachian foreland basin. Such source varieties were presumably located east and south of the foreland basin, but were evidently shielded or 'masked' by the intervening ridges raised by Middle and Late Palaeozoic collision.

Western Wyoming foreland basin (Jackson 'exogeosyncline')

One of the best exposed and most completely preserved foreland basin sequences occurs in and around Jackson, Wyoming (Schwab, 1967, 1969a, b). This western Wyoming foreland basin straddles the former 'hinge line' that separated the western continental margin of North America from the 'Cordilleran

Table 4. Mean detrital modes for sandstones, Appalachian foreland basin*

Formation or unit	No.	Q %	F %	L %	No.	Qm %	F %	Lt %
Upper Pennsylvanian sandstones	11.	79	3	18	10.	76	3	21
Pennsylvanian sandstones, Appalachian basin	10.	92	5	3	9.	63	5	32
Pennsylvanian sandstones, central Pennsylvania	9.	61	0	39	8.	60	0	40
Pennsylvania (Pottsville) ss., south-western W.Va.	8.	95	0	5	7.	82	0	18
	7.	73	2	25	6.	61	2	37
Mississippian Mauch Chunk Fm., Pennsylvania	6.	90	4	6	5.	73	4	23
Upper Devonian units, N.Y.	5.	72	2	23	4.	40	3	57
Upper Devonian Catskill delta, N.Y.	4.	58	4	38				
Lower Palaeozoic Taconic molasse	3.	78	3	19	3.	77	3	20
Ordovician Martinsburg Fm., Va.-Pa.	2.	65	7	28	2.	59	7	34
Ordovician Tourelle Fm., Gaspe, Que.	1.	68	12	20	1.	62	12	26
Chilhowee Group, eastern Tenn.	a.	85	13	2				
Precambrian or Cambrian Antietam Fm. (Chilhowee Gp.), Va.	b.	94	6	0	b.	90	6	4
Precambrian or Cambrian Harpers Fm. (Chilhowee Gp.), Va.	c.	93	6	1	c.	88	6	6

* All data from Dickinson *et al.* (1983, appendix).

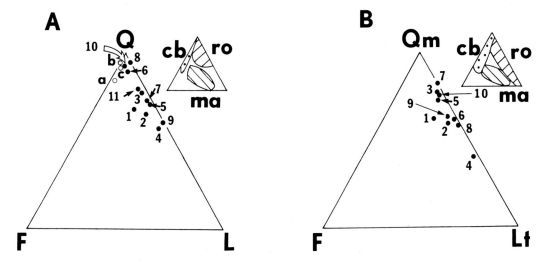

Fig. 3. Sandstone framework mineralogy, Appalachian foreland basin, and implied provenance types (also see Table 4). (A) QFL plot. (B) QmFLt plot. Number and letter keys for formations are listed in Table 4. The smaller adjacent triangles show the three main provenance area fields shown in Fig. 1: cb = continental block; ma = magmatic arc; and ro = recycled orogen.

miogeosyncline'. The 10–11 km thick composite section consists of: (1) Palaeozoic–Triassic sedimentary rocks deposited across a broad continental shelf-coastal belt, (2) a Jurassic–Cretaceous clastic wedge assemblage (the foreland basin itself) derived from rising source areas approaching from the west, and (3) intermontane basin deposits.

Table 5 and Fig. 4 summarize the details of sandstone framework mineralogy for the foreland basin assemblage. The early dominance of (easterly)

Table 5. Sandstone framework mineralogy, western Wyoming foreland basin (Jackson 'exogeosyncline') (Schwab, 1967, 1969a, b)

Stratigraphic unit (and age)	Q %	F %	L %	Qm %	F %	Lt %	Qp %	Lv %	Ls %	Qm %	P %
17. Hoback Fm. (Palaeocene–Eocene)	79·1	1·0	19·8	71·3	1·1	27·6	28	13	59	98	1
16. Mesaverde Fm. (White Ss.) (Upper Cretaceous)	66·8	0·7	32·6	60·9	0·7	38·3	16	10	74	99	0·5
15. Lenticular Ss., and Shale (Upper Cretaceous)	65·5	1·1	33·3	56·7	1·1	42·1	21	5	74	98	1
14. Coaly Sequence (U. Cretaceous)	65·5	2·6	31·9	54·8	2·6	42·5	26	0·5	74	95	4
13. Bacon Ridge Fm. (U. Cretaceous)	67·3	6·5	26·2	57·1	6·5	36·4	29	6	65	90	7
12. Frontier Fm. (U. Cretaceous)	69·7	9·2	21·1	60·8	9·2	30·0	30	12	58	87	12
11. Aspen Fm. (Lower Cretaceous)	66·5	6·7	26·8	55·5	6·7	37·7	29	9	62	89	7
10. Bear River Fm. (Lower Creta.)	87·7	0·7	11·7	73·5	0·7	25·8	56	4	40	99	1
9a. Gannet Group (Jackson, Wyo.) (Jur.)	81·8	1·4	16·8	57·6	1·4	41·1	53	1	46	98	2
9b. Gannet Group (Afton, Wyo.) (Jur.)	81·9	0·7	17·4	48·7	0·7	50·6	63	4	33	99	1
8. Tygee–Rusty Beds (Jurassic)	95·7	1·3	2·7	92·8	1·3	5·7	57	—	43	98	1
7. Cloverly–Morrison Fms. (Jur.)	92·4	5·0	2·6	88·6	5·0	6·4	60	7	33	95	4
6. Stump Fm. (Jurassic)	87·0	4·4	8·6	82·1	4·3	13·5	36	2	62	95	3
5. Preuss Fm. (Jurassic)	79·0	5·0	16·0	73·4	5·0	21·6	26	4	70	94	5
4. Nugget Sandstone (Jurassic)	96·5	3·3	0·2	92·0	3·3	4·7	95	2·5	2·5	95	3
3. Chugwater Fm. (Triassic)	94·5	4·3	1·1	91·7	4·3	3·9	70	5	25	95	3
2. Phosphoria Fm. (Permian)	90·5	—	9·5	86·8	—	12·9	45	5	50	100	—
1. Tensleep Fm. (Pennsylvanian)	96·5	1·7	1·6	90·5	1·7	7·8	77	—	23	98	1
Mean 1–4	94·5	2·3	3·1	90·3	2·3	23·5	71·8	3·1	25·1	97	2
Mean 5–10	86·5	2·6	10·8	73·8	2·6	23·5	50·1	3·1	46·7	97	3
Mean 11–17	68·6	4·0	27·4	59·6	4·0	36·4	25·6	7·9	66·5	94	5

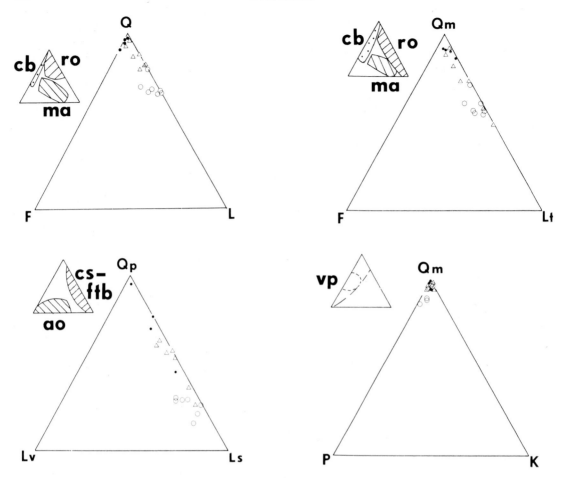

Fig. 4. Sandstone framework mineralogy and implied provenance types, Western Wyoming foreland basin (Jackson 'exogeosyncline'). Individual formations can be grouped into sequences dominated by low-lying easterly source areas (solid dots, units 1–4), mountainous westerly orogenic highland sources (open triangles, units 5–10), and mixed orogenic highland sources (open circles, units 11–17). Also see Table 5. The smaller triangles duplicate the main provenance fields shown in Fig. 1 and listed in Table 2: cb = continental block; ma = magmatic arc; ro = recycled orogen; ao = arc orogen; cs-ftb = collision suture and fold-thrustbelt sources; vp = Circum-Pacific volcanic-plutonic sources.

low-lying source rocks (mainly platform sedimentary cover) is clearly indicated (units 1–4). The development of recycled orogen sources (units 5, 9a, 9b, 10, 11, 12–17) occurs gradually upward through the succession, with some interplay between orogenic highland sources and the earlier continental block source rocks.

Noteworthy upward trends include a progressive increase in the ratio of chert to quartz (reflecting the replacement of basement and platform rocks with uplifted foldbelt-thrust belt sediments) and a progressive increase in rock fragments at the expense of quartz and feldspar (supracrustal sources replacing

basement rocks). Magmatic arc rocks were never apparently important sources of detritus, either because they were located too far to the west, or perhaps because intervening uplifted welts trapped or sedimentologically 'shielded' such sources. A complete absence of material indicative of subduction complex sources is also suggested.

Western (French–Italian) Alps

Throughout the Alps, classical late orogenic foreland basin sequences consist of thick, mainly continental molasse deposits and immediately subjacent older, deeper water flysch sequences. Dickinson (1974) and others have related this general sequence to an earlier episode of rapid depression of the foreland during the earliest stage of collision (foredeep stage), and a later increase in sediment supply relative to rate of subsidence, producing an upward change from lower marine and shoreline deposits to upper fresh-water (continental) molasse. Füchtbauer (1967) was even able to document a series of coarsening-upward megacycles related to phases of Alpine deformation and the successive unroofing of various source areas (in particular, the progressive elevation of advancing nappe fronts). This paper considers detailed aspects of the foreland assemblage in the Western (French–Italian) Alps (Schwab, 1981 and unpublished).

(1) *Total stratigraphic sequence, Western Alps (Fig. 5 (A–D) and Table 6)*

The post-orogenic foreland basin area straddles the earlier passive, rifted, progressively subsiding Atlantic-type margin developed during Late Palaeozoic to Jurassic time. Uplift and erosion of deep-seated basement crustal block units (horsts?) generated feldspar-rich sand detritus (unit 1). Subsequent erosional stripping of the sedimentary cover and low-lying basement of the European craton generated quartz-rich detritus (units 2–5) with some local unroofing perhaps of more basement (unit 6), indices of a classical continental block provenance. Westward (?) subduction resulted in collision that thrust and

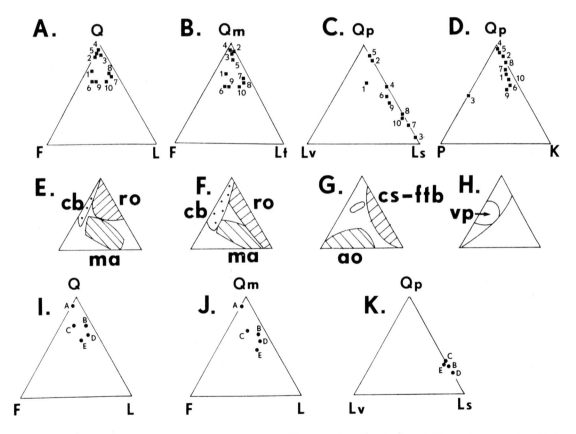

Fig. 5. Sandstone framework mineralogy and provenance, Western Alps: (A, B, C) and (D), sandstone modes, total stratigraphic sequence (also see Table 6); (E, F, G) and (H), main provenance fields shown as Fig. 1 and listed in Table 2 (cb = continental block; ma = magmatic arc; ro = recycled orogen; ao = arc orogen; cs-ftb = collision suture and fold-thrustbelt sources; vp = Circum-Pacific volcanic-plutonic sources; (I, J) and (K), sandstone modes, Western Alps foreland basin assemblages only (also see Table 7).

Table 6. Western (French–Italian) Alps: composite (Late Palaeozoic to Cenozoic) basin assemblage (Schwab, 1981 and unpublished)

Unit description	Q %	F %	L %	Qm %	F %	L %	Qp %	Lv %	Ls %	Qp %	P %	K %
10. Cretaceous–Tertiary molasse, Dauphinois Z.	63·8	13·8	22·4	55·7	13·8	30·5	27·3	1·5	71·2	65·3	6·8	27·9
9. Cretaceous–Tertiary flysch, Brianconnais, Sub-brianconnais, Valais, and Dauphinois Z.	63·7	23·3	13·1	57·8	23·3	19·0	40·9	7·3	51·8	57·0	10·0	33·0
8. Late Cretaceous flysch, Subalpine belt	70·9	6·0	23·1	62·5	6·0	31·5	27·8	—	72·2	81·6	—	18·4
7. Late Cretaceous sandstone and breccia, Valais Zone (Flysch de Tarentaise)	69·2	5·9	24·9	64·5	5·9	29·6	19·1	—	80·9	74·1	4·8	21·1
6. Late Cretaceous Helmenthoid flysch	62·5	26·1	11·3	55·8	26·1	18·0	47·6	6·6	45·9	58·5	8·3	33·2
5. Deep-water Early Cretaceous Helmenthoid flysch, Piedmont Z.	93·0	5·7	1·3	84·3	5·7	10·0	89·4	0·8	9·8	94·0	0·2	5·8
4. Deep-water Middle Jurassic Dauphinois Z. (foredeep).	95·6	3·3	1·1	94·0	3·3	2·7	58·1	—	41·9	94·6	1·1	4·2
3. Early Jurassic sandstone, Dauphinois Z.	91·7	3·3	4·9	91·6	3·3	5·0	3·0	—	97·0	47·6	52·4	—
2. Triassic terrigenous shelf sandstones	90·0	10·0	0·2	93·2	0·7	6·1	84·8	0·9	14·3	88·7	0·1	11·5
1. Late Palaeozoic molasse	75·2	19·8	5·0	69·7	19·8	10·5	60·9	18·3	20·8	71·7	5·1	23·1

Table 7. French–Italian (Western) Alps: Subalpine Foredeep and Foreland basin (Schwab, 1981 and unpublished)

Unit description	Q %	F %	L %	Qm %	F %	Lt %	Qp %	Ly %	Ls %
E. Upper Miocene molasse deposits	57·3	16·5	26·2	46·9	16·5	36·6	29·9	3·5	66·7
D. Lower Miocene molasse deposits	62·7	8·8	28·4	55·1	8·8	36·0	21·1	—	78·9
C. Oligocene molasse deposits	70·8	16·0	13·2	65·2	16·0	18·8	32·4	1·2	66·5
B. Upper Cretaceous, initial flysch deposits, Subalpine belt	70·9	6·0	23·1	62·5	6·0	31·5	27·8	—	72·2
A. Upper Cretaceous, basal Senomanian conglomerate	92·9	7·1	—	92·3	7·7	—	—	—	—

folded pre-existing sediments of the continental margin upward immediately adjacent to and actually within the foreland basin. The mineralogy of units 6–10 suggests such a classical recycled orogen source during Cretaceous and Cenozoic time. No obvious secular trends can be detected in the nature of this recycled orogen detritus, but magmatic arc and subduction complex sources are not apparent based on sandstone framework mineralogy.

(2) Foreland basin assemblage, sensu stricto (Fig. 5 (I–K) and Table 7)

If the data that document the earlier, pre-orogenic episode is deleted, the dominance of recycled orogen source areas is further underlined, but again, no clear upward systematic changes in type of orogenic

highland source varieties can be implied. Lumping of sandstone mineral modes conceivably obscures diagnostic 'souce rock indicator suites', a possibility that must be tested elsewhere, as follows.

SECULAR (VERTICAL) SHIFTS IN RECYCLED OROGEN PROVENANCE— IS THERE DETAILED EVIDENCE?

The above analyses indicate only a crude shift in the provenance of foreland basin sediments, from initial continental block sources to recycled orogen sources. No detailed unroofing of specific types of recycled orogen source rocks is obvious. How well do detailed sandstone framework mineralogy analyses document the evolution of individual recycled orogen source areas?

To answer such a question, it is necessary to focus in even greater detail on a particular foreland basin assemblage. Rapson (1965) quantitatively analysed the sedimentary petrology of the westward-thickening foreland basin clastic wedge assemblage in the Canadian Cordilleran belt. This assemblage was derived from the developing Nevadan–Laramide fold-thrustbelt located in the southern Canadian Cordillera of Alberta and British Columbia. The units which make up this clastic wedge are Jurassic and Cretaceous in age. They consist of a thick series of lithic arenites, conglomerates, and shales assigned to the Fernie Formation, Kootenay Formation and Blairmore Group.

Rapson did very detailed point counts of 46 sandstone thin sections randomly collected from the composite stratigraphic column (Table 8). She identified 12 different detrital components: monocrystalline quartz, feldspar, stable accessories like zircon and tourmaline, chert, detrital carbonate rock fragments, sandstone-siltstone clasts, phosphorite and bituminous material, shale-argillite clasts, metamorphic rock fragments, volcanic rock fragments, and miscellaneous detrital grains. Rapson was able to group these individual rock and mineral components into four different suites that she linked with four distinct source terranes:

(1) *A metamorphic suite* (essentially basement rock detritus) derived from the Selkirks metamorphic province located 100 km west of the foreland basin in British Columbia.
(2) *A stable Palaeozoic detrital suite* derived from Palaeozoic (mainly Mississippian and Pennsylvanian carbonates) sedimentary rocks that immediately overlie the Selkirks metamorphic terrane.
(3) *A pyroclastic suite* eroded from Pennsylvanian and Permian volcanic rocks (Cache Creek Volcanics) located even further west, beyond the Selkirk–Palaeozoic carbonate cover terranes that generated suites 1 and 2.
(4) *A phosphatic chert-bitumen-carbonate suite* derived from latest Palaeozoic (Permian), Triassic and Jurassic sedimentary rock units immediately adjacent to the foreland basin.

On the basis of diagnostic petrographic characteristics and independent structural and stratigraphic analyses, Rapson worked out a basin evolution model. The model required the principal source terranes

(keyed to suites 1–4 above) to be unroofed systematically as deforming cordillera emerged west of the foreland basin and eventually impinged upon it. Rapson predicted that the sedimentary sequence should be successively dominated from the base up by sandstone suite (4), then (3), then (2), and in the last stages, by sandstone suite (1), thanks to systematic source unroofing. In other words, she proposed that erosion of emerging welts immediately adjacent to the foreland basin would initially generate sands from source terrane (4) above. A westward shift of source areas inferred next would produce an influx of material derived from source terrane (3). This was interpreted to be followed by the progressive encroachment of uplifted source rocks more proximal to the foreland basin, with erosion first unroofing the late Palaeozoic carbonate cover (source terrane 2), and finally the underlying Selkirk metamorphic complex (source terrane 1).

However, Rapson unhappily conceded that this hypothesized systematic unroofing history could not be substantiated by sandstone framework mineralogy. In other words, the sedimentary sequence was not successively dominated from the base up by sandstone suites (4), then (3), then (2), and in the last stages, suite (1). Rapson explained this lack of correlation with several reasons: (i) late mixing of sand-sized detritus from a confusingly wide variety of sources; (ii) erosional removal of an underlying Triassic sequence which, if studied, would in fact reveal the unroofing history; and (iii) the overriding influence of sand grain size rather than source terrane composition on sandstone mineralogy.

Table 8 and Fig. 6 underscore this inability of sandstone petrology to definitively pinpoint an evolving provenance. Rapson's raw petrographic data are grouped into QFL, QmFLt, and QpLvLs data sets (Rapson did not discriminate plagioclase and K-feldspar because neither was abundant). Table 9 shows means calculated for the top (D), third (C), second (B), and bottom (A) quartile portions of the composite sequence. What are the salient features of these plots?

(1) All three plots show quite clearly that the southern Canadian Cordilleran foreland basin fill (like that of the Ouachitas, Appalachians, Western Wyoming, and Western Alps) was derived mainly from recycled orogen sources.
(2) Two prominent source terranes are suggested. (a) In the earlier stages, a continental block provenance (particularly the sedimentary cover, rather

Table 8. Rapson (1965). Late Jurassic–Early Cretaceous, southern Canadian Rocky Mountains (secular trends)

Sample number (top to bottom)	Q %	F %	L %	Qm %	F %	Lt %	Qp %	Lv %	Ls %
D—top quartile									
2A	80·7	1·6	17·1	72·7	1·6	25·7	31·1	15·5	53·4
2B	63·4	—	36·6	36·4	—	63·4	42·4	10·7	46·9
4	90·8	—	9·2	84·2	—	15·8	41·7	28·3	30·0
5	79·2	—	20·8	69·8	—	30·2	31·0	14·7	54·3
6B	56·9	—	43·1	16·5	—	83·5	48·4	20·6	31·1
7	83·5	—	16·5	73·4	—	26·6	38·1	6·8	55·1
8	55·9	—	44·1	36·0	—	64·0	31·2	15·1	53·7
9	77·3	0·1	22·6	49·7	0·1	50·2	54·9	1·1	44·0
10	73·6	0·3	26·2	49·4	0·2	50·3	37·8	4·9	57·3
11	81·3	1·0	17·6	63·1	1·0	35·9	46·2	9·7	44·1
12	92·3	2·2	5·5	80·3	2·2	17·5	70·2	4·3	25·5
C—third quartile									
13	92·0	4·3	3·8	84·9	4·3	11·3	50·0	23·3	26·7
14	93·6	3·2	3·2	82·8	3·2	14·0	65·3	25·0	9·7
15	88·4	1·7	9·9	77·4	1·7	20·9	52·8	13·0	34·3
16	74·5	—	25·5	24·4	—	75·5	65·3	3·2	31·5
16X	59·4	—	40·6	40·1	—	59·9	38·4	3·5	58·2
16X	44·7	—	55·3	11·2	—	88·8	53·5	3·3	43·3
17	67·3	1·1	31·7	37·4	1·1	61·6	45·9	6·9	47·3
18	77·5	—	22·5	33·8	—	66·2	66·8	3·9	29·4
18X	82·5	—	17·5	73·3	—	26·7	32·7	13·8	53·5
19	95·8	0·7	3·5	87·7	0·7	11·7	65·7	13·3	21·0
19X	70·9	—	29·1	37·7	—	62·3	63·2	8·3	38·5
B—second quartile									
20	55·4	—	44·6	21·8	—	78·2	43·0	21·0	36·0
21	99·7	0·3	—	99·1	0·3	0·6	100·0	—	—
22	69·8	—	30·2	31·5	—	68·5	55·9	12·7	31·5
23	87·2	—	12·8	72·0	—	28·0	54·3	23·7	22·0
24	70·6	—	29·4	13·2	—	86·8	66·1	3·1	30·8
25	69·6	1·2	29·2	37·5	1·2	61·4	52·4	6·3	41·3
27	100·0	—	—	93·4	—	6·6	100·0	—	—
30	65·2	0·4	34·5	34·1	0·4	65·5	40·5	5·1	54·5
31	71·2	—	28·8	32·0	—	68·0	57·7	—	42·3
32	64·1	—	35·9	32·7	—	67·3	46·1	3·5	50·4
34	58·6	—	41·4	34·1	—	65·9	37·1	6·0	56·9
A—bottom quartile									
35	51·6	—	48·4	5·2	—	94·8	49·0	2·6	48·5
37/38	85·8	—	14·2	77·2	—	22·8	37·8	3·3	58·9
39	77·7	0·7	21·5	69·9	0·7	29·3	25·3	4·8	69·9
40	75·1	—	24·9	23·3	—	76·7	67·6	0·3	32·1
41	100·0	—	—	98·8	—	1·2	100·0	—	—
42	91·8	1·6	6·7	85·6	1·6	12·9	48·0	6·0	46·0
43	90·4	2·4	7·2	88·0	2·4	9·6	25·0	25·0	50·0
45	74·5	0·4	25·2	60·2	0·4	39·5	36·3	2·0	61·8
46A	93·9	—	6·1	90·0	—	10·0	38·5	—	61·5
46B$_I$	74·1	—	25·9	62·2	—	37·8	31·4	9·1	59·5
46B$_{II}$	57·5	—	42·5	24·2	—	75·8	44·0	14·1	41·9
47a	74·7	—	25·3	62·1	—	37·9	35·6	5·9	58·4
47b	72·2	—	27·8	60·1	—	39·9	29·5	7·0	63·6

than underlying basement) predominates. (b) In the later stages, a recycled orogen provenance predominates.

(3) A magmatic arc provenance is not evident, despite the real existence of detritus that can be at least generally linked with a Pennsylvanian–Permian pyroclastic source rock terrane.

(4) Mean sandstone modes (Table 9 and plotted in

Table 9. Based on Rapson (1965). Late Jurassic–Early Cretaceous sandstone modes, southern Canadian Rocky Mountains, calculated means

	Q %	F %	L %	Qm %	F %	Lt %	Qp %	Lv %	Ls %
D. Top quartile (2A–12)	75·9	0·5	23·6	57·4	0·3	42·1	43·0	12·0	45·0
C. Third quartile (13–19X)	69·7	1·0	22·1	53·7	1·0	45·5	54·5	10·7	34·9
B. Second quartile (20–34)	73·8	0·2	26·1	45·6	0·2	54·3	59·4	7·4	33·2
A. Bottom quartile (35–47b)	78·4	0·4	21·2	62·1	0·4	37·6	43·7	6·2	50·1

Fig. 6) for the uppermost (D), third (C), second (B), and bottom (A) quartiles of the composite section show no statistically significant trends.

The inability, at least in this case, of QFL, QmFLt, and QpLvLs plots to pinpoint vertical, through time changes in provenance is enlightening, though disturbing. Obviously collective grouping of sandstone mineral modes into such 'Dickinson' ratios is not always beneficial. On the other hand, such critical ratios have proven useful in other localities for documenting source area changes through time (see, e.g. Yagishita, 1985).

BULK CHEMISTRY—FORELAND BASINS VERSUS OTHER BASIN TYPES

The bulk chemical composition of the complete, composite sedimentary assemblage in various types of basins can be estimated using representative analyses of the average chemical composition of the major sedimentary rock types, or where possible, actual published compositions of individual formations. The chemistry of different lithological components are weighted proportional to thickness

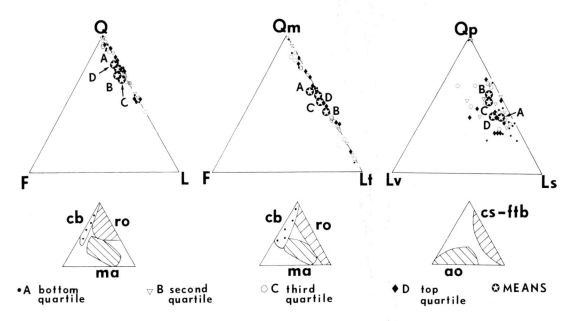

Fig. 6. QFL, QmFLt and QpLvLs plots of sandstone modal analyses of Rapson (1965). Sandstone modes are categorized by quartile but show no discernible trends through time in dominant provenance type. See also Tables 8 and 9. Smaller triangles show main provenance fields of Table 2 and Fig. 1: cb = continental block; ma = magmatic arc; ro = recycled orogen; ao = arc orogen; cs-ftb = collision suture and fold-thrustbelt sources.

Table 10. Bulk chemical composition of various sedimentary rock assemblages

	1 Foreland basin, Wyoming[1] %	2 Sediment cover, all platforms[2] %	3 Sediment fill geo-synclines[2] %	4 Continental shelf and rise[2] %	5 Abyssal ocean floor[2] %	6 Ancient miogeoclines[3] %	7 Ancient eugeoclines[3] %	8 Aulacogens, rift valleys %	9 Forearc basin (Franciscan[4]) %	10 Abyssal plain pelagic red clay[5] %
SiO_2	61.0	48.0	50.8	50.2	36.8	41.3	65.1	71.9	69.3	60.9
TiO_2	0.4	0.7	0.7	0.7	0.5	0.3	0.5	0.5	0.5	0.9
Al_2O_3	11.1	10.5	13.4	12.7	9.0	6.0	13.0	13.3	14.6	18.4
Fe_2O_3	3.6	2.7	2.6	2.7	4.7	1.7	2.1	2.2	1.7	8.6
FeO	1.8	2.2	3.2	2.9	1.0	1.0	3.1	2.4	3.8	
MnO	—	0.1	0.1	—	0.9	0.2	0.1	0.2	0.1	2.2
MgO	3.0	4.1	3.2	3.5	2.5	7.2	2.8	1.6	2.8	3.8
CaO	8.7	15.1	13.2	13.8	23.7	18.7	5.2	2.7	2.2	0.8
Na_2O	1.1	1.4	1.7	1.6	2.6	0.6	2.3	1.7	3.6	1.4
K_2O	2.1	2.4	2.0	2.1	1.4	1.2	2.3	3.7	1.7	3.0
CO_2	7.4	12.9	9.0	10.1	16.6	21.3	3.4	0.5	—	—
Total (%)	100.0	100.1	99.9	100.3	99.7	99.5	99.9	100.7	100.3	101.6

[1] Schwab (1969). [2] Ronov (1983). [3] Schwab (1971). [4] Bailey, Irwin & Jones (1964). [5] Wedepohl (1971).

Table 11. Bulk composite chemistry of various sedimentary rock assemblages (modified from Table 10)

	1 Foreland basin, Wyoming %	2 Platform cover %	3 Geosynclinal fill %	4 Shelf-rise sediments %	5 Abyssal plain cover %	6 Ancient miogeoclines %	7 Ancient eugeoclines %	8 Aulacogens rift valleys %	9 Forearc basin %	10 Pelagic red clay %
SiO_2	66.9	53.0	56.7	55.6	41.5	43.7	72.7	80.8	77.3	74.7
Al_2O_3	12.2	11.6	15.0	14.1	10.2	6.3	14.5	14.9	16.3	22.6
$CaO+MgO+CO_2$	20.9	35.4	28.3	30.3	48.3	49.9	12.7	4.3	6.5	2.7
Total (%)	100.0	100.0	100.0	100.0	100.0	99.9	99.9	100.0	100.1	100.0

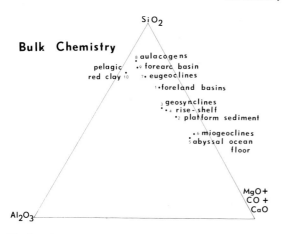

Bulk Chemistry

Fig. 7. Bulk chemistry of foreland basin fill versus other basin types (see also Tables 10 and 11).

percentage in the total stratigraphic column (see Schwab, 1969a, 1971).

The fill of a typical foreland basin (represented by the Jackson, Wyoming succession) most closely resembles the average fill of ancient eugeoclines (eugeosynclines) and the bulk composition for geosynclines estimated by Ronov (1983) (Table 10). Modern continental shelf-slope-rise prisms and the adjacent but thinner sedimentary veneer of modern abyssal plain regions are considerably richer in carbonate and depleted in SiO_2. Deposition of thick, carbonate-rich sediment on the abyssal ocean floor is probably exclusively a post-Early Jurassic phenomenon, dependent on the Jurassic evolution of floating, carbonate-secreting pelagic organisms.

Table 11 and Fig. 7 show the bulk chemistry of the various basin assemblages recalculated in terms of SiO_2, Al_2O_3 (mainly mudrock), and $CaO + MgO +$ CO_2 (mainly carbonate). Contrasts and similarities mainly arise from the secular variation in the locus of carbonate sedimentation.

COMPARATIVE SEDIMENT ACCUMULATION RATES

Table 12 and Fig. 8 summarize the mean accumulation rates (in m/1000 yr) for the sedimentary successions in a variety of ancient and modern sedimentary basins (based largely on Schwab, 1976). The accumulation rate for individual basin assemblages was calculated using (1) the published overall thickness (in metres) of the 'average' or 'typical' sedimentary section as determined directly by field measurement (or in a few cases, generally estimated), and (2) the estimated interval of time (in millions of years) during which an individual basin persisted as both a geographical and tectonic entity.

Cratonic basins (e.g. the Michigan and Williston basins) and normal cratonic shelf areas (whether continental or on the ocean floor) show the lowest accumulation rates (1). Ancient miogeoclines (2) and their modern analogue, continental terraces (3) together with eugeoclines (4) and modern continental rise-slope areas (5) show similar, but somewhat higher (by a factor of 3–4) rates of accumulation. Foreland basins (6) and the 'average exogeosyncline' (7) have sediment accumulation rates 5–30 times higher than basin types 1–5! A random sampling of separate flysch (8) and molasse (9) sediment accumulation rates selected from category (6) shows surprising results. Trenches (10) and ancient aulacogen-modern rift valleys (12) exhibit high sediment accumulation rates like foreland basins. Successor basins (11), essentially

Table 12. Accumulation rates for selected sedimentary basin types (Schwab, 1976)

Basin type description	Data base no.	Mean accumulation rate (m/1000 yr)	Range (m/1000 yr) Low	High
1. Cratonic basins (abyssal plain areas, continental cratonic basins like the Michigan basin, etc.)	11	0·0066	0·001	0·024
2. Ancient miogeoclines (continental shelves)	8	0·022	0·012	0·031
3. Modern continental terraces	7	0·036	0·029	0·057
4. Ancient eugeoclines (continental rises)	5	0·034	0·029	0·037
5. Modern slope-rise assemblages	3	0·028	0·015	0·044
6. Various composite foreland basins	6	0·186	0·044	0·927
7. Typical 'exogeosyncline'	1	0·180		
8. Flysch deposits in foreland basins	4	0·345	0·15	0·927
9. Molasse deposits in foreland basins	3	0·20	0·10	0·40
10. Trench fill	6	0·17	0·001	0·50
11. Successor basins	9	0·113	0·077	0·20
12. Rift valleys and aulacogens	11	0·169	0·071	0·40

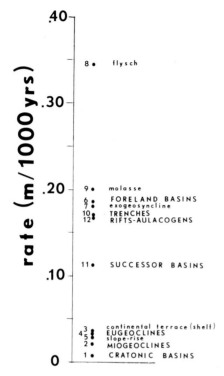

Fig. 8. Mean accumulation rates for foreland basins and other major basin types (see also Table 12).

late orogenic intermontane basins like the Uinta, Big Horn and Powder River basins, accumulate sediment at semi-fast rates (higher than 1–5; lower than 6–10; 12).

ACKNOWLEDGMENTS

I thank Peter Homewood, Philip Allen and Graham Williams for providing a forum to which the above synthesis might contribute. Gian G. Zuffa's comments helped to improve the manuscript.

REFERENCES

BAILEY, E.H., IRWIN, W.P. & JONES, D.L. (1964) Franciscan and related rocks, and their significance in the geology of western California. *Bull. Calif. Div. Mines Geol.* **183**, 177 pp.

DICKINSON, W.R. (1984) Interpreting provenance relations from detrital modes of sandstones. In: *Provenance of Arenites* (Ed. by G. G. Zuffa), pp. 333–361. Reidel, Dordrecht.

DICKINSON, W.R., BEARD, L.S., BRAKENRIDGE, G.R., ERJA-VEC, J.L., FERGUSON, R.C., INMAN, K.F., KNEPP, R.A., LINBERG, F.A. & RYBERG, P.T. (1983) Provenance of North American Phanerozoic sandstones in relation to tectonic setting. *Bull. geol. Soc. Am.* **94**, 222–235.

DICKINSON, W.R. & SUCZEK, C.A. (1979) Plate tectonics and sandstone compositions. *Bull. Am. Ass. Petrol. Geol.* **63**, 2164–2182.

DICKINSON, W.R. & VALLONI, R. (1980) Plate settings and provenance of sands in modern oceans. *Geology*, **8**, 82–86.

FUCHTBAUER, H. (1967) Die sandsteine in der Molasse nordlich der Alpen. *Geol. Rdsch.* **56**, 266–300.

RAPSON, J.E. (1965) Petrography and derivation of Jurassic-Cretaceous clastic rocks, southern Rocky Mountains, Canada. *Bull. Am. Ass. Petrol. Geol.* **49**, 1426–1452.

RONOV, A.B. (1983) The Earth's sedimentary shell: quantitative patterns of its structure, compositions, and evolution. *Am. Geol. Inst. Reprint Series*, **V**, 80 pp.

SCHWAB, F.L. (1967) *An analysis of Phanerozoic sedimentation through time, western Wyoming.* Unpublished Ph.D. Thesis, Harvard University, 341 pp.

SCHWAB, F.L. (1969a) Geosynclinal compositions: what contribution to the crust? *J. sedim. Petrol.* **39**, 150–158.

SCHWAB, F.L. (1969b) Cyclical geosynclinal sedimentation: a petrographic examination. *J. sedim. Petrol.* **39**, 1325–1343.

SCHWAB, F.L. (1971) Geosynclinal compositions and the New Global Tectonics. *J. sedim. Petrol.* **41**, 928–938.

SCHWAB, F.L. (1976) Modern and ancient sedimentary basins: comparative accumulation rates. *Geology*, **4**, 723–727.

SCHWAB, F.L. (1981) Evolution of the western continental margin, French-Italian Alps: sandstone mineralogy as an index of plate tectonic setting. *J. Geol.* **89**, 349–368.

WEDEPOHL, K.H. (1971) *Geochemistry.* Holt, Rhinehart & Winston, New York, 231 pp.

YAGISHITA, K. (1985) Evolution of a provenance as revealed by petrographic analyses of Cretaceous formations in the Queen Charlotte Islands, British Columbia, Canada. *Sedimentology*, **32**, 671–684.

Spec. Publs int. Ass. Sediment. (1986) **8**, 411–423

Compositional trends within a clastic wedge adjacent to a fold-thrust belt: Indianola Group, central Utah, U.S.A.

TIMOTHY F. LAWTON

Department of Earth Sciences, New Mexico State University, Las Cruces, New Mexico 88003, U.S.A.

ABSTRACT

The Indianola Group, a coarse-grained clastic wedge at the western margin of the Cretaceous Cordilleran foreland basin in central Utah, was derived from the unroofing of the Sevier orogenic belt lying to the west. Compositional trends in Indianola Group conglomerates and sandstones confirm that the clastic detritus was eroded from an uplifted miogeoclinal and cratonal section of Precambrian to Jurassic age. The miogeoclinal prism consists of Precambrian to Cambrian quartzite and argillite units and middle to late Palaeozoic carbonate strata, while the Mesozoic cratonal section is dominated by sandstone. Thrusting and uplift of the section resulted in deposition of carbonate-rich detritus eroded from the Palaeozoic section in the lower part of the Indianola Group and quartzose detritus from the Precambrian and Cambrian section in the upper part. The upsection enrichment in quartz is reflected in both conglomerate-clast populations and detrital modes of sandstones. Chert grains are an important derivative of the carbonate provenance and provide durable evidence of a carbonate source even in rocks lacking detrital carbonate grains.

The combination of compositional trends and Indianola depositional patterns suggest that influxes of contrasting detritus may be tied to major ramp uplift on two thrust systems. Deposition of the initial carbonate-rich wedge occurred during ramping and uplift of the Canyon Range thrust in late Albian time. Deposition of alluvial-fan deposits in the overlying quartzose wedge resulted from uplift during ramping of the Pavant thrust. Almost all of the Indianola detritus, however, was derived from the Canyon Range plate, first during uplift above the active Canyon Range thrust and second as the plate rose passively above the younger Pavant system in late Santonian to Campanian time. Frontal structures developed during late Campanian thrusting folded the Indianola Group and terminated subsidence along the basin margin.

INTRODUCTION

The Cretaceous Indianola Group forms the proximal part of a synorogenic clastic wedge within the central Utah part of the Cordilleran foreland basin. The presence of thick sequences of conglomerate in the Indianola Group led Spieker (1946) to postulate the presence of an orogenic terrane to the west. Documented ages within the section range from late Albian to late Campanian and indicate that deposition of the Indianola Group occurred during emplacement of thrust allochthons in the Sevier orogenic belt (Armstrong, 1968a). An upsection decrease in the relative abundance of limestone cobbles in Indianola conglomerates of the Gunnison Plateau (Spieker, 1949) was interpreted by Armstrong (1968a) to

represent an unroofing sequence created as upper Palaeozoic carbonate clasts were initially stripped from uplifted thrust plates before lower Palaeozoic and upper Precambrian quartzites could be eroded. Similar inverted stratigraphies have been reported from the northern Utah–Wyoming–Idaho thrust belt (Armstrong & Oriel, 1965; Royse, Warner & Reese, 1975) and the Alberta foreland basin (Price & Mountjoy, 1970).

This paper presents compositional data from Indianola conglomerates and sandstones which document a secular trend in detrital modes within the stratigraphic section. The compositional trend is interpreted to have resulted from the unroofing of a single thrust plate in two episodes. The first period of erosion occurred as the thrust plate initially ramped upward from deeper structural levels, while the second episode occurred when the allochthon rose during ramping of a younger, structurally deeper thrust.

Geological setting

Indianola Group rocks crop out in a region which lies immediately east of ramp-style thrust faults but west of undeformed strata within the foreland basin (Fig. 1). Outcrops generally have homoclinal moderate to steep dips both to the east and west, although most major panels dip east. This deformation resulted from uplift of the section over the easternmost interpreted thrust ramp immediately west of the Gunnison Plateau, as well as from folding and faulting above a

triangle zone west of the line indicating easternmost thrust deformation in Fig. 1 (Lawton, 1985). At the westernmost exposure of the Indianola Group, the Canyon Range thrust plate overlies a thick conglomeratic section called the Canyon Range fanglomerate by Armstrong (1968a).

Ramp-style thrust faults which lie to the west of Indianola outcrops occur in a Precambrian and Palaeozoic miogeoclinal section and a Triassic through Jurassic cratonal section. The miogeoclinal section west of the Gunnison Plateau consists of Precambrian and lower Cambrian quartzite and argillite overlain by a Cambrian to Mississippian carbonate-dominated section (Armstrong, 1968b). A thin Pennsylvanian to Permian section thickens dramatically northward within the area of Fig. 1 and in the Wasatch Mountains consists of 7925 m of

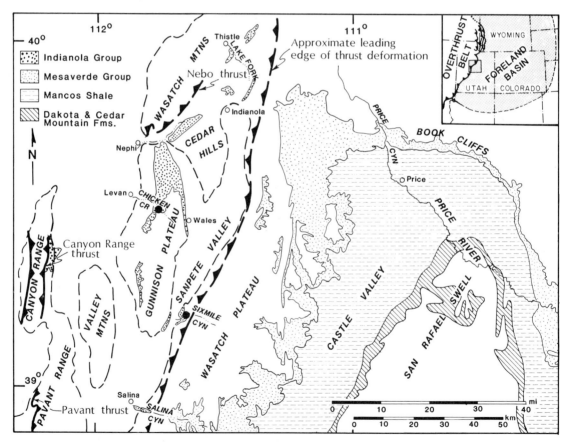

Fig. 1. Map of study area in central Utah and geographical names used in this report. Indianola outcrops occur within the Sevier orogenic belt, the eastern extent of which is indicated by the dashed line. Other units indicated, the Cedar Mountain and Dakota formations, the Mancos Shale, and the Mesaverde Group, are distal equivalents of the Indianola Group in the foreland basin. Solid dots indicate locations of sections studied for this report.

quartzite and carbonate strata in the upper plate of the Nebo thrust (Baker, 1947). The Mesozoic section consists of sandstone, shale and subordinate limestone.

Two thrust plates, the Canyon Range and Pavant allochthons, have been identified west of the Gunnison Plateau (Burchfiel & Hickcox, 1972). The structurally higher Canyon Range thrust emplaced Precambrian quartzite and argillite on Cambrian to Devonian carbonate strata. The lower Pavant thrust emplaced the entire Palaeozoic section on strata as young as Jurassic. The Canyon Range plate has been extensively stripped away by erosion and forms a large klippe of Precambrian strata resting on the Pavant plate (Christiansen, 1952).

Indianola Group stratigraphy

Upper Cretaceous rocks which crop out in the thrust belt region of central Utah west of the Wasatch Plateau (Fig. 1) were assigned to the Indianola Group by Spieker (1946). The most complete Indianola section in the Cedar Hills is approximately 4000 m thick (Jefferson, 1982). The equivalent but thinner section within the little-deformed foreland basin to the east includes the Cedar Mountain and Dakota Formations, the Mancos Shale, and the Mesaverde Group. This distal sequence is dominated by marine shale and interbedded delta-front sandstones, but contains fluvial sandstone and siltstone in the Cedar Mountain Formation and upper part of the Mesaverde Group (Fisher, Erdman & Reeside, 1960).

The Indianola Group was divided into four formations by Spieker (1946), with lithotypes in the Sixmile Canyon area. These formations in ascending order are the Sanpete Formation, Allen Valley Shale, Funk Valley Formation, and Sixmile Canyon Formation (Fig. 2). The lower three units were provisionally recognized in other parts of the Wasatch Plateau by Spieker (1946). Correlation of formations within the Indianola Group in the Gunnison Plateau and Cedar Hills using new fossil data and physical stratigraphy has subsequently confirmed Spieker's stratigraphy (Jefferson, 1982; Lawton, 1982).

Deposition of the Indianola Group ranged from late Albian to late Campanian (Fig. 2). Fluvial rocks near the base of the unit are difficult to date, but palynomorphs collected 600 m above the base of the Indianola Group on the Gunnison Plateau are late Albian (Aspen Shale equivalent) in age (Standlee, 1982). Cenomanian fission-track ages (96.2 ± 5.0 Ma; 90.6 ± 4.8 Ma; 90.3 ± 4.8 Ma; Willis, 1986) on zircon grains have been acquired from claystone beds in a conglomeratic section overlying known Middle Jurassic (Callovian) strata in Salina Canyon. These dates appear to corroborate the palynomorph data. The marine upper part of the Sanpete Formation, the Allen Valley Shale and the Funk Valley Formation

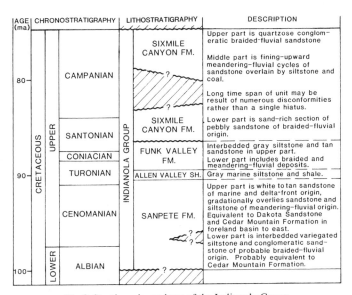

Fig. 2. Stratigraphy and age of the Indianola Group.

are well dated by marine fossils (Fouch *et al.*, 1983). Palynomorphs collected from the upper part of the fluvial Sixmile Canyon Formation have yielded a late Campanian age (Fouch *et al.*, 1983).

The recent age determinations from fluvial rocks of probable Indianola affinity but very low in the section have raised a nomenclature problem. These ages indicate the presence of approximately a 60 Myr disconformable hiatus between Middle Jurassic strata and Cretaceous strata. However, Spieker (1946) tentatively assigned these Albian to Cenomanian beds to the Morrison(?) Formation of Late Jurassic age based on stratigraphic position and lithologic similarity to Morrison beds elsewhere. These beds have been more recently called Cedar Mountain Formation after the time-equivalent unit on the Colorado Plateau (Standlee, 1982; Lawton & Willis, 1986), but this practice mixes nomenclature from the thrust belt and foreland basin. In this paper, I provisionally include the Albian strata in the Sanpete Formation pending further studies and consider them to represent the initial molasse deposit of the foreland basin.

Depositional trends

The Indianola Group generally comprises an upward-coarsening megasequence at all localities where significant sections are exposed. The trend appears to have resulted from an interaction of eustatic and tectonic factors which affected the western margin of the foreland basin. The following overview of Indianola depositional environments summarizes more complete descriptions and interpretations from previously published work (Lawton, 1982).

The lower half of the Indianola section, encompassing the Sanpete Formation, Allen Valley Shale and Funk Valley Formation, consists of interfingering deposits of non-marine and marine origin. The lower part of the Sanpete Formation is characterized by a 2300 m thick section of conglomeratic sandstone and conglomerate beds 5–25 m thick, interbedded with thick variegated shale and siltstone beds. The cross-bedded conglomerate and sandstone beds are composed of individual and stacked channelform units. Each channel unit fines upward and is truncated by the overlying channel or grades to siltstone. At localities in the Wasatch Plateau these channelform beds of fluvial origin decrease in average grain size upsection. In addition, shale content of the section increases. Coal is present in some shale sequences and oyster-bearing lenticular beds of burrowed sandstone underlie fossiliferous upward-coarsening shoreface

sequences at the top of the Sanpete Formation. The overlying Allen Valley shale consists of 150–200 m of grey thin-bedded siltstone, mudstone and sandstone. These prodelta or marine-shelf deposits represent a maximum transgression of marine conditions in mid-Turonian time.

The thickness and environmental diversity of the Sanpete Formation varies considerably along strike. In the northern part of Fig. 1 along Lake Fork and in the Cedar Hills, the Sanpete Formation consists of fluvial conglomerate and sandstone overlain by a few metres of transgressive marine sandstone, in contrast to the thick deltaic and marginal marine deposits described above and found in Salina and Sixmile Canyons. This variability is interpreted to be a result of valley backfilling during transgression. Weimer (1984) has described similar fluvial and deltaic rocks of late Albian age deposited in river valleys that were backfilled as transgression proceeded. Although located 500 km to the NE in the foreland basin, the valley-fill deposits described by Weimer appear to match the sequence of the lower part of the Indianola section, which was deposited at the culmination of the same transgression. Thus, although the Indianola sediments were deposited within 80 km of the thrust front, sea-level change appears to have had a strong effect on patterns of deposition.

The overlying Funk Valley Formation is characterized by a lower 350 m sequence of sandstone and siltstone deposited in shoreface, delta-plain, and fluvial environments and an upper 600 m marine sequence of prograding shoreface deposits. In northern locations, cobble conglomerates were deposited above strata of possible lagoonal origin in the lower part of the Sanpete Formation suggesting the possibility of fan-delta development east of the thrust front.

The upper half of the Indianola section in Sixmile Canyon, included in the Sixmile Canyon Formation, is composed of 1350 m of sandstone, pebbly sandstone, and conglomerate deposited in meandering-fluvial and braided-fluvial environments. In more western outcrops, strata equivalent to the Sixmile Canyon Formation are composed of cobble and boulder conglomerates deposited in an alluvial-fan environment. The Sixmile Canyon Formation, while containing significant grain-size and environmental variability, represents an overall coarsening of grain size within the section as depositional environments proximal to the orogenic terrane migrated eastward ahead of the thrust front. Thus, orogenic processes appear to have dominated sedimentation within the upper part of the Indianola section.

PETROGRAPHY OF THE INDIANOLA GROUP

Methods

Compositional data were collected from two Indianola sections, one at Chicken Creek in the Gunnison Plateau and the other at Sixmile Canyon in the Wasatch Plateau (Fig. 1). The Chicken Creek section is dominated by conglomerate, while the Sixmile Canyon section, located farther from the major thrust plates, is dominated by sandstone. Clast data were obtained by counting all pebble and cobble-sized clasts within a delineated rectangle on conglomeratic outcrop faces and recording each lithic type, usually to yield a minimum population of 300 clasts. The data thus obtained are frequency, rather than volumetric, counts. Sandstone samples were thin sectioned, and a minimum of 400 framework grains counted for each sample on a 0.6×0.6 mm grid. A total count of 400 modal points ensures a two-sigma confidence range of 5% or less of the whole rock for any calculated modal percentage (Van der Plas & Tobi, 1965). Standard QtFL (Dickinson, 1970) and QmFLt (Graham, Ingersoll & Dickinson, 1976) triangular diagrams were plotted to illustrate compositional characteristics of Indianola Group sandstones. Data from both thin-section and conglomerate counts were plotted stratigraphically to determine compositional trends within the stratigraphic section.

Definition of grain types

Clast and grain types for conglomerates and sandstones were defined operationally at the outset of the study to insure counting consistency. Brief descriptions of the lithologic groupings are outlined in Table 1 for conglomerates and Table 2 for sandstones. Clast-count data and point-count results are shown in Tables 3 and 4, respectively.

Conglomerate clast types

The following clast types were distinguished during clast counts at conglomeratic outcrops:

(1) Quartzite. Quartzite clasts are compositional quartz arenites with quartz cement. Plain and banded, red, pink and purple quartzites were discriminated from white, tan and grey quartzites because the red, pink and purple clasts may be tied to a specific source

Table 1. Lithic types discriminated in clast counts of Indianola Group conglomerates, and calculated conglomerate parameters plotted in Fig. 5

(a) Clast types
  (1) Quartzite (Qz)
    (a) Red, pink and purple quartzite (Qzrp)
    (b) Quartzite of other colours (Qzo)
  (2) Carbonate (CO_3)
    (a) Limestone (Ls)
    (b) Dolomite (Dol)
  (3) Chert (Ch)
  (4) Other clastic
    (a) Sandstone (Ss)
    (b) Mudstone, siltstone (Ms)
  (5) Miscellaneous
    (a) Vein quartz
    (b) Silicified bone and plant fragments
(b) Calculated conglomerate parameters
  (1) Qz = Qzrp + Qzo = total percentage quartzite
  (2) CC = CO_3 + Ch = percentage (carbonate + chert)

Table 2. Sandstone grain categories used in calculating QtFL and QmFLt plots

(1) Qt: Total framework quartz (Qt = Qm + Qp)
    (a) Qm: Monocrystalline quartz
    (b) Qp: Polycrystalline quartz
      (1) Chert
      (2) Polycrystalline quartz of sedimentary, igneous, metamorphic origin
      (3) Aggregate quartz of indeterminate origin
(2) F: Total framework feldspar (F = K + P)
    (a) K: Potassium feldspar
    (b) P: Plagioclase feldspar
(3) L: Framework lithic fragments (for QtFL plot; L = Ls + Lv)
    (a) Ls: Sedimentary lithic fragments
      (1) Argillite—shale
      (2) Very fine grained feldspathic sandstone
      (3) Detrital carbonate
    (b) Lv: Volcanic and hypabyssal lithic fragments
(4) Lt: Total framework lithic fragments (for QmFLt plot: Lt = L + Qp)

lithology, the Upper Precambrian Mutual Formation presently exposed in the upper plate of the Canyon Range thrust to the west (Christie-Blick, 1982; Sprinkel & Baer, 1982).

(2) Carbonate. Carbonate clasts consist of a wide range of colours and textures. They are most commonly grey and tan, and range from micritic limestone to coarse saccharoidal dolomite. Limestone and dolomite were discriminated during counts using a stain prepared with alizarin red-S solution (Dickson, 1966). The carbonate category includes clasts partly replaced by chert (Fig. 3).

Table 3. Conglomerate clast data, Chicken Creek, Gunnison Plateau, with counts listed in stratigraphic order above the base of the Indianola Group shown in Fig. 5. Clast size is listed in centimetres for maximum and average long-axis measurements. Clast type fractions are listed as percentages of the total count (*n*). Grain parameters are as listed in Table 1

Interval (m)	Clast size (cm) Max	Average	Quartzite Qzrp	Qzo	Carbonate Ls	Dol	Chert	Other clastic Ss	Ms	Misc	*n*
2220	40	3·0–4·0	12·1	79·6	0	0	1·5	6·8	0	0	264
2010	15	1·5–2·0	8·6	70·7	0	0	7·7	11·9	0·8	0·3	362
1775	30	4·0–6·0	10·6	72·1	0	0	5·0	10·6	1·8	0	283
1655	100	2·5–4·0	12·3	72·2	0	0	1·4	11·6	2·5	0	277
1595	15	2·0–3·0	5·2	37·2	33·6	3·3	8·8	10·0	1·8	0	330
1380	10	1·0	12·6	43·5	21·4	12·2	0·7	2·2	7·4	0	271
1380	10	1·5–2·0	9·5	53·1	18·7	8·2	3·9	6·2	0·3	0	305
1320	5	1·0	0	5·4	0·5	89·7	0·9	3·6	0	0	223
1275	10	1·5–3·0	4·7	20·4	60·4	10·2	1·8	1·8	0·7	0	275
1080	10	0·8–1·0	0·3	19·9	43·4	8·3	16·2	1·8	10·1	0	327
1005	26	8·0	5·1	66·7	0	1·1	3·7	11·4	11·7	0·3	351
730	45	5·0	1·7	16·4	32·8	45·6	2·6	0·6	0·3	0	353
610	60	5·0	1·2	7·6	23·9	62·4	2·3	1·8	0·6	0·3	343
525	60	2·5–4·0	1·7	5·4	18·6	72·3	1·0	1·0	0	0	296
370	27	n.d.	3·9	18·9	12·8	60·2	0·9	1·6	1·6	0	312
315	12	n.d.	4·8	3·3	2·7	86·9	2·1	0·3	0	0	335
180	15	n.d.	3·0	51·2	35·6	2·6	1·7	5·6	0	0·3	303
120	n.d.	n.d.	2·4	37·3	39·7	15·5	2·1	0·9	0	2·1	330
75	25	n.d.	0·3	41·3	41·7	9·7	1·3	1·3	0	4·3	300

Table 4. Mean modal compositions of Indianola Group sandstones. Numbers in parentheses are standard deviations

Unit	*n*	Qt	F	L	Qm	Lt	K	P	Qp	Ls	Lv	Detrital Co₃
Indianola Group	8	74·7	0·3	25·0	71·5	28·2	0·3	0·1	3·2	25·0	0	20·7
Undifferentiated		(19·4)	(0·3)	(19·6)	(19·8)	(20·0)	(0·3)	(0·1)	(1·5)	(19·6)		(22·9)
Sanpete Formation	5	77·1	2·8	20·2	75·5	21·7	2·7	0·1	1·5	20·1	0	17·4
		(10·3)	(1·6)	(11·8)	(9·2)	(10·7)	(1·5)	(0·2)	(1·4)	(11·8)		(11·5)
Funk Valley	10	71·3	3·4	25·3	69·0	27·6	3·4	0	2·3	25·2	0	24·7
Formation		(8·4)	(3·7)	(11·1)	(8·7)	(10·3)	(3·7)		(2·0)	(11·1)		(11·5)
Sixmile Canyon	9	84·9	0·1	15·0	80·3	19·6	0·1	0	4·6	15·0	0	13·9
Formation		(15·7)	(0·2)	(15·6)	(14·2)	(14·1)	(0·2)		(4·7)	(15·6)		(16·5)

(3) Chert. This category includes cryptocrystalline siliceous pebbles that are tan, grey, white, red, and rarely, black. The chert occasionally contains thin laminations or silicified invertebrate fossils, indicating an origin by replacement of limestone or dolomite rather than deep-basin deposition.

(4) Other clastic. The clastic category includes a wide range of detrital rock types including red to tan feldspathic and sublitharenitic sandstone and white, brown, and grey mudstone and siltstone.

Miscellaneous categories include white vein or bull quartz, silicified wood fragments, and bone material.

Sandstone grain types

Grain parameters discriminated in point counts are summarized in Table 2. Lithic grain types are in general defined following the descriptions of Graham *et al.* (1976). However, there are some differences in the classification used here; hence, a brief description of the dominant grain types follows:

(1) Argillite-shale: murky, fine-grained siliceous or argillaceous fragments, many containing silt-sized quartz and feldspar grains.

(2) Chert: microcrystalline aggregates of equant silica grains, with most domains less than 0·03 mm.

(3) Detrital carbonate: limeclasts of variable texture, ranging from micrite through mosaic microspar and pseudospar to coarse-grained or monocrystalline spar.

(4) Volcanic-hypabyssal: fine-grained felsitic fragments with aphanitic to microporphyritic or mosaic textures and rare microlitic and flow-banded siliceous grains.

(5) Aggregate quartz: fine-grained polycrystalline

Fig. 3. Laminated dolomite clast partly replaced by chert, Indianola Group, Chicken Creek. Marks on staff are 10 cm long.

quartz, including chalcedony, probably chiefly of vein origin.

(6) Feldspathic sandstone: detrital quartzofeldspathic aggregates of silt to very fine sand grains.

Discrimination of the above lithic types is sometimes difficult and in some rocks occasionally subjective. Extraformational argillite grains cannot always be discriminated texturally from intraformational mudstone clasts, which rarely exceed 1% in the samples counted and must be recognized by their anomalously large grain sizes. Fine-grained argillites and chert compose end members of a range of siliceous mudstones with variable amounts of included argillaceous and opaque material. Most, if not all, chert originated by diagenetic replacement of limestone and dolomite prior to erosion of the clastic grains. The replacement of micrite frequently resulted in the inclusion in impure chert grains of very fine detrital material, including occasional silt grains, generally quartz but rarely feldspar, and aphanitic material that imparts a dark grey smoky texture to the chert. Moreover, silicified invertebrate fossils are common in chert grains. In general, grains containing more

than 10% argillaceous or opaque inclusions were counted as argillite, although pebbles of the same material in conglomerates would undoubtedly be counted as chert. In general, high confidence is attributed to the relative proportions of the major grain types, (Qm, Qp, F, Ls, and Lv) the subcategories of which are listed in Table 2. The potentially greatest source of inconsistency rests in the discrimination of chert and argillite, which may affect the relative proportions of Qp and Ls, respectively.

Detrital mineralogy of Indianola Group

Indianola sandstones are compositional litharenites, sublitharenites, and quartz arenites, using the classification system and nomenclature proposed by McBride (1963). QtFL and QmFLt compositions are plotted in Fig. 4. The dominant lithic grain type in Indianola sandstones is detrital carbonate, which ranges in abundance from 0 to 52%, with a mean of 19·5% (Table 4). Argillite-shale grains are next in abundance, ranging from 0 to 16%, with a mean of 2·5%. Chert grains range from 0 to 13%, with a mean of 2%.

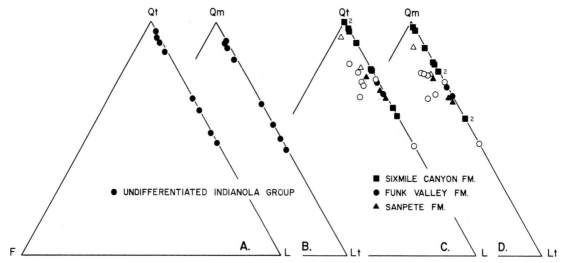

Fig. 4. QtFL and QmFLt plots for Indianola Group sandstones from Chicken Creek (triangles A, B) and Sixmile Canyon (triangles C, D). Open symbols indicate samples from marine facies. Numerals adjacent to symbols indicate multiple coincident observations.

Diagenetic effects

The major diagenetic effect noted in Indianola sandstones is development of sparry calcite cement and the recrystallization of detrital carbonate grains to form coarse mosaic spar. Consequently, detrital carbonate grains often occur as relicts, recognizable by the presence of impurities in tracts of sparry calcite. In sandstone samples that do not contain detrital carbonate grains, calcite cement is not present. The observed relationships suggest that calcite cement in the Indianola sandstones was formed in part from dissolution of detrital carbonate grains. In addition, sandstones of the Indianola section at Chicken Creek contain an average of 5·4% less total carbonate (cement plus grains) than the conglomerates, suggesting that loss of carbonate occurs during grain-size reduction and diagenesis. Thin rims of haematite cement often occur on detrital carbonate grains, but are poorly developed to absent on siliciclastic grains.

Compositional trends

For comparison of conglomerate and sandstone compositional trends through the Indianola section, frequency data are plotted with respect to stratigraphic position at both Chicken Creek (Fig. 5) and Sixmile Canyon (Fig. 6). Parameters plotted for conglomerates include pink and purple quartzite (Qzrp), total

quartzite (Qz), and carbonate plus chert (CC). Sandstone grain parameters plotted are Qm (in place of Qz), carbonate plus chert (CC), and total feldspar (F) at Sixmile Canyon only. The textural evidence in both conglomerate clasts and sandstone grains for formation of chert through carbonate silicification indicates that a single compositional parameter of combined chert and carbonate should be a sensitive indicator of the Palaeozoic carbonate sequence in the thrust terrane. The combined parameter has the additional advantage in the sandstone analyses of recording the presence of mechanically and chemically durable chert grains even when carbonate grains were lost through transport and dissolution. Thus, the carbonate provenance may be interpreted even in the absence of detrital carbonate grains.

Least squares curves fitted to the data indicate important compositional trends in both conglomerates and sandstones for several parameters:

(1) Quartzite (Qz) and monocrystalline quartz (Qm). The monocrystalline quartz and quartzite content clearly increases upsection in both the Chicken Creek and Sixmile Canyon sections. A dramatic increase in quartz content occurs at the top of the Chicken Creek section in the proximal alluvial-fan deposits.

(2) Pink and purple quartzite (Qzrp). Banded quartzite clasts derived from the Precambrian Mutual Formation increase in frequency upsection. A small popula-

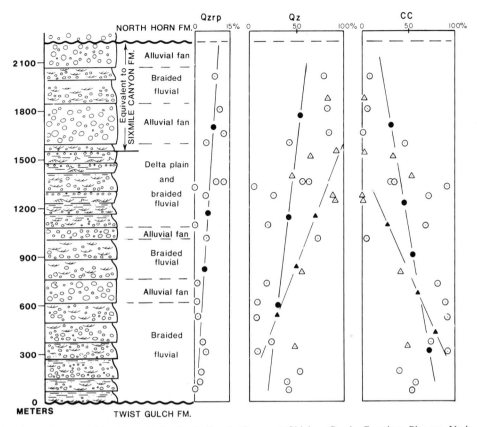

Fig. 5. Stratigraphic compositional trends for the Indianola Group at Chicken Creek, Gunnison Plateau, Utah. Column headings refer to clast and grain types described in Tables 1 and 2, with Qm substituting for Qz in sandstone samples. Key to symbols: circles, conglomerate clast data; triangles, sandstone modal data. Solid symbols indicate best-fit curves for each data category.

tion of distinctive red and purple quartzite clasts is present in the stratigraphically lowest conglomerate beds.

(3) Carbonate and chert (CC). Clast and grain types derived from carbonate source rocks decrease in frequency upsection. A rapid decrease in the percentage of carbonate and chert clasts occurs at metre 1325 of the Chicken Creek section (Fig. 5), coincident with the base of boulder and cobble conglomerates equivalent to the Sixmile Canyon Foundation in Sixmile Canyon.

(4) Feldspar (F). The total feldspar frequency in the Sixmile Canyon section does not display a characteristic stratigraphic trend. Measurable feldspar percentages are restricted to marine units of the Sanpete and Funk Valley Formations. The marine sandstones tend to be very fine grained, while associated non-marine sandstones range from fine grained to coarse-grained (Lawton, 1982).

DISCUSSION

Interpretation of compositional data

The compositional data from Indianola sandstones and conglomerates provide insight into the structural evolution and palaeogeography of the thrust belt. Sandstone modal plots and clast compositions simply corroborate the assumption that the thrust belt served as a source of Indianola clastic debris. Indianola sandstones plot within the field for recycled orogenic provinces (Dickinson & Suczek, 1979) on the QtFL and QmFLt triangular plots of Fig. 4. The almost exclusive presence of sedimentary lithic grains indicates that the detritus was derived from the thrusted sedimentary strata of the Sevier orogenic belt. This observation is further substantiated by the presence of only sedimentary lithic clasts in Indianola conglomerates.

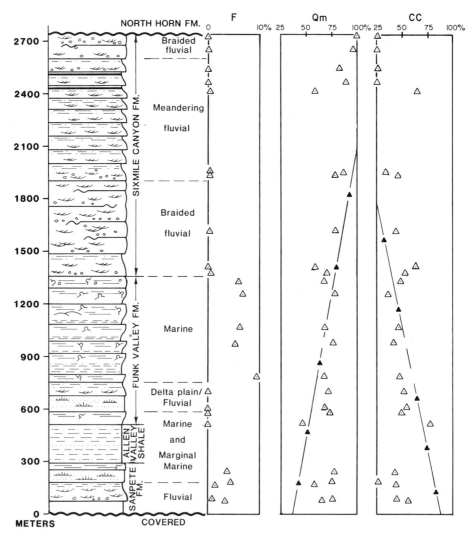

Fig. 6. Stratigraphic compositional trends for sandstones of the Indianola Group at Sixmile Canyon, Wasatch Plateau, central Utah. Column headings refer to grain types described in Table 2. Triangles represent sandstone modal data. Solid symbols indicate best-fit curves for data.

However, the stratigraphic compositional trends of Figs 5 and 6 permit more detailed interpretation of the nature of the source terrane than do the standard compositional plots. Several aspects of the trends pertain to detailed structural development within the thrust belt. Stratigraphic plots for both sandstones and conglomerates indicate a secular evolution from carbonate-rich compositions to quartzose compositions. Thus, the Indianola section does contain a simple record of unroofing of the miogeoclinal sequence to the west as suggested by Armstrong (1968a). Moreover, although significant scatter exists

in the stratigraphic plots, it appears that the carbonate to quartzite cycle occurs only once in the Gunnison Plateau area. Thus, apparently only one miogeoclinal section was stripped by erosion, indicating that the section was not repeated several times by numerous thrust faults. The mapped relations of the Pavant and Canyon Ranges (Christiansen, 1952; Burchfiel & Hickcox, 1972) support this interpretation. The Canyon Range plate, eroded to Precambrian rocks, rests on the Pavant plate which carries an uneroded section ranging in age from presumed Precambrian to Jurassic.

Although an inverted clast stratigraphy is present in Indianola conglomerates, clasts from the Precambrian Mutual Formation (Qzrp category of Table 1) occur in the lowermost part of the section. This indicates that drainage systems in the thrust terrane had early access to strata low in the miogeoclinal section. This may have occurred by rapid downcutting of canyons transverse to what was probably a strike-dominated drainage network within the thrust belt. Alternatively, early exposure of Precambrian strata may have occurred at positions of along-strike structural relief formed by tear faults or lateral ramps. Such a feature, the Leamington Canyon Fault, exists immediately north of the Canyon Range (Sprinkel & Baer, 1982), but it is unclear if clasts shed from that area would have been deposited as far south as Chicken Creek. In either case, the early appearance of Precambrian clasts suggests that structural relief sufficient to expose Precambrian strata to erosion was developed synchronously with uplift. Because folds with several kilometres of amplitude are absent in the area, I suggest that both structural and erosional relief

were developed by ramping and duplexing of the exposed thrust plates. This is probably the most common uplift mechanism associated with thrust-fault tectonics (Boyer & Elliott, 1982).

Interaction of basin-margin tectonics and basin fill

A comparison of regional stratigraphic and structural relations with the compositional data discussed earlier indicates that the structural geology of the thrust belt may be tied to depositional and compositional trends in the basin to form a unified tectonic scenario for basin development. The basin may be considered to consist of two major compositional wedges, a lower carbonate-rich wedge and an upper quartzose wedge (Fig. 7). The quartzose wedge is displaced eastwards with respect to the carbonate-rich wedge, and is overlain in turn by a smaller, more eastern clastic wedge (Price River Formation).

The lower carbonate-rich wedge encompasses the 'Canyon Range fanglomerate' of Armstrong (1968a), and the Sanpete, Allen Valley and Funk Valley

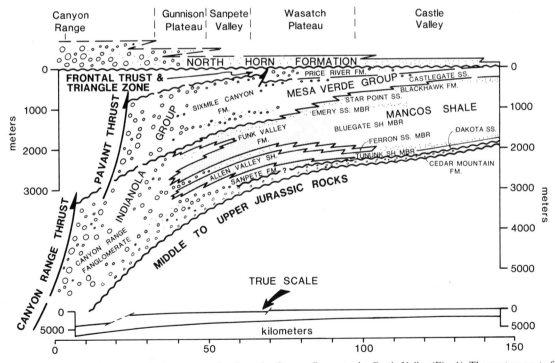

Fig. 7. Reconstructed east–west basin cross-section from the Canyon Range to the Castle Valley (Fig. 1). The western part of the foreland basin forms three offlapping wedges, each tied westward to the major thrust event which provided a source for the clastic detritus. The lowermost wedge is rich in detrital carbonate in the conglomerate and sandstone fractions. The middle wedge (indicated by light stipple) is dominated by quartzite-clast conglomerate and quartzarenite. The upper wedge (Price River Formation), not discussed in this paper, consists of litharenite and sublitharenite (Lawton, 1983).

Formations and their equivalents. The most distal deposits of the lower wedge in Utah are dominated by the Mancos Shale. Because thrust deformation and basin subsidence were only incipiently developed in the west, possibly coupled with rising sea-level beginning in late Albian time (Vail, Mitchum & Thompson, 1977), the depositional systems were strongly influenced by fluctuating shoreline conditions. The lower wedge is structurally bounded on the west by the Canyon Range thrust, from which the clastic material was derived and which ultimately overrode the basin margin. The juxtaposition of thrust plate and conglomerates suggests that the late thrust deformation occurred close to, or at, the synorogenic erosion surface.

The upper quartzose wedge encompasses the Six-mile Canyon Formation and its more proximal equivalents in the Gunnison Plateau. The upper wedge grades eastward into quartzarenites of the Castlegate Sandstone (Lawton, 1983) which extend 200 km from the thrust front before pinching out into marine siltstone (Van de Graaff, 1972). The quartzose wedge is not bounded by a thrust fault on the west in present exposures, but deposition occurred prior to thrust ramping which folded the Indianola outcrops concomitant with the development of easternmost thrust structures (Lawton, 1985). I suggest that the uplift which provided the source of the quartzose detritus was created by ramping of the Pavant thrust beneath the Canyon Range plate. The Canyon Range plate was then eroded more deeply and removed over most of the area. The appearance of large numbers of sandstone clasts (other clastic category, Table 3) contemporaneously with the flood of quartzose detritus indicates that the Mesozoic sandstone units beneath the Canyon Range thrust also contributed clastic material. The abrupt shift from pebbly braided-fluvial deposits to cobble and boulder conglomerates of alluvial-fan origin in the Chicken Creek section coincides with the base of the quartzose wedge and further documents the proximal uplift related to ramping of the Pavant thrust.

The third clastic wedge, although not discussed here, consists of the Price River Formation (Fig. 7). This clastic wedge is displaced eastward with respect to the quartzose wedge and consists of litharenites higher in sedimentary lithic fragments than the quartzose wedge. It was probably deposited during the formation of the easternmost structures of the thrust terrane (Lawton, 1985).

In the Cedar Hills, north of the sections studied for this report, the Indianola Group does not show an increased quartz content upsection (Jefferson, 1982). Instead, quartzite clasts are abundant throughout the section. This contrast with the results reported here is interpreted to reflect a difference in the stratigraphic section of the source area. As mentioned earlier, the Pennsylvanian to Permian section thickens dramatically northward and contains abundant quartzite (Baker, 1947). Detritus shed from the initial uplift would thus have been rich in upper Palaeozoic quartzite, and a trend reflecting later contributions of Precambrian quartzite would be masked. Jefferson's data do not include a separate red quartzite category to allow interpretation of possible trends in clast types derived from known Precambrian strata.

CONCLUSIONS

The Indianola Group forms a coarsening-upward megasequence whose detrital compositions record unroofing of the Sevier orogenic belt immediately to the west. The sequence consists of two offlapping clastic wedges, the lower one relatively enriched in detrital carbonate in both conglomeratic and sandy facies, the upper one relatively rich in quartzite clasts and quartzose sandstone. Each clastic wedge is interpreted to have resulted from major uplift related to thrust ramping. Although the two major ramping episodes may be attributed to two different thrust systems, the Canyon Range and Pavant thrusts, the dominant source of basin clastics was the Canyon Range plate, first as it actively ramped upsection and second as it underwent passive uplift above the Pavant thrust system. Ages of the synorogenic wedge indicate that uplift above the Canyon Range thrust occurred in late Albian time, followed by major uplift above the Pavant thrust probably in early Santonian time. Indianola deposition terminated in late Campanian time as thrust deformation affected the foreland basin margin, folding Indianola strata above a triangle zone at the eastern limit of thrusting. The record of this late-stage deformation is found in the Price River Formation, which lies immediately east of the frontal structural belt.

ACKNOWLEDGMENTS

I gratefully acknowledge discussions of this topic with J. W. Collinson, W. R. Dickinson, and W. S. Jefferson and a review by S. A. Graham. The research reported here was supported by Earth Sciences

Division, National Science Foundation Grant EAR-7926379. Drafting and manuscript support was provided by Sohio Petroleum Company. I thank Leta Smith and Peggy Montgomery for drafting and Judy Reed and Maggie Olmstead for typing the manuscript.

REFERENCES

ARMSTRONG, R.L. (1968a) Sevier orogenic belt in Nevada and Utah. *Bull. geol. Soc. Am.* **79**, 429–458.

ARMSTRONG, R.L. (1968b) The Cordilleran miogeosyncline in Nevada and Utah. *Bull. Utah geol. mineral. Surv.* **78**, 58 pp.

ARMSTRONG, F.C. & ORIEL, S.S. (1965) Tectonic development of the Idaho–Wyoming thrust belt. *Bull. Am. Ass. Petrol. Geol.* **49**, 1847–1886.

BAKER, A.A. (1947) Stratigraphy of the Wasatch Mountains in the vicinity of Provo, Utah. *U.S. geol. Surv. Oil Gas Invest. Prelim. Chart* 30.

BOYER, S.E. & ELLIOTT, D. (1982) Thrust systems. *Bull. Am. Ass. Petrol. Geol.* **66**, 1196–1230.

BURCHFIEL, B.C. & HICKCOX, C.W. (1972) Structural development of central Utah. In: *Plateau-Basin and Range Transition Zone, central Utah* (Ed. by J. L. Baer and E. Callaghan), pp. 55–73. *Utah geol. Ass. Publ.* **2**.

CHRISTIANSEN, F.W. (1952) Structure and stratigraphy of the Canyon Range, central Utah. *Bull. geol. Soc. Am.* **63**, 717–740.

CHRISTIE-BLICK, NICHOLAS (1982) Upper Proterozoic and Lower Cambrian rocks of the Sheeprock Mountains, Utah: regional correlation and significance. *Bull. geol. Soc. Am.* **93**, 735–750.

DICKINSON, W.R. (1970) Interpreting detrital modes of graywacke and arkose. *J. sedim. Petrol.* **40**, 695–707.

DICKINSON, W.R. & SUCZEK, C.A. (1979) Plate tectonics and sandstone compositions. *Bull. Am. Ass. Petrol. Geol.* **63**, 2164–2182.

DICKSON, J.A.D. (1966) Carbonate identification and genesis as revealed by staining. *J. sedim. Petrol.* **36**, 491–505.

FISHER, D.J., ERDMAN, C.E. & REESIDE, J.B. (1960) Cretaceous and Tertiary Formations of the Book Cliffs, Carbon, Emery, and Grand counties, Utah, and Garfield and Mesa counties, Colorado. *Prof. Pap. U.S. geol. Surv.* **332**, 80 pp.

FOUCH, T.D., LAWTON, T.F., NICHOLS, D.J., CASHION, W.B. & COBBAN, W.A. (1983) Patterns and timing of synorogenic sedimentation in Upper Cretaceous rocks of central and northeast Utah. In: *Mesozoic Paleogeography of West-Central United States* (Ed. by M. W. Reynolds & E. D. Dolly), pp. 305–336. Society of Economic Palaeontologists and Mineralogists, Rocky Mountain Section.

GRAHAM, S.A., INGERSOLL, R.V. & DICKINSON, W.R. (1976) Common provenance for lithic grains in Carboniferous sandstones from Ouachita Mountains and Black Warrior basin. *J. sedim. Petrol.* **46**, 620–632.

JEFFERSON, W.S. (1982) Structural and stratigraphic relations of Upper Cretaceous to lower Tertiary orogenic sediments of the Cedar Hills, Utah. In: *Overthrust Belt of Utah* (Ed. by D. L. Nielson), pp. 65–80. *Utah geol. Ass. Publ.* **10**.

LAWTON, T.F. (1982) Lithofacies correlations within the Upper Cretaceous Indianola Group, central Utah. In: *Overthrust Belt of Utah* (Ed. by D.L. Nielson), pp. 199–213. *Utah geol. Ass. Publ.* **10**.

LAWTON, T.F. (1983) Late Cretaceous fluvial systems and the age of foreland uplifts in central Utah. In: *Rocky Mountain Foreland Basins and Uplifts* (Ed. by J. D. Lowell), pp. 181–199, Rocky Mountain Association of Geologists.

LAWTON, T.F. (1985) Style and timing of frontal structures, thrust belt, central Utah. *Bull. Am. Ass. Petrol. Geol.* **69**, 1145–1159.

LAWTON, T.F. & WILLIS, G.C. (1986) The geology of Salina Canyon. In: Centennial Field Guide—Rocky Mountain Section, *DNAG Guidebook Series* (Ed. by J. K. Rigby). Geological Society of America.

McBRIDE, E.F. (1963) A classification of common sandstones. *J. sedim. Petrol.* **33**, 664–669.

PRICE, R.A. & MOUNTJOY, E.W. (1970) Geologic structure of the Canadian Rocky Mountains between Bow and Athabasca Rivers—a progress report. *Spec. Pap. geol. Ass. Can.* **6**, 7–25.

ROYSE, F., JR, WARNER, M.A. & REESE, D.L. (1975) Thrust belt structural geometry and related stratigraphic problems, Wyoming–Idaho–northern Utah. *Rocky Mountain Ass. Geol. Guidebook*, pp. 479–486.

SPIEKER, E.M. (1946) Late Mesozoic and early Cenozoic history of central Utah. *Prof. Pap. U.S. geol. Surv.* **205-D**, 117–161.

SPIEKER, E.M. (1949) The transition between the Colorado Plateaus and the Great Basin in central Utah. *Utah geol. Soc. Guidebook to the Geology of Utah*, no. 4, 106 pp.

SPRINKEL, D.A. & BAER, J.L. (1982) Overthrusts in the Canyon and Pavant ranges. In: *Overthrust Belt of Utah* (Ed. by D. L. Nielson), pp. 303–313. *Utah Geol. Ass. Publ.* **10**.

STANDLEE, L.A. (1982) Structure and stratigraphy of Jurassic rocks in central Utah: their influence on tectonic development of the Cordilleran fold and thrust belt. In: *Geologic Studies of the Cordilleran Thrust Belt* (Ed. by R. B. Powers), pp. 357–382. Rocky Mountain Associaton of Geologists.

VAIL, P.R., MITCHUM, R.M., JR, & THOMPSON, S., III (1977) Seismic stratigraphy and global changes of sea level, part 4: global cycles of relative changes of sea level. *Mem. Am. Ass. Petrol. Geol.* **26**, 83–97.

VAN DE GRAAFF, F.R. (1972) Fluvial-deltaic facies of the Castlegate Sandstone (Cretaceous), east-central Utah. *J. sedim. Petrol.* **42**, 558–571.

VAN DER PLAS, L. & TOBI, A.C. (1965) A chart for judging the reliability of point counting results. *Am. J. Sci.* **263**, 87–90.

WEIMER, R.J. (1984) Relation of unconformities, tectonics, and sea-level changes, Cretaceous of Western Interior, U.S.A. *Mem. Am. Ass. Petrol. Geol.* **36**, 7–35.

WILLIS, G.C. (1986) Geologic map of the Salina Quadrangle, Sevier County, Utah. *Utah geol. mineral. Surv. Map Series*, **83**, 16 pp. 1 : 24,000.

Spec. Publs int. Ass. Sediment. (1986) **8**, 425–436

Provenance modelling as a technique for analysing source terrane evolution and controls on foreland sedimentation

S. A. GRAHAM,[1] R. B. TOLSON,[1] P. G. DECELLES,[1] R. V. INGERSOLL,[2]
E. BARGAR,[3] M. CALDWELL,[4] W. CAVAZZA,[2] D. P. EDWARDS,[5]
M. F. FOLLO,[6] J. F. HANDSCHY,[7] L. LEMKE,[8] I. MOXON,[1] R. RICE,[9]
G. A. SMITH[10] *and* J. WHITE[11]

[1] *School of Earth Sciences, Stanford University, Stanford, California 94305, U.S.A.;* [2] *Department of Earth and Space Sciences, University of California, Los Angeles, California 90024, U.S.A.;* [3] *SOHIO Petroleum, 5420 LBJ Freeway, Dallas, Texas 75240, U.S.A.;* [4] *Department of Geology, Western Michigan University, Kalamazoo, Michigan 49008, U.S.A.;* [5] *Department of Geology, Northern Arizona University, Flagstaff, Arizona 86001, U.S.A.;* [6] *Department of Geological Sciences, Harvard University, Cambridge, Massachusetts 02138, U.S.A.;* [7] *Department of Geology and Geophysics, Rice University, Houston, Texas 77251, U.S.A.;* [8] *Department of Geosciences, University of Arizona, Tuscon, Arizona 85719, U.S.A.;* [9] *Department of Geology, McMaster University, Hamilton, Ontario L8S 4M1, Canada;* [10] *Department of Geology, Oregon State University, Corvallis, Oregon 97331, U.S.A.;* [11] *Department of Geological Sciences, University of California, Santa Barbara, California 93106, U.S.A.*

ABSTRACT

Tectonism and climate are widely viewed as principal controls on alluvial-fan sedimentation in foreland and other fault-bounded basins. For example, upward coarsening and thickening sedimentary sequences often are interpreted as responses to cratonward propagation of thrust-fault systems in foreland basins. Source-rock lithology is recognized as influencing gross fan morphology (e.g. sandy versus gravelly fans), but the impact of time-varying provenance has not been assessed. The Sphinx Conglomerate of southwestern Montana clearly displays the controlling role of changing provenance in an eroding foreland thrust belt on sedimentary style in the adjacent foreland basin. The Maastrichtian Sphinx Conglomerate crops out atop the 3315 m high Sphinx Mountain in the Madison Range as an erosional remnant, over 1000 m thick, of the proximal realm of a Laramide foreland basin. Conglomerate clasts define an unroofing sequence: recognizable Cretaceous to Cambrian clasts appear progressively upward in the Sphinx. Conglomerate units increase upward in abundance and thickness, and the Sphinx Conglomerate is overridden by a foreland thrust fault, implying that conglomerate distribution is a progradational sedimentary response to faulting. However, the Mesozoic section exposed in the Madison Range is largely muddy and contains few units capable of producing conglomerate clasts. In contrast, the middle and lower Palaeozoic section consists largely of carbonates capable of generating great volumes of cobbles in a temperate climate. Thus, the inverted stratigraphy seen in clasts, apparent overall upward coarsening, and increasing abundance and thickness of conglomerate beds suggest that sedimentary style in the Sphinx is determined largely by the lithology of units exposed at specific times in the encroaching thrust plate. To test this hypothesis, we tabulated thicknesses of resistant lithologies capable of yielding gravel clasts for each Phanerozoic unit exposed in the Madison Range, as compared to total unit thickness. Conglomerate clast counts of the Sphinx Conglomerate permitted the identification of units eroding in the adjacent thrust plate at specific times. Comparison of the composition of hypothetical conglomerates eroded from particular suites of source units with the actual composition of Sphinx Conglomerate samples yielded striking results. Modelled compositions closely match actual compositions in several instances. These results suggest that changing provenance can produce bedding trends in foreland basins that are often otherwise attributed to tectonic or climatic controls. Provenance modelling can aid in evaluating these alternative interpretations.

INTRODUCTION

For many decades, it has been apparent that tectonic setting strongly influences the depositional style and lithologic composition of sediments of foreland basins. Studies of tectonically well constrained, ancient foreland basin deposits like the northern Alpine Molasse and observations of modern forelands like the Indo–Gangetic plain adjacent to the Himalaya give rise to a composite view of the foreland basin as a cross-sectionally asymmetric sediment prism that is elongate parallel to an adjacent fold-thrust mountain belt. Floored by continental crust, such basins are generally characterized by shallow-marine to alluvial sediments, rather than deep-marine strata (Dickinson, 1976). Facies typically reflect transverse supply of detritus from adjacent tectonic uplands. Basinward, deposystems are orthogonally deflected to become longitudinal with respect to the axes of the basin and mountain belt (Graham, Dickinson & Ingersoll, 1975). Clastics of the basin bear the compositional imprint of the adjacent mountainous sediment source terrane. Although petrographic details vary among individual foreland basins, foreland sandstones, for instance, stand as a distinctive petrographic class when compared to sandstones from other tectonic settings (Dickinson & Suczek, 1979).

A key element of the foreland basin is its syntectonic character. The greatest thickness of the foreland-basin fill borders the fold-thrust belt, reflecting enhanced subsidence through the combined agencies of sediment and thrust-sheet loading (Jordan, 1981). Furthermore, the proximal margin of the basin progressively becomes involved in the propagating fold-thrust belt with time. Sediments shed from the rising thrust-belt are eroded and resedimented in the foreland basin, only to be recycled again with basinward propagation of folds and faults. This process results in time-progressive shifts in the locus of deposition and subsidence and gives rise to the concept of the 'migrating foredeep' (Bally, Gordy & Steward, 1966).

Regions of foreland basins most proximal to mountain fronts are often characterized by alluvial fan sedimentation. Details of sedimentary style in this setting vary widely, controlled in a complicated manner by obvious factors of gradient (implicitly including rates of uplift, sea-level fluctuations in marine basins, etc.), climate and drainage area. Less obvious and less often discussed is the significance of the lithology of sediment source areas on foreland sedimentation. For example, in simplest terms, it seems clear that monolithologic shaley source terranes yield detritus that forms mud-rich fans. But what is the impact on foreland sedimentation of a lithologically diverse source terrane, whose heterogeneity varies with time and depth of erosional stripping?

The problem is complex, and the literature provides no easy answer. In fact, studies of the provenance of foreland-basin sediments are numerous, but quite limited in character. Often cited are studies relating foreland sandstones to global tectonic petrographic schemes (e.g. Dickinson & Suczek, 1979). Studies at the basin level are also common. These are valuable, but generally they simply document the adjacent fold-thrust belt as the sediment source, or at best, broadly track compositional changes in sandstone composition with unroofing of uplifting source terranes. Studies of the latter sort have long been used world-over qualitatively to track the timing of thrusting recorded by associated clastic wedges in foreland basins. For instance, a study by Denson & Pipiringos (1969) of Palaeogene coarse clastics of the Wyoming foreland revealed compositional changes reflecting progressive unroofing of miogeoclinal Palaeozoic sedimentary rocks from the Precambrian cratonal basement that cores foreland fault blocks in the region. It is easy to imagine that the progressive change from detritus derived from Palaeozoic carbonates, shales and quartz sandstones to detritus derived from Archaean granites and gneisses should have had a significant impact on styles of fan sedimentation in depocentres of the Palaeogene 'broken foreland' (terminology of Dickinson, 1976) of Wyoming, but such effects have not been assessed.

On a broader level, this question is relevant to alluvial-fan sedimentation in all types of fault-bounded basins. Cyclic behaviour (on the order of hundreds of metres) commonly observed in vertical successions of alluvial sediments is usually attributed to autocyclic fan mechanisms or allocyclic climatic or tectonic controls (e.g. Miall, 1970; Heward, 1978; Gloppen & Steel, 1981). Many times, sufficient data exist to support these interpretations. But how many times has the character of alluvial megasequences (Fig. 1) been attributed to fault movements, based on little more than preconception? Could time-varying source lithology play a greater role in shaping depositional style in the foreland basin than generally supposed? Furthermore, how could one even set about answering the question?

PROVENANCE MODELLING

Where circumstances permit, it may be possible to asses the role of time-varying source lithology on depositional style through provenance modelling. In this process, lithologic factors can be isolated by comparing well known source terranes with strati-

Fig. 1. Southern face of Sphinx Mountain, Madison Range, Montana, presents the appearance of an upward coarsening and thickening sequence, suggestive of tectonically triggered alluvial-fan progradation. However, maximum clast-size in the Sphinx Conglomerate, which makes up the 3315 m mountain, does not vary systematically, and the upward increase in conglomerate may instead illustrate the influence of source lithology on fan sedimentation.

graphically controlled samples of the derivative foreland basin fill (Fig. 2).

In our approach to provenance modelling, we focus on the conglomeratic components of the alluvial section. The distinct advantage in using conglomerate and gravel lies in the clarity of provenance and tectonic interpretations available from large clasts (e.g. Miall, 1970; Tanner, 1976). In contrast, the often non-diagnostic quartz, feldspar and clay that volumetrically make up most sandstones and mudrocks may be derived from a great many kinds of source terranes. Nevertheless, interpretation of conglomerate provenance is not as simple as it might seem (e.g. Blatt, 1982). Conglomerate compositions are strongly determined by the survivability of clasts during transport, which is, in turn, controlled by several variables such as petrophysical properties of source rocks, degree and spacing of jointing in source terranes, intensity of climatically controlled weathering processes. Despite these uncertainties, conglomerates remain the most diagnostic and abundant provenance indicators in alluvial sequences close to fault-bounded margins of foreland basins. Compositions of conglomerates occurring within perhaps 10 km of parent source terranes probably faithfully record their full provenance spectrum (fig. 5-1 of Blatt, 1982).

Where the thrust sheets that comprised source terranes have been fully removed by erosion, constraints sufficient for modelling are lacking. However, where the full range of source units is at least partially preserved for later inspection, it is possible to attempt provenance modelling using the approach outlined below and illustrated in concept and by example in Figs 2 and 3.

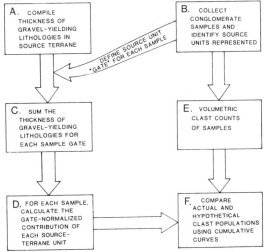

Fig. 2. Conceptual basis for provenance modelling of conglomerates.

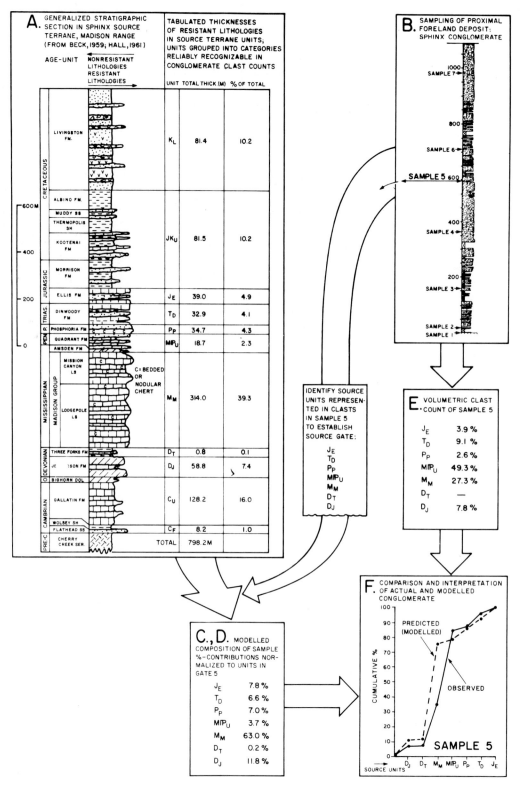

Fig. 3. Application of provenance modelling to the Sphinx Conglomerate: (A) shows the Phanerozoic section (modified from Beck, 1959 and Hall, 1961) that was recycled into the Sphinx Conglomerate, while (B–F) track sample 5 from the Sphinx Conglomerate through the modelling process.

(a) The first step in the process is to determine the relative proportion which each stratigraphic unit (n) in the source terrane is likely to contribute to the gravel fraction when the section is eroded. In practice, we compile (from field measurement or published measured sections) the thickness of erosion-resistant beds in each formation (Rtn) in the source region.

(b) The next step requires compositional analysis of foreland-basin conglomerates. Clast counts of stratigraphically controlled conglomerates reveal which stratigraphic units were exposed in the source region and recycled into the conglomerates at specific times (Fig. 4). For each sample, a 'window' or 'gate' of stratigraphic units of the source terrane represented in the clast counts is tabulated.

(c) Using the gate of source units determined for a particular sample, the thickness of resistant lithologies in the source unit gate (Rt_G) is derived by summing the resistant thickness for each formation (n_1 to n_G) in the gate:

$$Rt_G = \sum_{n1}^{nG} Rtn.$$

This is the total thickness of resistant lithologies contributing coarse detritus to the sample.

(d) The relative, gate-normalized contribution (C) of each source-terrane unit to the conglomerate,

$$C = \frac{Rtn}{RT_G},$$

is calculated for each formation in the gate. A more sophisticated approach might try to assess planimetric as well as vertical distribution of resistant source lithologies, or try to 'weight' differentially the transport durability of various clast lithologies.

(e) These hypothetical detrital contributions (C) are now compared with actual data. The analysis of foreland-basin gravel samples, mentioned in step b, consists of volumetric clast counts (much like thin-section modal analysis of sandstones). Clast-count categories with source-terrane significance are employed in order to assess the contribution of each source unit to the total clast population. In practice, unambiguous assignment of all clasts to source-unit categories may be difficult; if the proportion of unidentified clasts is relatively small, these clasts are excluded from final calculations by normalization.

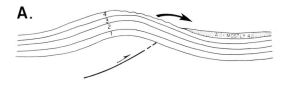

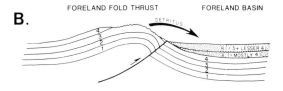

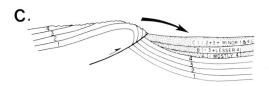

Fig. 4. Sequential block diagrams schematically illustrating the manner in which the emplacement and erosion of a foreland thrust plate gives rise to an unroofing sequence in the proximal realm of the adjacent foreland basin.

(f) Finally, the hypothetical compositions of gravels are compared with appropriate real samples. Histograms and pie diagrams could be used, but here we have chosen cumulative composition plots as a convenient visual means for making such comparisons. If hypothetical and actual curves closely coincide, the occurrence of the foreland-basin conglomerate is adequately explained by the distribution of source-terrane lithologies alone. Divergence of the curves requires other explanations.

APPLICATION TO THE SPHINX CONGLOMERATE

The technique described above was developed in the course of a field-based study of the Sphinx Conglomerate of southwestern Montana (DeCelles *et al.*, 1986). The Sphinx Conglomerate is a synorogenic deposit that crops out in a synclinal erosional remnant atop Sphinx Mountain and The Helmet in the Madison Range (Fig. 5). Although age limits are poorly constrained, the Sphinx is Maastrichtian in its lower middle portion, and is best considered an intermon-

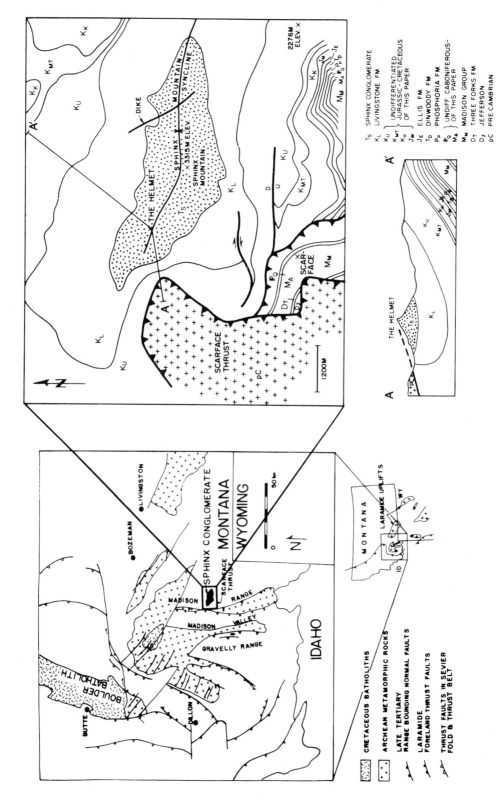

Fig. 5. Geological context of the Sphinx Conglomerate, a Laramide syntectonic molasse deposit in SW Montana. Local geologic map and cross-section modified from Beck (1959).

tane molasse deposit of the Laramide 'broken foreland' (terminology of Dickinson, 1976). The 1000+ m of preserved section consists largely of non-marine conglomerate beds, rich in limestone clasts, with lesser amounts of sandstone and mudstone (Fig. 6A).

Three factors make the Sphinx Conglomerate especially interesting:

(1) The unit crops out less than 1 km from the present erosional limit of the thrust plate from which the Sphinx sediments were derived (Fig. 5).
(2) The thrust plate consists of a nearly complete Phanerozoic sedimentary section resting atop Archean crystalline basement (Fig. 3A). Lithologies in this section are diverse, and for the most part, are easily distinguished, even in hand samples. This section is well displayed in outcrops in the Madison range (Fig. 5).
(3) The Sphinx Conglomerate is well exposed in three dimensions for its full vertical extent atop the two mountains (Fig. 5), where it gives the initial impression of an upward coarsening sequence (Fig. 1), whose apparent progradational character is related to movement on the nearby Scarface thrust fault.

Outcrop study quickly established several key relations. First, clast counts of conglomerates revealed that conglomerates low in the Sphinx section are dominated by clasts derived from upper Mesozoic units (Fig. 6A, D). Upward in the section, these clasts are progressively supplanted by lower Mesozoic and upper Palaeozoic clasts, and finally lower Palaeozoic clasts. The compositions of Sphinx sandstones suggest similar provenance relations (Ingersoll *et al.*, 1986). Lower Sphinx sandstones are rich in chert and siliciclastic rock fragments derived from Mesozoic and upper Palaeozoic units, whereas upper Sphinx sandstones are rich in carbonate grains derived from the Palaeozoic section (Fig. 6C). In sum, the Sphinx Conglomerate is an unroofing sequence (cf. Fig. 4), reflecting progressive erosion of the Phanerozoic section on the adjacent Scarface thrust plate.

Facies and palaeocurrent analysis support the interpretation that the Sphinx Conglomerate was derived from the Scarface thrust plate to the west (Fig. 5). Grain-size and facies trends indicate that the down-fan direction was generally to the ENE (DeCelles *et al.*, 1986). However, these data also show that the seemingly upward coarsening sequence that forms the dramatic upper cliff faces of Sphinx Mountain and The Helmet (Fig. 1) is more apparent than real. The lower third of the Sphinx Conglomerate

indeed shows an upward increase in maximum grain size and bed thickness (Fig. 6A, B), as well as increasingly proximal facies. Thus, fan progradation, possibly related to fault activity, likely controlled the character of the vertical section, at least in part. However, the markedly conglomeratic upper two-thirds of the Sphinx shows no particular vertical trends in facies or grain size (Fig. 6B), only an increase upward in the *abundance* of conglomerate beds in the section (Fig. 6A).

The latter result seems surprising. If fault tectonics simply controlled fan progradation, a systematic increase upward in grain size, especially over a section several hundred metres thick, would be expected. Instead, the entire upper two-thirds of the formation consists of proximal- to middle-fan facies of comparable sedimentary style and very coarse grain size (Fig. 6A, B). Key to the problem is the recognition of the message contained in the unroofing sequence documented by Sphinx conglomerate composition.

Following the provenance modelling procedure outlined above, we compiled the thickness of erosion-resistant (hence, gravel-forming) stratigraphic units of the Phanerozoic section that crops out in the Madison Range (Fig. 3A). The Madison Range currently resides in a semi-arid, mid-latitude position in the rain-shadow of the Cordillera of western North America. Thus, carbonate rocks commonly are major cliff-forming units in outcrop. In the absence of competing, durable quartz sandstones, carbonate clasts commonly comprise most of the gravel in modern streams draining Palaeozoic carbonate terranes of the Madison Range. Conditions in the latest Cretaceous and early Tertiary probably were at least broadly similar to the present. SW Montana was a mid-latitude region characterized by Laramide basin and range topography in the rain-shadow of the Cordillera. Although floral and faunal remains are rare in the Sphinx, pollen assemblages suggest a temperate climate (personal communication, Joyce Clark, Clark Geological Services), consistent with a regional trend toward cooler and more arid climate conditions across the Cretaceous–Tertiary boundary (Hall & Norton, 1967). Thus, the compositional character of the Sphinx, long noted for the abundance of carbonate clasts it contains (Lowell & Klepper, 1953; Beck, 1959), is consistent with the palaeoclimatic record.

For comparison against this source-terrane data base, we analysed the composition of Sphinx conglomerates at 11 stratigraphic positions in the sequence. In order to produce volumetric counts, we erected a grid

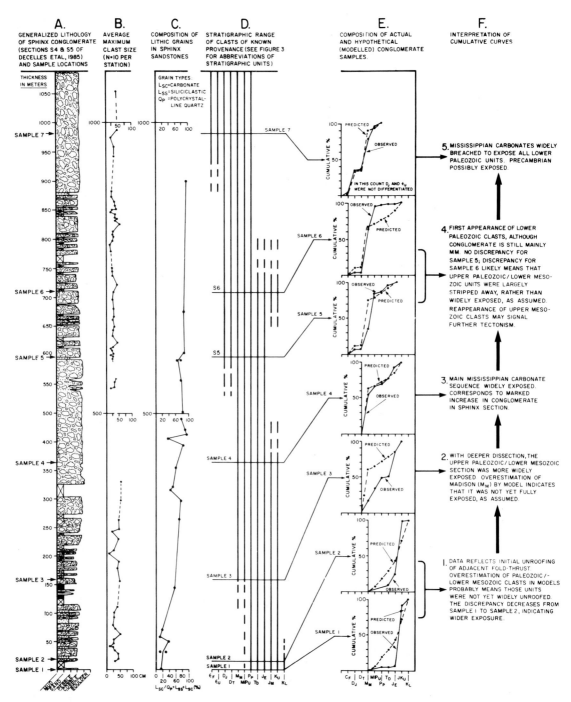

Fig. 6. Summary diagram comparing lithology/bedding sequence (A), grain-size trends (B), clast composition trends (C, D), and conglomerate modelling results (E, F) for the Sphinx Conglomerate.

on the outcrop, and recorded lithology and probable source unit for 100 clasts that corresponded to nodes in the 10 cm spaced grid.

These outcrop data were used in two ways. First, we let the counts define the 'gate' of source formations to be used in the modelling step (Fig. 3B–D). Second, we compared the actual clast counts with the modelled conglomerate compositions by plotting the cumulative-percentage contribution of each provenance unit to actual and modelled conglomerates (Fig. 3F).

Seven of the 11 analyses represent sampling of a specific vertical section (Fig. 6A). In going beyond the clast-type range chart (Fig. 6D), which qualitatively documents the unroofing of the Scarface thrust sheet, the sequence of cumulative composition curves (Fig. 6E) documents the manner in which the entire volume of conglomerates changes in composition with time. Significantly, the change from interbedded mudstone, sandstone and conglomerate observed in the lower Sphinx section to the limestone-clast-dominated conglomerate of the upper Sphinx coincides with an important change in clast composition. Samples 1, 2 and 3 in the lower Sphinx mostly include clasts eroded from relatively shale-rich upper Palaeozoic and Mesozoic formations (cf. the early phase of unroofing in Fig. 4A). However, samples 4 and higher are volumetrically composed almost entirely of Palaeozoic carbonate clasts. Apparently, by the time sample 4 was deposited, almost all of the Mesozoic sequence had been erosionally stripped from the adjacent thrust sheet, exposing the Palaeozoic carbonate sequence (cf. Fig. 4B, where Madison is Unit 3). The Mississippian Madison Group (Fig. 3A) is an especially thick, shale-poor, cliff-forming sequence in south-western Montana today. Its erosion in the Late Cretaceous yielded only gravel carbonate clasts to the proximal realm of Sphinx deposition, and only in latest Sphinx time was the thick Madison Group sufficiently unroofed to permit older Palaeozoic carbonates to contribute significantly to Sphinx gravels (cf. Fig. 4C). Thus, the overall upward increase in conglomerate in the Sphinx seems clearly related to the distribution of lithologies in the eroding source terrane.

Furthermore, when actual Sphinx samples are compared to the hypothetical composition of conglomerates generated by the modelling procedure discussed above, we find remarkable agreement. For instance, sample 4 consists of more than 50% clasts referrable to the Madison Group, with the balance evenly divided between other upper Palaeozoic and Mesozoic units. If the clast count is used to define the modelling gate as Madison Group through Upper Cretaceous,

and if it is assumed that the full thickness of each of those formations was exposed at the time, then the tabulation of resistant lithologies for those units can be used to determine the relative contribution of each formation in the Madison–Cretaceous gate to a hypothetical conglomerate. The result in the case of sample 4 is extraordinary agreement between actual and predicted composition. This suggests that source lithology strongly controlled the character of Sphinx conglomerate sample 4, and that the assumption that all of the formations represented by clasts were widely exposed is valid.

In fact, the correlation between actual and modelled compositions is remarkably strong for most of the 11 samples. Departures that do occur locally between the curves may yield additional insights into the evolution of the source terrane and deposit, however. For instance, sample 6 qualitatively shows that units as old as Devonian were exposed at the time of deposition. Lacking better information, we assumed in modelling the conglomerate that all units from the Jefferson through the Livingston were fully exposed. Clearly this was not true, however, because the observed abundance of upper Palaeozoic–Mesozoic clasts is far less than predicted under that assumption. Instead, probably only small patches of the younger units were left as erosional remnants at the time of deposition of sample 6, or perhaps younger units were still present only as steeply dipping beds on the overturned limb of a hanging-wall anticline (cf. Fig. 4C) or were briefly reintroduced into the source terrane along an imbricate thrust slice (cf. fig. 12 of DeCelles *et al.*, 1986). Both interpretations are consistent with the position of sample 6 in the upper portion of the Sphinx sequence.

Extending this interpretive method to the full sequence of comparative curves available for the Sphinx section shown in Fig. 6, it is possible to track the unroofing of the Scarface thrust sheet in detail. Lacking better resolution, the modelling method assumes that if any clasts derived from a particular source unit are present in the Sphinx, then the entire unit was fully exposed and contributing detritus. However, this assumption cannot be true in general, so discrepancies between modelled and observed compositional curves may have real significance in terms of extent of unroofing of source-terrane units, as argued above for sample 6. Thus, curves for samples 1–3 indicate early unroofing with only limited exposure of upper Palaeozoic–lower Mesozoic units (Fig. 6E, F). The close correspondence of modelled and actual curves for samples 4 and 5 suggests that by the onset of deposition of the main Sphinx conglomerate sequence, the middle Palaeozoic–lower Mesozoic source section was widely exposed. Through time, most of the lower Mesozoic section was stripped from the thrust plate (sample 6), until only lower to middle Palaeozoic units remained in the source terrane in late Sphinx time (sample 7).

Finally, the only arguably upward coarsening in the Sphinx Conglomerate occurs in the lower 300 m of the section. This upward coarsening, better illustrated in sections other than that shown in Fig. 6(A), could be taken to represent fan progradation, perhaps related to thrust movement. An upward change through this interval from organized conglomerates-sandstones-mudstones to poorly organized conglomerates provides independent evidence for fan progradation. However, even here it can be argued that the clear-cut unroofing of the Madison Group carbonates seen upward from samples 1 to 4 (Fig. 6C, D, E) would result in an attendant increase in maximum grain size and decrease in overall mud content in the Sphinx. Moreover, the increased 'proximality' of facies may simply reflect the control of grain size on facies type. Thus, it seems likely that source lithology played a considerable, if not dominant, role in shaping the character of the entire Sphinx section.

It is important, however, to remember that in some cases the departures between modelled and observed curves highlight limitations of the method. Clearly, the one-dimensional vertical section analysis presented here cannot rigorously reproduce the vagaries of an eroding, three-dimensional source terrane. Secondly, our conglomerate-clast analyses are modal analyses analogous to thin-section point counts of sandstones, and as such are subject to the same measures of statistical reliability (Van der Plas & Tobi, 1965). The 100 clasts/sample counts we used in this study permit a 2σ confidence of only about $\pm 10\%$ for clasts in abundances of 30–70% and about $\pm 5\%$ for clasts comprising 7% of the population. Thus, a portion of the discrepancies seen in Fig. 6(E) are an artefact of sampling.

More fundamentally, however, these statistics of sampling presuppose a perfect knowledge of the provenance of every gravel clast counted. We have no precise estimate of the uncertainty in our clast identifications, although we think they are generally accurate. One possible measure, suggested to us by Timothy Lawton, would be to compare the average of the seven counts that sample the entire Sphinx section (Fig. 6E) with modelled contributions from the entire source section (Fig. 3A), under the assumption that if

basin fill reflects source lithology, the values will match. The results, plotted as histograms (Fig. 7), show that our counts match model expectations within a few per cent for half of the counting categories. The most notable exceptions lie in the upper Palaeozoic chert-bearing carbonate units, reflecting the difficulty in identifying clasts of massive carbonate and chert. However, the sum of observed Madison and Amsden–Quadrant clasts closely matches the sum of the two modelled categories. Thus, the discrepancy in sample 5 (Fig. 6) probably largely reflects misidentification of Madison and Amsden–Quadrant clasts in the field. On the other hand, the underrepresentation of Cambrian–Devonian carbonates as clasts in the total population (Fig. 7) probably reflects incomplete unroofing of the lower Palaeozoic source section and erosion of the top of the Sphinx section, rather than a failure of the modelling technique or clast misidentification.

However, this error analysis is itself limited by failures of the model, such as potential errors in extrapolating outcrop weathering characteristics to gravel composition (Fig. 3A). For instance, the third major discrepancy in Fig. 7, the observed occurrence of Jurassic–Cretaceous (undifferentiated) clasts in amounts above that predicted by the model, remains unexplained, but may reflect an incomplete understanding of detrital contributions from Mesozoic strata. A better documented example of this problem lies in the analysis of sample 2, where an overestimation of the contribution of upper Palaeozoic units probably indicates that the bold outcrops of chert tabulated in the compilation of resistant lithologies (Fig. 3A) do not directly translate into an abundance of chert clasts. A quick inspection of outcrops of the

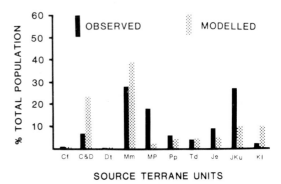

Fig. 7. Comparison of average abundance of clast types in the Sphinx Conglomerate (averages of samples 1–7, Fig. 6E) and modelled clast contributions from the source section (Fig. 3A).

Phosphoria Formation, the major upper Palaeozoic chert-bearing unit, confirms that, when tectonized, the chert is often fractured into domains smaller than pebble size.

Clearly, error analysis in provenance modelling is fraught with difficulty. Nevertheless, close scrutiny is necessary to distinguish geological factors such as tectonic effects and source terrane lithology effects from methodological artefacts.

CONCLUSIONS

The fill of a foreland basin reflects the evolution of the fold-thrust belt to which it is genetically linked. Unfortunately, sandstones and mudstones, for the most part, contain detrital components too cosmopolitan to be of more than general use in tectonic interpretations. Although volumetrically much less significant, conglomerates, with their less equivocal provenance message, offer great interpretive potential for basin areas proximal to fold-thrust belts. Such conglomerates have generally been used in a qualitative way to associate the timing of clastic wedges with fault movements and to suggest the presence or absence of particular formations in eroding source terranes.

However, it is possible in favourable circumstances to model the erosional stripping of thrust sheets and recycling of detritus into nearby foreland-basin deposits more rigorously by comparing actual conglomerate clast populations with hypothetical conglomerates contrived by assessing the contribution of transport-durable lithologies that might be eroded from various stratigraphic units in the source terrane (Figs 2 and 3). For example, the Upper Cretaceous Sphinx Conglomerate of the Laramide foreland of SW Montana is best interpreted using the provenance-modelling approach. A striking increase upward in conglomerate content in the Sphinx can be seen to be related to the unroofing of a thrust sheet of lithologically diverse Phanerozoic formations (Fig. 6). Instead of reflecting rejuvenated or accelerated fault activity, as many widely used models would predict, the upward increasing occurrence of conglomerate can be seen, through provenance modelling, principally to reflect the lithologic composition of the source terrane as it varied through time with erosion. The main conglomeratic element of the Sphinx Conglomerate represents the time needed to erode through a very thick section of erosion-resistant Mississippian carbonate rocks contained in the thrust-sheet adjacent to the Sphinx depocentre.

Relatively strict input constraints, such as fairly complete knowledge of the character of source formations and uncomplicated sediment dispersal pathways, probably limit the general applicability of the method. Nevertheless, where such conditions are met, provenance modelling of conglomerates may be used to assess the role of source lithology on the development of thrust-related sedimentary sequences in the proximal realms of foreland basins. Furthermore, the provenance modelling technique described here may be a useful basin-analysis tool in studies of fault-bounded basins characteristic of other tectonic settings, as well.

ACKNOWLEDGMENTS

The authors of this study comprise the faculty and students of the 1984 Graduate Field Seminar at the Indiana University Geologic Field Station. We gratefully acknowledge the support of Indiana University, and particularly, the Director of the Field Station, Lee J. Suttner, in the course of our study of the Sphinx Conglomerate. Additional funding was generously provided by the Tenneco Foundation. We also thank Christopher J. Schmidt for helpful discussion and comment, and Timothy Lawton for a thoughtful review of the manuscript.

REFERENCES

BALLY, A.W., GORDY, P.L. & STEWARD, G.A. (1966) Structure, seismic data and orogenic evolution of southern Canadian Rocky Mountains. *Bull. Can. Petrol. Geol.* **14**, 337–381.

BECK, F.M. (1959) *Geology of the Sphinx Mountain area, Madison and Gallatin Counties, Montana.* M.A. Dissertation, University of Wyoming, 65 pp.

BLATT, H. (1982) *Sedimentary Petrology.* W. H. Freeman, San Francisco, 564 pp.

DECELLES, P.G. *et al.* (1986) Laramide thrust-generated alluvial-fan sedimentation, Sphinx Conglomerate, southwestern Montana. *Bull. Am. Ass. Petrol. Geol.* In press.

DENSON, N.M. & PIPIRINGOS, G.N. (1969) Stratigraphic implications of heavy-mineral studies of Paleocene and Eocene rocks of Wyoming. In: *Wyoming Geological Association 21st Field Conference Guidebook* (Ed. by J. A. Barlow, Jr), pp. 9–18.

DICKINSON, W.R. (1976) Plate tectonic evolution of sedimentary basins. *Am. Ass. Petrol. Geol. Continuing Education Course Note Ser.* **1**, 1–62.

DICKINSON, W.R. & SUCZEK, C.A. (1979) Plate tectonics and sandstone compositions. *Bull. Am. Ass. Petrol. Geol.* **63**, 2164–2182.

GLOPPEN, T.G. & STEEL, R.J. (1981) The deposits, internal structure and geometry in six alluvial fan-fan delta bodies (Devonian–Norway)—a study in the significance of bedding sequence on conglomerates. *Spec. Publ. Soc. econ. Paleont. Mineral.* **31**, 49–69.

GRAHAM, S.A., DICKINSON, W.R. & INGERSOLL, R.V. (1975) Himalayan–Bengal model for flysch dispersal in the Appalachian–Ouachita system. *Bull. geol. Soc. Am.* **86**, 273–286.

HALL, J.W. & NORTON, N.J. (1967) Palynological evidence of floristic change across the Cretaceous–Tertiary boundary in eastern Montana. *Palaeogeogr. Palaeoclim. Palaeoecol.* **3**, 121–131.

HALL, W.B. (1961) *Geology of the upper Gallatin River basin, southwestern Montana.* Ph.D. Dissertation, University of Wyoming, 239 pp.

HEWARD, A.P. (1978) Alluvial fan sequence and megasequence models: with examples from Westphalian D–

Stephanian B coal fields, northern Spain. *Mem. Can. Soc. Petrol. Geol.* **5**, 669–702.

INGERSOLL, R.V. (1986) Provenance of impure calclithites in the Laramide Foreland of southwestern Montana. *J. sedim. Petrol.* (submitted).

JORDAN, T.E. (1981) Thrust loads and foreland basin evolution, Cretaceous, western United States. *Bull. Am. Ass. Petrol. Geol.* **65**, 2506–2520.

LOWELL, W.R. & KLEPPER, M.R. (1953) Beaverhead formation, a Laramide deposit in Beaverhead County, Montana. *Bull. geol. Soc. Am.* **64**, 235–244.

MIALL, A.D. (1970) Devonian alluvial fans, Prince of Wales Island, Arctic Canada. *J. sedim. Petrol.* **40**, 556–571.

TANNER, W.F. (1976) Tectonically significant pebble types: sheared, pocked and second-cycle examples. *Sediment. Geol.* **16**, 69–83.

VAN DER PLAS, L. & TOBI, A.C. (1965) A chart for judging the reliability of point counting results. *Am. J. Sci.* **263**, 87–90.

Spec. Publs int. Ass. Sediment. (1986) **8**, 437–443

Reconstruction of patterns of differential subsidence using an episodic stratigraphic model

E. J. ANDERSON, PETER W. GOODWIN *and* PETER T. GOODMANN

Department of Geology, Temple University, Philadelphia, PA 19122, U.S.A.

ABSTRACT

Applying a general model of episodic stratigraphic accumulation, the Hypothesis of Punctuated Aggradational Cycles (PACs), it is possible to reconstruct patterns of differential subsidence within short time intervals in the Appalachian Foreland Basin. As small-scale (1–5 m thick) time-stratigraphic shallowing-upward cycles, PACs provide the necessary chronologic and palaeoenvironmental control for discriminating the amount of stratigraphic thickness attributable to differential subsidence. Correlation of PACs in the shallow carbonate facies of the Keyser Formation of central Pennsylvania permits precise determination of significant differential subsidence between localities only 60 km apart. Over this distance an 11·5 m difference in stratigraphic thickness (30 m versus 18·5 m) can be attributed to differential subsidence by establishing correlative horizons which were topographically level surfaces at the time of deposition. Correlation is accomplished by dividing the entire section into PACs at each locality, tracing individual PACs between pairs of localities and by correlating three major deepening events, distinguished at all localities by marked facies changes. Two PACs, one at the bottom of the sequence studied and one in the middle of the Keyser Formation aggraded to sea-level and were selected for subsidence analysis. The tops of each of these PACs once represented essentially horizontal synchronous surfaces. Difference in stratigraphic thickness between these surfaces can only be explained by differential subsidence. Differential subsidence (11·5 m in 60 km of distance) was progressively greater toward the south and was introduced into the stratigraphic record non-uniformly (more in some PACs than in others). Only by applying a model of small-scale episodic accumulation is it possible to recognize the synchronous palaeoisotopographic surfaces necessary for interpreting basin dynamics at this scale. In that PACs are predicted to occur in all facies affected by relative sea-level fluctuations, this method has potential for very detailed reconstruction of the patterns and amounts of differential subsidence in the stratigraphic record, especially in nearshore facies.

INTRODUCTION

Differences in stratigraphic thickness at the systemic scale across a depositional basin imply differential tectonic subsidence. For example Suppe (1985, p. 423) showed that the combined Silurian and Devonian Systems in the Appalachian Foreland Basin are nearly an order of magnitude thicker than they are westward on the craton. This thickness differential records the long-term cumulative difference in subsidence between these two geographic areas.

In contrast, estimation of short-term differences in subsidence within a basin is complicated because topography, differences in sedimentation rates and the identification of sufficiently precise time-lines become significant factors in the analysis. These complications can be eliminated if closely spaced, synchronous, isotopographical surfaces can be identified in the stratigraphic record. Recognition of such surfaces permits isolation of the contribution of subsidence to stratigraphic thickness by eliminating the local effects of topography and sedimentation rates. We have recognized such surfaces in stratigraphic sequences in the Appalachian Basin as a consequence of applying an episodic model of stratigraphic accumulation, the Hypothesis of Punctuated Aggradational Cycles (Goodwin & Anderson, 1985).

The PAC Hypothesis is a comprehensive stratigraphic model which states that most stratigraphic accumulation occurs episodically as thin (1–5 m thick) time-stratigraphic shallowing-upward cycles separated by sharply defined surfaces marked by abrupt change to deeper facies (Fig. 1). In the hypothesis it is argued that these surfaces are created by geologically instantaneous basin-wide relative base-level rises; deposition occurs during the intervening periods of base-level stability. Thus all deposition occurs in aggradational episodes interrupted by short periods of non-deposition (punctuation events). From this perspective the stratigraphic record should consist of thin, laterally extensive rock units, at a scale larger than individual beds but usually thinner than 5 m, separated from each other by basin-wide synchronous surfaces. We refer to these time-stratigraphic rock units as PACs.

THE PAC HYPOTHESIS

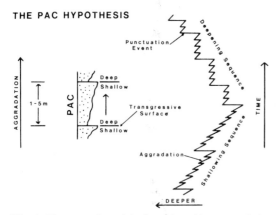

Fig. 1. The general model of stratigraphic accumulation applied in this paper. PACs are bounded by surfaces of abrupt change to deeper facies. Facies changes within PACs are gradual. At a larger scale, shallowing and deepening sequences consist entirely of PACs produced during periods of aggradation punctuated by deepening events (from Goodwin & Anderson, 1985).

ORIGINS OF THICKNESS VARIATION IN PACS

Given that the episodic component of abrupt base-level rise, whether eustatic or tectonic, is equal throughout the lateral extent of a single PAC, several local factors can contribute to thickness variation within that PAC. For example local topography on the top surface of the underlying PAC will result in more stratigraphic room for accumulation in low areas. Also lateral variation in sediment accumulation rates may produce thickness differences within the same PAC. For example the deposits in mound,

biohermal or shoal areas will likely be thicker than in those in nearby channels or lagoons. Although of negligible importance in the carbonate PACs of the Keyser Formation, diffferential compaction can produce thickness differences within the same PAC if lithologies within that PAC are very different laterally. Finally, differential subsidence can produce significant thickness differences within a PAC by creating more stratigraphic room in areas of greater subsidence. In order to isolate and quantify this variable (differential subsidence) for a single PAC or a sequence of PACs each of the other variables which may have contributed to differences in PAC thickness must be eliminated.

STRATIGRAPHIC DATA BASE

Consistent with the predictions of the PAC model, the Upper Silurian to Lower Devonian Keyser Formation at Tyrone, Pennsylvania, is divisible into small-scale shallowing-upward cycles (Figs 2 and 3). In every case the deepest facies within a PAC occurs at or near its base above which facies progressively shallow. PAC boundaries are placed at surfaces of abrupt change to deeper facies. Some PACs in this section (e.g. PACs 4–8 in the Lower Keyser Formation) are totally subtidal. However other PACs in the upper Keyser Formation and in the underlying Tonoloway Formation aggrade to sea-level (Fig. 2).

The magnitude of deepening events (punctuation events) is interpreted from facies analysis and expressed by means of a relative water depth curve (Fig. 2). Events which initiated deposition of the fourth and seventh Keyser PACs as well as the first New Creek Formation PAC just above the Keyser were relatively large events which produced pronounced facies changes. Recognition of large punctuation events, patterns of shallowing and deepening PAC sequences, and unique facies which appear to be developed locally only in one PAC provide the basis for correlation of PACs to other localities in the study area (Fig. 4). At all five localities, the stratigraphic sections are complete and divisible into PACs. Correlations are indicated by designating equivalent PACs with the same number at all localities (Fig. 4).

SUBSIDENCE ANALYSIS

The surfaces which bound a single PAC at all localities in the study area are synchronous surfaces

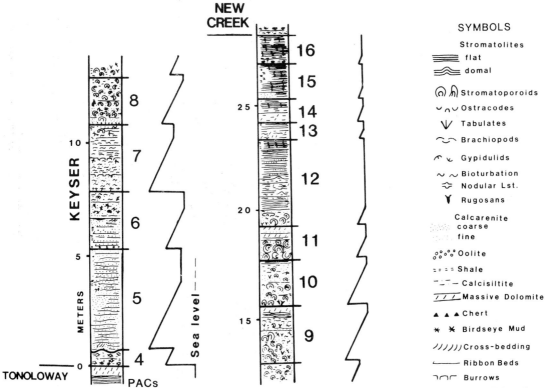

Fig. 2. Columnar section of the Keyser Formation at Tyrone, Pennsylvania. PACs numbered at the right side of the column are set off by heavy horizontal lines. Facies depicted with symbols are interpreted with a relative water-depth curve (deeper to the left). The top PAC in the Tonoloway Formation and PACs 9–16 all aggrade to sea-level. Keyser PACs 1–3 are missing in a disconformity at this locality.

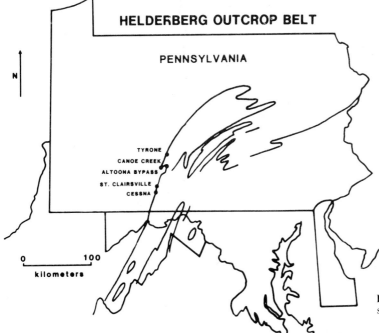

Fig. 3. Locality map of the five described sections in south-central Pennsylvania.

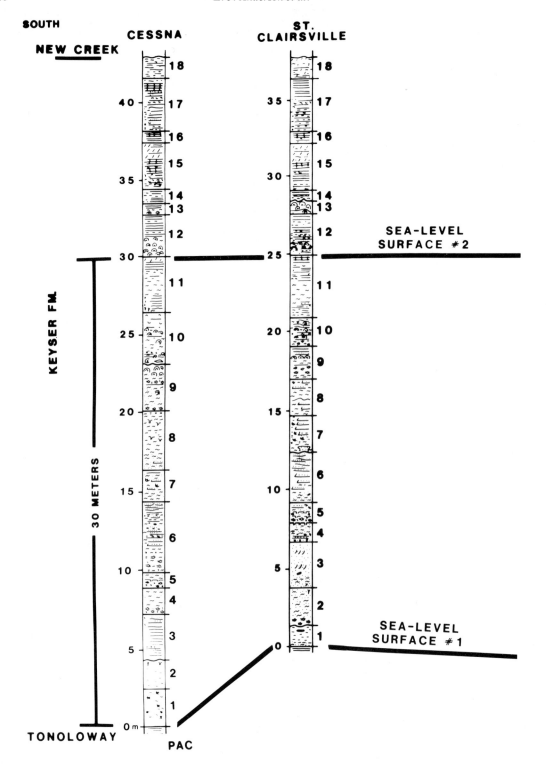

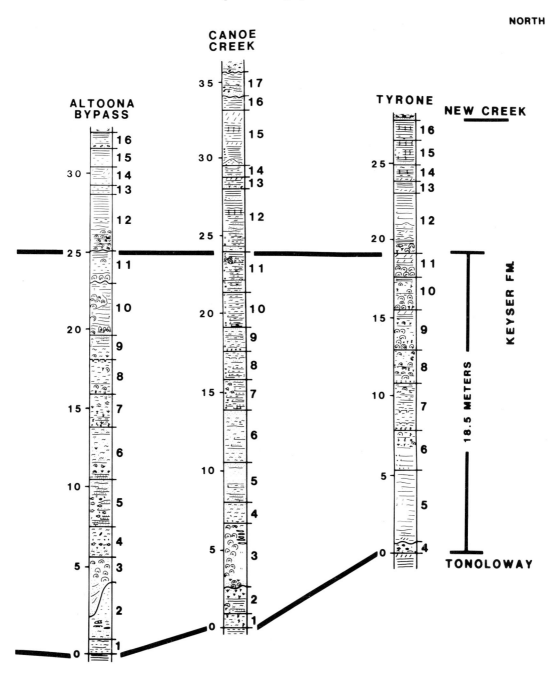

Fig. 4. Correlated columnar sections (Tyrone is the same as Fig. 2). Each section is totally divisible into PACs numbered on the right of the columns. PACs with the same number at each locality are correlative. The top of the last Tonoloway PAC and the top of PAC 11 each were once horizontal synchronous surfaces and are used in the subsidence analysis. The difference in thickness of 11·5 m between these surfaces from Tyrone to Cessna represents differential subsidence. Total distance between Tyrone and Cessna is 60 km.

because they are produced by rapid basin-wide events independent of the local sedimentary environment (i.e. base-level rises). These and all PAC boundaries meet the first requirement, that of synchroneity, for using surfaces in an analysis of subsidence.

Some of the PACs which aggraded to sea-level at Tyrone (Fig. 2) also aggraded to sea-level at the other localities studied (Fig. 4). The tops of these PACs, observed both in the upper Tonoloway and upper Keyser Formations, are characterized by high intertidal to supratidal facies. Because these PACs aggraded to approximately the level of mean high tide, their tops were essentially horizontal surfaces at the time that deposition was terminated by the next punctuation event.

When these two criteria (synchroneity and original horizontality) characterize closely spaced traceable surfaces in stratigraphic sections a basis is available for eliminating complicating variables and quantifying amounts of differential subsidence which may have occurred among these sections. At two localities the difference is stratigraphic thickness between any two surfaces, that were synchronous and horizontal at the time of their origin, is equal to the amount of differential subsidence which occurred between those localities during the time interval represented (Fig. 5).

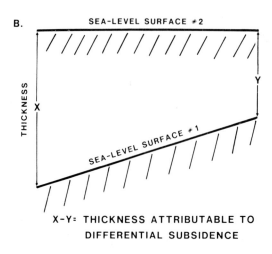

X-Y= THICKNESS ATTRIBUTABLE TO DIFFERENTIAL SUBSIDENCE

Fig. 5. Model for calculating differential subsidence based on recognition of once synchronous horizontal surfaces: (A) represents time 1 and (B) a later point in time.

RESULTS

To illustrate an example of an analysis of subsidence at a small scale, two upper PAC boundaries which aggraded to sea-level throughout the study area were selected, one near the top of the Tonoloway Formation and a second in the middle of the Keyser Formation (Fig. 4). The thickness of the stratigraphic interval between these two surfaces at Tyrone is 18·5 m. The thickness of the interval progressively increases to the south reaching 30 m at the locality near Cessna, Pennsylvania, 60 km from Tyrone. The difference of 11·5 m of stratigraphic thickness between these two surfaces at the two most distant localities represents the amount of differential subsidence that occurred between the times of origin of the two surfaces. If an average duration of cycles can be calculated and the number of cycles counted between the key surfaces used in the measurement an estimate of small-scale subsidence rates at each locality could be obtained.

A sea-level fall and subsequent episodic rise has resulted in a disconformity within this stratigraphic interval. As a result of onlap to the north there are more Keyser PACs (PACs 1–3) in the southern localities (Fig. 4). During the period of sea-level fall a southward dipping disconformable surface was produced by differential subsidence and then episodically overstepped (by onlap) to the north during the subsequent sea-level rises. The presence of the disconformity does not affect the calculation of the amount of differential subsidence because any thickness difference between these surfaces is the result of differential subsidence whether deposition was continuous or discontinuous.

CONCLUSIONS

Application of the Hypothesis of Punctuated Aggradational Cycles permits recognition of synchronous and originally horizontal surfaces within the Keyser Formation of central Pennsylvania. These surfaces, which can be traced to all localities in the study area, provide the basis for a calculation of differential subsidence. The amount of differential subsidence is equal to the difference in thickness between the selected palaeoisotopographic synchronous surfaces. Intervening disconformities which do not involve folding do not affect the analysis. Within a thin stratigraphic interval in the Keyser Formation 11·5 m of differential subsidence occurred over a linear distance of less than 60 km.

ACKNOWLEDGMENTS

We thank the secretarial and technical staffs of the Geology Departments at the University College of Swansea and Temple University for aid in preparing the manuscript. The project has been supported by National Science Foundation Grants EAR-8107690 and EAR-8305900.

REFERENCES

GOODWIN, P.W. & ANDERSON, E.J. (1985) Punctuated aggradational cycles: a general hypothesis of episodic stratigraphic accumulation. *J. Geol.* **93**, 515–533.

SUPPE, J. (1985) *Principles of Structural Geology.* Prentice-Hall, New Jersey.

Index

References to figures appear in *italic type*.
References to tables appear in **bold type**.